Polyimides and Other High Temperature Polymers, Volume 2

POLYIMIDES AND OTHER HIGH TEMPERATURE POLYMERS:

SYNTHESIS, CHARACTERIZATION AND APPLICATIONS

VOLUME 2

Editor:
K.L. Mittal

CRC Press
Taylor & Francis Group
Boca Raton London New York

CRC Press is an imprint of the
Taylor & Francis Group, an **informa** business

First published 2003 by VSP Publishing

Published 2018 by CRC Press
Taylor & Francis Group
6000 Broken Sound Parkway NW, Suite 300
Boca Raton, FL 33487-2742

First issued in paperback 2019

No claim to original U.S. Government works

ISBN 13: 978-0-367-44678-9 (pbk)
ISBN 13: 978-90-6764-378-8 (hbk)

This book contains information obtained from authentic and highly regarded sources. Reasonable efforts have been made to publish reliable data and information, but the author and publisher cannot assume responsibility for the validity of all materials or the consequences of their use. The authors and publishers have attempted to trace the copyright holders of all material reproduced in this publication and apologize to copyright holders if permission to publish in this form has not been obtained. If any copyright material has not been acknowledged please write and let us know so we may rectify in any future reprint.

Visit the Taylor & Francis Web site at
http://www.taylorandfrancis.com

and the CRC Press Web site at
http://www.crcpress.com

Contents

Polyimides and Other High Temperature Polymers, Vol. 2, pp. ix–x
Ed. K.L. Mittal
© VSP 2003

Preface

This volume documents the proceedings of the Second International Symposium on Polyimides and Other High Temperature Polymers: Synthesis, Characterization and Applications held under the auspices of MST Conferences in Newark, New Jersey, December 3-6, 2001. Because of the interest evinced in this topic, the third symposium in this series is scheduled to be held December 10-12, 2003 in Orlando, Florida.

As polyimides possess many desirable attributes, so this class of materials has found applications in many technologies ranging from microelectronics to high temperature adhesives to membranes. For quite some time there has been a brisk R&D activity in synthesizing new polyimides and/or in ameliorating/modifying the existing materials. All signals indicate that interest in polyimides and other high temperature polymers will continue unabated.

The first symposium on this topic held in 1999 was quite successful and was well received by the community, and its proceedings were properly documented in a hard-bound book [1]. This second symposium comprised 44 papers covering many and varied aspects of polyimides and other high temperature polymers. The technical program also included a special session on metallized polyimides and other polymers. This symposium provided a forum for discussion of latest developments, and there were lively discussions among the participants through the Symposium.

As for this volume, it contains a total of 32 papers (others are not included for a variety of reasons) addressing many aspects and new developments regarding polyimides and other high temperature polymers. It must be recorded that all manuscripts were rigorously peer reviewed and suitably revised (some twice or thrice) before inclusion in this volume. So this volume is not a mere collection of unreviewed papers – which is generally the case with many symposia proceedings – rather it reflects information which has passed peer scrutiny.

This volume is divided into two parts. Part 1 "Synthesis, Properties and Bulk Characterization"; and Part 2 "Surface Modification, Interfacial or Adhesion Aspects and Applications". The topics covered include: Synthesis and characterization of a variety of polyimides; photoalignable polyimides; high-modulus poly(p-phenylenepyromellitimide) films; structure-property relationships in polyimides; aromatic benzoxazole polymera; polybenzobisthiazoles; polyimide L-B films; transport of water in high T_g polymers; surface modification of polyimides; adhesion of metal films to polyimide and other polymers; investigation of interfacial interactions between metals and polymers; polyimide film surface properties; applications of polyimides in microelectronics, as membranes for gas separation, as composite films; fabrication of thin-film transistors on polyimide films; polyimide modified with fullerenes; semicrystalline polyimides for advanced composites; and wear performance of polyetherimide composite.

I sincerely hope this and its predecessor volume [1] addressing many aspects and recent developments in the domain of polyimides and other high temperature polymers will be of interest to anyone interested (centrally or tangentially) in this topic. It is further hoped that these two volumes containing bountiful information will serve as a commentary on the current R&D activity in this arena.

Acknowledgements

Now comes the pleasant task of thanking those who helped to bring this project to fruition. First, my sincere thanks are extended to my friend and colleague, Dr. Robert H. Lacombe, for taking care of the myriad details necessary in organizing such a symposium. Second, I am most appreciative of the time and efforts of the unsung heroes (reviewers) for providing valuable comments which are a desideratum to main the highest standard of a publication. I am profusely thankful to the authors for their interest, enthusiasm and contribution without which this book would not have seen the light of day. In closing, my appreciation goes to the staff of VSP (publisher) for giving this book a body form.

K.L. Mittal
P.O. Box 1280
Hopewell Jct., NY 12533

1. K.L. Mittal (Ed.), *Polyimides and Other High Temperature Polymers: Synthesis, Characterization and Applications.* Vol. 1. VSP, Utrecht (2001).

Part 1

Synthesis, Properties and Bulk Characterization

Polyimides and Other High Temperature Polymers, Vol. 2, pp. 3–35
Ed. K.L. Mittal
© VSP 2003

Poly(amic acid)s and their ionic salt solutions: Synthesis, characterization and stability study

ROHIT H. VORA,[1, 2, 3] * P. SANTHANA GOPALA KRISHNAN,[1]
S. VEERAMANI[1] and SUAT HONG GOH[2]

[1]*Institute of Materials Research and Engineering (IMRE), 3, Research Link, Singapore 117602*
[2]*Department of Chemistry, National University of Singapore (NUS), 10 Kent Ridge Crescent, S8-03, Science Drive 3, Singapore 117543*
[3]*Department of Materials Science, National University of Singapore (NUS), 10 Kent Ridge Crescent, S7-03-01, Science Drive 4, Singapore 117543*

Abstract—Two series of co-poly(amic acid)s (Co-PAAs) containing various mole percents of *ortho*-Tolidine (o-TDA) and 1,3-bis(3-aminopropyl) tetramethyl disiloxane (SiDA) were prepared by reacting with 3,3',4,4'-biphenyl tetracarboxylic dianhydride (BPDA) or 3,3',4,4'-benzophenone tetracarboxylic dianhydride (BTDA) in N-methyl-2-pyrrolidone (NMP) solvent at room temperature. Subsequently, from these Co-PAAs, we have formulated two series each of the ionic bond type photodefinable co-poly(amic acid)s (Co-PDPAA), i.e. ionic salts, using N,N-diethylamino ethyl methacrylate (DEEM) and N,N-dimethylamino ethyl methacrylate (DMEM) respectively. Storage stability of these solutions at room temperature was monitored for a period of one month by gel permeation chromatography (GPC), potentiometry and viscometry. Bulk viscosity, inherent viscosity, acid number and molecular weight all decreased drastically as the storage time increased.

Rheological behaviour of Co-PAA and Co-PDPAA in solution was also investigated as a function of shear rate and temperature. At room temperature, all co-polymer solutions showed shear thinning behaviour in the shear rate range of 3.07-38.4 s^{-1}. The temperature dependence of bulk viscosity followed the Arrhenius equation. The activation energies of viscous flow of BTDA based Co-PAA and Co-PDPAA were in the range of 25-27 and 22-26 kJ/mol respectively, whereas for BPDA based Co-PAA and Co-PDPAA were in the range of 12-28 and 19-33 kJ/mol, respectively. Both the activation energy of viscous flow and ln [A], where A is pre-exponential factor, decreased with an increase in SiDA content in both Co-PAA and Co-PDPAA series. For a given composition, activation energy of viscous flow was comparable for both DMEM and DEEM based Co-PDPAA.

Thermal stability of precipitated solids of Co-PAA and Co-PDPAA was investigated using thermogravimetric analyzer (TGA). Both samples showed a two-step weight loss in air atmosphere, the first step due to imidization and the second step due to degradation. The activation energy for thermal degradation was determined using Coats-Redfern and Chang methods. The activation energy determined by the two methods was comparable.

Keywords: Co-poly(amic acid); photodefinable co-poly(amic acid) ionic salt; polyimide storage stability and degradation.

---

*To whom all correspondence should be addressed. Phone: (65) 6874 4299, Fax: (65) 6776 3604, E-mail: rohitvora@nus.edu.sg

1. INTRODUCTION

Polyimides, being versatile engineering polymers, have inherently high mechanical properties, good chemical resistance, low dielectric constant and high thermal stability. These have a wide range of diverse and potential applications in several major technologies. Currently, high performance polyimides are being widely used for several primary applications in the electronics area as:

(1) Fabrication aids such as photoresists, planarization layers and in implant masks;

(2) Passivation overcoats and interlevel insulators;

(3) Adhesives and underfill materials for microBGA packaging and flip chip technology;

(4) Substrate components.

These polymers, due to their unique properties, are also very suitable as materials for insulating and protective layers and coating formulations for aircraft, automotive and aerospace applications, as well as matrix resins for high performance light weight composites.

For high resolution photolithography, the use of photosensitive or photodefinable polyimides (PDPI) greatly simplifies the complex, multistep processing for microelectronic component fabrication requirement [1], but are often sold and used as solutions of the precursor, poly(amic acid) (PAA) and its ionic salt formulation (PDPAA) having photosensitive tertiary amines containing acryloyl groups for negative acting type system for the photolithography application. These precursors can undergo many reactions such as hydrolysis of amide bonds or terminal anhydride groups and cyclization with the elimination of amine and anhydride or water depending on the synthesis and processing steps shown in Figure 1 and storage conditions [2]. These reactions not only affect the properties of PAA/PDPAA or of PDPAA exposed to UV radiation followed by baking but also of PI which has been prepared either thermally or chemically from PAA. To prevent or to minimize the degradation of PAA/PDPAA during storage, it is necessary to have a clear understanding of the type of reaction taking place, whether hydrolysis or cyclization. Hydrolysis of PAA derived from [BPDA + PPD] [3] and [PMDA + ODA] [3, 4] and also the effect of moisture on the stability of [ODPA + m-Tolidine] and [ODPA + o-Tolidine] PAA and their model compounds have been reported [5-6].

In this work we have synthesized and characterized two amic acid compounds and two series of co-poly(amic acid)s containing various mole percents of *ortho*-Tolidine (o-TDA) and 1,3-bis(3-aminopropyl) tetramethyl disiloxane (SiDA) prepared by reacting with 3,3,4,4'-biphenyl tetracarboxylic dianhydride (BPDA) or 3,3',4,4'-benzophenone tetracarboxylic dianhydride (BTDA) in N-methyl pyrrolidone (NMP). Subsequently, using these Co-PAAs, two series each of negative type photodefinable Co-PDPAA were made from N,N-diethylamino ethyl methacrylate (DEEM) and N,N-dimethylamino ethyl methacrylate (DMEM).

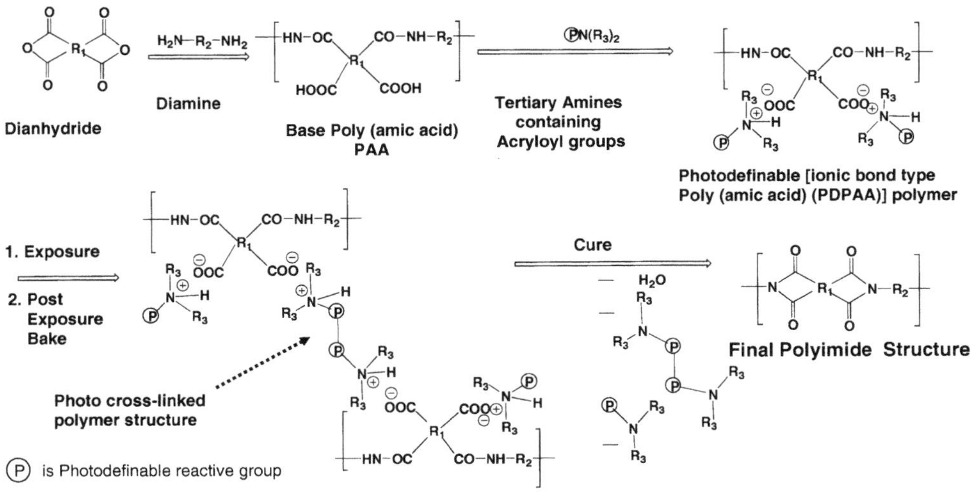

Figure 1. Poly(amic acid) and its photosensitive formulation (PDPAA) having photosensitive tertiary amines containing acryloyl groups for negative acting type system, and photolithography processing steps.

NMR studies were carried out on model compounds both in the absence and presence of added water at room temperature using ¹H NMR.

Any changes, whether increase or decrease, in molecular weights, bulk and inherent viscosities and acid number during storage will highlight the degradation path of PAA, Co-PAA and Co-PDPAA. Therefore, we have monitored molecular weights by gel permeation chromatography (GPC) [a.k.a. size exclusion chromatography], bulk and inherent viscosities by viscometry and acid number by potentiometry, as a function of storage time. This study will help to understand what type of reaction is taking place predominantly at the storage temperature which accounts for the degradation of Co-PAA and Co-PDPAA. In addition, rheological behaviour of Co-PAA and Co-PDPAA in solution was also investigated as a function of shear rate and temperature. We have determined the rate constants at various temperatures from the activation energy using the Arrhenius equation for viscous flow of BTDA and BPDA based Co-PAA and Co-PDPAA. Also, we have investigated the thermal stability behaviour of Co-PAA and Co-PDPAA using a thermogravimetric analyzer (TGA) and the results are reported here.

2. EXPERIMENTAL

2.1. Materials

Phthalic anhydride (PA) was purchased from BDH Laboratory, London, UK; BPDA and BTDA were obtained from Chriskev Co., USA; (3,3'-dimethyl benzidine) i.e. (o-TDA) was received from Wakayama Seika Kogyo Co. of Japan;

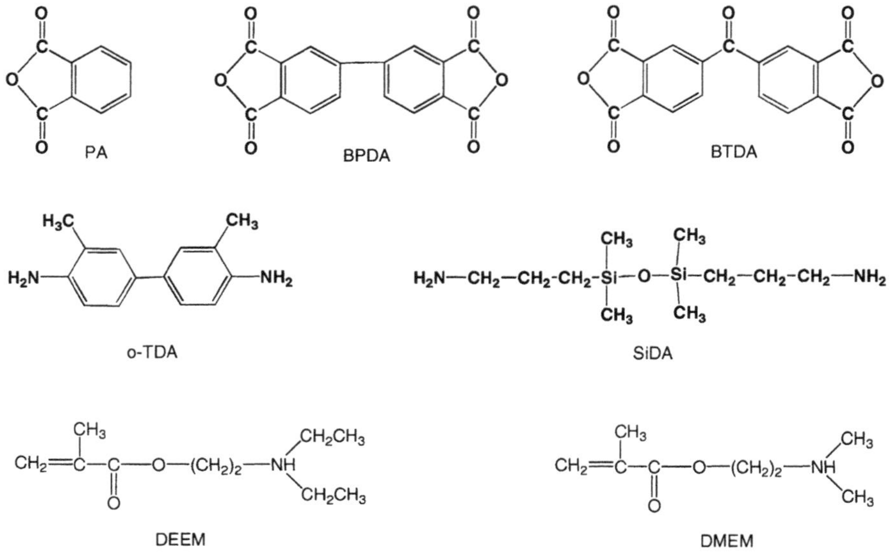

Figure 2. Structure of monomers.

1,3-bis(3-aminopropyl) tetramethyl disiloxane (SiDA) was obtained from Tokyo Kasei Organic Chemicals Co. of Japan; N-methyl-2-pyrrolidone (NMP) was purchased from Lab-Scan of Singapore: N,N-diethylamino ethyl methacrylate (DEEM), N,N-dimethylamino ethyl methacrylate (DMEM), Dichloromethane (DCM), acetone, tetrahydrofuran (THF), tetramethyl ammonium hydroxide (~ 25% in methanol) (TMAH) and potassium hydrogen phthalate were obtained from Aldrich Chemical Company of USA.

Both dianhydrides were dried in an oven at $150 \pm 2°C$ and o-TDA was dried at $55 \pm 2°C$ overnight. PAA, Co-PAA and Co-PDPAA were powdered and dried in an air oven overnight at $110 \pm 2°C$. Other chemicals were used as received. The chemical structures of monomers used are shown in Figure 2.

2.2. Synthesis

2.2.1. (a) Poly(amic acid) (PAA) and co-poly(amic acid) (Co-PAA)

Several methods for the preparation of polyimides have been reported in the literature [7-14]. The most common procedure used in this investigation is a simplified one-pot two-step polymerization synthesis process used by Vora [15] as per the polymerization scheme shown in Figure 3.

Typically, in the case of the synthesis of a poly(amic acid) (PAA) for example [BPDA + o-TDA] based on biphenyl dianhydride and 3,3'-dimethyl benzidine, 29.42 g (0.10 mole) of accurately weighed solid BPDA powder was added to an equimolar amount of o-TDA diamine (21.23 g), pre-dissolved in freshly distilled NMP to make 20 weight % solids concentration. For a random co-poly(amic acid)

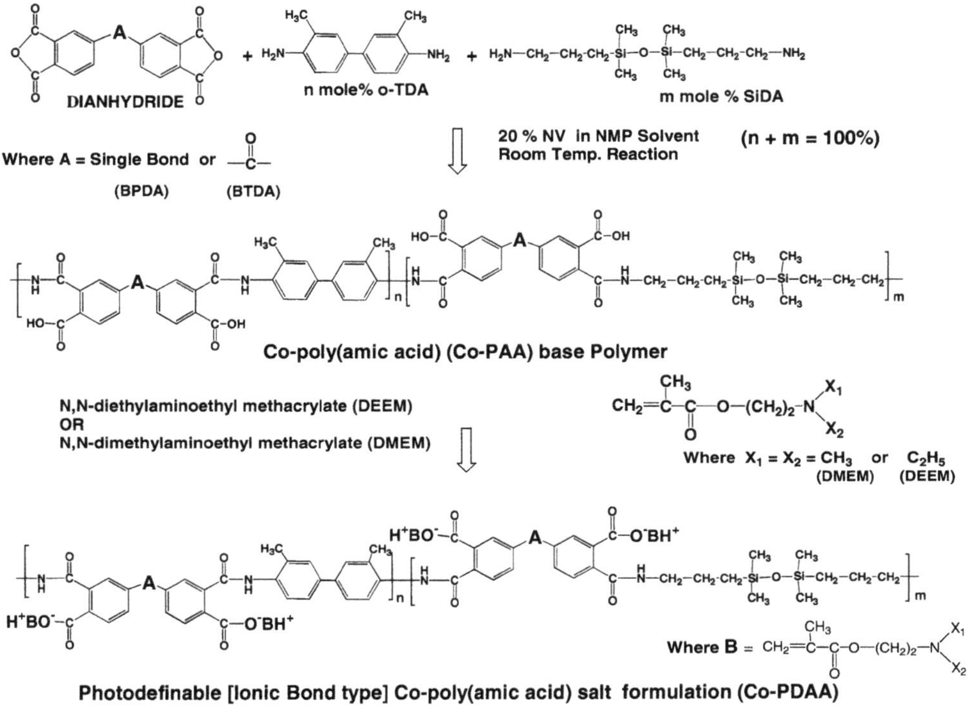

Figure 3. Synthesis schematic of base poly(amic acid) (PAA), co-poly(amic acid) (Co-PAA) and their ionic bond type photodefinable poly(amic acid) (PDPAA)s.

(Co-PAA), for example, [BPDA + o-TDA (90%) + SiDA (10%)], 38.25 g (0.13 mole) of solid BPDA powder was added to an equimolar amount of diamines mixture [consisting of 24.84 g of o-TDA (0.117 mole) and 3.23 g of SiDA (0.013 mole)] pre-dissolved in freshly distilled NMP to make 20% solid concentrations. The reaction mixture was stirred under a nitrogen atmosphere at room temperature overnight to make viscous polyamic or co-poly(amic acid) solutions. Then it was filtered and packed in a polypropylene bottle under an argon environment.

2.2.1. (b) Photodefinable ionic salt formulation (PDPAA) from poly(amic acid) (PAA) and co-poly(amic acid) (Co-PAA)

Two series of photodefinable ionic salt formulations (PDPAA) were synthesized from the BTDA and BPDA based poly(amic acid) (PAA) and co-poly(amic acid) (Co-PAA) with DEEM and DMEM. A stoichiometric amount of DEEM or DMEM was added based on the calculation of theoretical mole % of free carboxylic acid groups on the polymer chain from the chemical structural repeat unit. The solution was stirred in a nitrogen environment at room temperature for two hours and then packaged in 60 mL polypropylene bottles under an argon environment for further study.

2.2.2. Model compounds synthesis and characterization

We believe that the most common possibility of polymer degradation activity would be in the orthocarboximide (-CO-NH-) region at various parts of the polymer chain, e.g. proximity to the aromatic ring, or the proximity to the (-CH$_2$ CH$_2$ CH$_2$-Si-O-Si-CH$_2$ CH$_2$ CH$_2$-) group. Therefore, to understand the viscosity drift of PAA, Co-PAA and its ionic salt solution [i.e. photosensitive poly(amic acid) (PDPAA)] upon storage at room temperature, two model compounds, **A** and **B**, were synthesized whose chemical structures were chosen after careful inspection of the chemical structure of Co-PAA (Figure 4).

2.2.2. (a) Synthesis of model compound A (Bisphthalamic acid of 3,3'-dimethyl-4,4'-diaminobiphenyl)

The synthesis of model compound **A** was straightforward. The synthesis scheme is shown in Figure 5.

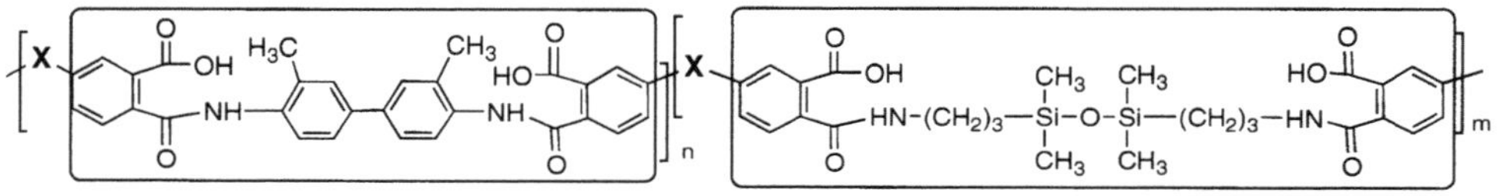

Model compound A   **Model Compound B**

Figure 4. Model compounds selected based on the chemical structure of Co-PAA repeat unit.

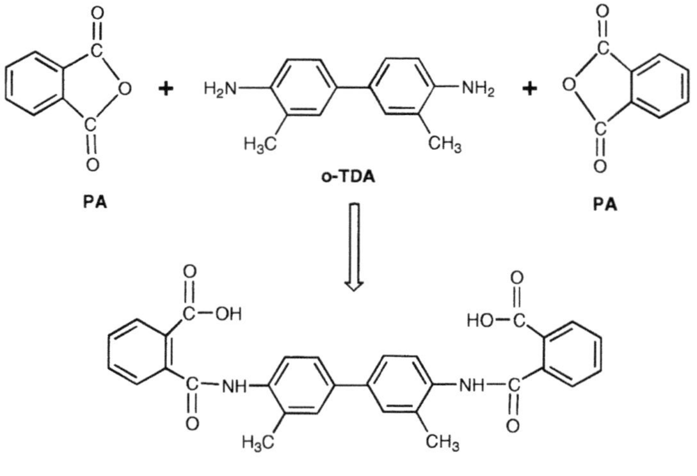

Model compound A

Bisphthalimic acid of 3,3'-dimethyl-4,4'-biphenyl

Figure 5. Synthesis scheme for model compound **A**.

In a 500 mL round bottom flask with a magnetic stirrer were introduced o-TDA (10 g, 0.0472 moles) in 400 mL DCM (total conc. ~ 6%). The contents were stirred until the diamine dissolved. Phthalic anhydride (PA) (13.974 g, 0.0943 moles, 2 mole equivalent) was gradually added to the solution and the mixture was stirred overnight at RT. The compound was analysed by ^{1}H NMR and the chemical structure was confirmed.

^{1}H NMR (DMSO); δ 10.36 (s, COOH); 7.9-7.8 (d, J = 6.6 Hz, Aromatic); 7.70-7.51 (m, Aromatic); 7.0 (d, J = 8.2 Hz, Aromatic); 2.02 (s, CH$_3$).

At one point a dense white precipitate began to form indicating the formation of the amic acid. After 24 h the solution was filtered and the white powder was left to dry. 22.12 g of model compound **A** was obtained after drying (Yield 92.3%).

2.2.2. (b) Synthesis of model compound B (Bisphthalamic acid of 1,3-bis(3-propyl) tetramethyl disiloxane)

The synthesis of model compound was sluggish due to reactivity of SiDA. The synthesis scheme is shown in Figure 6.

23 mL of THF was added to a 50 mL single necked round bottomed flask fitted with a calcium chloride guard tube. Then 2 g of SiDA was introduced and dissolved using a magnetic stirrer. 2.6 g of freshly dried phthalic anhydride (PA) was gradually added into the above solution in instalments. It took about 5-10 minutes for the PA to dissolve completely. The flask was fitted with the calcium chloride guard tube. The solution was stirred overnight at room temperature (25 ± 2°C). The next morning, the solution was precipitated from hexane and the mixture was kept in a refrigerator.

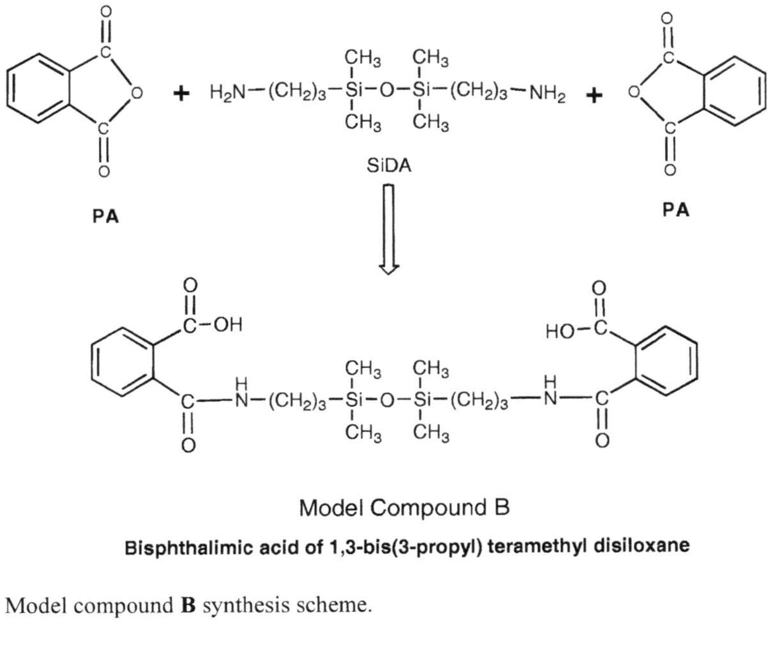

Model Compound B

Bisphthalimic acid of 1,3-bis(3-propyl) teramethyl disiloxane

Figure 6. Model compound **B** synthesis scheme.

White crystals had formed at the bottom. The upper solution layer was decanted slowly taking care not to lose the crystals formed. The trace amount of trapped solvent in crystalline material was removed by applying vacuum at 40°C in an air oven for a few hours. 3.6 g of bisphthalamic acid of siloxane diamine (**B**) was obtained (Yield 81.7% (wt/wt). The compound was analysed by ^{1}H NMR and the chemical structure was confirmed.

^{1}H NMR (DMSO); 0.1(a, singlet), 0.55(d, multiplet), 1.0(b, triplet), 1.5(c, multiplet), 2.5(DMSO, multiplet), 3.3(Water, singlet), 7.5-7.8(Aromatic, multiplet), 8.3(e, triplet), and 12.9 ppm (COOH, broad singlet).

2.3. Characterization

2.3.1. ^{1}H NMR
NMR studies were conducted using ACF 300 MHz NMR spectrometer at RT using DMSO-d6 as a solvent. The chemical shift (δ) is given in ppm. Tetramethylsilane (TMS) was used as an internal reference. NMR samples were prepared at 0% 10% and 20% by weight water concentration and were studied at RT, and NMR was run after 8 hours of water treatment at RT. Then, NMR was run again for the same sample after 1 week at RT to see whether any degradation had taken place due to presence of added water.

2.3.2. Inherent viscosity
Inherent viscosity of PAA, Co-PAA and Co-PDPAA was determined according to ASTM 2515 / D446 using a Schott-Gerate model AVS360, and capillary viscometer type DIN Ubbelohde at 25°C in NMP.

2.3.3. Bulk viscosity
Bulk viscosity of PAA, Co-PAA and Co-PDPAA was determined as a function of time, shear rate, and temperature using a Brookfield Programmable Rheometer Model DV-III (with Rheocalc software and Brookfield water bath model TC-200/500) at 5 rpm using CP42 spindle. About 1 mL of bubble-free sample was used. Prior to analysis, the sample was allowed to reach equilibrium for 1 minute before taking readings. An average of 6 readings was taken for plotting graphs.

2.3.4. Gel permeation chromatography (GPC)
The molecular weights of PAA, Co-PAA and Co-PDPAA were measured using Waters GPC system, sample concentration was 5 mg/mL and injection volume used was 200 μL, at the flow rate of 1 mL/min through a GPC column 'Gelpack GL-S300MDT-5 X 3 of size: 8 mm X 300 mm. Data were acquired from the Waters 2410 RI detector using Millennium 32 software. The relative molecular weights (M_n, M_w and polydispersity) were calculated against standard polystyrene using a mixture of THF/DMF = 1:1 by volume containing H_3PO_4 (0.06 M) and LiBr (0.06 M) as a mobile phase.

2.3.5. Potentiometric analysis. Determination of acid number in PAA, Co-PAA and Co-PDPAA solutions

Tetramethyl ammonium hydroxide (TMAH) was used as the titrant for determining the % carboxylic acid in the poly(amic acid) samples. One mole of TMAH reacts with one mole of carboxylic acid to produce its salt and water. A typical reaction of TMAH with the poly(amic acid) is given in Figure 7.

The acid number was determined by titrating with TMAH [16-17] using a TitroLine alpha (TZ 2055) (Schott-Gerate GmbH) automatic potentiometric titrator fitted with a magnetic stirrer (TM125). The electrode used was SAN 6480 (filled with LiCl/glacial acetic acid solution). Here the acid number is reported as mg of TMAH per gram of poly (amic acid) (PAA), Co-PAA or Co-PDPAA solution. Acid number was calculated using the following equation [17].

$$\text{Acid Number} = \frac{V \times N \times 91}{W}$$

where V = Volume of TMAH (ml), N = Normality of TMAH (N), W = Weight of poly (amic acid) solution taken (g).

The acid number determined for Co-PDPAA i.e. the ionic salt formulation of co-poly(amic acid) (Co-PAA) containing DEEM or DMEM is not only for the free acid present but also for the carboxylic acid attached to DEEM or DMEM. However, it is not possible to determine whether the carboxylic acid group present is in ionic form or in unionised form using this method. This is the limitation of this method.

2.3.6. Thermogravimetric analysis

Thermal decomposition temperatures (5% wt. loss) of polymer films were determined using dynamic TGA [Perkin Elmer model TGA-7 with Pyris software]. Scans were run at a heating rate of 10°C/min in a flowing air atmosphere (10 mL/min).

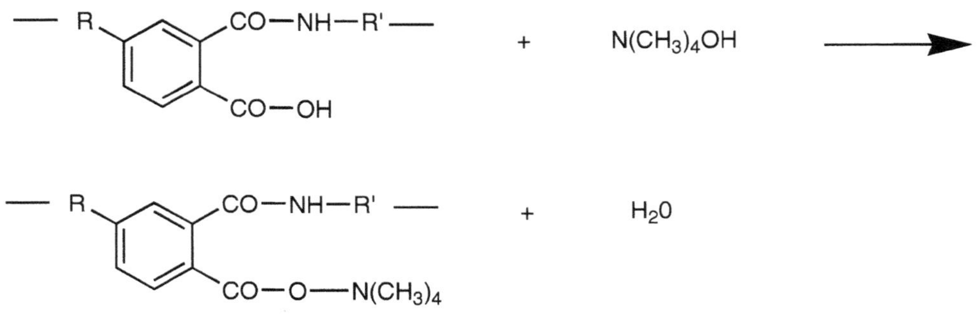

Figure 7. Reaction of TMAH with the o-carboxyamide group in poly(amic acid).

3. RESULTS AND DISCUSSION

3.1. NMR spectroscopic analysis of model compounds

NMR spectra indicated that the amic acid model compound **A** was sufficiently stable under the experimental conditions and that no chemical reactions were occurring to cause any structural changes detectable by NMR. Since the spectra (Figures 8-9) of the water treated model compound **A** remained the same as the untreated one, it can be inferred that more drastic conditions other than presence of water are needed to bring about any perceptible change.

In the NMR spectra (Fig. 10 top) of siloxane diamine (SiDA), the amine protons appear at 2.6 ppm. However, in the spectrum of model compound **B** this peak was not present at room temperature when it was synthesized, which confirmed that the prepared compound was devoid of unreacted amine and was pure. However, after the water treatment and storage for 8 hours, there was a change in the NMR spectra of **B** at 10 and 20% added water at RT. At the end of 8 hours study, a peak corresponding to an amino group was observed at 2.8 ppm both at 10 and 20% water contents.

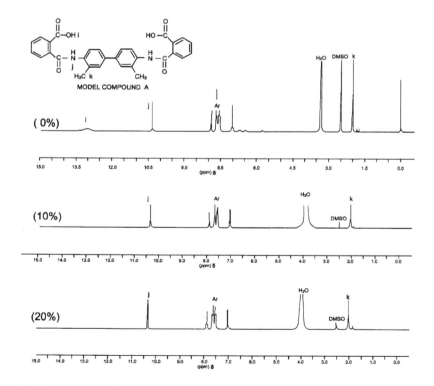

Figure 8. NMR study of model compound **A**: Effect of 0, 10 and 20% water content at RT.

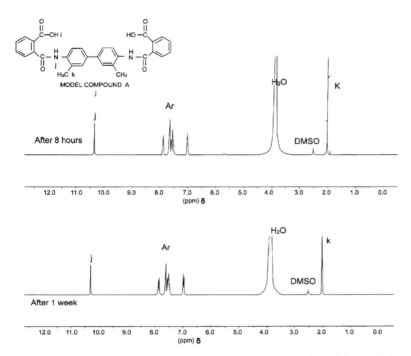

Figure 9. NMR study of model compound **A** containing 10% water: Effect of time at RT.

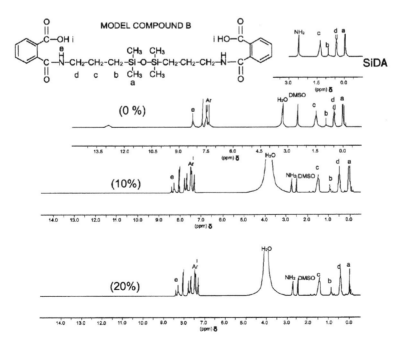

Figure 10. NMR study of model compound **B**: Effect of 0, 10 and 20% water content at RT.

This suggests that the model compound **B** was not stable at room temperature when compared to model compound **A** in the presence of water after 8 hours of storage. The model compound **A** was found to be stable at RT in presence of water even after 1 week (Figure 9), whereas the NMR spectrum of model compound **B** containing SiDA showed amine peak even at room temperature in presence of water in just 8 hours (Figure 10). Therefore, it suggests that model compound **B** is following a reverse order reaction (degradation) through hydrolysis in the presence of water to give amine which appeared as a peak at 2.8 ppm. This confirms that the model compound **B** is not stable at room temperature and starts decomposing into amine (Figure 10 bottom two spectra).

3.2. Monitoring changes in bulk viscosity, inherent viscosity, molecular weight and acid number

The inherent viscosity, molecular weight and acid number were determined before and after one month of storage at RT. Bulk viscosity, inherent viscosity, molecular weight and acid number all decreased irrespective of the chemical composition and chemical structure of Co-PAA and Co-PDPAA (Tables 1 and 2).

The change in bulk viscosity, inherent viscosity, molecular weight and acid number of PAA, Co-PAA and Co-PDPAA were studied for 1 month during storage at RT. For bulk viscosity, the sample was tested once a week. The results are shown in Figures 11 and 12.

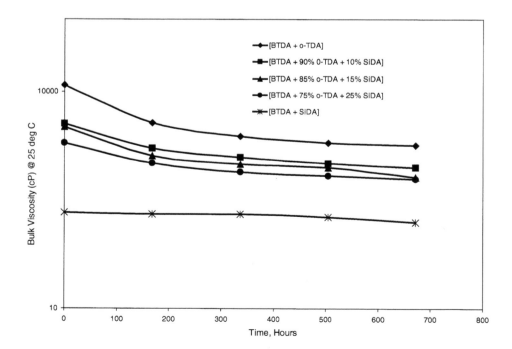

Figure 11. Bulk viscosity as a function of storage time at 25°C for BTDA based polymer systems.

Table 1.
Solution properties of PAA, and PDPAA (molecular weight, bulk viscosity, and inherent viscosity)

Polymer composition and Formulation	Time interval between analysis (hr)												
	0			672			0	168	336	504	672	0	672
	Molecular weights by GPC						Bulk viscosity (cP)					Inh. Visc. (dl/g)	
	Mw	Mn	PD	Mw	Mn	PD							
[BTDA + o-TDA]	71844	42767	1.68	36404	25688	1.42	12186	3615	2376	1905	1772	0.87	0.51
[BTDA + o-TDA] PAA + DEEM	73655	44483	1.66	45970	30200	1.52	19994	9564	6328	3881	3031	0.53	0.36
[BTDA + o-TDA] PAA + DMEM	74109	46042	1.61	48142	31464	1.53	18790	8714	5796	3707	2867	0.57	0.35
[BPDA + o-TDA]	64385	38612	1.67	49874	31664	1.58	21069	20378	19891	19200	18509	0.89	0.79
[BPDA + o-TDA] PAA + DEEM	61458	37549	1.64	43969	28407	1.54	36352	20403	15898	11750	10667	0.53	0.44
[BPDA + o-TDA] PAA + DMEM	62347	37838	1.65	45896	29513	1.56	32870	19866	15479	11913	9506	0.56	0.43
[BTDA + SiDA]	16378	12519	1.31	11893	9828	1.07	204	195	195	180	153	0.22	0.17
[BPDA +SiDA] PAA + DEEM	14096	11183	1.26	10228	8594	1.16	225	205	195	170	163	0.12	0.10
[BPDA +SiDA] PAA + DMEM	14385	11339	1.27	10658	8881	1.19	307	270	240	180	143	0.12	0.09
[BPDA +SiDA]	16886	12992	1.30	11692	9583	1.17	204	195	190	170	163	0.21	0.15
[BPDA +SiDA] PAA + DEEM	14923	11795	1.26	10143	8897	1.08	184	175	170	170	163	0.12	0.10
[BPDA +SiDA] PAA + DMEM	15257	11971	1.27	10534	9003	1.12	220	200	180	177	174	0.14	0.13

PD: Polydispersity

Table 2.
Solution properties of Co-PAA, and Co-PDPAA (molecular weight, bulk viscosity, inherent viscosity and acid number) at the nth number of hours of monitoring

Polymer composition and Ionic salt formulation	Time interval between analysis (hr)									
	0		672		0	672	0	672	0	672
	Molecular Weight by GPC				Bulk Viscosity (cP)		Inh. Visc. (dl/g)		Acid Number*	
	Mw	Mn	Mw	Mn						
BTDA based Co-PAA										
[90% oTDA + 10 % SiDA]	64656	43499	39107	28321	3543	870	0.62	0.42	72.3	62.7
[85% oTDA + 15 % SiDA]	60457	41602	36423	26735	3197	645	0.60	0.39	72.7	62.2
[75% oTDA + 25 % SiDA]	55334	38253	33793	25398	1894	604	0.56	0.38	68.9	60.2
BTDA based Co-PAA + DEEM										
[90% oTDA + 10 % SiDA]	53220	37456	36731	27072	1802	727	0.30	0.24	61.6	50.4
[85% oTDA + 15 % SiDA]	47243	33989	34199	25479	1321	634	0.28	0.23	57.5	50.7
[75% oTDA + 25 % SiDA]	40270	29841	30777	23210	1004	419	0.27	0.21	60.6	51.2
BTDA based Co-PAA + DMEM										
[90% oTDA + 10 % SiDA]	48198	34601	35369	26237	7167	2949	0.30	0.23	60.4	48.1
[85% oTDA + 15 % SiDA]	46710	33164	27903	20990	1475	583	0.27	0.22	60.0	47.2
[75% oTDA + 25 % SiDA]	38537	28129	27903	20990	655	368	0.27	0.19	62.4	48.2
BPDA base Co-PAA										
[90% oTDA + 10 % SiDA]	55723	39026	36406	26531	2796	1024	0.55	0.43	73.7	63.2
[85% oTDA + 15 % SiDA]	46402	32880	36076	26373	1167	675	0.45	0.39	71.0	62.6
[75% oTDA + 25 % SiDA]	37801	27448	28399	21443	532	337	0.35	0.30	69.1	63.6
BPDA based Co-PAA + DEEM										
[90% oTDA + 10 % SiDA]	51634	36964	33828	24967	3062	1126	0.31	0.25	59.7	55.0
[85% oTDA + 15 % SiDA]	47963	34642	31903	23512	1618	655	0.28	0.22	60.8	57.9
[75% oTDA + 25 % SiDA]	38941	28382	27913	21101	675	399	0.27	0.20	59.8	56.3
BPDA based Co-PAA + DMEM										
[90% oTDA + 10 % SiDA]	53139	37596	37293	27323	1792	706	0.31	0.24	58.5	51.8
[85% oTDA + 15 % SiDA]	47736	34221	34904	25912	1290	593	0.28	0.22	64.8	51.6
[75% oTDA + 55 % SiDA]	40572	29670	31383	23498	952	471	0.25	0.21	59.7	54.4

*measured as mg of TMAH/g of the solution

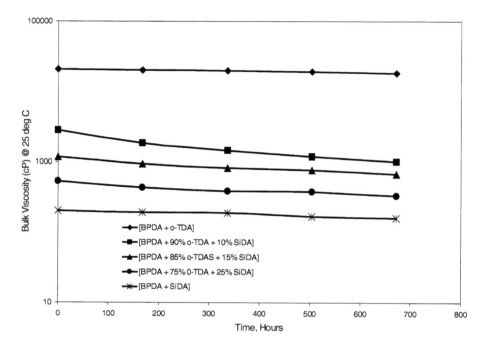

Figure 12. Bulk viscosity as a function of storage time at 25°C for BPDA based polymer systems.

It is known that a poly(amic acid) solution can undergo chemical transformation upon extended storage. During storage of freshly synthesized PAA at ambient temperature, in addition to propagation (forward) reaction, depolymerization, amic acid hydrolysis as well as some imidization reaction can also take place [18]. Some PAA solutions can also undergo pseudogelation, which could be due to the physical association of the chains in the solution and polymer-solvent interaction [19-20]. If pseudogelation had taken place in our polymers, then bulk viscosity would have increased to well above 100,000 cP. However, this was not the case for the polymers in our study as viscosity and molecular weight values drifted to lower values.

We observed that the initial viscosity of [BTDA + o-TDA] PAA and [BPDA + o-TDA] PAA based PDPAA (with DEEM) increased in the range of about 18 and >28%, respectively, than using DMEM. This could be attributed to the larger molecular dimensional volume of DEEM than that of DMEM and ionic bonding of methacrylate moieties to the carboxylic groups on the PAA. We also noted that the average rate of bulk viscosity drift of BTDA based Co-PAA system was 2.84 cP/hr, and 1.62 orders of magnitudes higher than that of BPDA based Co-PAA system. However, the average rate of bulk viscosity drift of both BTDA and BPDA based Co-PDPAA systems was around 1.10 cP/hr.

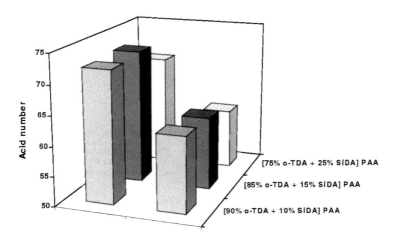

Figure 13. Acid number of BPDA based co-poly(amic acid) (Co-PAA) after 1 month storage at RT.

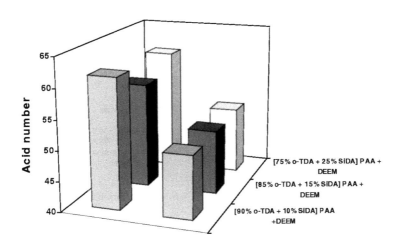

Figure 14. Acid number of BTDA based Co-PDPAA [ionic salt formulation of co-poly(amic acid) (Co-PAA) with DEEM] after 1 month storage at RT.

Even though bulk viscosity, inherent viscosity and molecular weight decreased over a period of time in our study, the acid number was also shown to decrease rather than increase (Figures 13-17). We used BTDA with an E_a (electron affinity) value of 1.57 eV, BPDA with an E_a value of 1.38 eV and a very weak diamine, SiDA, as co-diamine in the synthesis of Co-PAA. It has been reported that most reactive dianhydrides with high E_a values are most susceptible to premature hydrolysis during storage with an increase in acid number and are most likely to undergo side reactions with an amide solvent [21, 22].

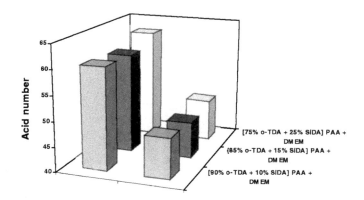

Figure 15. Acid number of BPDA based Co-PDPAA [ionic salt formulation of co-poly(amic acid) (Co-PAA) with DMEM] after 1 month storage at RT.

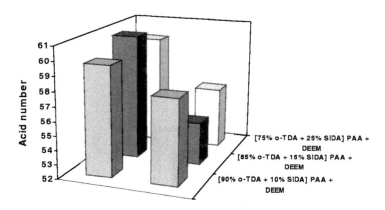

Figure 16. Acid number of BTDA based Co-PDPAA [ionic salt formulation of co-poly(amic acid) (Co-PAA) with DEEM] after 1 month storage at RT.

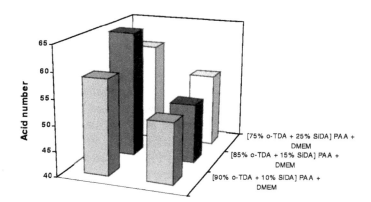

Figure 17. Acid number of BPDA based Co-PDPAA [ionic salt formulation of co-poly(amic acid) (Co-PAA) with DMEM] after 1 month storage at RT.

This explanation holds true evidently due to higher viscosity drift of BTDA than BPDA based Co-PAA systems. On analysis of our results, we could further explain that in our synthesis of Co-PAA involving the less reactive, i.e. the less basic diamine SiDA, there would be a lower ratio of forward reaction/reverse reaction. As a result, once the degradation of poly (amic acid) has begun, the starting dianhydride and diamine are regenerated (as shown in Figure 18).

These regenerated monomers may not immediately undergo further repolymerization in the presence of less reactive diamine and absence of stirring/shaking (kinetic energy) especially in the case when poly(amic acid) is stored undisturbed at ambient temperature. This would lead to further degradation of polymer's molecular weight and hence viscosity. It is also clear from the NMR spectra and acid number values that in our polymer system, the main weak link is located at the o-carboxyamide connected to SiDA as show in Figures 3, 6 and 10.

Even after the initial synthesis period for a poly(amic acid) is over, a molecular equilibration continues for a long time and simultaneously the o-carboxyamide group generated during reaction of amine with anhydride can undergo several reactions at ambient temperature. One such reaction is the hydrolysis via formation of an anhydride intermediate as shown in Figure 18. The rate constant for the reaction of dianhydride with the diamine is five to six orders of magnitude higher that that of the reverse reaction (depolymerization) but only one order of magnitude higher than the rate constant of dianhydride hydrolysis with water.

As per the literature [18, 23-24] the following types of reactions (Figure 19) might be taking place at RT in our PAA, Co-PAA and Co-PDPAA containing silicone diamine (SiDA).

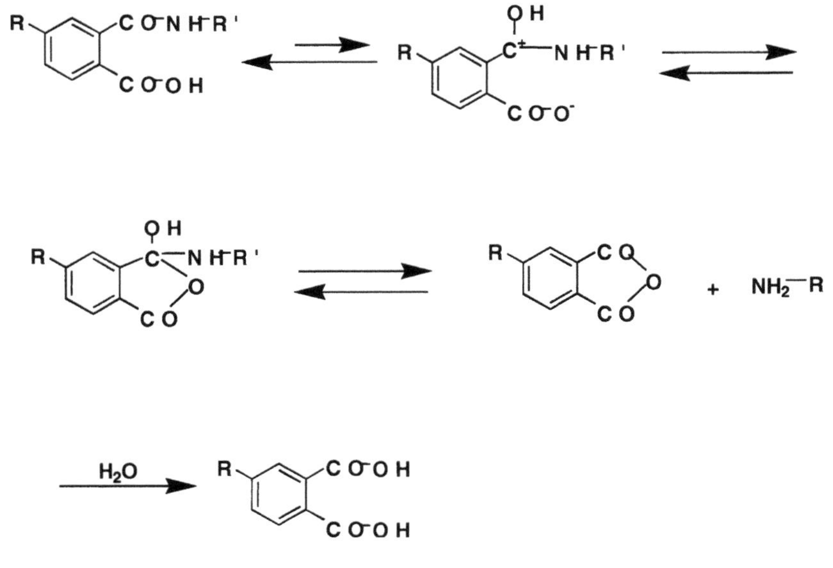

Figure 18. Hydrolysis mechanism of amic acid as reported by Harris [23].

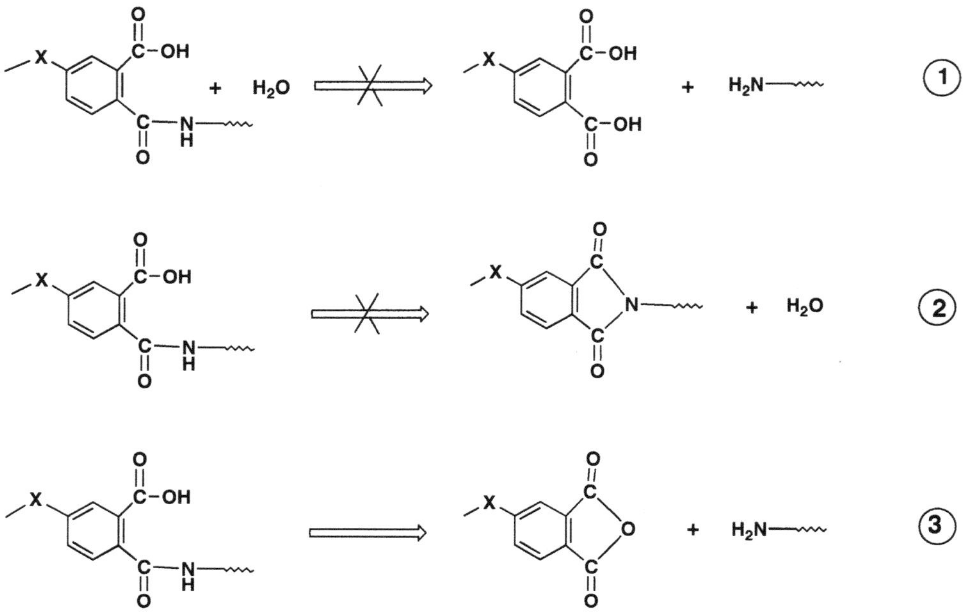

Figure 19. Potential reaction mechanisms of PAA, Co-PAA and Co-PDPAA degradation at RT.

Out of above three potential poly(amic acid) degradation reactions shown in Figure 19, the first two are ruled out as per the following reasoning: the first reaction is ruled out, since in our study the acid number was shown to decrease rather than increase (Figures 13-17) which would have been had it undergone hydrolysis reaction as per the first reaction. Large scale imidization is also less likely as its rate constant is two or three orders of magnitude less than that for the depolymerization reaction as well as that for amic acid hydrolysis. Hence the second reaction is also ruled out. Therefore the only feasible reaction is the anhydride formation, the third reaction.

3.3. Rheological study of PAA, Co-PAA and Co-PDPAA

It is known from a laboratory study that the rheological properties of polymer solutions depend on the bulk viscosity of the solution [25]. In our study, the bulk viscosity was not extremely high but was sufficiently high to be determined by the Brookfield viscometer. We have studied the rheology (shear rate and temperature dependence of viscous flow) and calculated the activation energy of the viscous flow behaviour of the PAA, Co-PAA and Co-PDPAA.

3.3.1. Bulk viscosity as a function of shear rate at constant temperature
A shear rate dependence of bulk viscosity was found in most polymer solutions. In a dilute solution, the randomly extended chains can deform and orient with the

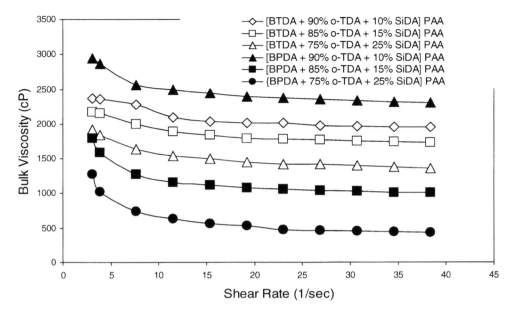

Figure 20. Shear rate dependence of bulk viscosity for both series of BTDA and BPDA based Co-PAAs.

applied shear and thus offer less resistance to shear [26]. Figures 20, 21 and 22 show the comparison of shear rate dependence of bulk viscosity of both BPDA and BTDA based series of Co-PAA and Co-PDPAA with DEEM and DMEM. All the curves show similar trends, and a sharp shear thinning behaviour is observed early on, in the experiment at shear rates from 3 to 15 sec^{-1}. than from about 20 to 38 sec^{-1} indicating that the bulk viscosity remained nearly stable.

The decrease in viscosity with increasing shear rate could be interpreted in terms of stretching and disentangling of polymer chains in the viscous mass. At low shear rates, the polymer chain entanglements impede shear flow and, therefore, viscosity is high. As the shear rate increases, this allows polymer chains to orient and slip through each other in the direction of flow and disentangle from one another and the viscosity decreases. In other words, the decrease of bulk viscosity with increasing shear rate was attributed to the gradual progressive disentangling of polymer chains due to shearing of the solution.

Within the range of measurements, all shear rate dependent curves of bulk viscosity for both series of BPDA and BTDA based Co-PDPAAs with DEEM as well as DMEM showed two regions with quite different curvatures. When the shear rate was less than 20 sec^{-1}, the bulk viscosity decreased more with the increase in shear rate. Then, the bulk viscosity remained stable from 20 to 38 sec^{-1}. However, interestingly, for both BPDA and BTDA based series of Co-PDPAAs with DMEM there was a sharper drop in viscosity in the shear rate range of 3 to 20 sec^{-1}.

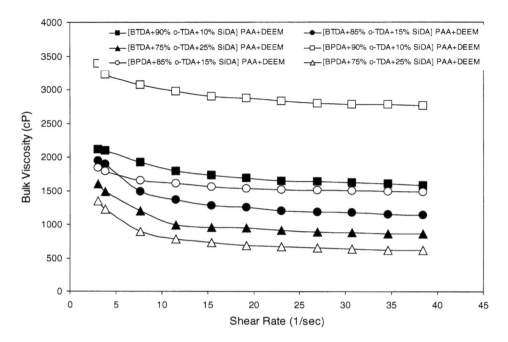

Figure 21. Shear rate dependence of bulk viscosity for both series of BTDA and BPDA based Co-PDPAAs with DEEM.

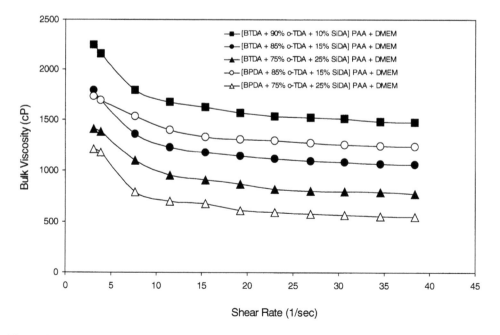

Figure 22. Shear rate dependence of bulk viscosity for both series of BTDA and BPDA based Co-PDPAAs with DMEM.

3.3.2. Bulk viscosity as a function of temperature at constant shear rate

The viscosity of a polymer solution usually varies with shear temperature [25] according to an exponential Arrhenius type relation given by

$$\eta = A \ e^{\left(E_a/RT\right)}$$

where η viscosity

A pre-exponential factor

E_a activation energy

R gas constant and

T absolute temperature.

Over limited ranges of temperatures a plot of log η versus $[1/T]$ will produce a straight line and the activation energy, E_a, will be given by the slope:

$$\ln \eta = \ln A + \left(E_a/RT\right)$$

It is well known that the flow of a fluid must overcome the energy barrier. For the small molecule based fluids this energy barrier arises from the friction force between the molecules. However, for the polymer solution, especially for a concentrated solution, this energy barrier primarily arises due to the sufficiently larger entangled polymer chains as well as interaction between polymer chains and solvent. Strong solvating NMP would swell the polymer as well as create hydrogen bonding with the polymer chain and thus increase the activation energy [25].

Figures 23 and 24 show that the temperature dependence of bulk viscosity for both series of BPDA and BTDA based Co-PDPAAs with DEEM as well as DMEM followed the Arrhenius equation. The activation energy of flow for BPDA and BTDA based Co-PAAs, and Co-PDPAAs with DEEM and DMEM calculated from the slope of ln (Viscosity) vs. 1/T in the plot of ln η vs. $[1/T]$ are shown in Figures 25 to 32 and are reported in Tables 3 and 4.

Table 3.
Activation energy of series of BTDA and BPDA based Co-PAAs and Co-PDPAAs with DEEM and DMEM

o-TDA (mole %)	SiDA (mole %)	E_a (kJ/mole)					
		BTDA			BPDA		
		Co-PAA	DEEM	DMEM	Co-PAA	DEEM	DMEM
90	10	27.34	26.28	25.91	28.44	30.74	32.91
85	15	27.04	25.38	24.18	22.81	25.49	24.81
75	25	25.59	24.28	22.58	12.55	20.65	19.17

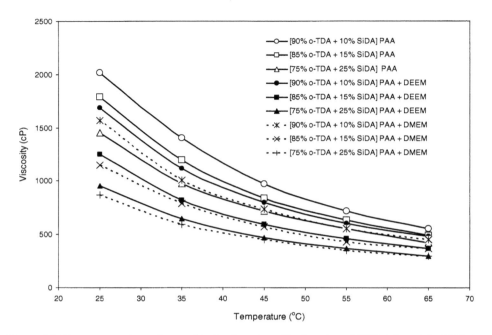

Figure 23. Temperature dependence of bulk viscosity of both series of BTDA based Co-PDPAAs with DEEM and DMEM.

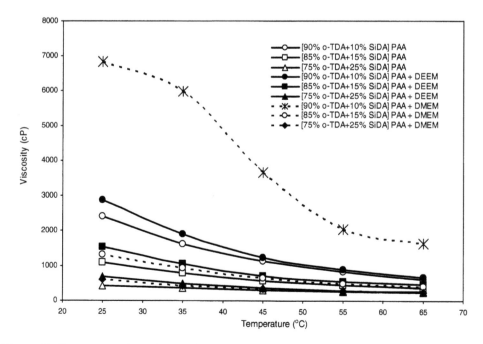

Figure 24. Temperature dependence of bulk viscosity of both series of BPDA based Co-PDPAAs with DEEM and DMEM.

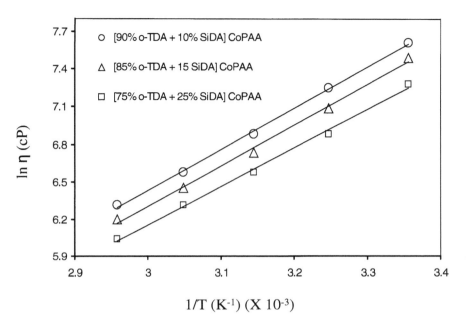

Figure 25. ln η vs. [1/T] plot for series of BTDA based Co-PAAs.

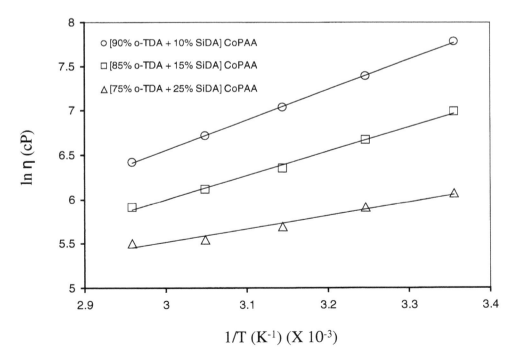

Figure 26. ln η vs. [1/T] plot for series of BPDA based Co-PAAs.

R.H. Vora et al.

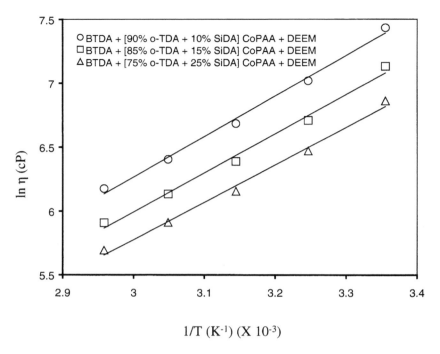

Figure 27. ln η vs. [1/T] plot for BTDA based Co-PDPAAs with DEEM.

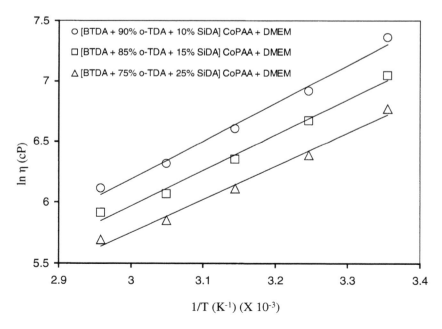

Figure 28. ln η vs. [1/T] plot for BTDA based Co-PDPAAs with DMEM.

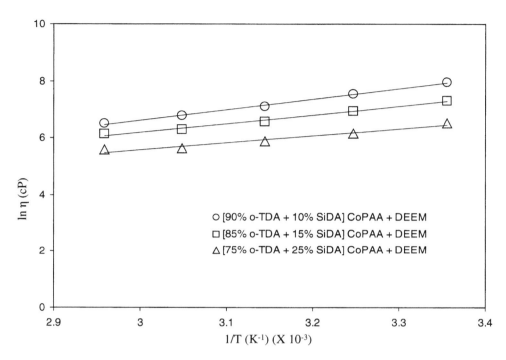

Figure 29. ln η vs. [1/T] plot for BPDA based Co-PDPAAs with DEEM.

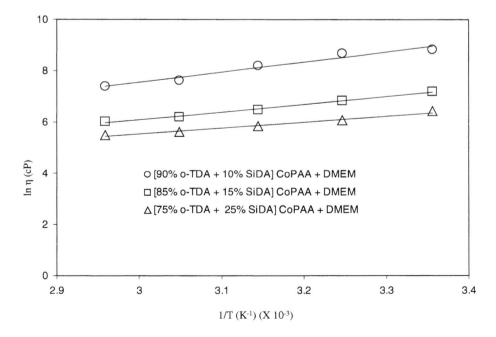

Figure 30. ln η vs. [1/T] plot for BPDA based Co-PDPAAs with DMEM.

R.H. Vora et al.

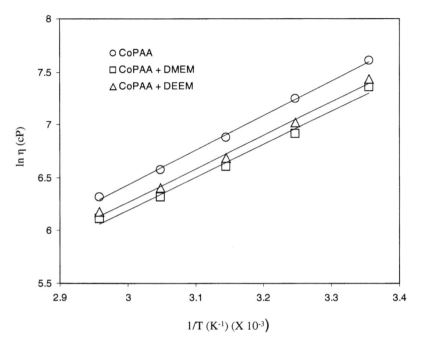

Figure 31. ln η vs. [1/T] plot for 10 mole % SiDA containing BTDA based Co-PDPAAs with DEEM and DMEM.

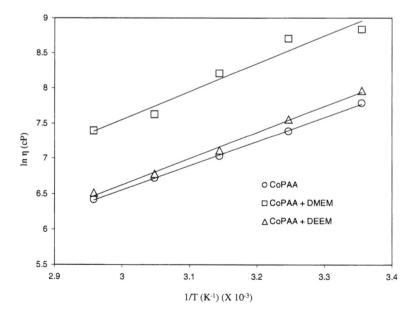

Figure 32. ln η vs. [1/T] plot for 10 mole % SiDA containing BPDA based Co-PDPAAs with DEEM and DMEM.

Table 4.
Activation energy of BTDA and BPDA based Co-PAAs and Co-PDPAAs with DEEM and DMEM

o-TDA (mol %)	SiDA (mol %)	Formulation with	E_a (kJ/mole)	
			BTDA	BPDA
90	10	–	27.34	28.44
90	10	DEEM	26.28	30.74
90	10	DMEM	25.91	32.91

The activation energies for viscous flow of BTDA based Co-PAAs and Co-PDPAAs are in the range of 25-27 and 22-26 kJ/mol respectively; whereas for BPDA based Co-PAAs and Co-PDPAAs are in the range of 12-28 and 19-33 kJ/mol, respectively. Both the activation energy for viscous flow and ln [A] decreased with increase in SiDA content in both Co-PAA and Co-PDPAA series.

For a given composition, the activation energy for viscous flow is comparable for Co-PDPAAs based on both DMEM and DEEM. The higher the measurement temperature, the more obvious the change in the bulk viscosity was observed.

While comparing the temperature dependence of the bulk viscosity for both series of BPDA and BTDA based Co-PAAs having 10 mole % SiDA as well as their respective Co-PDPAAs with DEEM and DMEM, the changes in activation energy was found to be in the reverse order as shown in Table 4.

4. THERMAL STABILITY STUDY OF POLYIMIDES (PI), CO-POLYIMIDES (CO-PI)

As a part of the overall project, a study was undertaken to determine the thermal stability of solid PI and CoPI derived from BTDA and BPDA based PAA, Co-PAA and Co-PDPAA, via in-situ thermal imidization using a thermogravimetric analyzer (TGA) at a heating rate of 10°C/min in flowing air (10 cc/min).

First, solid PAA and Co-PAA were precipitated from their solutions in NMP by reverse precipitation in de-ionised water at room temperature in a high speed blender chopper. These were washed several times with fresh DI-Water, allowed to air dry overnight and further dried in a vacuum desiccator overnight.

Similarly the PDPAA and Co-PDPAA were precipitated from their solutions in NMP by reverse precipitation in acetone at room temperature in a high speed blender chopper. They were washed several times with fresh acetone, allowed to air dry overnight and further dried in a vacuum desiccator overnight.

Samples were then analysed for their thermal stability by TGA. The activation energy for thermal degradation was determined using Coats-Redfern [27] and Chang methods [28].

R.H. Vora et al.

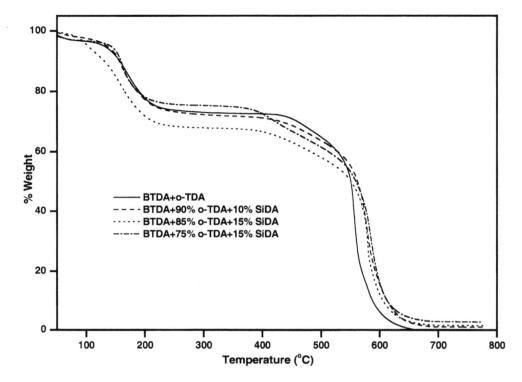

Figure 33. TGA thermograms of BTDA based Co-PDPAA [ionic salt formulation of co-poly(amic acid) with DEEM].

It is obvious that when PAA is heated it undergoes imidization and converts to a polyimide. A similar process occurs when a film of ionic salt of a poly(amic acid) on silicon wafer is thermally cured after photolithography process in the actual IC-chip fabrication process as shown in Figure 1. During the thermal curing of Co-PDPAA, in addition to water, the photosensitive acrylate compound is also liberated and degraded during the process. Hence the shapes of the curves in Figure 33 are different from those of pure polyimides in Figure 34.

From the slicing data of the thermograms of BTDA and BPDA based PI's and Co-PIs, we have calculated the thermal degradation kinetic energy using Coats and Redfern as well as Chang methods.

Coats and Redfern developed the following equation relating α with T

$$\ln\left[\frac{-\ln(1-\alpha)}{T^2}\right] = \ln\left(\frac{A\,R}{\phi\,E_a}\right)\left[1-\left(\frac{2RT}{E_a}\right)\right] - \left(\frac{E_a}{RT}\right)$$

where α is the fraction decomposed at temperature T

ϕ is heating rate

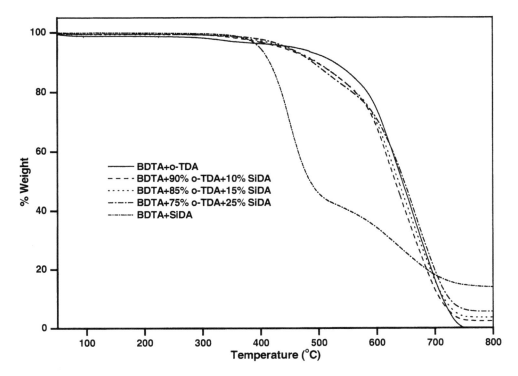

Figure 34. TGA thermogram of BTDA based series of polyimides and copolyimides derived from solid PAA and Co-PAA.

E_a is activation energy for the decomposition reaction

R is universal gas constant

A is Arrhenius frequency factor

When the order of reaction is one, then the plot of ln [-ln $(1-\alpha)$ / T^2] vs. 1/T gives a straight line with slope equivalent to $(-E_a / R)$. The results are shown in Table 5.

Whereas from the Chang method:

$$\ln\left[\frac{\left(d\alpha/dt\right)}{\left(1-\alpha\right)^n}\right] = \ln(A) - \frac{E_a}{RT}$$

where $\left(d\alpha/dt\right)$ is the rate of fraction decomposed

α is fraction decomposed at temperature T

E_a is activation energy for the decomposition reaction

R is universal gas constant

A is Arrhenius frequency factor

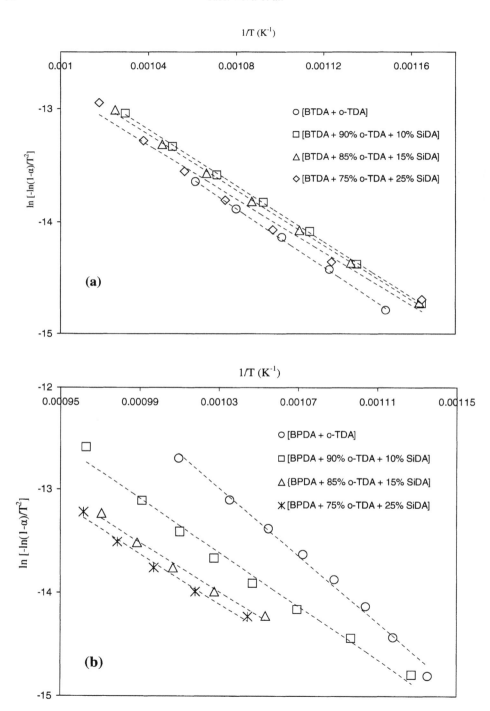

Figure 35. Activation energy (E_a) calculated by the Coats and Redfern method for both (a) BTDA and (b) BPDA based Co-PIs derived from solid Co-PDPAAs.

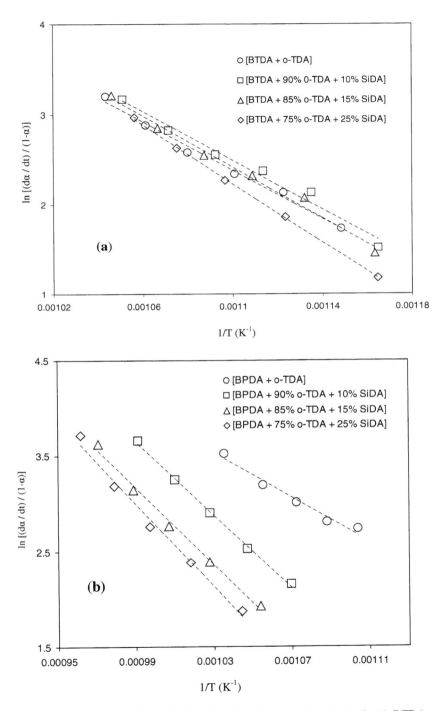

Figure 36. Activation energy (E_a) calculated by the Chang method for both (a) BTDA and (b) BPDA based copolyimides derived from solid Co-PDPAA.

R.H. Vora et al.

Table 5.
Activation energy for thermal degradation of BTDA and BPDA based copolyimides

o-TDA (mole %)	SiDA (mole %)	E_a (kJ/mole) BTDA based copolyimide		BPDA based copolyimide	
		Coats-Redfern method	Chang method	Coats-Redfern method	Chang method
90	10	106.7	117.8	93.2	92.2
85	15	101.6	112.4	73.2	76.4
75	25	98.2	106.4	65.6	74.95

A plot of ln $[(d\alpha / dt) / (1-\alpha)^n]$ against $1 / T$ yields a straight line if the decomposition order n is selected correctly. A straight line was obtained when n is equal to one. The slope and intercept of this line provide the $(-E_a/R)$ and ln (A) values, respectively. The results are shown in Table 5.

The thermograms of both series of materials showed a two-step weight loss in air atmosphere. The first step was obviously due to imidization to polyimide and loss of water produced as a by-product and the second step was due to degradation of polyimide. The activation energies determined by Coats-Redfern as well as Chang methods decreased as the silicon content in the polymer increased, indicating that less energy was required to degrade the polymer with silicon diamine (SiDA). The extent of decrease in activation energy was comparable.

5. CONCLUSIONS

For both series of PAA, Co-PAA and Co-PDPAA, the bulk viscosity decreased upon storage at room temperature for a period of one month, irrespective of the chemical composition or the chemical structure of the Co-PAA or Co-PDPAA. Also molecular weights, inherent viscosity and acid number decreased as a function of time at room temperature, which indicated that the most likely reaction mechanism was anhydride formation. The weak linkage was the amide bond between the dianhydride moiety and the siloxane diamine (SiDA) moiety. Ionic salt formulation (Co-PDPAA) was less stable than PAA. BPDA based PAA showed less viscosity drift than BTDA based PAA. This behaviour can be explained on the basis of E_a values of BPDA and BTDA. The viscosity drift of Co-PDPAA formulation with DEEM was lower than with DMEM.

Shear thinning was observed in both PAA and Co-PDPAA. The shear temperature dependence of the bulk viscosity of PAA and Co-PDPAA followed the exponential Arrhenius relationship.

Thermal stability behaviour of both DEEM and DMEM systems showed similar trends and distinctly showed two-step weight losses. Imidization occurred from 50 to 350°C (removal of water, and DEEM or DMEM) followed by thermal degradation in the second step.

Acknowledgements

The authors would like to express their sincere gratitude for the materials (o-TDA diamine) support provided by the Wakayama Seika Kogyo Co. Ltd. Japan. Also thanks are due to Dr. Pramoda Kumari, Engineering Fellow for useful discussion on DSC analysis.

REFERENCES

1. L.M. Sonnett and T.P. Gannett, in: *Polyimides: Fundamentals and Applications*, M.K. Ghosh and K.L. Mittal (Eds.), Ch.6, pp. 151-185, Marcel Dekker, New York (1996).
2. V.I. Kolegov, V.P. Sklizkova, V.V. Kudryavtsev, B.G. Belen'kii, S.Y. Frenkel and M.M. Koton, Dokl. Akad. Nauk. USSR (Engl. Translation) **232**, 848 (1977).
3. N.G. Bel'nikevich, V.M. Denisov, L.N. Korzhavin and S.Y. Frankel, Vysokomol. Soedin. Ser. A, **23**, 1268 (1981).
4. K. Matsumoto and H. Saitoh, Kobunshi Ronbunshu **48**, 711 (1991).
5. P.S.G. Krishnan, R.H. Vora and T-S. Chung, Polymer, **42**, 5165 (2001).
6. Y. Tong, S. Veeramani and R.H. Vora, Paper presented at the 6th World Congress of Chemical Engineering, Melbourne, AUSTRALIA (Sep. 2001).
7. W. Volksen and P.M. Cotts, in: *Polyimides: Synthesis, Characterization and Applications*, K.L. Mittal (Ed.) Vol. 1, pp. 163, Plenum, New York (1984).
8. A. Berger and R.H. Vora, US Patent 4,681 928 (1987).
9. A.L. Landis and A.B. Naselow, US Patent 4, 645 824 (1987).
10. D.M. Stoakley, A.K. St. Clair and R.M. Baucom, SAMPE Quarterly, **4**, 3 (1989).
11. P.M. Hergenrother and S.J. Havens, in: *Polyimides: Materials Chemistry and Characterization*, C. Feger, M.M. Khojasteh and J.E. McGrath (Eds.), p. 45, Elsevier, New York (1989).
12. R.H. Vora, US Patent 4,933 132 (1990).
13. Y.S. Negi, Y-I. Suzuki, I. Kawamura, T. Hagiwara, Y. Takahashi, M. Iijima, M-A. Kakimoto and Y. Imai, J. Polym. Sci.: Part A: Polym. Chem., **30**, 2281 (1992).
14. J.D. Summers, B.C. Auman, M.J. Grovola and M.A. Guidry, Presented at 6th Int. Conf. on Polyimides and Other Low K Dielectrics, held at McAfee, New Jersey (1997).
15. R.H. Vora, P.S.G. Krishnan, S.H. Goh and T.-S. Chung, Adv. Funct. Mater., **11**, 361 (2001).
16. T. Yilmaz, H. Guclu, O. Ozarslan, E. Yildiz, A. Kuyulu, E. Ekinci and A. Gungor, J. Polym. Sci., Part A, **35**, 2981 (1997).
17. Y.J. Kim, T.E. Glass, G.D. Lyle and J.E. McGrath, Macromolecules, **26**, 1344 (1993).
18. M.I. Bessonov, M.M. Koton, V.V. Kudryavtsev and L.A. Laius, *Polyimides: Thermally Stable Polymers*, Ch. 1, pp. 1-96, Plenum, New York (1987).
19. R.A. Dine-Hart and W.W. Wright, J. Appl. Polym. Sci. **11**, 609 (1967).
20. J.I. Jones, F.W. Ochynski and F.A. Rackley, Chem. Ind., 1686 (1962).
21. L.W. Frost and I. Kesse, J. Appl. Polym. Sci. **8**, 1039 (1964).
22. M.L. Bender, Y.L. Chow and F. Chluoek, J. Am. Chem. Soc. **80**, 5380 (1958).
23. F.W. Harris, in: *Polyimides*, D. Wilson, H.D. Stenzenberger and P.M. Hergenrother (Eds.), Ch. 1, pp. 1-37, Chapman and Hall, New York (1990).
24. T. Takekoshi, in: *Polyimides: Fundamentals and Applications*, M.K. Ghosh and K.L. Mittal (Eds.), Ch. 2, pp. 7-48, Marcel Dekker, New York (1996).
25. D.W. Van Krevelen, *Properties of Polymers*, Ch. 16, pp. 499-520, Elsevier, New York (1990).
26. J.R. Fried, *Polymer Science and Technology*, Ch. 11, pp. 373-426, Prentice Hall (1995).
27. A.W. Coats and J.W. Redfern, J. Polym. Sci. Part C Polym. Lett. **3**, 917 (1965).
28. W.L. Chang, J. Appl. Polym. Sci., **53**, 1759 (1994).

Polyimides and Other High Temperature Polymers, Vol. 2, pp. 37–45
Ed. K.L. Mittal
© VSP 2003

A new interpretation of the kinetic model for the imidization reaction of PMDA-ODA and BPDA-PDA poly(amic acid)s

CHANG-CHUNG YANG, KUO HUANG HSIEH and WEN-CHANG CHEN[*]

Department of Chemical Engineering, National Taiwan University, Taipei 106, Taiwan

Abstract—In this study, a new interpretation is given for the imidization of poly(amic acid)s based on the Seo model: $k(t) = b \times \mathrm{sech}(-at)$. The parameters a and b in the original Seo model are the rate constant of the imidization reaction and the invariant constant, respectively. However, according to the equation $k(t) = b \times \mathrm{sech}(-at)$, $k(t)$ approaches b at the initial time, $t \to 0$. The assumption that the parameter b is independent of temperature conflicts obviously with the fact that the rate constant of first-order reaction $k(t)$ is a function of temperature. In this study, the activation energies of a and b were proposed as the energy barriers for the transition state and the rate constant of the imidization reaction, respectively. The new interpretation was tested by imidizing two commercially available poly(amic acid)s: PMDA-ODA and BPDA-PDA from 120 to 225°C. The extent of imidization was determined by a comparison of the intensity of the 1380 cm^{-1} peak in the FTIR spectra. The experimental results suggested that the activation energy from the new interpretation was similar to the literature reports. However, the activation energy from the original Seo model was significantly different from the literature values reported.

Keywords: Poly(amic acid); imidization; kinetic model.

1. INTRODUCTION

Thermally stable polyimides have been used as a dielectric material since 1970 [1-6]. Generally, poly(amic acid)s were prepared first followed by spin coating and curing to complete the imidization reaction, as shown in eq. (1):

$$(1)$$

Kinetic studies on the imidization reaction of poly(amic acid)s have been extensively carried out [7-23]. A first-order reaction was always used to describe the

---

*To whom all correspondence should be addressed. Phone: 886-2-23628398, Fax: 886-2-23623040, E-mail: chenwc@ms.cc.ntu.edu.tw

imidization reaction since it was regarded as an intramolecular reaction. Kreuz *et al.* [7] considered that the imidization reaction involved two consecutive first-order reactions. They found that the ratio of the rate constant of the first stage to that of the second stage was 5.4. Besides, the activation energies of the first and second stages were 109 and 96.64 kJ/mole, respectively. They concluded that the decreasing rate constant with curing time was due to the stiffness of molecular chain segments caused by the transformation of the soft poly(amic acid)s to rigid polyimides. Bessonov *et al.* [8] obtained an average value of activation energy for the first-order reaction at the first stage of 102.1 kJ/mole for many different poly-imides. However, Lauver [9] and Sacher [10] regarded the imidization of poly(amic acid)s as two second-order irreversible reactions because of the influence of aprotic solvents. Pyun *et al.* [11] considered the self-catalytic effect of poly(amic acid)s and proposed a second-order reaction at an early stage and a first-order reaction at a later stage.

Although it was considered that rate constants did not vary, the experimental results showed that the rate constant of imidization decreased at a later stage. Such decrease in rate constant is possible due to many effects, including the non-equivalent activities of poly(amic acid)s [7-8, 12-14], solvent evaporation [7, 12, 15-20], and the increase in the stiffness of polymer chains [12, 21, 22]. In order to explain the decreasing rate constant at a later stage of imidization, Seo and co-workers [21, 22] considered the difference in the activities of poly(amic acid)s and proposed a first-order rate constant (k) as $k(t) = b \times \text{sech}(-at)$, where t is time and a, b are parameters. They tested the proposed model with the poly(amic ester)s synthesized by themselves. They concluded that the activation energy for the imidization reaction obtained from the parameter a was 102.1 kJ/mole. The parameter, b, was found to be constant in their study.

In this study, a different explanation for the parameters a and b of the Seo model is proposed. Since $k(t) = b \times \text{sech}(-at)$ was based on the Seo model, $k(t)$ approaches b at the initial time, $t \rightarrow 0$. As the first-order rate constant $k(t)$ is a function of temperature, b should not be an invariant constant independent of temperature. Besides, parameter a represents the transition between the activated state and the de-activated state of the poly(amic acid)s, the corresponding activation energy calculated from a should be the energy barrier for the transition state instead of the energy barrier for the imidization reaction. On the other hand, parameter b indeed represents the rate constant of the reaction because $k(t) \rightarrow b$ for $t \rightarrow 0$. The new interpretation of the imidization kinetics of the Seo model was tested with two commercially available poly(amic acid)s: PMDA-ODA (pyromellitic dianhydride-*co*-4,4'-oxydianiline) and BPDA-PDA (3,3',4,4'-biphenyltetracarboxylic dianhydride-*co*-1,4-phenylenediamine). The imidization of the poly(amic acid)s was carried out at a temperature ranging from 120 to 225°C, as shown in Figure 1. The degree of imidization at various temperatures and times was evaluated using FTIR spectroscopy. The Arrhenius equation was used for both parameters a and b to obtain the corresponding pre-exponential factor and activation energy.

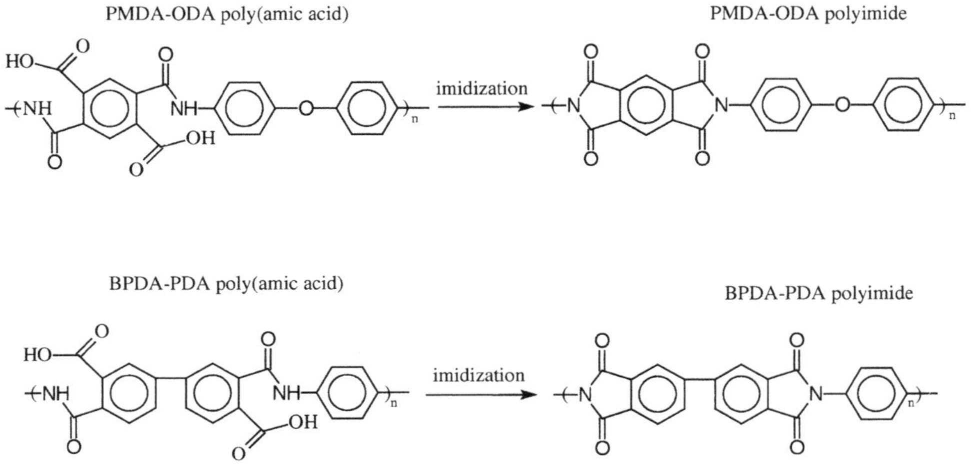

Figure 1. Imidization reactions of two poly(amic acid)s: PMDA-ODA and BPDA-PDA.

2. EXPERIMENTAL

2.1. Materials and processing

The poly(amic acid)s of pyromellitic dianhydride-*co*-4,4'-oxydianiline (PMDA-ODA, 14 ± 1 wt%, in NMP/aromatic hydrocarbon, viscosity: 10-12 poises) and 3,3',4,4'-biphenyltetracarboxylic dianhydride-*co*-1,4-phenylenediamine (BPDA-PDA, in NMP, 10.5 ± 1 wt%, viscosity: 25-30 poises) were purchased from Aldrich Co. and used without further purification. After standard RCA cleaning procedures on 4-in. silicon wafers, the wafers were spin-coated with poly(amic acid)s to form uniform thin films. The spin-coating was carried out at 500 rpm for 30 sec followed by 4000 rpm for another 60 sec. These films were then heat treated for imidization kinetics study.

2.2. Degree of imidization (x)

The FTIR spectra of cured polyimides were obtained using a Bio-rad QS300 FTIR spectrophotometer. Conversion of the poly(amic acid)s to polyimides was monitored by a comparison of the peak intensity at 1380 cm^{-1} between the curing temperature and 400°C. Here, it is assumed that the poly(amic acid)s are fully converted to polyimides after curing at 400°C for two hours. To eliminate the difference in absorption due to variation of film thickness after curing, the peak at 1500 cm^{-1} was used as an internal standard for comparison. This peak represents vibration of the benzene ring of the diamine part in the polyimide and this part would not participate in the imidization reaction. The degree of imidization of poly(amic acid)s to polyimides was determined using eq. (2):

$$Degree\ of\ imidization(x) = \frac{(peak\ area\ at\ 1380cm^{-1})_{time=t}}{(peak\ area\ at\ 1380cm^{-1})_{400°C}} \Bigg/ \frac{(peak\ area\ at\ 1500cm^{-1})_{time=t}}{(peak\ area\ at\ 1500cm^{-1})_{400°C}} \quad (2)$$

2.3. Evaluation of the parameters of the Seo model

The Seo model was used to correlate the degree of imidization with time [21-22]. In Seo's approach [21], the rate constant was proposed as $k(t) = b \times sech(-at)$. Inserting into a first-order rate equation, the relationship between the degree of imidization and curing time is obtained as

$$-\ln(1-x) = -\frac{2b}{a}\tan^{-1}e^{-at} + \text{constant} \quad (3)$$

The constant in eq. (3) was obtained by fitting with the experimental data to the original Seo model [21]. However, the constant in eq. (3) was found to be 0.785 by fitting our experimental data at the initial condition ($t = 0$, $x = 0$). Hence, the kinetic equation was expressed as eq. (4).

$$-\ln(1-x) = -\frac{2b}{a}(\tan^{-1}e^{-at} - 0.785) \quad (4)$$

The parameters a and b were expressed in the form of Arrhenius expressions in this study, which are shown as eqs. (5) and (6).

$$a = A_a \exp(-E_a/RT) \quad (5)$$

$$b = A_b \exp(-E_b/RT) \quad (6)$$

The constants A_a, A_b, E_a, and E_b were calculated from the Arrhenius plots of the parameters a and b.

3. RESULTS AND DISCUSSION

Figure 2 shows the FTIR spectra for the PMDA-ODA poly(amic acid) when cured at 165°C for various curing times of (a) 20 sec, (b) 4 min, (c) 110 min, and (d) at 400°C for two hours. As the curing progresses, the intensities of the absorption peaks of polyimide increase while those of the poly(amic acid) and NMP solvent decrease. Hence, it is observed in Figure 2 that the intensities of the peaks at 1680 cm^{-1} (carbonyl vibration of NMP solvent), 1640–1680 cm^{-1} (amide I of poly(amic acid)), and 1550 cm^{-1} (amide II of poly(amic acid)) decrease with increasing curing time. On the other hand, the peak intensities at 1780 cm^{-1}, 1720 cm^{-1}, 725 cm^{-1} and 1380 cm^{-1} representing carbonyl symmetric stretching, asymmetric stretching, bending vibration, and C-N vibration of the imide group, respectively, show an increasing trend with curing time.

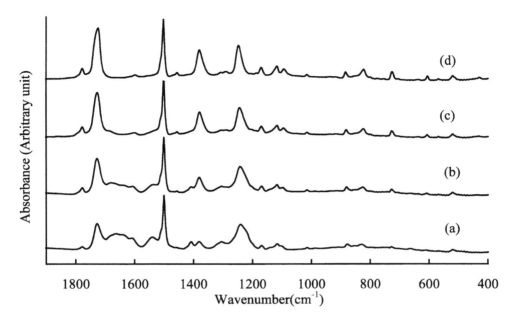

Figure 2. FTIR spectrum of PMDA-ODA poly(amic acid) cured at 165°C for different times (a) 20 sec, (b) 4 min, (c) 110 min, and (d) 400°C for two hours.

In order to quantify the degree of imidization reaction, the peak at 1380 cm^{-1} was used to represent the conversion from poly(amic acid)s to polyimides. Figure 3 shows the FTIR absorption peak at 1380 cm^{-1} for PMDA-ODA poly(amic acid) at 165°C at different curing times and for the fully imidized PMDA-ODA. It can be observed that the intensity of the peak increases with curing time, suggesting the progress of the imidization reaction. The imidization of the poly(amic acids) was studied at different curing temperatures and times.

Figures 4 and 5 show the degrees of imidization of PMDA-ODA and BPDA-PDA poly(amic acid)s, respectively, at different curing temperatures ranging from 120 to 225°C with the curing time. The degree of imidization was obtained from eq. (2) by comparing the peak areas at 1380 cm^{-1}. It can be observed from these two figures that the degree of imidization increases rapidly at a short curing time and then reaches a steady value. The higher the curing temperature, the shorter the curing time for reaching the steady value. The solid lines in Figures 4 and 5 represent the fitting curves for the experimental results based on Seo model (eq. (4)). Figure 6 shows the rate constant (k) of imidization of the PMDA-ODA poly(amic acid) at different curing temperatures. As can be observed from this figure, the rate constant k drops rapidly with increasing curing time and temperature. The parameters a and b were obtained from eq. (4) and Figures 4-6. Figure 7 shows the variation of ln(a) and ln(b) with 1/T for the imidization of PMDA-ODA and BPDA-PDA poly(amic acid)s. From the Arrhenius plots, the activation energies

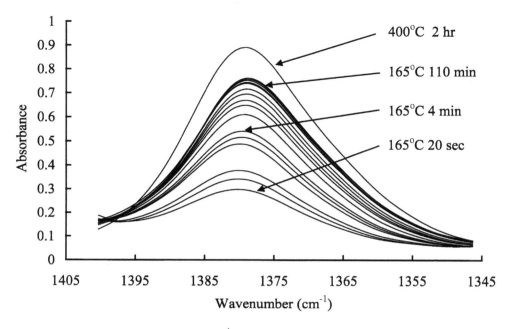

Figure 3. FTIR absorption peak at 1380 cm⁻¹ of PMDA-ODA poly(amic acid) cured at 165°C at different times and the fully imidized form by curing at 400°C for two hours.

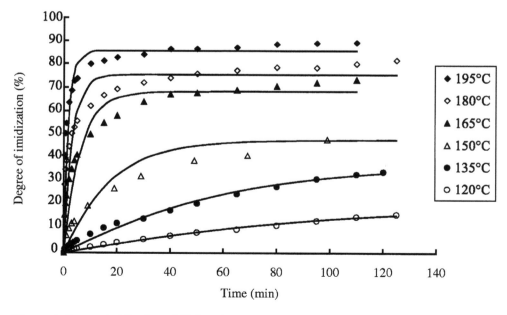

Figure 4. Percent imidization of PMDA-ODA poly(amic acid) as a function of time at different temperatures (symbols represent experimental results while solid lines represent results fitted with eqs. (4), (5), and (6).

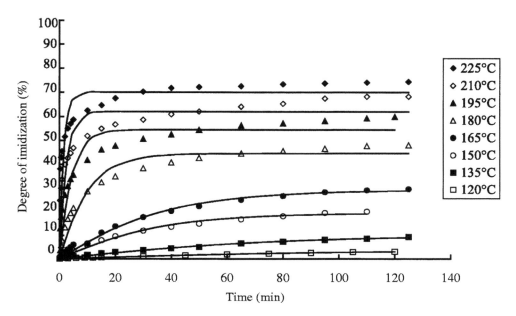

Figure 5. Percent imidization of BPDA-PDA poly(amic acid) as a function of time at different temperatures (symbols represent experimental results while solid lines represent results fitted with eqs. (4), (5), and (6).

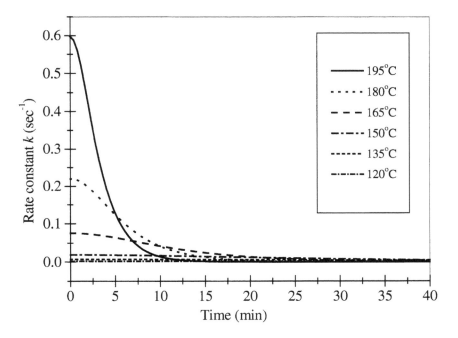

Figure 6. The first-order rate constant $k(t)$ for the imidization of PMDA-ODA poly(amic acid) as a function of time at different temperatures.

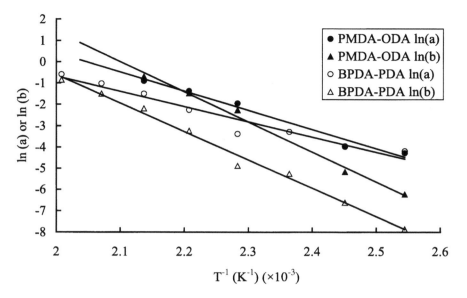

Figure 7. Arrhenius plots of kinetic parameters a and b in eqs. (5) and (6) for the poly(amic acid)s of PMDA-ODA and BPDA-PDA.

(E_a, E_b) and pre-exponential factors (A_a, A_b) were obtained. The E_a, E_b, A_a and A_b for the imidization reaction of the PMDA-ODA poly(amic acid) were found to be 74.69 kJ/mole, 117.54 kJ/mole, 9.65×10^7 s^{-1} and 7.87×10^{12} s^{-1}, respectively. For the case of BPDA-PDA poly(amic acid), the E_a, E_b, A_a and A_b are 59.37 kJ/mole, 110.88 kJ/mole, 8.09×10^5 s^{-1} and 2.10×10^{11} s^{-1}, respectively. The values of E_b for both polyimides are close to the literature value ($\sim$ 100-110 kJ/mole) [7, 8] and those of E_a from our study are much lower than those reported [7, 8]. In this study, it is proposed that the corresponding activation energy calculated from parameter a should be the energy barrier for the transition state and parameter b represents the rate constant of the imidization reaction. If the original Seo model was used to fit our experimental data according to eq. (3), the activation energies for the imidization of PMDA-ODA and BPDA-PDA poly(amic acid)s would be 64.33 and 45.16 kJ/mole, respectively. These values show a large discrepancy with the previous report by Seo *et al.* [21] or other literature reports [7, 8]. Hence, the results show that the new interpretation of the parameters a and b is more suitable to describe the Seo model than the previous reports [21, 22].

4. CONCLUSION

In this study, a new interpretation was given for the imidization of poly(amic acid)s based on the Seo model: $k(t) = b \times \mathrm{sech}(-at)$. The assumption that the parameter b in the original Seo model is independent of temperature conflicts obviously with the fact that the rate constant of first-order reaction $k(t)$ is a function of temperature since $k(t)$ approaches b at the initial time, $t \rightarrow 0$. In this study, the corresponding activation energy obtained from parameter a should be the energy barrier for the transition state and parameter b represents the rate constant for the imidization reaction. The new interpretation was successfully verified by imidization of two commercially available poly(amic acid)s: PMDA-ODA and BPDA-PDA.

Acknowledgement

The authors thank the National Science Council of Taiwan for financial support of this work.

REFERENCES

1. C.E. Sroog, in: *Polyimides: Fundamentals and Applications*, M.K. Ghosh and K.L. Mittal (Eds.), p. 1, Marcel Dekker, New York (1996).
2. P. Lakshmanan, in: "High Temperature Nanofoams Based on Ordered Polyimide Matrices", PhD Thesis, Chap. 2, p. 6, Virginia Polytechnic Institute, Blacksburg, Virginia, USA (1995).
3. H. Satou, H. Suzuki and D. Makino, in: *Polyimides*, D. Wilson, H.D. Stenzenberger and P.M. Hergenrother (Eds.), pp. 227-251, Blackie & Son, Glasgow (1990).
4. K.R. Carter, *Mater. Res. Soc. Symp. Proc.*, **476**, 87 (1997).
5. F.W. Harris, S.O. Norris, L.H. Lanier, B.A. Reinhardt, R.D. Case, S. Varaprath, S.M. Padaki, M. Torres and W.A. Feld, in: *Polyimides: Synthesis, Characterization, and Applications*, K.L. Mittal (Ed.), Vol. 1, pp. 3-14, Plenum, New York (1984).
6. F.W. Harris, in: *Polyimides*, D. Wilson, H.D. Stenzenberger and P.M. Hergenrother (Eds.), p. 1, Blackie & Son, Glasgow (1990).
7. J.A. Kreuz, A.L. Endrey, F.P. Gay and C.E. Sroog, *J. Polym. Sci.: Part A-1* **4**, 2607 (1966).
8. M.I. Bessonov, M.M. Koton, V.V. Kudryavtsev and L.A. Laius, *Polyimides: Thermally Stable Polymers*, 2nd ed., p. 57, Plenum, New York (1987).
9. R.W. Lauver, *J. Polym. Sci.: Polym. Chem.*, **17**, 2529 (1979).
10. E. Sacher, *J. Macromol. Sci. Phys.*, **B25**, 405 (1986).
11. E. Pyun, R.J. Mathisen and C.S.P. Sung, *Macromolecules*, **22**, 1174 (1989).
12. L.A. Laius, M.I. Tsapovetsky, in: *Polyimides: Synthesis, Characterization, and Applications*, K.L. Mittal (Ed.), Vol. 1, p. 295, Plenum, New York (1984).
13. M. Navarre, in: *Polyimides: Synthesis, Characterization, and Applications*, K.L. Mittal (Ed.), Vol. 1, p. 429, Plenum, New York (1984).
14. M.J. Brekner and C. Feger, *J. Polym. Sci. Polym. Chem.*, **25**, 2005 (1987).
15. A.I. Baise, *J. Appl. Polym. Sci.*, **32**, 4043 (1986).
16. R. Ginsburg and J.R. Susko, in: *Polyimides: Synthesis, Characterization, and Applications*, K.L. Mittal (Ed.), Vol. 1, p. 237, Plenum, New York (1984).
17. D. Kumar, *J. Polym. Sci.: Polym. Chem.*, **19**, 795 (1981).
18. P.D. Frayer, in: *Polyimides: Synthesis, Characterization, and Applications*, K.L. Mittal (Ed.), Vol. 1, p. 273, Plenum, New York (1984).
19. J.C. Johnston, M.A.B. Meador and W.B. Alston, *J. Polym. Sci. Part A, Polym. Chem.*, **25**, 2175 (1987).
20. S.I. Numata, K. Fujisaki and N. Kinjo, in: *Polyimides: Synthesis, Characterization, and Applications*, K.L. Mittal (Ed.), Vol. 1, p. 259, Plenum, New York (1984).
21. Y. Seo, S.M. Lee, D.Y. Kim and K.U. Kim, *Macromolecules*, **30**, 3747 (1997).
22. Y. Seo, *Polym. Eng. Sci.*, **37**, 772 (1997).

Polyimides and Other High Temperature Polymers, Vol. 2, pp. 47–70
Ed. K.L. Mittal
© VSP 2003

Synthesis and characterization of new organosoluble poly(ether imide)s derived from various novel bis(ether anhydride)s

DER-JANG LIAW,* WEN-HSIANG CHEN and CHING-CHENG HUANG

Department of Chemical Engineering, National Taiwan University of Science and Technology, Taipei 106, Taiwan

Abstract—A series of new bis(ether anhydride)s was prepared in three steps starting from nitro-displacement reaction of new diols with 4-nitrophthalonitrile to form bis(ether dinitrile)s, followed by alkaline hydrolysis of the bis(ether dinitrile)s and subsequent dehydration of the resulting bis(ether diacid)s. A series of new poly(ether imide)s was prepared from the bis(ether anhydride)s and various diamines by a conventional two-stage synthesis involving polyaddition and subsequent chemical cyclodehydration. All the polymers showed typical amorphous diffraction patterns. All of the poly(ether imide)s showed excellent solubility and were readily dissolved in various solvents such as *N*-methyl-2-pyrrolidinone, *N,N*-dimethylacetamide (DMAc), *N,N*-dimethylformamide, pyridine, cyclohexanone, tetrahydrofuran and even in chloroform. These polymers had glass transition temperatures in the range of 226–262°C. Thermogravimetric analysis showed that all polymers were stable up to 440°C, with 10% weight loss recorded in the range of 441–535°C in nitrogen. Transparent, tough and flexible polymer films could be obtained by solution casting from the DMAc solution. These polymer films had tensile strength in the range of 80–116 MPa and tensile modulus in the range of 1.7–2.7 GPa.

Keywords: Organosoluble; poly(ether imide)s; bis(ether anhydride).

1. INTRODUCTION

Aromatic polyimides are widely used in the semiconductor and electronic packaging industries because of their outstanding thermal stability, good insulation properties with low dielectric constant, good adhesion to common substrates, and superior chemical stability [1, 2]. However, their applications are limited in many fields because the early polyimides were insoluble and intractable. Therefore, considerable research has been undertaken in order to identify new ways to circumvent these limitations. The introduction of flexible groups and/or bulky units in the polymer backbone has been a general approach to alter the chemical struc-

---
*To whom all correspondence should be addressed. Phone: 886-2-27376638 or 886-2-27335050, Fax: 886-2-23781441 or 886-2-27376644, E-mail: liaw@ch.ntust.edu.tw, liaw8484@yahoo.com.tw

ture of polyimides [3–11]. Poly(ether imide)s were developed as a result of research interests in aromatic nucleophilic displacement chemistry combined with a perceived marketplace for high performance polymers which could be readily fabricated by standard plastics extrusion and injection molding processes. An important example is Ultem 1000® developed and commercialized by General Electric Co. [12], which exhibits reasonable thermal stability and good mechanical properties together with good moldability.

The incorporation of 2,2'-disubstituted biphenylene in a para-linked polymer chain reduced the interactions between polymer chains. The phenyl rings are forced by the 2,2'-disubstitution into a noncoplanar conformation, decreasing the intermolecular forces between the polymer chains. The crystallization tendency is markedly lowered and the solubilities are significantly enhanced [13–18]. On the other hand, another effective approach to obtain organosoluble polyimides is the incorporation of substituted methylene linkages, such as isopropylidene [19–22], hexafluoroisopropylidene [23–27] and diphenylmethylene [28] units which provide kinks between the rigid phenyl rings in the backbone and lead to enhanced solubility of the polymer. The incorporation of these flexible linkages such as isopropylidene into the polymer backbone is expected to reduce the crystallinity, and enhance the solubility and melt-moldability of the poly(ether imide)s [19–22]. The improved solubility of the polymers is ascribed to the presence of the kink units in the polymer backbone that lower the chain rigidity. It was observed that the polymers with diphenylmethylene unit showed better thermal stability than those containing isopropylidene and hexafluoroisopropylidene [28]. Therefore, the incorporation of noncoplanar 2,2'-dimethyl-4,4'-biphenylene and kink diphenylmethylene in the poly(ether imide) backbone was expected to provide organosoluble poly(ether imide)s with good thermal stability. The introduction of cardo (Latin meaning loop) groups into the backbone of polymers is another approach for improving solubility and, thereby, processability. Cardo polymers exhibit a valuable set of properties: the combination of an increased thermal stability with an increased solubility in organic solvents because of the specific contribution of the cardo groups [29–32]. In our previous works, we have found several means for the introduction of cardo groups such as cyclododecylidene [33], adamantane [34], norbornyl [35] and tricyclo[5.2.1.0$^{2.6}$]decane [36] groups in the polymer backbone. In these attempts, the solubility of polyimide was enhanced while high glass transition temperature and thermal stability were maintained [33–36]. In continuation of these studies, we were interested in the potential usefulness of a *tert*-butylcyclohexylidene group as a bulky pendent group in the polymer backbone. The cardo group such as *tert*-butylcyclohexylidene could be considered as a bulkier pendent group as compared with other pendent groups mentioned above (such as cyclododecylidene, adamantane, norbornyl and tricyclo[5.2.1.0$^{2.6}$]decane groups [33–36]. The bulkier groups will possibly contribute to an enhanced solubility of the polymers [37].

Our group has reported the preparation of new diols such as 2,2'-dimethyl-4,4'-dihydroxybiphenyl (**1A**), bis(4-hydroxyphenyl)diphenylmethane (**1B**), 1,1-bis(4-

hydroxyphenyl)-4-*tert*-butylcyclohexane (**1C**), 1,1-bis[4-hydroxyphenyl]cyclodo-decane (**1E**) and 1,1-bis(4-hydroxyphenyl)-4-phenylcyclohexanon (**1F**) (Scheme I) [5–8, 38–43]. The commercial diol, 2,2-bis(4-hydroxy-3,5-dimethylphenyl)pro-pane (**1D**), was purchased from TCI. The present study is concerned with the syn-thesis and characterization of a series of new highly soluble poly(ether imide)s based on new bis(ether anhydride)s such as 2,2'-dimethyl-4,4'-bis[4-(3,4-dicarboxyphenoxy)]biphenyl dianhydride (**4A**, Scheme II) bearing noncoplanar 2,2'-dimethyl-4,4'-biphenylene unit, bis[4-(3,4-dicarboxyphenoxy)phenyl]diphenyl-methane dianhydride (**4B**) containing kink diphenylmethylene linkage, 1,1-bis[4-(4-dicarboxyphenoxy)phenyl]-4-*tert*-butylcyclohexane dianhydride (**4C**) bearing bulky pendent *tert*-butylcyclohexylidene unit, 3,3',5,5'-tetramethyl-2,2-bis[4-(4-dicarboxyphenoxy)phenyl]propane dianhydride (**4D**) bearing both flexible ether and isopropylidene bridges between the phenylene units and tetramethyl substitu-ents on the phenylene unit, 1,1-bis[4-(3,4-dicarboxyphenoxy)phenyl] cyclodo-decane dianhydride (**4E**) bearing a bulky pendent alicyclic cyclododecane unit, and 1,1-bis[4-(4-dicyanophenoxy)phenyl]-4-phenylcyclohexane dianhydride (**4F**) bearing a 4-phenyl cyclohexylidene unit [38–43]. The effects of different units on the polymer properties such as solubility, thermal and mechanical properties will be discussed here [38–43].

2. EXPERIMENTAL

2.1. Materials [38–43]

The materials, 2,2-bis(4-hydroxy-3,5-dimethylphenyl)propane (**1D**, Scheme II, from TCI) and 4-nitrophthalonitrile (from TCI) were used without further purifi-cation. *N,N*-dimethylformamide (DMF), *N,N*-dimethylacetamide (DMAc) and pyridine were purified by distillation under reduced pressure over calcium hy-dride before use. Acetic anhydride was purified by vacuum distillation.

2.2. Synthesis of diols, bis(ether dinitrile)s, bis(ether diacid)s and bis(ether anhydride)s (Schemes I and II)

2.2.1. Synthesis of new diols (1A–1C, 1E and 1F, Scheme I)
The new diols (**1A–1C**, **1E** and **1F**, Scheme I) were synthesized by the procedures in the previous studies as shown in Scheme I [5–8, 38–43].

2.2.2. Synthesis of bis(ether dinitrile)s (2A–2F, Scheme II) [38–43]
A typical procedure for synthesis of **2A** (Scheme II) was as follows [38, 39]: In a flask, 7.39 g (34.5 mmol) of 2,2'-dimethylbiphenyl-4,4'-diol (**1A**) and 12.2 g (70 mmol) of 4-nitrophthalonitrile were dissolved in 80 mL of dry DMF. Anhydrous potassium carbonate (10.1 g, 73 mmol) was added and the suspension was stirred at 170°C for two days. The reaction mixture was then poured into 500 mL of wa-ter to give a pale-yellow solid product that was washed repeatedly with water and

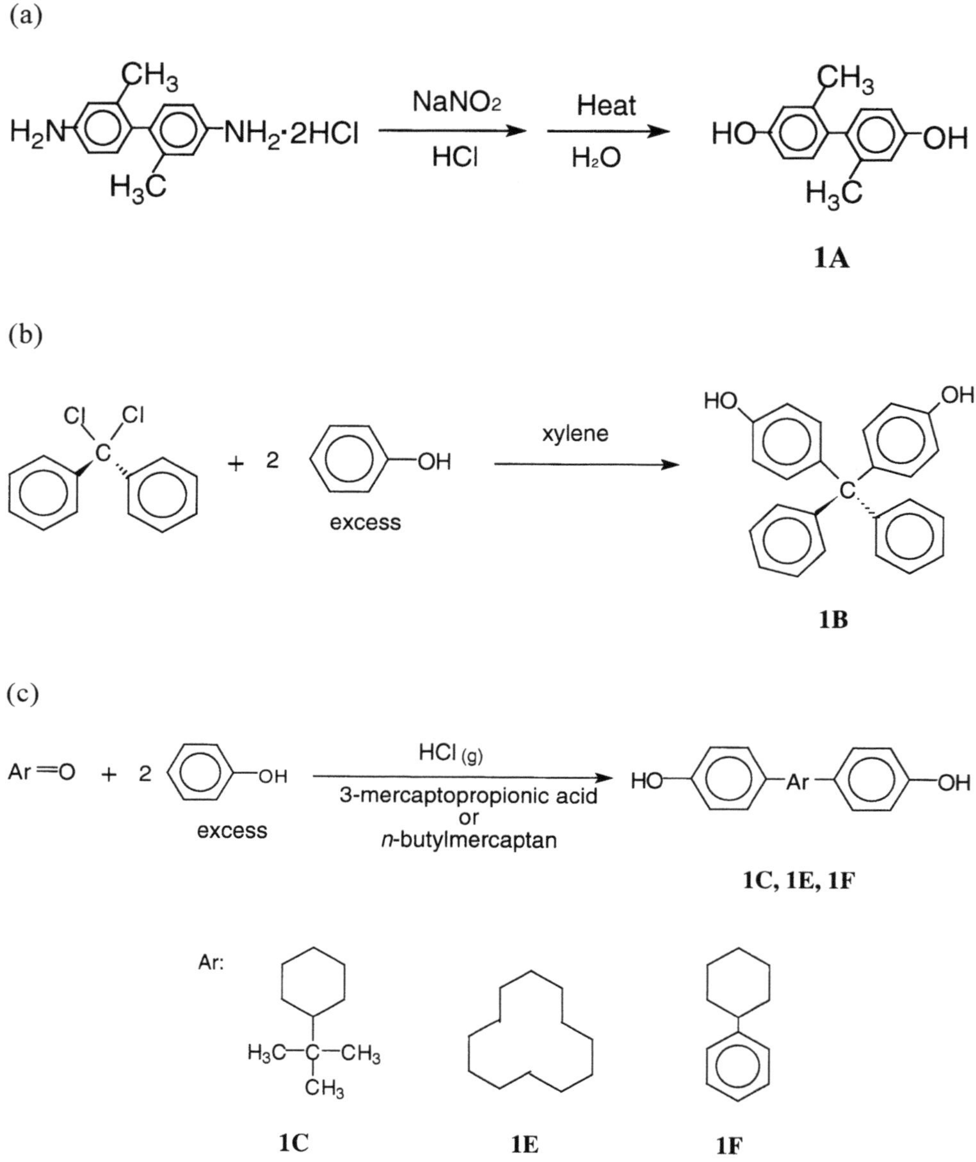

(a)

1A

(b)

1B

(c)

1C, 1E, 1F

Ar:

1C   **1E**   **1F**

Scheme I. Synthesis of various diols (**1A–1C**, **1E** and **1F**). (From Refs. 38–43.)

methanol, filtered out and dried. The crude product was then recrystallized from acetonitrile twice to obtain yellow crystals of bis(ether dinitrile) (**2A**, Scheme II), in 83% yield [38, 39].

The synthesis methods for bis[4-(3,4-dicyanophenoxy) phenyl]diphenyl-methane (**2B**), 1,1-bis[4-(4-dicyanophenoxy) phenyl]-4-*tert*-butylcyclohexane

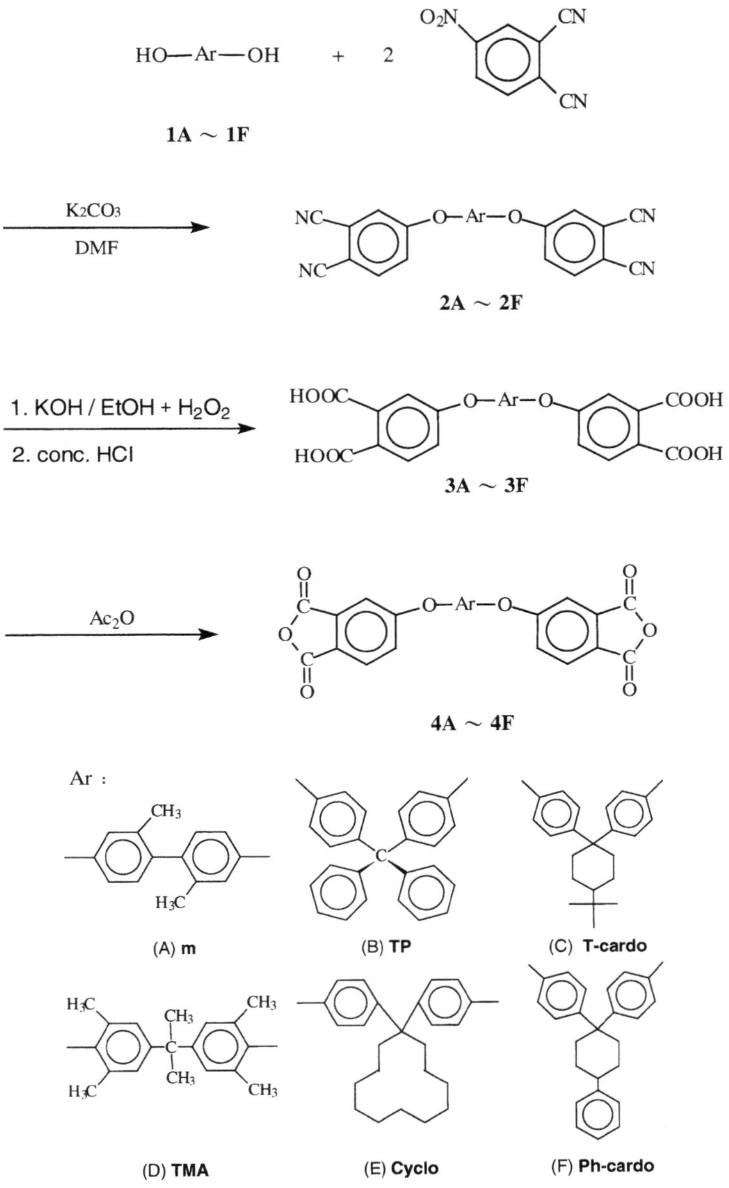

Scheme II. Synthesis of the new bis(ether anhydride)s (**4A–4F**). (From Refs. 38–43.)

(**2C**), 3,3',5,5'-tetramethyl-2,2-bis[4-(4-dicyanophenoxy)phenyl] propane (**2D**), 1,1-bis[4-(3,4-dicyanophenoxy)phenyl]cyclododecane (**2E**), and 1,1-bis[4-(4-dicyanophenoxy)phenyl]-4-phenylcyclohexane (**2F**) were the same as that for bis(ether dinitrile) **2A** [38–43].

2.2.3. Synthesis of bis(ether diacid)s (3A–3F, Scheme II) [38–43]

A typical procedure for synthesis of **3A** (Scheme II) was as follows [38, 39]: In a flask, 5.41 g (11.6 mmol) of bis(ether dinitrile) **2A** were suspended in a solution of 12.9 g (0.23 mol) potassium hydroxide in 40 mL water and 40 mL ethanol. The solid bis(ether dinitrile) could be dissolved within one hour. Refluxing was continued for two days until the evolution of ammonia had ceased. After filtration and removal of the residual ethanol under reduced pressure, the cooled filtrate was diluted with 200 mL of water and acidified by concentrated HCl. The precipitated bis(ether diacid) **3A**, was filtered off and washed thoroughly with distilled water until the filtrate was neutral. The yield was 92%. The compound had an endothermic peak due to thermal cyclodehydration at about 165°C (by DSC) [38].

The synthesis methods for bis[4-(3,4-dicarboxylphenoxy)phenyl]diphenyl-methane (**3B**), 1,1-bis[4-(4-dicarboxyphenoxy)phenyl]-4-*tert*-butylcyclohexane (**3C**), 3,3',5,5'-tetramethyl-2,2-bis[4-(4-dicarboxyphenoxy) phenyl] propane (**3D**), 1,1-bis[4-(3,4-dicarboxyphenoxy) phenyl] cyclododecane (**3E**), and 1,1-bis[4-(4-dicarboxyphenoxy) phenyl]-4-phenylcyclohexane (**3F**) were the same as that for **3A** [38–43].

2.2.4. Synthesis of bis(ether anhydride)s (4A–4F, Scheme II) [38–43]

A typical procedure for synthesis of **4A** (Scheme II) was as follows [38, 39]: In a flask, 3.7 g (6.8 mmol) of bis(ether diacid) **3A** was suspended in 35 mL of glacial acetic acid and 25 mL of acetic anhydride. The mixture was boiled under reflux for 24 hrs. Then, the mixture was filtered and left to crystallize overnight. The precipitated product was filtered off and further recrystallized from acetic anhydride. The brown crystals obtained were filtered off, washed with dry toluene and dried at 100°C for 24 hrs under vacuum to give bis(ether anhydride), 2,2'-dimethyl-4,4'-bis[4-(3,4-dicarboxyphenoxy)] biphenyl dianhydride **4A**. Yield: 81% [38, 39].

The synthesis methods for bis[4-(3,4-dicarboxylphenoxy)phenyl]diphenylmethane dianhydride (**4B**), 1,1-bis[4-(4-dicarboxyphenoxy)phenyl]-4-*tert*-butylcyclohexane dianhydride (**4C**), 3,3',5,5'-tetramethyl-2,2-bis[4-(4-dicarboxyphenoxy)phenyl]propane dianhydride (**4D**), 1,1-bis[4-(3,4-dicarboxyphenoxy)phenyl] cyclododecane dianhydride (**4E**) bearing a bulky pendent alicyclic cyclododecane unit, and 4-phenyl-cyclohexane dianhydride (**4F**) were the same as that for **4A** [38–43].

2.3. Synthesis of diamines (Scheme IV) [17, 48]

2.3.1. Synthesis of 1,4-bis(4-aminophenoxy)2-tert-butylbenzene (BATB) [48]

The dinitro compound 1,4-bis(4-nitrophenoxy)-2-*tert*-butylbenzene (**BNTB**, Scheme IV) was synthesized by reaction of *tert*-butylhydroquinone (41.5 g, 0.25 mol) and *p*-chloronitrobenzene (81.9 g, 0.52 mol) in the presence of potassium carbonate (79.4 g, 0.57 mol) and *N,N*-dimethylformamide (300 mL) at 160°C for 8 hrs. The mixture was cooled and poured into 600 mL of ethanol-water mixture (1:1 by volume). The crude product was recrystallized from glacial acetic acid to provide brown needles (m.p. 154°C) in 85% yield. The IR spectrum

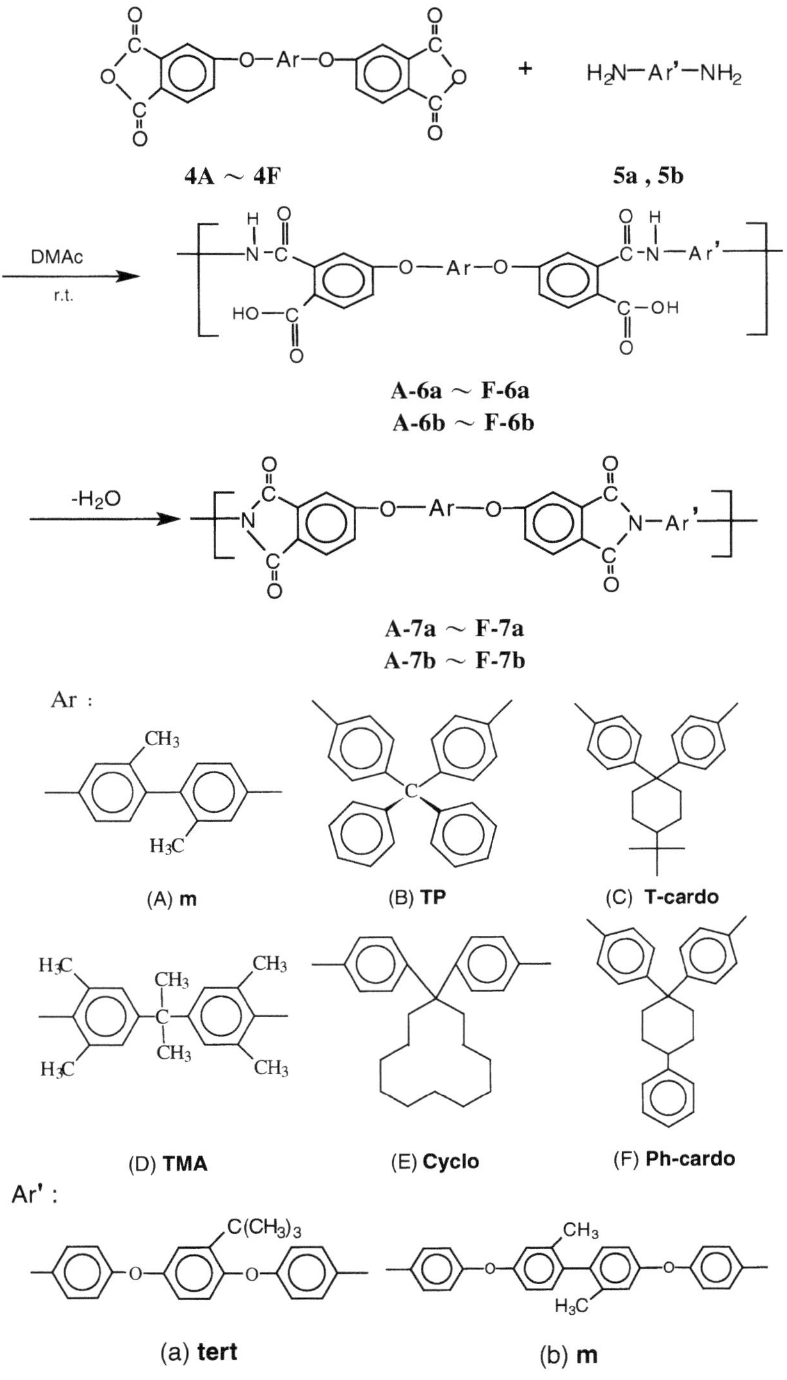
4A ~ 4F
5a , 5b
$H_2N-Ar'-NH_2$
DMAc
r.t.
A-6a ~ F-6a
A-6b ~ F-6b
-H_2O
A-7a ~ F-7a
A-7b ~ F-7b
Ar :
(A) m
(B) TP
(C) T-cardo
(D) TMA
(E) Cyclo
(F) Ph-cardo
Ar' :
(a) tert
(b) m

Scheme III. Synthesis of various poly(ether imide)s. (From Refs. 38–43.)

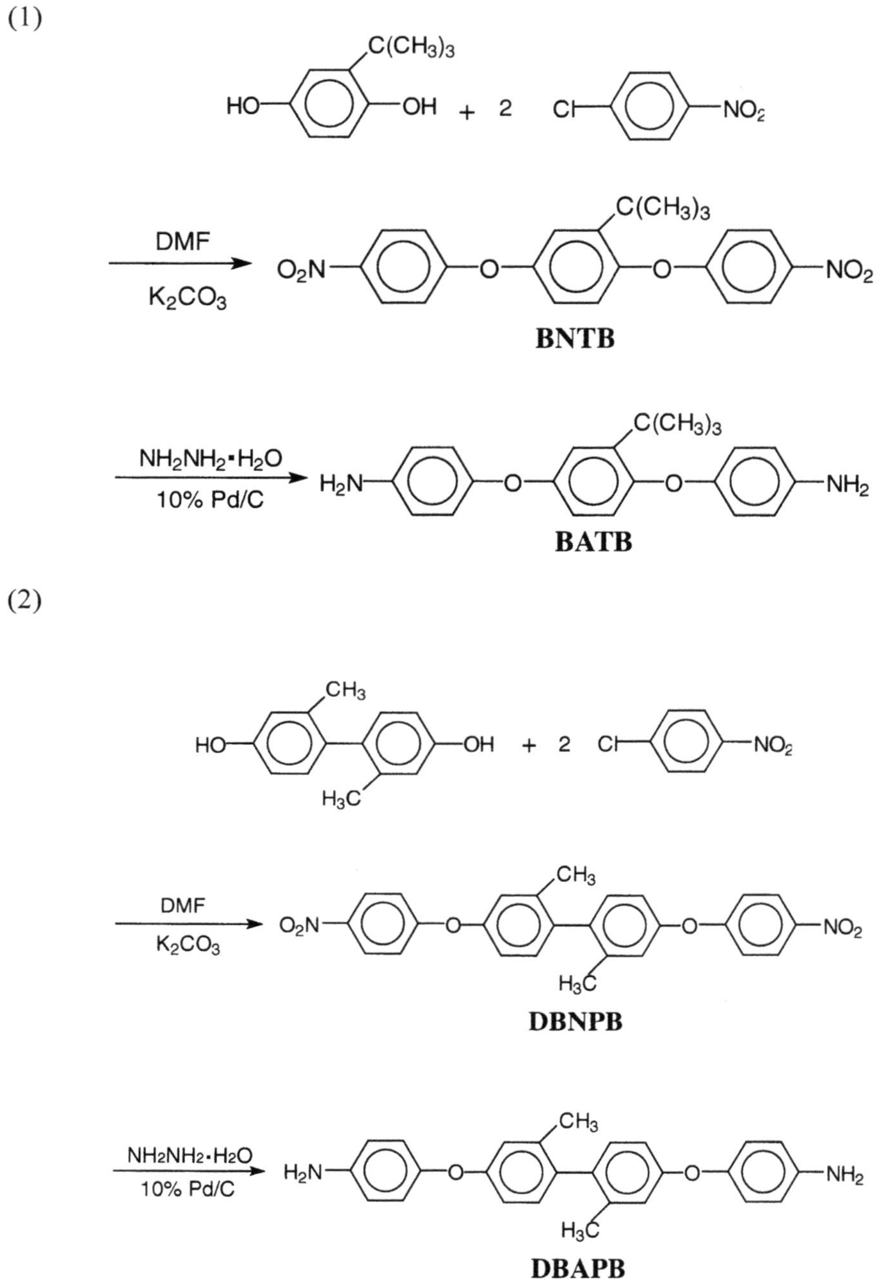

(1)

(2)

Scheme IV. Synthesis of 1,4-bis(4-aminophenoxy) 2-*tert*-butylbenzene (**BATB**) and 2,2'-dimethyl-4,4'-bis(4-aminophenoxy)biphenyl (**DBAPB**).

(KBr) exhibited absorptions at 1509 and 1335 cm^{-1} (NO$_2$), and 1233 cm^{-1} (C-O-C). ^{1}H NMR (CDCl$_3$): δ (ppm) = 1.33 (s, 9H), 6.91 (s, 2H), 7.03 (dd, 4H), 7.19 (s, 1H), 8.19 (dd, 4H). ^{13}C NMR (CDCl$_3$): δ (ppm) = 29.91, 34.91, 116.86, 117.21, 119.08, 120.13, 123.06, 125.87, 125.95, 142.59, 144.49, 150.58, 151.26, 163.09, 163.15. Elemental analysis calcd for C$_{22}$H$_{20}$O$_6$N$_2$: C, 64.70%; H, 4.94%; N, 6.86%; found: C, 64.52%; H, 4.79%; N, 6.42%. The obtained dinitro compound [1,4-bis(4-nitrophenoxy)-2-*tert*-butylbenzene (**BNTB**, Scheme IV)] (57.1 g, 0.14 mol), 0.3 g of 10% Pd/C, and 400 mL ethanol were introduced into a three-necked flask to which 130 mL of hydrazine monohydrate was added dropwise over a period of 1 hr at 85°C. After the addition was complete, the reaction was continued at reflux temperature for an additional 24 hrs. The mixture was then filtered to remove Pd/C. After cooling, the precipitated needle crystals were isolated by filtration and recrystallized from ethanol in 83% yield (m.p. 134°C). The diamine containing *tert*-butyl group 1,4-bis(4-aminophenoxy)2-*tert*-butylbenzene (**BATB**, Scheme IV) was obtained. The IR spectrum (KBr) exhibited absorptions at 3364 and 3440 cm^{-1} (N-H), and 1203 cm^{-1} (C-O-C). ^{1}H NMR (CDCl$_3$): δ (ppm) = 1.40 (s, 9H), 3.55 (s, 4H), 6.61–6.67 (m, 6H), 6.81 (dd, 4H), 7.05 (d, 1H). ^{13}C NMR (CDCl$_3$): δ (ppm) = 29.85, 34.74, 115.31, 116.23, 116.30, 116.91, 119.73, 120.12, 141.42, 141.72, 141.91, 149.50, 150.21, 151.85, 153.06. Elemental analysis calcd for C$_{22}$H$_{24}$O$_2$N$_2$: C, 75.83%; H, 6.94%; N, 8.04%; found: C, 75.44%; H, 6.87%; N, 7.70%.

2.3.2. Synthesis of 2,2'-dimethyl-4,4'-bis(4-aminophenoxy)biphenyl (*DBAPB*, Scheme IV) [48]

The dinitro compound 2,2'-dimethyl-4,4'-bis(4-nitrophenoxy)biphenyl (**DBNPB**, Scheme IV) was synthesized by the reaction of 2,2'-dimethylbiphenyl-4,4'-diol (30 g, 0.14 mol) and *p*-chloronitrobenzene (47.2 g, 0.3 mol) in the presence of potassium carbonate (48.7 g, 0.35 mol) and 250 mL DMF at 160°C for 8 h. The mixture was then cooled and poured into methanol-water mixture (1:1 by volume). The crude product was recrystallized from glacial acetic acid to provide brown needles (m.p. 142–144°C) in 83% yield. The IR spectrum (KBr) exhibited absorptions at 1580 and 1339 cm^{-1} (NO$_2$), 1238 cm^{-1} (C-O-C). ^{1}H-NMR (CDCl$_3$): δ (ppm) = 2.09 (s, 6H), 6.96 (d, 4H), 7.01 (s, 2H), 7.07 (d, 2H), 7.17 (d, 2H), 8.22 (d, 4H). ^{13}C-NMR (CDCl$_3$): δ (ppm) = 19.98, 117.24, 117.53, 121.61, 125.91, 131.15, 137.68, 138.62, 142.72, 154.00, 163.27. Elemental analysis calcd for C$_{26}$H$_{20}$O$_6$N$_2$: C, 68.42 %; H, 4.42%; N, 6.14%; found: C, 67.98%; H, 4.59%; N, 6.22%.

The dinitro compound obtained 2,2'-dimethyl-4,4'-bis(4-nitrophenoxy)biphenyl (**DBNPB**, Scheme IV) (45.6 g, 0.1 mol), 0.3 g 10% Pd/C, and 300 mL ethanol were introduced into a three-necked flask to which hydrazine monohydrate (100 mL) was added dropwise over a period of 0.5 hr at 85°C. After the addition was complete, the reaction was continued at reflux temperature for another 24 hrs. The mixture was then filtered to remove Pd/C. After cooling, the precipitated crystals were isolated by filtration and recrystallized from ethanol and dried in vacuum.

The diamine containing noncoplanar 2,2'-dimethyl-4,4'-biphenylene group (**DBAPB**, Scheme IV) was obtained. The yield was 80%; m.p. 138–139°C. The IR spectrum (KBr) exhibited absorptions at 3324 and 3406 cm^{-1} (N-H), and 1226 cm^{-1} (C-O-C). Elemental analysis calcd for $C_{26}H_{24}O_2N_2$: C, 78.76%; H, 6.10%; N, 7.07%; found: C, 78.41%; H, 6.34%; N, 7.10%.

2.4. Polymerization procedures (Schemes III and V) [17, 38–43, 48]

2.4.1. Synthesis of poly(ether imide)s [38–43]

To a stirred solution of diamine (Such as **5a** or **5b**, Scheme III) (1.1 mmol) in DMAc (5 mL), bis(ether anhydride) (Such as **4A**, **4B**, **4C**, **4D**, **4E** or **4F**, Scheme III) (1.1 mmol) was gradually added. The mixture was stirred at room temperature for 2 hrs under argon atmosphere to form the poly(ether amic acid) precursor (Such as **A-6a**, **B-6a**, **C-6a**, **D-6a**, **E-6a**, **F-6a**, **A-6b**, **B-6b**, **C-6b**, **D-6b**, **E-6b** or **F-6b**, Scheme III). Chemical imidization was carried out by adding an extra 3 mL of DMAc, 1 mL of acetic anhydride and 0.5 mL pyridine into the above-mentioned poly(ether amic acid) precursor (Such as **A-6a**, **B-6a**, **C-6a**, **D-6a**, **E-6a**, **F-6a**, **A-6b**, **B-6b**, **C-6b**, **D-6b**, **E-6b** or **F-6b**, Scheme III) solution with stirring at room temperature for 1 h, and then heating at 100°C for 3 hrs. The homogeneous solution was subsequently poured into methanol and the yellow solid precipitate was filtered off, washed with methanol and hot water, and then dried at 100°C for 24 hrs to afford poly(ether imide)s (Such as **m-tert**, **TP-tert**, **Tcardo-tert**, **TMA-tert**, **Cyclo-tert**, **Phcardo-tert**, **m-m**, **TP-m**, **Tcardo-m**, **TMA-m**, **Cyclo-m** or **Phcardo-m**, Scheme III) [38–43].

2.4.2. Synthesis of polyimides derived from various commercial dianhydrides [17, 48]

To a stirred solution of 1.25 mmol of diamine (Such as **tert** and **m**, Scheme V) in 8 mL of DMAc, 1.25 mmol of dianhydride (Such as **PMDA**, **BPDA**, **ODPA**, **BTDA**, **6FDA** and **SDPA**, Scheme V) was gradually added. The mixture was stirred at room temperature for 2–4 hrs under argon atmosphere to form the poly(amic acid). From this solution a film was cast onto a glass plate and heated (8 hrs at 80°C, 1 hr at 120°C, 1 hr at 150°C, 2 hrs at 200°C, 1 hr at 250°C) to convert the poly(amic acid) into polyimide film. Chemical cyclodehydration was also carried out by adding DMAc, and an equimolar mixture of acetic anhydride and pyridine into the above-mentioned poly(amic acid) solution with stirring at room temperature for 1 hr, and then heating at 100°C for 3 hr. The polymer solution was poured into methanol. The precipitate was collected by filtration, washed thoroughly with methanol and hot water, and dried at 100°C under vacuum. A series of polyimides derived from various commercial dianhydrides (Such as **PMDA-tert**, **BPDA-tert**, **ODPA-tert**, **BTDA-tert**, **6FDA-tert**, **SDPA-tert**, **PMDA-m**, **BPDA-m**, **ODPA-m**, **BTDA-m**, **6FDA-m** and **SDPA-m**, Scheme V).

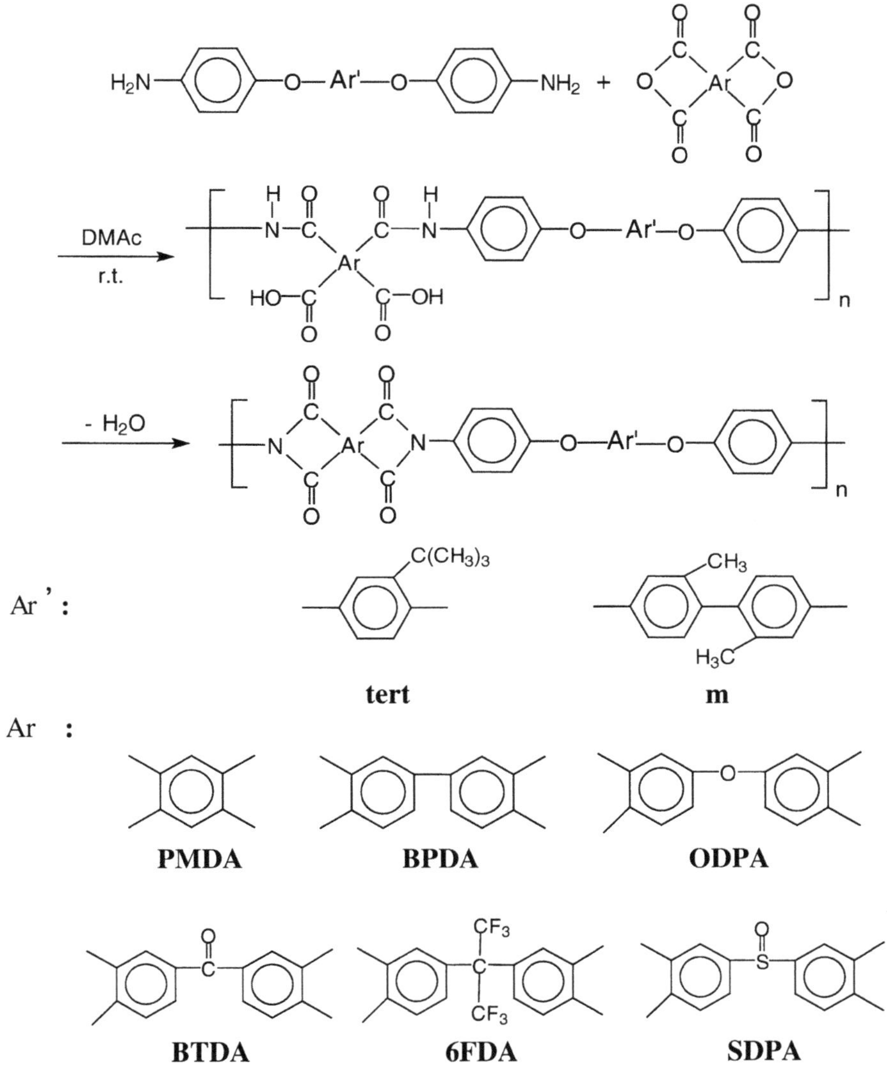

Scheme V. Preparation of polyimides derived from various commercial dianhydrides.

2.5. Characterization

The ^{13}C and ^{1}H NMR spectra were obtained using a JEOL EX-400 instrument operating at 100.40 MHz for carbon and 399.65 MHz for proton. The inherent viscosities of all poly(ether imide)s were measured using Ubbelohde viscometer. Weight-average ($\overline{Mw}$) and number-average molecular weights ($\overline{Mn}$) were determined by gel permeation chromatography (GPC). Four Waters (Ultrastyragel) columns were used for GPC analysis with tetrahydrofuran (THF) (1 mL·min^{-1}) as the eluent. The eluents were monitored with a UV detector (Gilson model 116) at 254 nm. Polystyrene was used as the standard. Thermogravimetric data were obtained on a DuPont 2100 equipment in flowing nitrogen or air (60 cm^3·min^{-1}) at a heating rate of 20°C·min^{-1}. Differential scanning calorimetry (DSC) analysis was performed on a DuPont 2000 differential scanning calorimeter.

3. RESULTS AND DISCUSSION

3.1. Monomer synthesis

As shown in Scheme II, all the bis(ether anhydride)s were prepared by a three-stage synthesis procedure starting from the nucleophilic nitrodisplacement reaction of diols (**1A–1F**) and 4-nitrophthalonitrile in dry DMF in the presence of potassium carbonate at room temperature. It was preferable to carry out the nitrodisplacement reaction at low temperature (at room temperature) than at higher temperature (higher than 100°C) since the products (**2A–2B**) always showed dark color when they were obtained at high temperature. After the nitrodisplacement reaction of diols (**1A–1F**) and 4-nitrophthalonitrile, a series of new bis(ether nitrile)s (**2A–2F**) were obtained. The resulting bis(ether dinitrile)s (**2A–2F**) were then hydrolyzed in an alkaline solution in the presence of hydrogen peroxide to obtain the corresponding bis(ether diacid)s (**3A–3F**). Nitriles can be hydrolyzed to give either amides or carboxylic acids. Although the amide is being formed initially, carboxylic acid is the most common product since amides are hydrolyzed with an acid or base. When carboxylic acid is desired, the reagent of choice is aqueous KOH containing about 6 to 12% hydrogen peroxide, although acid-catalyzed hydrolysis is also carried out frequently. The hydrolysis of **2A–2F** (except **2B**) was performed in two days. However, the hydrolysis of bis(ether dinitrile) **2B** needed longer time due to the poorer solubility of **2B** than **2A**. After complete hydrolysis, the solution became clear. Before acidification by aqueous HCl, the removal of the residual ethanol was necessary. The presence of the residual ethanol in the aqueous solution always resulted in viscous product during acidification. The bis(ether diacid)s were then cyclodehydrated to bis(ether anhydride)s (**4A–4F**) using dehydrating agent such as a mixture of acetic anhydride and glacial acetic anhydride. The structures of these compounds were confirmed by elemental analysis, IR and NMR. For instance, the cyano group (C $\equiv$ N) of compound **2A** was evident from the peak at 2222 cm^{-1} in the IR spectrum. However, in the IR spectrum of **3A**, the cyano stretching vibration was absent, although a broad C(O)O-H absorption appeared in the region 2500–3600 cm^{-1} and a C=O stretching absorption appeared at 1690 cm^{-1}. Furthermore, the IR spectrum of the bis(ether anhydride) **4A** shows characteristic cyclic anhydride absorptions at 1837 and 1767 cm^{-1} attributed to the asymmetrical and symmetrical stretching vibrations, respectively, of the carbonyl group. The NMR spectra data are listed in Tables 1–3. The NMR spectra agree satisfactorily with the proposed structure.

Table 1.
Melting point, IR and NMR results on various bis(ether dinitrile)s (**2A–2F**, Scheme II)

Monomer code	IR (KBr) (cm⁻¹)	NMR	m.p. (°C)
2A	(C≡N) 2222 (C-O) 1240	^{1}H-NMR (CDCl$_3$): δ (ppm) = 7.80 (d, 2H); 7.37 (d, 4H); 7.29 (t, 2H); 7.05 (s, 2H); 7.01 (d, 2H); 2.15 (s, 6H). ^{13}C-NMR (CDCl$_3$): δ (ppm) = 162.8, 153.8, 139.9, 139.1, 136.2, 132.2, 122.4, 122.3, 122.0, 118.3, 118.1, 115.9, 115.5, 109.2, 19.4.	227–228
2B	(C≡N) 2224 (C-O) 1244	^{1}H-NMR (DMSO-d_6): δ (ppm) = 8.06 (d, 2H), 7.78 (s, 2H), 7.41 (dd, 2H), 7.32–7.20 (m, 14H), 7.10 (d, 4H). ^{13}C-NMR (DMSO-d_6): δ (ppm) = 162.3, 153.3, 147.5, 145.0, 137.5, 133.8, 131.6, 129.1, 127.4, 123.9, 123.3, 120.4, 117.8, 116.9, 116.4, 109.3, 64.1.	219–220
2C	(C≡N) 2224 (C-O) 1245	^{1}H-NMR (DMSO-d_6): δ (ppm) = 8.10–8.04 (dd, 2H); 7.70 (d, 2H); 7.52 (d, 2H); 7.35–7.28 (m, 4H), 7.14 (d, 2H), 7.04 (d, 2H), 2.82 (d, 2H), 1.82–1.10 (m, 7H), 0.73 (s, 9H). ^{13}C-NMR (DMSO-d_6): δ (ppm) = 162.5, 152.8, 149.8, 143.6, 137.6, 137.5, 131.1, 129.2, 123.7, 123.0, 122.9, 121.1, 120.1, 117.7, 117.6, 116.8, 116.3, 109.0, 108.9, 47.2, 45.0, 36.5, 31.9, 27.1, 23.1.	211–212
2D	(C≡N) 2224 (C-O) 1246	^{1}H-NMR (DMSO-d_6): δ (ppm) = 8.04 (d, 2H), 7.55 (s, 2H), 7.14–7.09 (m, 6H), 2.00 (s, 12H), 1.65 (s, 6H). ^{13}C-NMR (DMSO-d_6): δ (ppm) = 162.1, 149.4, 148.5, 137.7, 130.6, 128.8, 121.5, 120.6, 117.9, 116.9, 116.4, 108.3, 41.9, 30.5, 15.7.	197–198
2E	(C≡N) 2224 (C-O) 1247	^{1}H-NMR (CDCl$_3$): δ (ppm) = 7.71(d, 4 H), 7.25(d, 4 H), 7.24(s, 4 H), 7.22(d, 4 H), 6.97(d, 4 H), 1.98–0.95 (m, 22H). ^{13}C-NMR (CDCl$_3$): δ (ppm) = 162.7, 152.1, 148.1, 136.1, 130.3, 122.0, 121.8, 120.3, 117.9, 115.8, 115.4, 109.0, 47.7, 32.6, 25.6, 25.4, 21.4, 21.1, 19.2.	211–212
2F	(C≡N) 2222 (C-O) 1244	^{1}H NMR (DMSO-d_6): δ (ppm) = 8.10–8.04 (t, 2H), 7.75 (d, 2H), 7.75 (d, 2H), 7.40~7.06 (m, 13H), 2.90–1.81 (m, 9H). ^{13}C NMR (DMSO-d_6): δ (ppm) = 162.5, 162.4, 153.0, 152.9, 149.6, 148.0, 143.4, 137.6, 137.5, 139.4, 129.3, 127.6, 127.0, 124.0, 123.8, 123.1, 122.9, 121.2, 120.9, 117.7, 117.6, 116.9, 116.4, 109.1, 109.0, 44.9, 43.0, 36.3, 30.0.	254–255

Source: Refs. 38–43.

Table 2.
Melting point, IR and NMR results on various bis(ether diacid)s (**3A–3F**, Scheme II)

Monomer code	IR (KBr) (cm^{-1})	NMR	m.p. (°C)
3A	(C(O)OH) 2500–3600 (C=O) 1690 (C-O-C) 1271	^{1}H-NMR (DMSO-d_6): δ (ppm) = 7.79 (d, 2H); 7.15–6.98 (m, 10H); 2.02 (s, 6H). ^{13}C-NMR (DMSO-d_6): δ (ppm) = 170.2, 169.2, 160.8, 155.7, 139.4, 137.9, 137.8, 132.7, 132.3, 129.9, 121.9, 120.0, 117.9, 117.8, 19.4.	a
3B	(C(O)OH) 2500–3600 (C=O) 1710 (C-O-C) 1227	^{1}H-NMR (DMSO-d_6): δ (ppm) = 7.85–7.67 (m, 2H), 7.33–7.00 (m, 24H). ^{13}C-NMR (DMSO-d_6): δ (ppm) = 169.9, 169.2, 160.3, 154.8, 147.6, 143.9, 138.0, 136.9, 133.6, 131.6, 129.1, 127.9, 127.4, 120.2, 119.8, 118.9, 64.0.	168–170
3C	(C(O)OH) 2500–3600 (C=O) 1705 (C-O-C) 1225	^{1}H-NMR (DMSO-d_6): δ (ppm) = 8.10–8.04 (dd, 2H); 7.70 (d, 2H); 7.52 (d, 2H); 7.35–7.28 (m, 4H), 7.14 (d, 2H), 7.04 (d, 2H), 2.82 (d, 2H), 1.82–1.10 (m, 7H), 0.73 (s, 9H). ^{13}C-NMR (DMSO-d_6): δ (ppm) = 162.5, 152.8, 149.8, 143.6, 137.6, 137.5, 131.1, 129.2, 123.7, 123.0, 122.9, 121.1, 120.1, 117.7, 117.6, 116.8, 116.3, 109.0, 108.9, 47.2, 45.0, 36.5, 31.9, 27.1, 23.1.	179–181
3D	(C(O)OH) 2500–3600 (C=O) 1696 (C-O-C) 1220	^{1}H-NMR (DMSO-d_6): δ (ppm) = 8.03 (d, 2H), 7.28 (s, 2H), 7.06 (s, 4H), 6.87 (d, 2H), 2.01 (s, 12H), 1.67 (s, 6H). ^{13}C-NMR (DMSO-d_6): δ (ppm) = 169.6, 169.2, 160.4, 149.4, 148.7, 138.8, 135.5, 130.9, 128.6, 127.9, 117.0, 41.8, 30.5, 16.0.	245–246
3E	(C(O)OH) 2500–3600 (C=O) 1706 (C-O-C) 1228	^{1}H-NMR (DMSO-d_6): δ (ppm) = 8.01(d, 4 H), 7.45(s, 2 H),7.17(d, 4 H), 7.05(d, 4 H), 6.97(d, 4 H), 2.01–0.87 (m, 22H, cyclododecane). ^{13}C-NMR (DMSO-d_6): δ (ppm) = 170.0, 169.5, 160.8, 154.2, 146.9, 138.5, 134.9, 130.4, 128.8, 120.2, 47.7, 32.7, 26.1, 25.9, 21.8, 21.6, 19.7.	152–153
3F	(C(O)OH) 2500–3600 (C=O) 1706 (C-O-C) 1225	^{1}H NMR (DMSO-d_6): δ (ppm) = 7.82–7.78 (t, 2H), 7.50 (d, 2H), 7.31 (d, 2H), 7.22–6.97 (m, 13H, aromatic and 4-phenyl group), 2.84–1.50 (m, 9H, cyclohexane group). ^{13}C NMR (DMSO-d_6): δ (ppm) = 170.3, 169.3, 160.9, 160.7, 154.4, 154.2, 148.8, 148.2, 142.5, 138.0, 133.1, 130.9, 129.6, 129.1, 127.7, 127.5, 127.3, 127.1, 120.8, 119.9, 118.4, 118.1, 44.8, 43.2, 36.4, 30.2.	200–202

a. The compound had an endothermic peak due to thermal cyclodehydration at about 165°C (by DSC).
Source: Refs. 38–43.

Table 3.
Melting point, IR, NMR and elemental analysis results on various bis(ether anhydride)s (**4A–4F**, Scheme II)

Monomer code (Formula)	IR (KBr) (cm^{-1})	NMR	m.p. (°C)	Elemental analysis (%)	
				C	H
4A (C$_{30}$H$_{18}$O$_8$)	(C=O) 1837, 1767 (C-O) 1272	^{1}H-NMR (DMSO-d_6): δ (ppm) = 7.79 (d, 2H); 7.18–6.97 (m, 10H), 2.02 (s, 6H). ^{13}C-NMR (DMSO-d_6): δ (ppm) = 170.3, 169.2, 160.8, 155.8, 139.5, 138.0, 137.9, 132.7, 132.3, 126.9, 122.0, 120.0, 118.0, 117.9, 19.9.	217–218	Calcd 71.15 Found 70.79	3.58 3.77
4B (C$_{41}$H$_{24}$O$_8$)	(C=O) 1842, 1764 (C-O) 1262	^{1}H-NMR (DMF-d_7): 8.00 (dd, 2H), 7.61 (d, 2H), 7.53 (s, 2H), 7.39–7.13 (m, 18H). ^{13}C-NMR (DMF-d_7): δ (ppm) = 171.1, 170.0, 166.6, 155.0, 148.7, 146.2, 136.3, 134.9, 136.7, 129.8, 129.7, 128.3, 127.0, 126.9, 121.4, 114.6, 65.3.	262	Calcd 76.39 Found 75.95	3.75 4.05
4C (C$_{38}$H$_{32}$O$_8$)	(C=O) 1841, 1768 (C-O) 1278	^{1}H-NMR (CDCl$_3$): δ (ppm) = 7.89–7.84 (m, 2H), 7.44–7.21 (m, 8H), 7.05 (d, 2H), 6.93 (d, 2H), 2.72 (d, 2H), 2.02–1.17 (m, 7H), 0.76 (s, 9H). ^{13}C-NMR (CDCl$_3$): δ (ppm) = 166.2, 166.0, 163.6, 163.5, 163.1, 152.7, 152.5, 149.9, 143.7, 134.5, 134.4, 130.8, 128.9, 128.2, 128.1, 125.4, 124.7, 124.6, 120.9, 120.7, 113.0, 112.7, 47.5, 45.0, 37.0, 31.7, 26.7, 22.9.	211–212	Calcd 74.00 Found 73.45	5.23 5.45
4D (C$_{35}$H$_{28}$O$_8$)	(C=O) 1838, 1764 (C-O) 1278	^{1}H-NMR (CDCl$_3$): δ (ppm) = 7.86 (d, 2H), 7.32 (d, 2H), 7.10 (d, 2H), 6.95 (s, 4H), 2.00 (s, 12H), 1.64 (s, 6H). ^{13}C-NMR (CDCl$_3$): δ (ppm) = 165.7, 163.9, 163.2, 149.3, 148.6, 134.8, 130.5, 128.5, 128.4, 124.2, 123.9, 110.7, 41.8, 30.3, 15.6.	164–165	Calcd 72.90 Found 73.00	4.89 5.16
4E (C$_{40}$H$_{36}$O$_8$)	(C=O) 1840, 1767 (C-O) 1266	^{1}H-NMR (DMSO-d_6): δ (ppm) = 7.74 (d, 4 H), 7.21–7.05 (m, 8H), 6.99 (d, 4 H), 2.05–0.88 (m, 22H, cyclododecane). ^{13}C-NMR (DMSO-d_6): δ (ppm) = 170.3, 169.3, 161.0, 154.1, 147.0, 137.9, 132.7, 130.4, 126.9, 120.3, 119.7, 117.7, 47.7, 32.6, 26.0, 25.8, 21.7, 21.5, 19.6.	212–213	Calcd 74.52 Found 74.25	5.63 5.68
4F (C$_{40}$H$_{32}$O$_8$)	(C=O) 1839, 1765 (C-O) 1264	^{1}H NMR (DMSO-d_6): δ (ppm) = 7.80–7.76 (t, 2H), 7.49 (d, 2H), 7.30 (d, 2H), 7.22–6.96 (m, 13H, aromatic and 4-phenyl group), 2.83–1.50 (m, 9H, cyclohexane group). ^{13}C NMR (DMSO-d_6): δ (ppm) = 170.3, 169.4, 169.3, 160.9, 160.8, 154.4, 154.3, 148.8, 148.2, 142.5, 127.4, 127.2, 127.1, 120.8, 120.5, 120.1, 119.9, 118.3, 118.0, 44.8, 43.2, 36.4, 30.2.	217–218	Calcd 75.46 Found 74.40	4.43 4.41

Source: Refs. 38–43.

3.2. Preparation of poly(ether imide)s

The poly(ether imide)s were prepared by the conventional two-step polymerization method, as shown in Scheme III, involving ring-opening polyaddition forming poly(ether amic acid)s following by chemical cyclodehydration. In general, the thermal cyclodehydration of the poly(amic acid) film is also performed under reduced pressure at high temperature (about 300°C). However, the high temperature thermal cyclodehydration may result in poorer solubility of the polymer than chemical cyclodehydration [9]. Since the aim of this study was to prepare organosoluble poly(ether imide)s, the chemical cyclodehydration was adopted here. The poly(ether amic acid) precursors were prepared by adding the bis(ether anhydride)s (**4A–4F**, Scheme II) to the diamine solution gradually. To the obtained viscous poly(ether amic acid) solutions were then added dehydrating agents such as a mixture of acetic anhydride and pyridine to yield various poly(ether imide)s. As shown in Table 4, the resulting poly(ether imide)s had inherent viscosities 0.50–0.81 dL·g^{-1}. GPC measurements revealed that the polymers exhibited number-average molecular weight ($\overline{Mn}$) and weight-average molecular weight ($\overline{Mw}$) up to 57,000 and 130,000, respectively. The structures of the polymers were identified by IR and NMR spectroscopies. Specifically, the IR spectra of the poly(ether imide) (Such as **cyclo-m**, Scheme III) revealed that the characteristic bands around 1768 and 1716 cm^{-1} were commonly attributed to the asymmetric

Table 4.
Inherent viscosities and average molecular weights of various poly(ether imide)s

Polymer code (Ar-Ar')	Inherent viscosity [a] (dL·g^{-1})	$\overline{Mn} \times 10^{-4}$ [b]	$\overline{Mw} \times 10^{-4}$ [b]
m-tert	0.81	4.5	7.9
TP-tert	0.56	2.3	5.2
Tcardo-tert	0.51	3.7	8.5
TMA-tert	0.54	3.3	6.1
Cyclo-tert	0.70	3.3	7.8
Phcardo-tert	0.64	5.7	13.0
m-m	0.65	1.7 [c]	3.1 [c]
TP-m	0.65	2.7	6.2
Tcardo-m	0.79	4.3	7.6
TMA-m	0.71	4.6	9.4
Cyclo-m	0.50	1.6	3.2
Phcardo-m	0.73	2.5	5.7

a. Inherent viscosity measured in DMAc at a concentration of 0.5 g·dL^{-1} at 30°C.
b. Measured by GPC in THF, polystyrene was used as standard.
c. The polymer was only partially soluble in THF.
Source: Refs. 38–43.

and symmetric stretches of carbonyl group of imides, respectively. The C-N band at 1369 cm^{-1} verified the formation of the imide structure. The NMR data on the polymer (**cyclo-m**, Scheme III) were obtained and the resonance peaks at 168.0 and 167.9 ppm were ascribed to the carbon of the imide group. The elemental analysis data on these polyimides were generally in good agreement with the respective structures. [For example, calcd. for **cyclo-m** (Scheme III) ($C_{66}H_{56}N_2O_8$): C, 78.86%; H, 5.62% N, 2.79% and found: C, 78.47%; H, 5.65%; N, 2.79%]. Films of all the polymers could be obtained by solution casting from their DMAc solution. All polymer films showed tough, transparent and flexible nature.

3.3. Polymer characterization [38–43]

The crystallinity of the polymers was examined by using wide-angle X-ray diffraction technique. All polymers showed amorphous patterns in the region of $2\theta = 8°$ and $40°$. This observation is reasonable. For example, the presence of non-coplanar conformation of 2,2'-dimethyl-substituted biphenylene unit and bulky and kink phenyl gorups on the diphenylmethylene linkage decrease the intermolecular forces between the polymer chains, causing a decrease in crystallinity. In general, the presence of the biphenylene unit in the polymer backbone leads to a rigid-rod polymer with high crystallinity and poor solubility. However, the incorporation of 2,2'-dimethyl substituent on the 4,4'-biphenylene unit effectively reduces the packing of the polymer chains [5–6, 10–13]. It should be noted that the polymer chain containing symmetric substituents always leads to better packing of a polymer chains [14, 15]. However, the diphenylmethylene linkage will be present as a kink conformation, thus the rigidity of polymer chain is lowered [38]. The crystallinity of the polymer containing a kink linkage was thus reduced [38]. The bulky pendent group, flexible ether and isopropylidene groups also inhibited a close packing of the polymer chains.

The solubilities of polyimides derived from various commercial dianhydrides are listed in Table 5. Polyimides (**PMDA-tert**, **BPDA-tert**, **ODPA-tert** and **BTDA-tert**, Scheme V) were almost insoluble in organic solvents but dissolved in concentrated sulfuric acid. Polyimides (**SDPA-tert** and **6FDA-tert**, Scheme V) containing hexafluoroisopropylidene and sulfone linkages exhibited excellent solubility toward test solvents. For comparison, polyimide (**SDPA-BAPB**) was prepared by condensation of **SDPA** with 1,4-bis(4-aminophenoxy)butyl-benzene (**BAPB**) through chemical cyclodehydration. The results from *tert*-butyl substituted polyimide (**SDPA-tert**) and analogous polyimide (**SDPA-BAPB**) show that *tert*-butyl substituent incorporated into the polyimide indeed increased its solubility. Also, polyimides (**PMDA-m**, **BPDA-m**, **ODPA-m**, **BTDA-m**, Scheme V) and polyimide (**SDPA-BAPB**) were almost insoluble in the solvents tested but were soluble in concentrated sulfuric acid. However, polyimides (**SDPA-m** and **6FDA-m**, Scheme V) derived from dianhydrides **SDPA** and **6FDA** were highly soluble in organic solvents such as NMP, DMAc, DMF and *m*-cresol. This result revealed that the solubility of polyimide (**SDPA-m**) was

Table 5.
Solubility of polyimides derived from various commercial dianhydrides[a]

Polymer code	Solvent[b]						
	NMP	DMSO	DMAc	DMF	m-Cresol	Pyridine	Conc. H$_2$SO$_4$
PMDA-tert	−	−	−	−	−	−	+ +
BPDA-tert	+ −	−	−	−	+ −	−	+ +
ODPA-tert	+ −	−	−	−	+	−	+ +
BTDA-tert	+	−	−	−	+	−	+ +
6FDA-tert	+ +	+ +	+ +	+ +	+ +	+ +	+ +
SDPA-tert	+ +	+	+ +	+ +	+ +	+ +	+ +
PMDA-m	−	−	−	−	−	−	+ +
BPDA-m	−	−	−	−	+	−	+ +
ODPA-m	+ −	−	−	−	+	−	+ +
BTDA-m	−	−	−	−	−	−	+ +
6FDA-m	+ +	+	+ +	+ +	+ +	+ +	+ +
SDPA-m	+ +	+ −	+ +	+ +	+ +	+ −	+ +
SDPA-BAPB[c]	−	−	−	−	−	−	+ +

a. Solubility: ++, soluble at room temperature; +, soluble on heating at 70°C; + −, partially soluble; −, insoluble.

b. Abbreviations: NMP: *N*-methyl-2-pyrrolidinone; DMSO: dimethylsulfoxide; DMAc: *N,N*-dimethylacetamide; DMF: *N,N*-dimethylformamide; THF: tetrahydrofuran.

c. Analogous polyimide **SDPA-BAPB**:

Source: Refs. 17 and 48.

considerably higher than that of analogous polyimide (**SDPA-BAPB**) in all organic solvents, indicating that the solubility of polyimide was enhanced by introducing noncoplanar conformation of 2,2'-dimethyl substituted biphenylene unit in the polymer backbone.

The solubilities of these poly(ether imide)s in several organic solvents at 5.0% (w/v) are also summarized in Table 6. Almost all of the poly(ether imide)s were soluble in the test solvents including *N*-methyl-2-pyrrolidinone, *N,N*-dimethylacetamide (DMAc), pyridine, cyclohexanone, tetrahydrofuran and even in chloroform at room temperature. The good solubility of these poly(ether imide)s was possibly imparted by the presence of flexible ether, isopropylidene group, bulk pendent group, noncoplanar biphenylene and kink linkage, which reduced the polymer chains interaction and rigidity. A comparison of the solu-

Table 6.
Solubility of various new poly(ether imide)s [a]

Polymer code	Solvent [b]						
	NMP	DMAc	DMF	Pyridine	Cyclohexanone	THF	Chloroform
m-tert	++	++	++	++	+	++	++
TP-tert	++	++	++	++	++	++	++
Tcardo-tert	++	++	++	++	++	++	++
TMA-tert	++	++	++	++	++	++	++
Cyclo-tert	++	++	+	−	++	+	++
Phcardo-tert	++	++	++	++	+	++	++
m-m	++	++	++	++	+	+−	+−
TP-m	++	++	++	++	++	++	++
Tcardo-m	++	++	++	++	++	++	++
TMA-m	++	++	++	++	++	++	++
Cyclo-m	++	++	+	−	++	++	++
Phcardo-m	++	++	++	++	+	++	++
BPDA-m [c]	+	+	+−	−	+	+−	−

a. Solubiltiy: ++, soluble at room temperature; +, soluble on heating at 70°C; + −, partially soluble; −, insoluble.

b. Abbreviations: NMP: *N*-methyl-2-pyrrolidinone; DMAc: *N,N*-dimethylacetamide; DMF: *N,N*-dimethylformamide; DMSO: dimethylsulfoxide; THF: tetrahydrofuran.

c. Analogous polyimide **BPDA-m**:

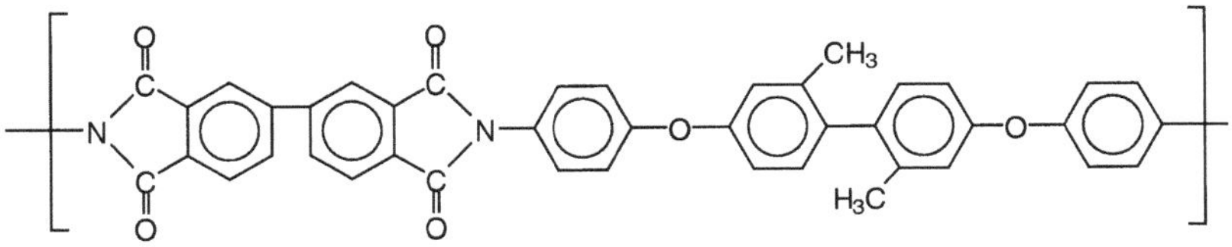

Source: Refs. 38–43.

bility between these poly(ether imide)s indicated that poly(ether imide)s containing the isopropylidene, kink and *tert*-butylcyclohexane groups shown slightly better solubility than the others. For comparison, polyimide **BPDA-m**, shown below, containing biphenylene without pendent cyclododecylidene was prepared from 4,4'- bis(4-aminophenoxy)biphenyl with dianhydride. It was observed that polyimide (**BPDA-m**) was nearly insoluble in the test solvents and showed poorer solubility than its analogous polymer (**Cyclo-m**, Scheme III). This may be due to the presence of the bulky pendent group which decreases the inter-chain interaction between rigid aromatic repeat units, resulting in enhanced solubility.

BPDA-m

The thermal properties of the poly(ether imide)s were determined using differential scanning calorimetry (DSC) and thermogravimetric (TG) analysis. The thermogravimetric (TG) analysis revealed that these poly(ether imide)s had good thermal stability. They were stable up to temperatures above 440°C. The temperature at 10% weight loss (Td_{10}) of these polymers reached 441–535°C in nitrogen atmosphere. It was observed that polymers with 2,2'-dimethyl-4,4'-biphenylene having biphenylene unit showed higher Td_{10} value than the others. For example, Td_{10} value of **m-tert** was observed to be 518°C in nitrogen, which is higher than that of **TP-tert** (501°C, in nitrogen), **Tcardo-tert** (504°C, in nitrogen), **TMA-tert** (474°C, in nitrogen), **Cyclo-tert** (441°C, in nitrogen) and **Ph-cardo-tert** (495°C, in nitrogen). This is because the polymers with biphenylene unit exhibit higher rigidity than the others. It was observed that polymers with 2,2'-dimethyl-4,4'-biphenylene unit showed higher Td_{10} values than those with diphenylmethylene linkage. On comparing the polymers derived from the diamines (**a** and **b**), it was observed that the polymer containing 2,2'-dimethyl-4,4'-biphenylene unit (**m-m**) exhibited a higher Td_{10} value than that bearing asymmetric *tert*-butyl substitution group (**m-tert**). That is, a noncoplanar conformation can impart a greater thermal stability to the polymer than the *tert*-butyl substitution [7, 18]. The solubility of polymer was enhanced by the noncoplanar conformation of 2,2'-dimethyl substituted biphenylene in the polymer backbone, which reduced the polymer chains interaction and rigidity [17]. The incorporation of the methyl substituent on the 2,2'-dimethyl-4,4'-biphenylene unit did not significantly decrease the thermal stability of the polymer but enhanced its processability [18]. The result of the TGA measurements on these polymers indicated that polymers with noncoplanar biphenylene groups exhibited higher thermal stability than those with kink diphenylmethylene linkages [38, 39]. The isothermal gravimetric analysis (IGA) results showed that these poly(ether imide)s had good thermo-oxidative stability. In general, the IGA results were in agreement with the TGA data. Specifically, following 20 hrs isothermal aging at 350°C in static air, the polymer showed weight losses of 7.0–10.5% (namely 7.0% for **m-tert** polyimide and 7.9% for **m-m** polyimide). Upon comparing the weight loss values, it is seen that polyimides containing 2,2'-dimethyl-4,4'-biphenylene unit (namely 7.0% for **m-tert** polyimide and 7.9% for **m-m** polyimide) were somewhat more thermally stable than those containing diphenylmethylene unit (namely 8.8% for **TP-tert** polyimide and 9.0% for **TP-m** polyimide) [38, 39].

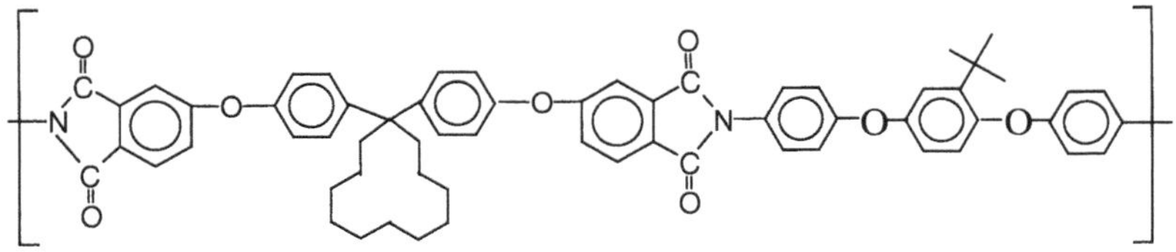

Cyclo-tert : Tg =262 °C

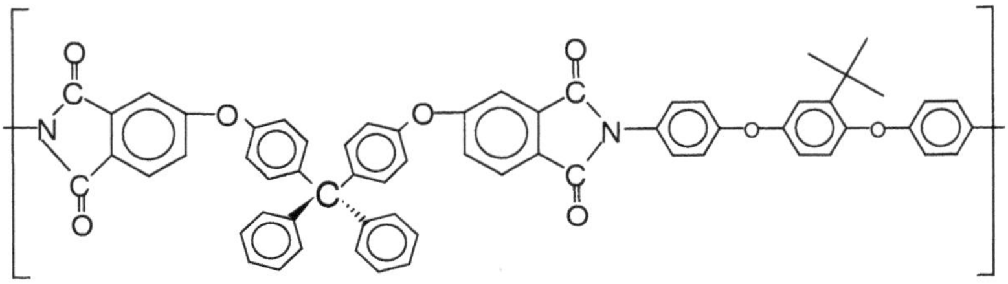

TP-tert : Tg =229°C

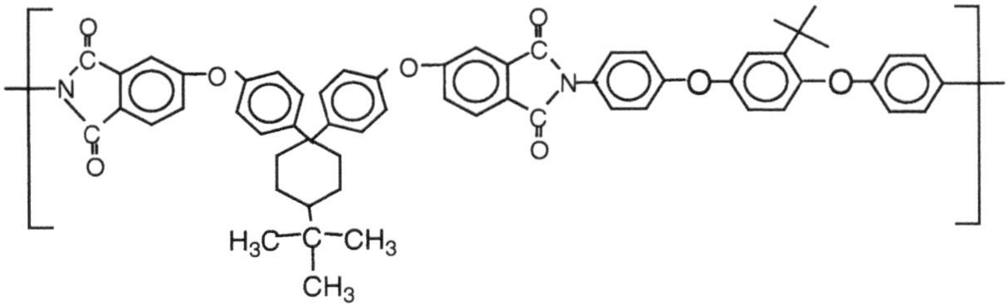

Tcardo-tert : Tg =233°C

The glass transition temperatures (Tg's) of the poly(ether imide)s were found to be in the range of 226–262°C. No melting endotherm peak was observed in the DSC traces. This also verified the amorphous nature of the poly(ether imide)s. On comparing the thermal properties, all the polyimides exhibited higher Tg values than the commercially available polyimide Ultem 1000 (Tg = 215°C). All the polymides reported here also showed better thermal properties than the results reported in our previous studies [35–48]. These polyimides could be considered as new processable high-performance polymeric materials [38–43]. On comparing the thermal properties of polyimides here with the alicyclic poly(ether imide)s reported in our previous study, it was observed that the poly(ether imide) (**Cyclo-tert**) with the introduction of cyclododecane group had a higher Tg value than **TP-tert** and **Tcardo-tert** containing kink [6] and *tert*-butylcyclohexane alicyclic [23] groups, respectively. This means that the introduction of cyclododecane group into the polymer backbone could enhance the thermal stability effectively.

Table 7.
Mechanical properties of various new poly(ether imide)s

Polymer code (Ar-Ar')	Tensile strength (MPa)	Elongation at break (%)	Initial modulus (GPa)
m-tert	103	7	2.5
TP-tert	84	12	1.9
Tcardo-tert	96	9	2.0
TMA-tert	83	6	1.9
Cyclo-tert	83	8	1.7
Phcardo-tert	91	8	2.1
m-m	116	6	2.7
TP-m	91	8	2.3
Tcardo-m	98	6	1.9
TMA-m	98	7	1.9
Cyclo-m	80	6	1.7
Phcardo-m	96	7	1.9

Source: Refs. 38–43.

The tensile properties of the poly(ether imide) films prepared by thermal treatment are summarized in Table 7. The mechanical properties of these poly(ether imide) films prepared by solution-casting from the DMAc solution are also summarized in Table 7. The films showed strong and tough nature. It was concluded that the polymer films with 2,2'-dimethyl-4,4'-biphenylene units were stronger than the others. It is quite reasonable that poly(ether imide) having the 4,4'-biphenylene unit showed rod-like nature and led to a higher rigidity of polymer chains than the others. These polymer films had tensile strength of 80–116 MPa and initial modulus of 1.7–2.7 GPa [38–43].

4. SUMMARY

Six new bis(ether anhydride)s: 2,2'-dimethyl-4,4'-bis[4-(3,4-dicarboxyphenoxy)]-biphenyl dianhydride (**4A**), bis[4-(3,4-dicarboxy phenoxy)phenyl]diphenylmethane dianhydride (**4B**), 1,1-bis[4-(4-dicarboxyphenoxy)phenyl]-4-tert-butylcyclohexane dianhydride (**4C**), 3,3',5,5'-tetramethyl-2,2-bis[4-(4-dicarboxyphenoxy)phenyl]propane dianhydride (**4D**), 1,1-bis[4-(3,4-dicarboxyphenoxy)phenyl] cyclododecane dianhydride (**4E**), and 1,1-bis[4-(4-dicarboxyphenoxy)phenyl]-4-phenylcyclohexane dianhydride (**4F**) were prepared in three steps starting from nitrodisplacement of 4-nitrophthalonitrile with corresponding bisphenols, followed by alkaline hydrolysis of the intermediate bis(ether dinitrile)s and subsequent dehydration of the resulting bis(ether diacid)s. A series of new highly organosoluble poly(ether imide)s were prepared from the bis(ether anhydride)s and various diamines by a conventional two-stage synthesis. The resulting poly(ether imide)s had inherent viscosities in the range of 0.50–0.81 dL·g^{-1}. GPC measurements revealed that the polymers exhibited number-average molecular weight and weight-average molecular weight up to 57,000 and 130,000, respectively. All the polymers showed typical amorphous diffraction patterns. Almost all of the poly(ether imide)s showed excellent solubility and readily dissolved in various solvents such as *N*-methyl-2-pyrrolidinone, *N,N*-dimethylacetamide (DMAc), *N,N*-dimethylformamide, pyridine, cyclohexanone, tetrahydrofuran and chloroform. These polymers had glass transition temperatures in the range of 226–262°C. Thermogravimetric analysis showed that all polymers were stable, with 10% weight loss recorded above 441°C in nitrogen. Tough and flexible polymer films could be easily obtained by solution casting from the DMAc solution. These polymer films had tensile strength of 80–116 MPa and tensile modulus of 1.7–2.7 GPa. The polyimides derived from six new dianhydrides have good solubility, thermal stability and mechanical properties.

REFERENCES

1. M. K. Ghosh and K. L. Mittal (Eds.), *Polyimides: Fundamentals and Applications*, Marcel Dekker, New York (1996).
2. D. Wilson, H. D. Stenzenberger and P. M. Hergenrother (Eds.), *Polyimides*, Chapman and Hall, New York (1990).
3. P. M. Hergenrother, N. T. Wakelyn and S. J. Havens, J. Polym. Sci., Part A: Polym. Chem. **25**, 1093 (1987).
4. T. Asanuma, H. Oikawa, Y. Okawa, W. Yamashita, M. Matsuo and A. Yamaguchi, J. Polym. Sci., Part A: Polym. Chem. **32**, 2111 (1994).
5. S. Matsuo and K. Mitsuhashi, J. Polym. Sci., Part A: Polym. Chem. **32**, 1969 (1994).
6. Y. Imai, N. N. Malder and M. Kakimoto, J. Polym. Sci., Part A: Polym. Chem. **22**, 2189 (1984).
7. D. J. Liaw, B. Y. Liaw, L. J. Li, B. Sillion, R. Mercier, R. Thiria and H. Sekiguchi, Chem. Mater. **10**, 734 (1998).
8. X. Sun, Y. K. Yang and F. Lu, Macromolecules **31**, 4291 (1998).
9. I. K. Spiliopoulos and J. A. Mikroyannidis, Macromolecules **31**, 515 (1998).
10. M. H. Yi, W. Huang, M. Y. Jin and K. Y. Choi, Macromolecules **30**, 5606 (1997).
11. D. J. Liaw and B. Y. Liaw, Polymer **40**, 3183 (1999).
12. J. G. Wirth, in *Discovery and Development of Polyetherimides in High Performance Polymers: Their Origin and Development*, R. B. Seymour and G. S. Kirshenbaum (Eds.), Elsevier, Amsterdam (1986).
13. M. Eashoo, Z. Wu, A. Zhang, D. Shen, C. Tse, F. W. Harris, S. Z. D. Cheng, H. K. Gardner and S. B. Hsiao, Macromol. Chem. Phys. **195**, 2207 (1994).
14. F. W. Harris and L. H. Lanier, in *Structure-Solubility Relationships*, F. W. Harris and R. B. Seymour (Eds.), p.183, Academic Press, New York (1977).
15. R. Sinta, R. A. Minns, R. A. Gaudiana and H. G. Rogers, Macromolecules **20**, 2374 (1987).
16. K. H. Becker and H. W. Schmidt, Macromolecules **25**, 6784 (1992).
17. D. J. Liaw, B. Y. Liaw and M. Q. Jeng, Polymer **39**, 1597 (1998).
18. D. J. Liaw, B. Y. Liaw, J. R. Chen and C. M. Yang, Macromolecules **32**, 6860 (1999).
19. D. J. Liaw, B. Y. Liaw and Y. S. Chen, Polymer **40**, 4041 (1999).
20. D. J. Liaw and B. Y. Liaw, Macromol. Chem. Phys. **199**, 1473 (1998).

21. D. K. Mohanty, Y. Sachdeva, J. L. Hedrick, J. F. Wolfe and J. E. McGrath, Polym. Prepr. **25(2)**, 19 (1984).
22. P. R. Sundararajan, Macromolecules **23**, 2600 (1999).
23. A. C. Misra, G. Tesoro, G. Hougham and S. M. Pendharkar, Polymer **33**, 1078 (1992).
24. J. W. Park, M. Lee, M. H. Lee, J. W. Liu, S. D. Kim, J. Y. Chang and S. B. Rhee, Macromolecules **27**, 3459 (1994).
25. Y. S. Negi, Y. I. Suzuki, I. Kawamur, M. A. Kakimoto and Y. Imai, J. Polym. Sci., Part A: Polym. Sci. **34**, 1663 (1996).
26. D. J. Liaw and K. L. Wang, J. Polym. Sci., Part A: Polym. Chem. **34**, 1209 (1996).
27. D. J. Liaw, B. Y. Liaw and C. Y. Chung, Acta Polym. **50**, 135 (1999).
28. D. J. Liaw, B. Y. Liaw and C. M. Yang, Macromolecules **32**, 7248 (1999).
29. M. Y. Jin and K. Y. Choi, Macromolecules **30**, 5606 (1997).
30. D. Ayala, A. E. Lozano, J. G. de la Campa and J. de Abajo, Polym. Prep. **39(1)**, 359 (1998).
31. V. V. Korshak, S. V. Vinogradova and Y. S. Vygodski, J. Macromol. Sci., Rev. Macromol. Chem. **C11**, 45 (1974).
32. N. Biolley, M. Gregoire, T. Pascal and B. Sillion, Polymer **32**, 3256 (1991).
33. D. J. Liaw and B. Y. Liaw, Polymer **40**, 3183 (1999).
34. D. J. Liaw and B. Y. Liaw, Macromol. Chem. Phys. **200**, 1326 (1999).
35. D. J. Liaw and B. Y. Liaw, Polym. J. **31**, 1270 (1999).
36. D. J. Liaw, B. Y. Liaw and C. Y. Chung, J. Polym. Sci., Part A: Polym. Chem. **37**, 2815 (1999).
37. D. J. Liaw, B. Y. Liaw and C. Y. Chung, Macromol. Chem. Phys. **201**, 1887 (2000).
38. D. J. Liaw, B. Y. Liaw, P. N. Hsu and C. Y. Hwang, Chem. Mater. **13**, 1811 (2001).
39. D. J. Liaw, C. Y. Hsu, I. W. Chen and B. Y. Liaw, Polymer Prepr. (Am. Chem. Soc Div. Polym. Chem.) **42**, 596 (2001).
40. D. J. Liaw, C. Y. Hsu and B. Y. Liaw, Polymer **42**, 7993 (2001).
41. D. J. Liaw, I. W. Chen, W. H. Chen and S. L. Lin, J. Polym. Sci., Part A: Polym. Chem **40**, 2556 (2002).
42. D. J. Liaw, I. W. Chen and M. C. Yang, Macromol. Chem. Phys. (In Press).
43. D. J. Liaw, C. Y. Hsu, P. N. Hsu and S. L. Lin, J. Polym. Sci., Part A: Polym. Chem. **40**, 2066 (2002).
44. R. A. Dine-Hart and W. W. Wright, Angew. Makromol. Chem. **153**, 237 (1972).
45. D. J. Liaw, B. Y. Liaw and M. Y. Tsai, Eur. Polym. J. **33**, 997 (1997).
46. D. J. Liaw and B. Y. Liaw, J. Polym. Sci., Part A: Polym. Chem. **35**, 1527 (1997).
47. D. J. Liaw and B. Y. Liaw, Eur. Polym. J. **33**, 1423 (1997).
48. D. J. Liaw and B. Y. Liaw, Polym. J. **28**, 970 (1996).

Polyimides and Other High Temperature Polymers, Vol. 2, pp. 71–90
Ed. K.L. Mittal
© VSP 2003

Synthesis of novel polyimides from dianhydrides with flexible side chains

YONG SEOK KIM,* KYUNG HOON LEE and JIN CHUL JUNG

Center for Advanced Functional Polymers/Polymer Research Institute, Department of Materials Science & Engineering, Pohang University of Science & Technology, San 31, Hyoja-dong, Pohang, 790-784, Korea

Abstract—Polyimides have been of great interest in engineering and microectronics for a number of applications because of their unique properties. In this study, we prepared a series of polyimides from pyromellitic dianhydrides having two 4-(n-alkyloxy)phenyloxy groups at their 3 and 6 positions. Also, another series of polyimides containing alicyclic units and alkyloxy side chains were prepared from 9,10-dialkyloxy-1,2,3,4,5,6,7,8-octahydro-2,3,6,7-anthracenetetracarboxylic 2,3,6,7-dianhydrides. Their physical properties such as thermal stability, phase transition and solution properties were investigated. WAXS studies revealed that substituted rigid-rod polyimides had typical well-developed layered structures, while non-rigid-rod polyimides had loosely packed layered structures.

In particular, uniform alignment layers, possessing high pretilt angles, have been obtained by mechanical rubbing of the polyimide film from **C$_n$-PMDA**, and their LC aligning property, such as pretilt angle and thermal stability of the LC director, were investigated with respect to the chemical architecture of polyimides. It is very interesting behavior that the pretilt angles in **C$_n$-OPIs** have been observed to possess both positive (θ_p) and negative ($-\theta_p$) angles depending on the side chain length.

Keywords: Polyimides; modification; side chain; thermal properties; pretilt angle.

1. INTRODUCTION

Wholly aromatic polyimides are a very important class of high performance polymers used in various applications requiring high thermal, mechanical and electrical characteristics [1]. Most of the aromatic polyimides, however, suffer from lack of processibility due to their infusible and insoluble nature, and must first be prepared as a soluble precursor, e.g., poly(amic acid) and then processed into the final form by thermal or chemical imidization methods. To overcome this drawback, a large number of structural modifications have been attempted such as

---

*To whom all correspondence should be addressed. Corresponding address: Advanced Materials Division, 1 Team, Korea Research Institute of Chemical Technology (KRICT), 100 Jang-dong, Yu-song, Daejon 305-343, Korea. Phone: +82-42-860-7304, Fax: +82-42-861-4151,
E-mail: yongskim@krict.re.kr

incorporation of thermally stable flexible or non-symmetrical linkages in the backbone [2, 3], introduction of polar or non-polar bulky substituents [4-6], or disruption of symmetry and recurring regularity through copolymerization. Another approach to enhance solubility is the insertion of alicyclic units into monomeric dianhydrides [7-9].

Among these structural modification methods, the appendage of flexible side chains to rigid backbone polymers is a particularly interesting approach because the entropy change on going from the crystalline to the nematic state is relatively low, and these polymers exhibit high melting and decomposition temperatures. As demonstrated by Majnusz *et al.* [10] and by Krigbaum *et al.* [11], the attachment of substituents of varying structure and length lowered the thermal transition temperature. A quantitative treatment of the phase transition behavior of rigid-rod polymers with flexible side chains has been elaborated recently [12-18]. This theory based on Flory's lattice model shows that the side chains act like a low molecular weight solvent.

Another very interesting property of the rigid-rod polymers with flexible side chains is their ability to form layered structures in crystals and liquid crystals when the flexible side chains reach a critical length, as shown in Figure 1. The layered phases are characterized by a segregated structure in which the rigid main chains form layers, while the flexible side chains occupy the space between the layers. Depending on the degree of molecular interaction between rigid main chain and flexible side chain, different layered structures are formed [19-21].

Recently, Matsumoto synthesized soluble polyimides using dianhydrides with polyalicyclic (cycloaliphatic) structures and aromatic diamines, the so-called

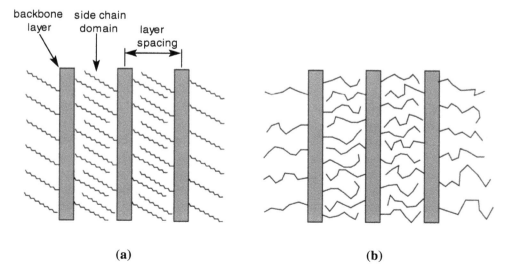

(a)   (b)

Figure 1. Schematic representation of main and side chains in a rigid-rod polymer having flexible side chains in the layered structure. (a) crystalline state (b) after side chain melting.

"semiaromatic polyimides" [22]. The introduction of an alicyclic unit into the polyimide backbone would reduce polymer-polymer interaction and enhance the solubility in organic solvents. These polyimides showed excellent thermal stability with no significant weight loss up to approximately 400°C and the 5% weight-loss temperature in a N_2 atmosphere was over 450°C; some of them had glass transition temperatures over 380°C. The high-temperature stability can also be explained by the introduction of an alicyclic structure, which would foster less probability of main-chain scission because of the presence of multibonds and would also increase main-chain rigidity. Of course, the use of alicyclic polymer constituents implies that the ultimate end use of such materials is targeted for applications with less stringent thermal requirements than those of aromatic polyimides. Alicyclic polyimides, however, have attracted much attention in recent years because of their potential applications including use as liquid crystal orientation layers, nonlinear optics buffer layers, or low dielectric materials.

On the other hand, there are only a few studies on polyimides containing various kinds of side groups as well as other functional groups such as alicyclic units. To achieve these aims, a proper design of new functionalized monomers is required. An effective functionalization of a monomer depends on its physical form, chemical reactivity and stability. These factors are crucial in the design of new materials and can be influenced by the conditions employed during their preparation.

This article reports on recent developments in polyimides having flexible side groups for heat resistance or LC alignable materials with particular emphasis on their synthesis and properties as well as the new structure-property relationships for novel polyimides.

2. SYNTHESIS OF POLYIMIDES FROM NEW DIANHYDRIDES HAVING ALKYLOXY PENDENT GROUP

2.1. Synthesis of new dianhydride monomers

Two different kinds of dianhydrides having various alkyl lengths were prepared. For easier comprehension, we have listed their chemical structures and codes in Figure 2.

First of all, as shown in Figure 3, the two (*n*-alkyloxy)phenyloxy side chains [23] were successfully introduced into pyromellitimide by nucleophilic substitution reaction of bromine atoms by (*n*-alkyloxy)phenoxide anions in pyridine at 0°C followed by hydrolysis of the imide groups. The dianhydrides blocked by aniline, 3,6-bis(*n*-alkyloxy)phenyloxy-*N*,*N*'-diphenylpyromellitimides, were hydrolyzed to obtain the corresponding tetracarboxylic acids. This hydrolysis was carried out by a two-step process: basic hydrolysis reaction using NaOH followed by acidic hydrolysis reaction through HCl/DMSO system. The resulting tetraacids were cyclodehydrated by treating with acetic anhydride. Irrespective of the side chain length, all the dianhydride monomers could be purified to polymerization grade by recrystallization from toluene.

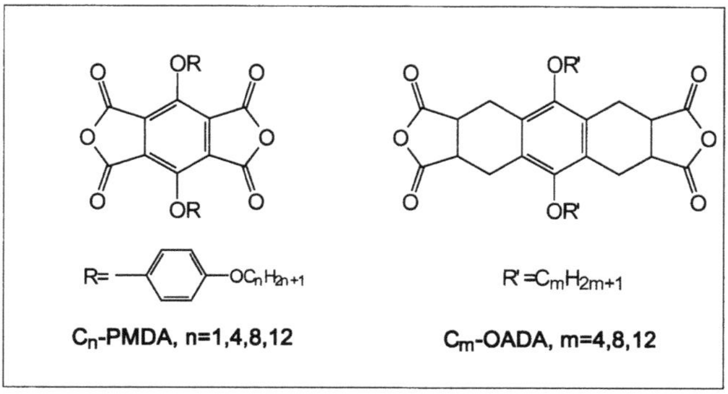

Figure 2. Chemical structures of new dianhydrides.

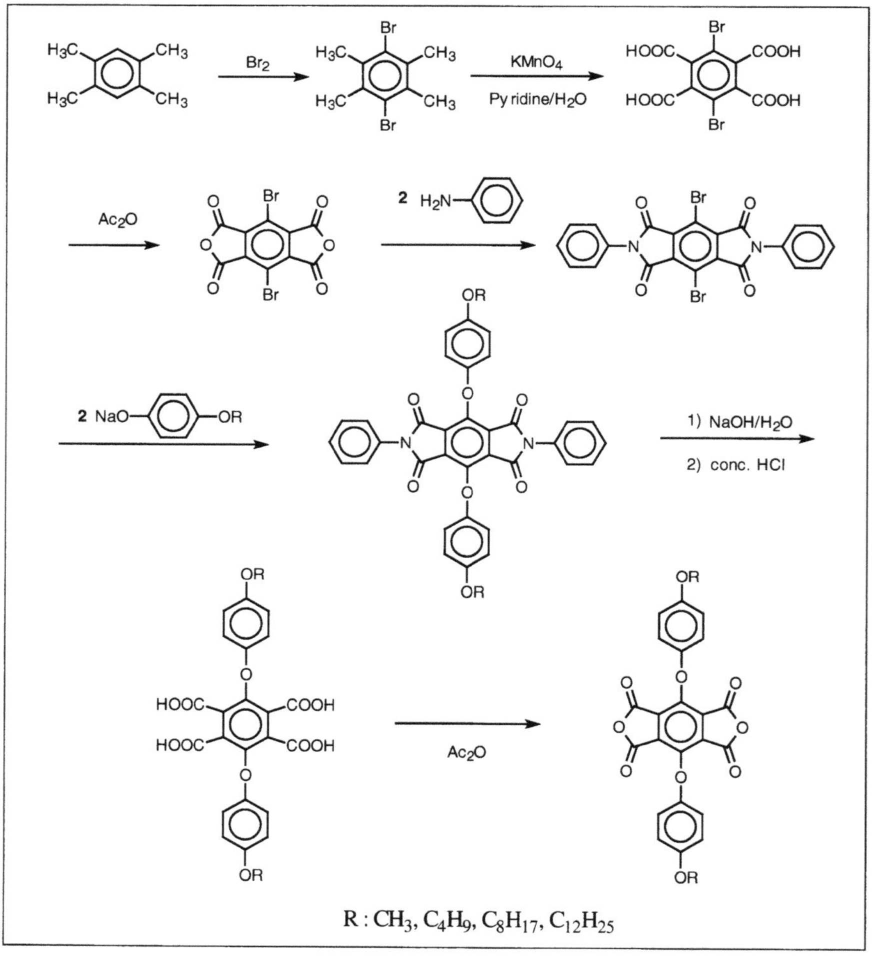

Figure 3. Synthesis route to **C$_n$-PMDA**.

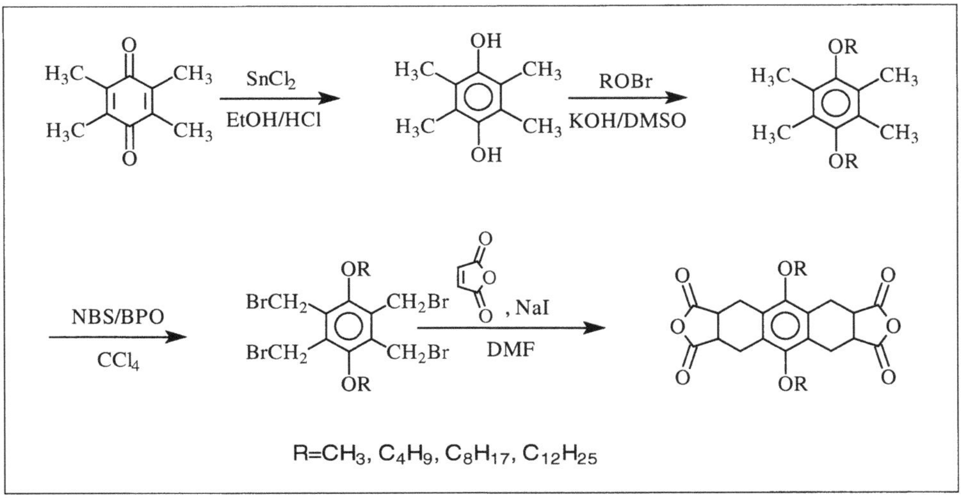

Figure 4. Synthesis route to **C$_m$-OADA**.

Other monomeric dianhydrides (**C$_m$-OADA**) [24] were prepared by bis-Diels-Alder reaction from maleic anhydride and 1,4-dialkyloxy-2,3,5,6-tetrakis (bromomethyl)benzene, which had been synthesized from duroquinone as a starting material *via* consecutive reduction, alkylation with *n*-bromoalkane and side-chain bromination of methyl groups using *N*-bromosuccinimide (NBS), as shown in Figure 4. Bis-Diels-Alder reaction was conducted to obtain monomeric dianhydrides below 80°C under a nitrogen flow. The reaction mixtures were dehydrated by treating with acetic anhydride, and then recrystallized from acetic anhydride or CH$_2$Cl$_2$/petroleum ether mixed solvent. After separation, crystalline materials were obtained with anhydride peaks at 1844, 1783 cm^{-1} in the FT-IR spectra. The NMR peaks at 0.9~1.8 ppm were regarded as the ones from alkyl side chains. And we also confirmed that doublets at ca. 2.5 and 3.5 ppm were split by the *geminal* hydrogen atom of the alicyclic unit in the monomers.

2.2. Polymerization

The polymerization reactions were carried out by one or two-step method and the chemical structure of new polyimides and their codes are listed in Figure 5.

The typical one-step method was as follows. To a solution of diamine in freshly distilled *N*-methylpyrrolidinone (NMP), equimolar dianhydride was added at room temperature under an argon atmosphere with 10% solid content (w/v). The reaction mixture was heated at 70~80°C for 2 hours. Then the solution temperature was slowly raised to 180°C and kept there for an additional 12 hours. In the two-step method, poly(amic acid)s were first prepared from reactions of dianhydride monomers with diamines in NMP and subsequently cyclized with triethylamine and acetic anhydride at room temperature.

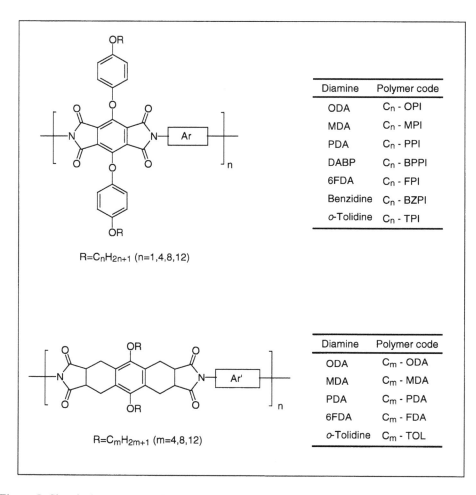

Figure 5. Chemical structures and codes of new polyimides.

In both methods, the precipitated polymers were washed several times with boiling methanol, and then dried at 100°C for 12 hours *in vacuo*. The reaction systems remained homogeneous in NMP. Later, spectroscopic characterizations confirmed that the cyclodehydration had proceeded completely.

3. PROPERTIES

3.1. Solution properties

The solution properties of polyimides derived from **C_n-PMDA** are summarized in Table 1. The polymers are highly soluble in NMP even at room temperature, but only slightly soluble in other polar solvents such as *m*-cresol, *N,N*-dimethyl-

Table 1.
Solution properties of polyimides from C_n-PMDA

Polymer code	n	$\eta_{inh.}$ [a] (dL/g)	Solubility [b]						
			NMP	DMSO	DMAc	DMF	HMPA	*m*-Cresol	H_2SO_4
C_n-OPI	1	0.51	+++	++	++	++	++	+	+
	4	0.68	+++	++	++	++	++	+	+
	8	0.66	+++	++	++	++	++	+	+
	12	0.46	+++	++	++	++	++	+	+
C_n-MPI	1	0.66	+++	++	++	++	++	+	+
	4	0.53	+++	++	++	++	++	+	+
	8	0.64	+++	++	++	++	++	+	+
	12	0.53	+++	++	++	++	++	+	+
C_n-TPI	4	0.56	+++	++	++	++	++	+	+
	8	0.66	+++	++	++	++	++	+	+
	12	0.56	+++	++	++	++	++	+	+
C_n-BZPI	1	–	++	++	++	++	++	++	+
	4	2.29	+++	+	+	+	+	–	++
	8	–	++	+	+	+	+	–	+
	12	–	++	+	+	+	+	–	+

[a] Inherent viscosities were measured at a concentration of 0.2 g/dL in NMP at 25°C
[b] Key: – Insoluble, + slightly soluble, ++ soluble at high temperature, +++ soluble at room temperature

acetamide (DMAc), concentrated sulfuric acid and hexamethylphosphoramide (HMPA). It is quite surprising that there is a great solubility difference between NMP and other highly polar aprotic or protic solvents. Inherent viscosities of the polymers measured from NMP are in the range of 0.46~2.29 dL/g, indicating that they have relatively high molecular weights.

From Table 2, it is observed that all the polyimides from C_8-OADA and C_{12}-OADA, even with a rigid diamine, PDA, are soluble in polar solvents such as NMP, *N,N*-dimethylformamide (DMF) as well as *m*-cresol. Apart from these solvents, the polyimides from C_{12}-OADA could also be dissolved in chloroform at high temperature. The improved solubility of polyimides clearly originates from the introduction of the long alkyl chain, which increases the free volume and reduces the chain packing of the polyimides. These results coincide with the trend of the polyimides having alkyl side chain [24]. The C_4-FDA, however, showed relatively low solubility. This is presumably because of its crystallinity due to strong dipolar interaction of hexafluoroisopropylidene groups, which was confirmed by X-ray diffractometry. Moreover, this result indicates that the increase in the chain flexibility of polyimide is not necessary for the increase of solubility.

Table 2.
Solution properties of polyimides from C_m-OADA

Polymer code	m	$\eta_{inh.}$ [a] (dL/g)	Solubility [b]				
			CHCl$_3$	DMAc	NMP	*m*-Cresol	H$_2$SO$_4$
C_m-ODA	4	0.84	−	++	+++	++	+++
	8	0.30	−	++	+++	+++	+++
	12	0.45	++	++	+++	+++	+++
C_m-MDA	4	0.42	−	++	+++	++	+++
	8	0.28	−	++	+++	+++	+++
	12	0.41	++	+++	+++	+++	+++
C_m-PDA	4	0.41	−	++	++	++	+++
	8	0.38	−	++	+++	+++	+++
	12	0.42	−	+++	+++	+++	+++
C_m-FDA	4	0.24 [c]	−	−	−	−	+++
	8	0.27	−	++	+++	+++	+++
	12	0.25	++	++	+++	+++	+++
C_m-TOL	4	0.42	−	++	+	++	+++
	8	0.38	−	++	+++	+++	+++
	12	0.53	++	++	+++	+++	+++

[a] Inherent viscosities were measured at a concentration of 0.2 g/dL in NMP at 25°C
[b] Key: − Insoluble, + slightly soluble, ++ soluble at high temperature, +++ soluble at room temperature
[c] Inherent viscosities were measured in 0.2 g/dL in conc. H$_2$SO$_4$ at 25°C

3.2. Thermal properties

The thermal stability of polymers was investigated by thermogravimetric analysis (TGA) in N$_2$ at 20°C/min heating rate and the results on thermal properties of polyimides are summarized in Tables 3 and 4. All the polyimides showed similar patterns in TGA thermograms. The thermal stability of the prepared polyimides was lowered due to the incorporation of the thermally fragile alkyloxy side chains. Typical TGA thermograms are shown in Figures 6 and 7. In Figure 6, all C_n-OPIs show a two-step pyrolysis behavior with increasing temperature. This behavior may reasonably be presumed to arise from the fact that side chains are degraded in the low-temperature range and the main chains in the high-temperature range. To determine which part of side chains splits away in the lower-temperature range, we calculated wt. %'s of *n*-alkyloxy group (W_a) and those of (*n*-alkyloxy)phenyloxy group (W_b) and compared with the measured wt. % values (W_m) in the thermograms. The observed W_m values were between the W_a and W_b values. This fact means that no side group selectively degrades in the low-temperature range. At this stage, it cannot be said whether the main chain also decomposes to some degree or not.

Table 3.
Thermal properties of polyimides from C_n-PMDA

Polymer code	n	$T_g(°C)^{a)}$	$T_m(°C)^{b)}$	Thermal stability in N_2			
				$T_{10}(°C)^{c)}$	$T_{d1}(°C)^{d)}$	$T_{d2}(°C)^{e)}$	$R_{900}(\%)^{f)}$
C_n-OPI	1	302	–	478	478	584	47
	4	279	345	480	463	591	41
	8	238	326	475	464	582	37
	12	217	316	458	457	577	33
C_n-MPI	1	308	314	488	453	591	46
	4	263	333	473	454	594	44
	8	192	–	457	448	598	37
	12	185	–	465	456	590	34
C_n-PPI	1	–	–	518	502	–	52
	4	–	–	496	480	–	45
	8	–	–	481	470	–	39
	12	–	–	480	472	–	32
C_n-BPPI$^{g)}$	1	301	–	496	455	582	52
	4	268	–	471	445	587	46
	8	258	312	470	458	594	39
	12	264	320, 375	458	452	579	36
C_n-FPI	1	307	–	511	451	–	48
	4	298	–	86	453	572	43
	8	196	–	476	461	579	37
	12	171	–	470	461	576	33
C_n-BZPI	1	–	–	498	460	602	50
	4	–	–	467	450	599	44
	8	–	–	473	462	604	39
	12	–	–	467	459	304	36
C_n-TPI	4	–	–	480	468	622	47
	8	–	–	471	463	621	40
	12	–	–	463	461	617	36

[a]From the second heating traces in DSC measurements conducted at a heating rate of 20°C/min in N_2 atmosphere; [b]Onset temperature of degradation; [c]1st, [d]2nd decomposition temperatures measured by TGA at a heating rate of 20°C/min under N_2; [e]10% weight loss temperatures measured by TGA at a heating rate of 20°C/min under N_2; [f]Residual weight % at 900°C measured by TGA at a heating rate of 20°C/min under N_2

Table 4.
Thermal properties of polyimides from C_m-OADA

Polymer code	T_g (°C) [a]	Thermal stability in N_2				
		T_o(°C) [b]	T_{d1}(°C) [c]	T_{d2}(°C) [d]	T_{10}(°C) [e]	R_{900}(%)[f]
C_4-PDA	292	439	472	702	459	47.58
C_8-PDA	176	422	497	680	440	40.76
C_{12}PDA	138	408	447	661	418	36.02
C_4-ODA	259	449	452	643	443	50.15
C_8-ODA	163	435	466	664	447	45.83
C_{12}-ODA	124	413	463	656	434	42.34
C_4-MDA	232	441	479	678	470	45.86
C_8-MDA	134	431	466	653	451	40.01
C_{12}-MDA	111	413	453	643	426	37.06
C_4-FDA	294	453	477	610	477	45.64
C_8-FDA	160(267)[g]	432	472	619	453	38.61
C_{12}-FDA	116	406	443	579	425	36.78
C_4-TOL	298	436	464	669	468	52.77
C_8-TOL	202	422	459	669	442	43.30
C_{12}-TOL	165	412	455	657	427	40.57

[a]From the second heating traces in DSC measurements conducted at a heating rate of 20°C/min in N_2 atmosphere; [b]Onset temperature of degradation; [c]1st, [d]2nd decomposition temperatures measured by TGA at a heating rate of 20°C/min under N_2; [e]10% weight loss temperatures measured by TGA at a heating rate of 20°C/min under N_2; [f]Residual weight % at 900°C measured by TGA at a heating rate of 20°C/min under N_2; [g]Melting transition measured by DSC

From Table 3, it is obvious that with increasing side chain length, the T_{d1} greatly decreases, indicating that thermal stability of the polyimides is greatly affected by the presence of the side chains. On the other hand, T_{d2} changed only slightly, which means that entire side chains had already degraded in the low-temperature range.

In case of polyimides having an alicyclic unit as well as an alkyloxy side chain, they were thermally stable to about 420°C, and then began to degrade drastically in the range of 443~497°C, followed by second slight degradation occurring in the 579~702°C range. Thus the degradation shows a two-step weight loss behavior. As shown in Table 4, the residual weight percent determined at 900°C varies with backbone structure of each polymer. When these values are compared with the theoretically calculated side chain content values, the coincidence falls within a ±4% range. These results indicate that the first drastic degradation originates mainly from the degradation of the side chains, while the second degradation arises from that of the polymer backbone containing alicyclic unit. All the residual weights showed higher values than 36.02% calculated on the basis of the structures of polyimides.

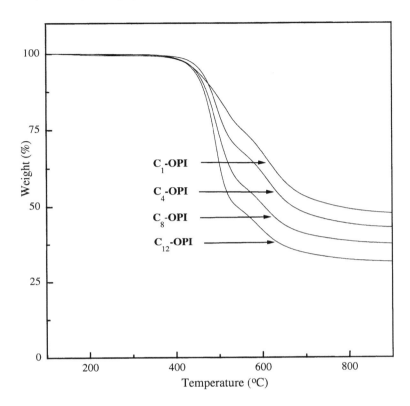

Figure 6. TGA thermograms of **C$_n$-OPIs**.

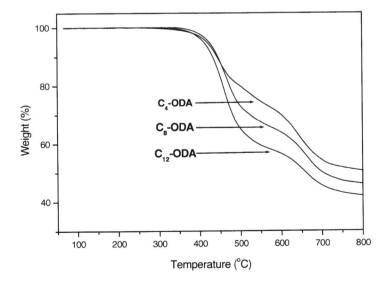

Figure 7. TGA thermograms of **C$_m$-ODAs**.

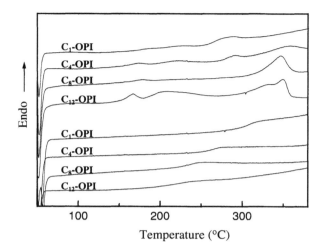

Figure 8. DSC thermograms of **Cₙ-OPIs** 1st (upper curves) and 3rd (lower curves) heatings.

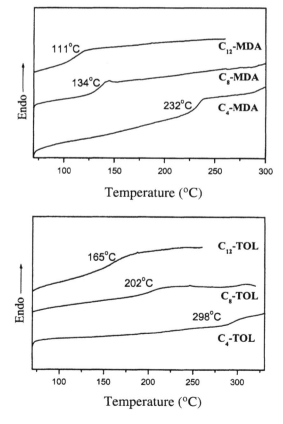

Figure 9. DSC thermograms of **Cₘ-MDAs** and **Cₘ-TOLs**.

DSC thermograms were used to study the phase transition behavior of the poly-imides. They were obtained from first and third heating scans of as-polymerized **Cₙ-OPIs** powder samples at 20°C/min scan rate and they are shown in Figures 8 and 9. DSC curves obtained from first heating scans of **C₈-** and **C₁₂-OPI** clearly show melting endotherms at 345 and 350°C, respectively, indicating that both have some crystallinity and also the presence of such crystallinity could be con-firmed by wide-angle X-ray scattering (WAXS). The lower T_g's reflect higher mobilities of the longer flexible side chains directly on the attached main chains.

For the series of polyimides from **Cₘ-OADA**, the glass transition temperatures (T_g's) of the polymers obtained ranged from 111 to 296°C depending on their structure. Generally, the T_g's of these polyimides were relatively lower than of polyimides which had no alicyclic unit, which means that the introduction of an alicyclic unit in the backbone induces non-coplanarity to lower the T_g. And also, T_g's highly depended on the length of side chain and the chemical structure of diamine. The T_g's of **C₄**-series were in the range 232~296°C, those of **C₈, C₁₂**-series were in the ranges 134~202°C and 111~165°C, respectively. With a given dianhydride family, the T_g's of the polyimides decreased with decreasing rigidity of the diamine in the order o-tolidine > PDA ≥ 6FDA > MDA. Interestingly, **C₈-FDA** showed T_m at 267°C as well as a T_g at 160°C, which reappeared with re-peated scans as shown in Figure 10. This means that **C₈-FDA** has a semi-crystalline structure and also it is confirmed by wide-angle X-ray diffraction. In case of **C₄-FDA** and **C₄-TOL**, their semi-crystalline nature is ascertained also by X-ray diffraction, but no melting transition was observed in the 70~350°C range. On the basis of these results, it is possible to adjust the T_g's from 110 to 300°C by the introduction of alkyloxy side chains having various lengths or using different kinds of diamines.

3.3. Structure

An attempt was made to evaluate the crystallinity of these polymers by means of X-ray diffraction experiment in a transmission mode. In the small and middle an-gle regions, intense peaks from the layer structure were observed. Especially in the diffractogram of **C₁₂-PPI**, 4th order reflection peak as well as 2nd and 3rd or-der reflections were observed. At wider angles they showed only amorphous halo originating from their very low crystallinity.

It is obvious from the increase of the scattering intensity that the internal order of the polyimide is enhanced with increasing side chain length. Also, the maxima in the diffractograms are shifted to smaller 2θ values, which means larger lateral dimensions. In the **Cₙ-PPIs** series, layer distances (*d*) were plotted versus the number of atoms in the alkyloxy side chain. A linear dependence was found with slope 1.24. The increment per methylene (CH_2) unit, 1.24 Å was similar to the theoretical value of 1.25 Å calculated from the fully extended and all-trans con-formation of alkyl group [25-27]. The *n*-alkyl groups of **Cₙ-PPIs**, therefore, are thought to be fully extended and located perpendicularly to the layer.

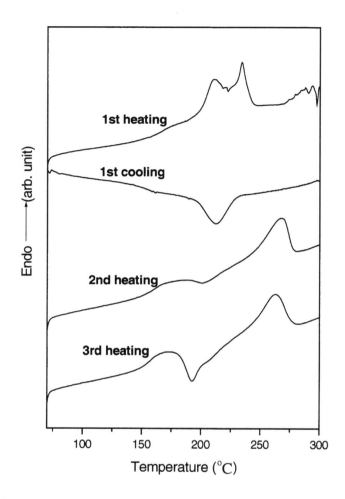

Figure 10. DSC thermograms of **C$_8$-FDA**.

The X-ray reflections of polyimides from **C$_m$-OADA** are summarized in Table 5. Most of the X-ray measurements agree in that only weak reflections show up in the typical wide-angle region ($2\theta > 10°$). On the other hand, the diffractograms exhibit at least one strong reflection in the middle angle region $2\theta = 2\sim8°$. This reflection is particularly strong and sharp in both **C$_{12}$-series** and **C$_8$-series** but is relatively broad in the case of **C$_4$-series**.

In completely amorphous **C$_m$-MDA** series, the resulting layer distances (d) were also plotted versus the number of carbon atoms in the alkyloxy side chain and its slope was 1.7. From these facts, it is thought that these polyimides have loose-layered structures because the layer distance (d in Table 5) is larger than the theoretical layer spacing calculated from the fully extended and completely interdigitated side chains.

Table 5.
X-ray reflections of polyimides from C_m-OADA

Polymer code	Middle angle[a]		WAXS 2θ (degree)
	2θ (degree)	d (Å)[b]	
C_4-PDA	–	–	15.0/18.3
C_8-PDA	4.35/8.59[c]	19.6/10.3	13.5/18.8/22.1
C_{12}-PDA	3.28	26.9	18.5
C_4-ODA	8.27	10.7	15.5/broad halo
C_8-ODA	4.86	18.7	16.4
C_{12}-ODA	3.34	24.5	15.5
C_4-MDA	8.07(broad)	10.9	18.2
C_8-MDA	4.57	19.3	18.8
C_{12}-MDA	3.53	24.7	19.2
C_4-FDA	8.15	10.8	11.5/13.8/15.7/17.8/22.9/24.8/28.3
C_8-FDA	4.71/8.26	18.7/10.7	17.8
C_{12}-FDA	3.48	25.4	18.4
C_4-TOL	5.86	15.1	15.0/17.9
C_8-TOL	5.46	16.6	18.4
C_{12}-TOL	3.57	24.7	19.0

[a]$2<2\theta<8$, [b]Layer distance is calculated via the Bragg equation; [c]multiple numbers indicate distinctive peaks

Figure 11 depicts the X-ray diffractograms of C_8-FDA obtained at various temperatures. As shown in Figure 10, a broad endothermic peak appears in the DSC thermogram of C_8-FDA. The diffractograms obtained at 175 and 220°C show no significant change in their diffraction patterns except that their small angle reflections at 4.71 degree are shifted to slightly smaller angle region. This means that at higher temperatures, the layer spacing is increased because of the thermal expansion of the polymer. The discernible X-ray reflections at 8.26 degree and higher angle peaks disappeared above 260°C, indicating that it had passed the mesophase transition [25]. But the remaining strong small-angle reflection indicates the layered-structure to be still present. In the cooling scans, the reflections were reproduced meaning that its transitions were reversible.

Moreover, a higher crystallinity of C_4-FDA was observed, presumably because of the strong dipolar interchain interaction of the hexafluoroisopropylidene groups and this result agrees with its poor solubility.

3.4. Liquid crystal (LC) alignment properties

The polyimides from C_n-PMDA coated onto glass substrates were rubbed with various rubbing densities (L/l) [28-30] using a laboratory rubbing machine

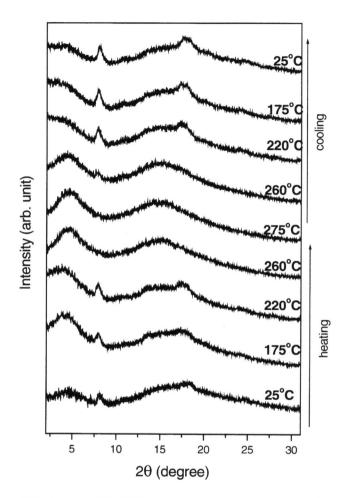

Figure 11. X-ray diffractograms of **C₈-FDA** at various temperatures (scan rate: 20°C/min).

(Wande Co., Dallas, TX) equipped with a roller covered by a rayon cloth (Yoshi-kawa Co., Okayama, Japan; YA-20-R). The rubbing density (L/l) was calculated according to equation (1):

$$\frac{L}{l} = N\left(\frac{2\pi r n}{60\nu} - 1\right) \tag{1}$$

where L (mm) is the total length of the rubbing cloth that contacts a fixed point of the polymer film, l (mm) is the contact length of the circumference of rubbing roller, N is the cumulative number of rubbings, n (rpm) is the speed of rubbing roller, r (cm) is the radius of roller, and ν (cm/s) is the velocity of the substrate stage.

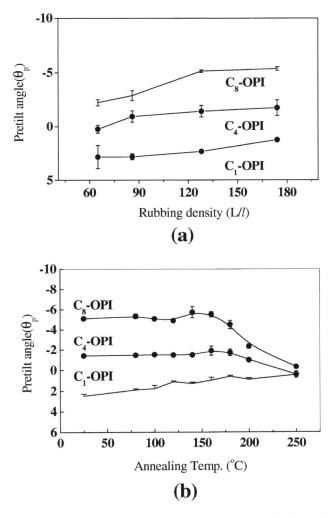

Figure 12. Pretilt angles of **C_n-OPIs** alignment layers as a function of rubbing density (a) and annealing temperature (b).

The LC material, 4-(n-pentyl)-4'-cyanobiphenyl (**5CB**), containing 1.0 wt% Disperse Blue 1 as a dichroic dye was injected into the empty cell in which the polyimide films were assembled into an antiparallel rubbing direction with 50 μm thick poly(ethylene terephthalate) film spacers. The pretilt angles (θ_p) were measured by the crystal rotation method and are plotted against the rubbing density for the LC cells fabricated with **C_n-OPIs** and **C_n-PPIs** in Figures 12 and 13, respectively.

For the rubbed surfaces of all polyimide films, microgrooves were observed in the rubbing direction by AFM (Atomic Force Microscopy). Also, polar diagrams

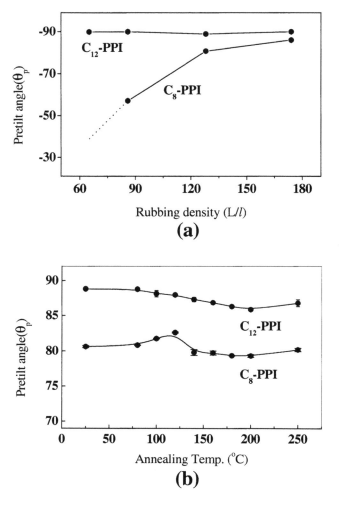

Figure 13. Pretilt angles of **C_n-PPIs** alignment layers as a function of rubbing density (a) and annealing temperature (b).

constructed after LC cell fabrications revealed that LC molecules were well aligned in the rubbing direction for all cases.

For the LC cells made from **C_n-OPIs**, θ_p's are in the range of $3 \sim -5°$, in which a negative angle indicates that LC molecules are oriented opposite to the rubbing direction. On the surface of **C_1-OPI**, the **5CB** molecules are aligned along the rubbing direction while **C_8-OPI** orients the **5CB** molecules against the rubbing direction, and in **C_4-OPI** both negative and positive angles are achieved depending on the rubbing density. **C_n-OPI** backbones are flexible in their structure. In order to examine the effect of the backbone flexibility on LC alignment, the pretilt angles were determined for the **C_n-OPIs** annealed at various temperatures for 2 hours (see Figure 12(b)). The pretilt angles gradually decrease with increasing annealing temperature and converge to zero at 250°C. Although this temperature (250°C) is not identical with T_g's of the **C$_n$-OPIs**, it can be shown that the backbone rigidity exerts a great effect on the thermal resistance of the rubbing director. All the other polyimides except **C$_n$-PPIs** demonstrated roughly similar LC-alignment and annealing behaviors as **C$_n$-OPIs**.

C$_n$-PPIs, which have the most rigid backbone structure, showed highly unique LC alignment behavior. As shown in Figure 13, **C$_{12}$-PPI** revealed homeotropic LC alignment independent of rubbing density. But in **C$_8$-PPI** the LC director changed from tilted LC alignment to homeotropic with increasing rubbing density ($\theta_p= -25{\sim}-55°$). This result is highly interesting since by adjusting the rubbing density, it is possible to control the pretilt angle. Figure 13(b) shows that the annealing below 250°C exerts practically no effect on the disappearance of the pretilt angles of **C$_n$-PPIs**. This means again that the rigidity of the polymer backbone strongly affects LC alignability of the polymer structure.

4. PROSPECTS

Polyimides having flexible side chains have been extensively investigated because of both scientific and commercial significances. In recent years, successful results have been reported and some of them have even achieved commercial development. Moreover, polyimides modified by the introduction of functional side groups have potential applications in electronic device materials like LC alignment layer, optical wave guide or photochromic materials. To obtain these functional materials, new and fine synthesis methodology is required. Due to the unique properties of polyimides, the demand for them will continue to expand steadily in the future. There is a need for acquiring of much more knowledge in this area.

REFERENCES

1. (a) M.K. Ghosh and K.L. Mittal (Eds.), *Polyimides: Fundamentals and Applications*, Marcel Dekker, New York (1996); (b) C.E. Sroog, A.L. Endery, S.V. Abramo, C.E. Berr, W.M. Edward and K.L. Oliver, *J. Polym. Sci.: A*, **3**, 1373 (1965).
2. Y.T. Chern, *Macromolecules*, **31**, 1898 (1998).
3. J. Yin, Y.F. Ye and Z.G. Wang, *Eur. Polym. J.*, **34**, 1839 (1998).
4. K.H. Lee and J.C. Jung, *Polym. Bull.*, **40**, 407 (1998).
5. F.W. Harris, Y. Sakaguchi, M. Shibata and S.Z.D. Cheng, *High Perform. Polym.*, **9**, 251 (1997).
6. J.A. Mikroyannidis, *J. Polym. Sci. Polym. Chem. Ed.*, **37**, 15 (1999).
7. S. Itamura, M. Yamada, S. Tamura, T. Matsumoto and T. Kurosaki, *Macromolecules*, **26**, 3490 (1993).
8. M. Kasuma, T. Matsumoto and T. Kurosaki, *Macromolecules*, **27**, 1117 (1994).
9. T. Matsumoto, *Macromolecules*, **32**, 4933 (1999).
10. J. Majnusz, J.M. Catala and R.W. Lenz, *Eur. Polym. J.*, **19**, 1043 (1983).
11. W.R. Krigbaum, H. Hakem and R. Kotek, *Macromolecules*, **18**, 965 (1985).

12. F.H. Metzmann, M. Ballauff, R.C. Schultz and G. Wegner, *Makromol. Chem.*, **190**, 985 (1989).
13. A.K. Whittaker, U. Falk and H.W. Speiss, *Makromol. Chem.*, **190**, 1603 (1989).
14. M. Ballauff, *Angew. Chem. Int. Ed. (Engl.)*, **28**, 252 (1989).
15. B.R. Harkness and J. Watanabe, *Macromolecules*, **24**, 6759 (1991).
16. P. Galda, D. Kistner, A. Martin and M. Ballauff, *Macromolecules*, **26**, 1595 (1993).
17. H.R. Kricheldorf, V. Linzer, M. Leland and S.Z.D. Cheng, *Macromolecules*, 30, 4828 (1997).
18. R.U. Zheng, E.Q. Chen and S.Z.D. Cheng, *Macromolecules*, **32**, 3574 (1999).
19. S.B. Park, H. Kim, W.C. Zin and J.C. Jung, *Macromolecules*, **26**, 1627 (1993).
20. J.C. Jung and S.B. Park, *Polym. Bull.*, **35**, 423 (1995).
21. J.C. Jung and S.B. Park, *J. Polym. Chem.: Polym. Chem.*, **34**, 357 (1996).
22. T. Matsumoto, *Macromolecules*, **32**, 4933 (1999).
23. K.H. Lee and J.C. Jung, *Polym. Bull.*, **40**, 407 (1998).
24. Y.S. Kim and J.C. Jung, *Polym. Bull.*, **45**, 311 (2000).
25. H.R. Kricheldorf and A. Domschke, *Macromolecules*, **29**, 1337 (1996).
26. C. Wutz, S. Thomsen, G. Schwarz and H.R. Kricheldorf, *Macromolecules*, **30**, 6127 (1997).
27. M. Dreja and W. Lennartz, *Macromolecules*, **32**, 3528 (1999).
28. T. Uchida, M. Hiramo and H. Sakai, *Liq. Cryst.*, **5**, 1127 (1989).
29. S.I. Kim, S.M. Pyo, M. Ree, M. Park and Y. Kim, Mol. *Crys. Liq. Cryst.*, **316**, 209 (1998).
30. J.H. Park, J.C. Jung, B.H. Sohn, S.W. Lee and M. Ree, *J. Polym. Sci.- Polym. Chem. Ed.*, **39**, 3622 (2001).

Polyimides and Other High Temperature Polymers, Vol. 2, pp. 91–98
Ed. K.L. Mittal
© VSP 2003

Polyimides based on rhenium(I) diimine complexes

LILLIAN SZE MAN LAM,[1] WAI KIN CHAN[*,1] and A.B. DJURIŠIĆ[2]

[1]*Department of Chemistry, The University of Hong Kong, Pokfulam Road, Hong Kong, China*
[2]*Department of Electrical and Electronic Engineering and Department of Physics,
The University of Hong Kong, Pokfulam Road, Hong Kong, China*

Abstract—Polyimides with significantly enhanced photoconductivity were synthesized by the polymerization of a rhenium tricarbonyl complex with various aromatic dianhydrides. The rhenium complex acted as the photosensitizer in the visible region. Electronic absorption spectra of the polymers showed a very strong intramolecular charge transfer transition and a large photocurrent response was observed. The photocharge generation process was studied by simulating the experimental quantum yield with the Onsager's theory, which indicated that the primary yield and the thermalization distance were of the order of 10^{-4} and 1.7 nm, respectively.

Keywords: Polyimide; rhenium diimine complex; photoconducting polymers; photosensitivity.

1. INTRODUCTION

Our research effort has been continuously focused on the synthesis of metal-containing polymers and their applications as opto-electronic materials. By incorporating different metal complexes into the polymer main chain, we were able to make use of the interesting photosensitivity/emissive properties of the metal complex moieties. The metal complexes can be attached to the polymer main chain or as the pendant group [1-6]. In our previous papers, we reported on the syntheses of different polyimides with bis(terpyridyl) ruthenium [7], diazacrown ether [8] (Figure 1), or rhenium diimine complexes [9]. Some preliminarly results on their photoconductivity and light emitting properties were also presented. In this paper, we report a detailed study of the photocharge generation process in a polyimide functionalized with chlorotricarbonyl 1,4-diaza-1,3-butadiene (DAB) rhenium(I) complex. Complexes of the type *fac*-[Re(X)(CO)$_3$(α-diimine)] (X = Cl, Br, CH$_3$) possess low lying excited states with metal-to-ligand charge transfer (MLCT) character [10-12]. They are very promising photooxidants, photosensitizers and photo- and electro-catalysts [13-16]. The fabrication of light emitting devices based on these complexes was also reported [17, 18]. Compared to other aromatic

---

*To whom all correspondence should be addressed. Phone: (852) 2859-8943,
Fax: (852) 2857-1586, E-mail: waichan@hkucc.hku.hk

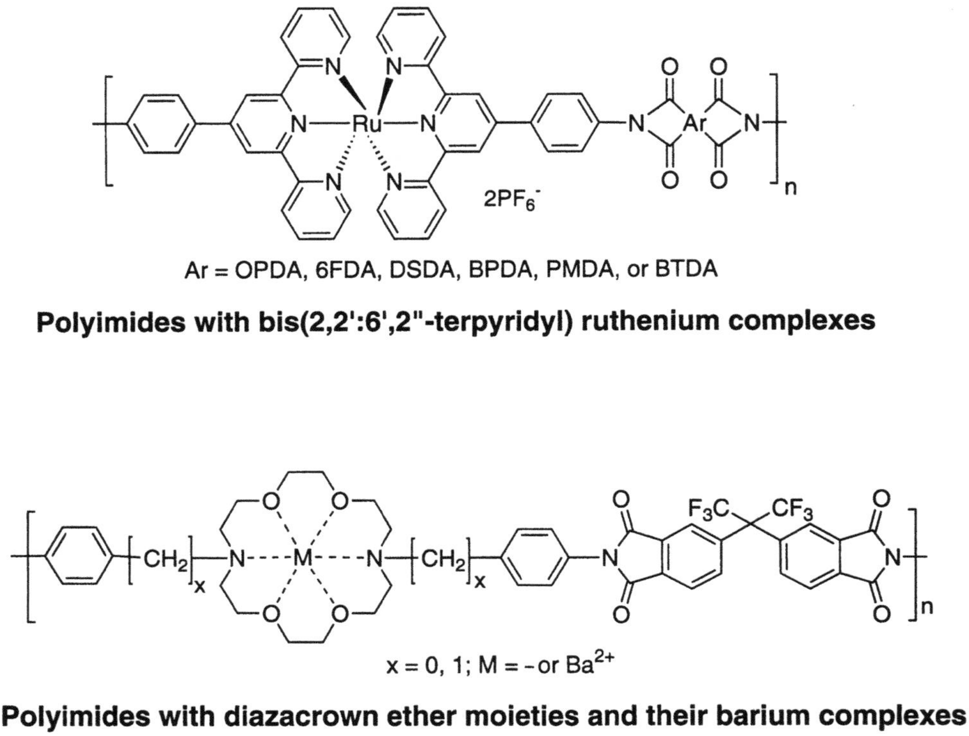

Ar = OPDA, 6FDA, DSDA, BPDA, PMDA, or BTDA

Polyimides with bis(2,2':6',2"-terpyridyl) ruthenium complexes

$x = 0, 1$; $M = -$or Ba^{2+}

Polyimides with diazacrown ether moieties and their barium complexes

Figure 1. Examples of some metal-containing polyimides.

heterocyclic diimines such as 2,2'-bipyridine and 1,10-phenanthroline, DAB and its derivatives exhibit lower π^* orbital energy and the rate of non-radiative decay is much faster [19]. As a result, these complexes are not emissive in nature and this may further enhance the photosensitivity because more excitons generated can be utilized for the photocharge generation process.

2. EXPERIMENTAL

2.1. Synthesis of polyimides

The polyimides were conveniently synthesized by the reaction between the diamino rhenium complex **1** with various dianhydrides in NMP followed by *in situ* imidization using acetic anhydride as the dehydrating agent and pyridine as the catalyst. Detailed synthesis procedures were reported in our previous paper [9]. Six different dianhydrides were used and the polyimide derived from 6FDA was found to have the best film quality and processability (Figure 2). Therefore, the **Re-6FDA** was used for detailed photoconductivity studies.

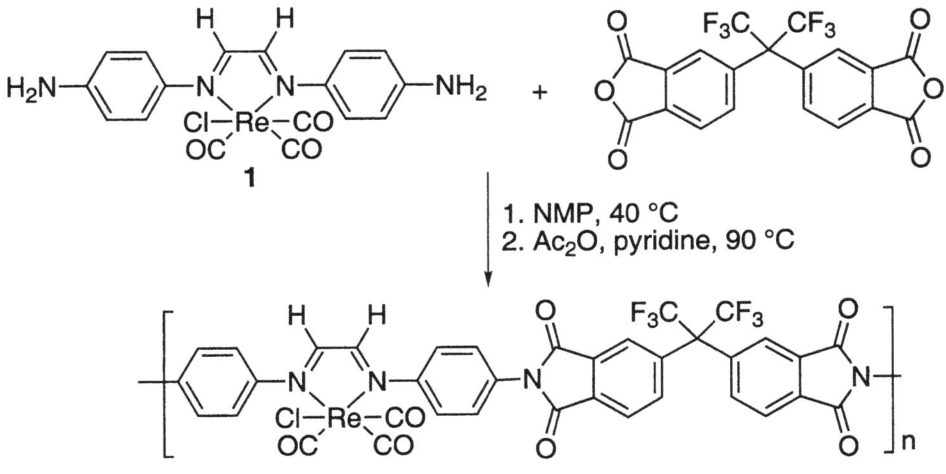

Figure 2. Synthesis of polyimide with rhenium diimine complexes.

2.2. Physical measurements

The polymer film (thickness = 0.6 to 1.0 μm) for photocurrent measurement was prepared by casting a polyimide solution in NMP on an indium-tin-oxide glass slide. A layer of aluminium electrode (10 nm) was deposited on the film surface by thermal evaporation. A Laser Physics Ar-ion laser (488 and 514 nm) was used as the light source. The incident laser light beam was modulated with a mechanical chopper and the photocurrent was measured with a Stanford Research System SR 510 lock-in amplifier.

3. RESULTS AND DISCUSSION

The TGA thermogram of **Re-6FDA** is shown in Figure 3. Two major weight losses at ca. 250 and 550°C were observed in the TGA thermogram. We suggest that the first process accounts for approximately 20% of weight loss and is probably due to the decomposition of the rhenium carbonyl moiety, while the second weight loss corresponds to the degradation of the polyimide main chain. The structure of the polyimide was confirmed by FTIR spectroscopy. Three very strong stretching bands corresponding to the characteristic C≡O stretching of the *facial* metal carbonyl ligands are observed at 2020, 1972 and 1914 cm^{-1}. In addition, other absorption bands at 1603 and 830 cm^{-1} are also observed, which are due to the diimine C=N stretching and the out-of-plane bending of the 1,4-disubstituted benzene, respectively. Other stretching bands observed at 1780, 1720, 1372 and 725 cm^{-1} are assigned to the C=O stretching, C-N stretching and

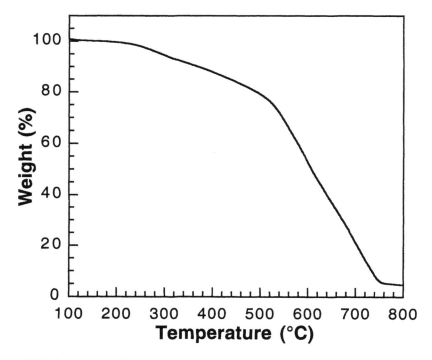

Figure 3. TGA thermogram of **Re-6FDA** in nitrogen atmosphere.

C=O bending of the diimide units, respectively. The absence of C=O stretching band due to poly(amic acid) at 1660 cm^{-1} also confirms a complete imidization in the second stage of polymerization. All these results prove the formation of polyimide and the presence of rhenium carbonyl.

Figure 4 shows the UV-vis absorption spectra of the monomer and **Re-6FDA** in both solid state and in DMF solution. A very broad and intense absorption band at ca. 560-570 nm is observed in the diamino substituted rhenium complex monomer, which is assigned to the MLCT transition between the rhenium and the diimine ligand. The polyimide in DMF solution has a strong peak centered at ca. 310 nm, while the polymer film showed a broad and featureless absorption band. This absorption band extends beyond 600 nm throughout the whole visible region. This electronic transition is tentatively assigned to the intramolecular charge transfer transition on the polyimide main chain [20]. This absorption completely overwhelms the MLCT band of the rhenium complex and thus its presence cannot be proven by UV-vis spectroscopy.

The photoconductivity of the polyimide was studied using an argon-ion laser as the light source. Figure 5 shows the photocurrent response of **Re-6FDA** as a function of the applied electric field at 514 and 488 nm. When the polymer was exposed to visible light irradiation, large photocurrent was detected. The photocurrent response is comparable to polyimides doped with aromatic amine donor [21]

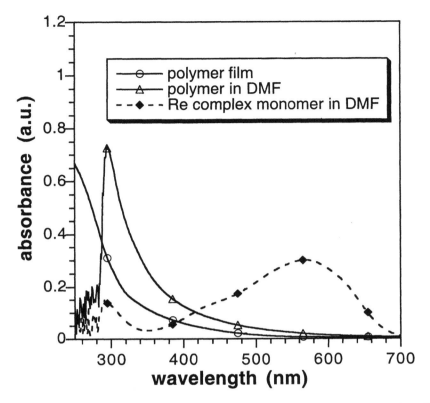

Figure 4. UV-vis absorption spectra of monomer **1** and **Re-6FDA** in both solid state and DMF solution.

and is also an order of magnitude higher than our previously reported polyimides using bisterpyridyl ruthenium complex as the photosensitizer [7].

The photocharge generation process was studied by irradiating the **Re-6FDA** film with an Ar-ion laser (488 nm) under an external electric field. Photoconduction quantum efficiency Φ is defined as the ratio of the number of charge carriers producing photocurrent generated by photoexcitations to the number of photons absorbed by the material and it can be expressed as

$$\Phi = J_{ph}/qI_0(1-T), \tag{1}$$

where J_{ph} is the photocurrent density, I_0 is the number of incident photons per unit area per second and T is the transmittance of the illuminated electrode. The field dependence of quantum efficiency Φ is well described by the germinate recombination theory of Onsager using a Gaussian distribution function for the initial pair radii [22]. The exciton formed by photoexcitation of a molecule may thermalize to form a bound electron-hole pair where the electron and hole are localized on the same or different molecules. The bound electron-hole pair would separate, in

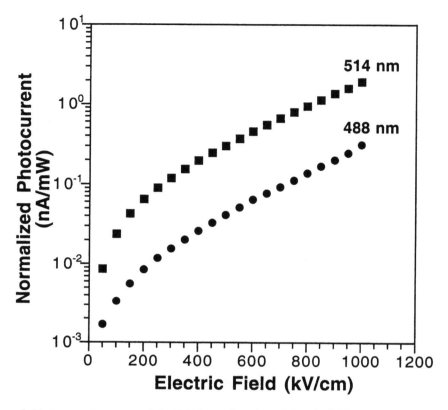

Figure 5. Photocurrent response of **Re-6FDA** as a function of electric field strength at 488 and 514 nm.

the presence of an external electric field, into a free electron and a hole. The fraction of absorbed photons that results in bound thermalized pairs is the primary quantum yield Φ_0. Under the assumption that primary quantum yield is independent of electric field and the distribution of bound pairs is spherically symmetric, the photogeneration efficiency can be expressed by:

$$\Phi(r_0, E) = \Phi_0 \left[1 - \left(\frac{eEr_0}{kT} \right)^{-1} \sum_{g=0}^{\infty} I_g \left(\frac{e^2}{4\pi\varepsilon_0 \varepsilon_r kTr_0} \right) I_g \left(\frac{eEr_0}{kT} \right) \right] \qquad (2)$$

where I_g is a recursive formula given by the equation:

$$I_{g+1}(x) = I_g(x) - x^{g-1} \exp(-x)/(g+1)! \qquad (3)$$

with $I_0(x) = 1 - \exp(-x)$, Φ_0 is the primary yield of thermalized bound pairs, r_0 is the initial thermalization separation between the bound charges, ε_0 is the permittivity of free space, ε_r is the relative permittivity and E is the applied electric field strength. Figure 6 shows the theoretical curves for $\Phi(r_0, E)$ as a function of E based on Eq. (2) and the experimental results for different complexes with incident light at 488 and 514 nm. The fitting results are summarized in Table 1. The polyimide exhibits a similar thermalization distance of approximately 1.7 nm, because the nature of the exciton is independent of the excitation light source. On the other hand, both the quantum yield and primary yield at 514 nm are higher than those polymer films excited at 488 nm. This cannot be attributed to the difference in absorbance at these two wavelengths because the polymer film absorbed stronger at 488 nm. However, monomer **1** exhibits a stronger absorption at 514 nm as it is closer to the peak of the MLCT absorption band (570 nm). Therefore, it is highly probable that the rhenium complex does play an active role in the photosensitization process by enhancing both the quantum efficiency and the primary yield.

Table 1.
Photosensitizing properties of **Re-6FDA**

Light source (nm)	Photoconductivity, σ $(10^{-12} \, \Omega^{-1} \, cm^{-1})$	Quantum efficiency, Φ (10^{-6})	Primary yield, $\Phi_0 (10^{-4})^a$	Thermalization distance, $r_0 \, (nm)^a$
488	1.3	3.9	3.4	1.75
514	1.1	4.7	5.6	1.68

^{a} Obtained by fitting the experimental results to the Onsager's equation.

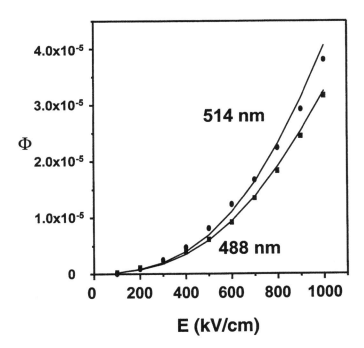

Figure 6. Photogeneration efficiencies (Φ) of **Re-6FDA** as functions of electric field strength at 488 and 514 nm (●). The solid lines show the curve fitting results from the Onsager's equation.

4. CONCLUSIONS

Polyimides functionalized with rhenium diimine complexes were synthesized and their photosensitizing properties were studied. It was found that the rhenium complex could act as an efficient photosensitizer. As a result, large photocurrent response was observed when the polymer film was irradiated with light. The field dependent quantum yield of photosensitization was simulated by the Onsager's theory and the primary quantum yield and thermalization distance were quite similar at 488 and 514 nm.

Acknowledgments

The work reported in this paper was substantially supported by The Research Grants Council of The Hong Kong Special Administrative Region, China (Project Nos. HKU and 7096/00P and 7075/01P). Partial financial support from the Committee on Research and Conference Grants and the Hung Hing Ying Physical Sciences Research Fund (University of Hong Kong) is also acknowledged.

REFERENCES

1. W. Y. Ng and W. K. Chan, *Adv. Mater.* **9**, 716 (1997).
2. P. K. Ng, X. Gong, W. T. Wong and W. K. Chan, *Macromol. Rapid Commun.* **18**, 1009 (1997).
3. S. C. Yu, X. Gong and W. K. Chan. *Macromolecules* **31**, 5639 (1998).
4. C. T. Wong and W. K. Chan, *Adv. Mater.* **11**, 455 (1999).
5. S. C. Yu, S. Hou and W. K. Chan, *Macromolecules* **32**, 5251 (1999).
6. S. C. Yu, S. Hou and W. K. Chan, *Macromolecules* **33**, 3273 (2000).
7. W. Y. Ng, X. Gong and W. K. Chan, *Chem. Mater.* **11**, 1165 (1999).
8. S. H. Chan, W. T. Wong and W. K. Chan, *Chem. Mater.* **13**, 4635 (2001).
9. L. S. M. Lam, S. H. Chan and W. K. Chan, *Macromol. Rapid Commun.* **21**, 1081 (2000).
10. D. J. Stufkens, *Comments Inorg. Chem.* **13**, 359 (1992).
11. K. S. Schanze, D. B. MacQueen, T. A. Perkins and L. A. Cabana, *Coord. Chem. Rev.* **63**, 122 (1993).
12. B. D. Rossenaar, C. J. Kleverlaan, M. C. E. van de Ven, D. J. Stufkens, A. Oskam, J. Fraanje and K. Goubitz, *J. Organomet. Chem.* **493**, 152 (1995).
13. R. Lin and T. F. Guarr, *Inorg. Chim. Acta.* **79**, 726 (1994).
14. W. B. Connick, A. J. Di Bilio, M. G. Hill, J. R. Winkler and H. B. Gray, *Inorg. Chim. Acta.* **240**, 169 (1995).
15. R.-J. Lin, K.-S. Lin and I.-J. Chang, *Inorg. Chim. Acta.* **242**, 179 (1996).
16. F. P. A. Johnson, M. W. George, F. Hartl and J. J. Turner, *Organometallics* **15**, 3374 (1996).
17. P. K. Ng, X. Gong and W. K. Chan, *Adv. Mater.* **10**, 1337 (1998).
18. W. K. Chan, P. K. Ng, X. Gong and S. Hou, *Appl. Phys. Lett.* **75**, 3920 (1999).
19. D. J. Stufkens and A. Vlček Jr., *Coord. Chem. Rev.* **177**, 127 (1998).
20. J. M. Salley and C. W. Frank, in: *Polyimides: Fundamentals and Applications*, M. K. Ghosh and K. L. Mittal (Eds.), p. 279, Marcel Dekker, New York (1996).
21. S. A. Lee, T. Yamashita and K. Horie, *Polym. J.* **29**, 752 (1997).
22. L. Onsager, *Phys. Rev.* **54**, 554 (1938).
23. S. Freilich, *Macromolecules* **20**, 973 (1987).

Polyimides and Other High Temperature Polymers, Vol. 2, pp. 99–112
Ed. K.L. Mittal
© VSP 2003

New highly phenylated bis(phthalic) and bis(naphthalic) anhydrides and polyimides therefrom

M.L. KESHTOV,[1] A.L. RUSANOV,[*,1] YU.I. FOGEL[1] and F.W. HARRIS[2]

[1] *A.N. Nesmeyanov Institute of Organoelement Compounds, Russian Academy of Sciences, ul. Vavilova 28, Moscow, 119991, Russia*
[2] *The University of Akron, Maurice Morton Institute and Department of Polymer Sciences, Akron, OH 44325-3909, USA*

Abstract—New highly phenylated bis(phthalic) and bis(naphthalic) anhydrides were synthesized by the Diels-Alder reactions of 4,4'-bis[(2,4,5-triphenylcyclopentadien-l-on-3-yl)]benzophenone with twofold molar amounts of maleic anhydride and 4-(phenylethynyl)naphthalic anhydride, respectively. The dianhydrides were used to prepare a series of new phenylated polyimides and poly-(naphthalmide)s that were soluble in organic solvents. The thermal, mechanical, electrical and optical properties of the polymers were determined.

Keywords: Polyimides; poly(naphthalimide)s; 4,4'-bis[(2,4,5-triphenylcyclopentadien-l-on-3-yl)]benzophenone; maleic anhydride; 4-(phenylethynyl)naphthalic anhydride; Diels-Alder reaction.

1. INTRODUCTION

Polyimides (PIs) [1-6] and poly(naphthalimide)s (PNIs) [7-9] belong to an important class of thermally stable polymers. However, the application of these polymers is frequently limited by their high glass transition temperatures and poor solubility in organic solvents. A successful approach to solving this problem involves the introduction of bulky substituents into the polymer chain [10-15]. This improves the solubility of the polymers without any appreciable deterioration of their thermal stability. Harris and coworkers [16-20] and Kirn and Hay [21] synthesized phenylated PIs possessing good solubility in organic solvents and high glass transition temperatures (near 300°C). Since such polymers have many potential applications, the objective of this work was to expand the previous work in the syntheses of new phenylated polyimides.

*To whom all correspondence should be addressed. Phone: +7(095)1356166, Fax: +7(095)1355085, E-mail: alrus@ineos.ac.ru

2. EXPERIMENTAL

The starting compounds and solvents were purified according to conventional techniques.

The ^1H and ^{13}C NMR spectra were recorded in $CDCl_3$ with a Bruker AMX-400 spectrometer operating at 400.13 and 100.61 MHz, respectively. Tetramethylsilane was used as an internal standard. The purity of the individual compounds was checked by thin-layer chromatography. Measurements were performed in benzene using Silufol-UV-245 plates, and they were developed under UV light. The IR and Raman spectra were measured with a Perkin-Elmer 1720X spectrophotometer. DSC and TGA experiments were carried out with DSC7 and TGA7 instruments, respectively. The heating rate was 20°C/min. The dielectric constants of the polymer films were measured with an E7-8 CRL digital meter at an electric field frequency of 1 kHz and a humidity of 0 and 50%.

2.1. Monomers

2.1.1. Synthesis of 4,4'-bis(2,3,6-triphenyl-4,5-dicarboxyphenyl) benzophenone dianhydride (II)

A two-necked flask equipped with a reflux condenser and a stirrer was charged with 4,4'-bis(2,4,5-triphenylcyclopentadien-1-on-3-yl)-benzophenone **1** (2.46 g, 3.1 mmol), maleic anhydride (0.61 g, 6.25 mmol), and bromobenzene (7.5 ml). The reaction mixture was boiled for 3 h, cooled, and a solution of bromine (0.55 ml) in bromobenzene (0.9 ml) was slowly added dropwise. After the mixture was boiled for another 3 h, it was then poured into hexane (150 ml). The precipitate was filtered off and dried under vacuum to provide compound **II** (2.51 g, 87%) as a white crystalline powder; M.p.= 361-362°C.

For $C_{65}H_{38}O_7$, Calcd. (%): C, 83.85; H, 4.11.

Found (%): C, 83.90; H, 4.15.

IR (v, cm^{-1}): 1779, 1843 (C=O of anhydride), 1661 (C=O of benzophenone), 1605 (C=C of aromatic ring).

^1H NMR ($CDCl_2$, δ, ppm): 6.76-6.77 (m, 4H), 6.85 (d, 4H, $J = 8.1$ Hz), 6.90-7.00 (m, 6H), 7.00-7.18 (m, 12H), 7.18-7.30 (m, 12H).

^{13}C NMR ($CDCl_3$, δ, ppm): 126.78 (2C), 126.85 (2C), 126.90 (2C), 127.11 (2C), 127.15 (2C), 127.24 (2C), 127.47 (2C), 127.55 (2C), 127.62 (2C), 127.65 (2C), 127.79 (2C), 127.89 (2C), 128.00 (2C), 128.02 (2C), 128.45 (2C), 128.59 (2C), 129.39 (2C), 129.45 (2C), 129.55 (2C), 130.19 (2C), 130.30 (2C), 133.81 (2C), 135.69 (2C), 140.84 (2C), 141.35 (2C), 149.47 (2C), 161.21 (-CO-dianhydride, 4C), 195.28 (-CO-benzophenone, 1C).

2.1.2. Synthesis of 4,4'-bis[2,3,5,6-tetraphenyl-(4,5-dicarboxynaphth-1-yl)phenyl-benzophenone dianhydride (III)

A three-necked flask equipped with a reflux condenser, a stirrer, and an inlet for feeding argon was charged with compound **II** (0.79 g, 1 mmol), 4-(phenylethynyl)naphthalic anhydride (0.5966 g, 2 mmol), and 1,2,4-trichlorobenzene (16 ml). The reaction mixture was boiled for 10 h, cooled, and poured into a 10-fold excess of methanol. The precipitate was filtered off, dried, and crystallized from isopropyl alcohol to provide compound **III** (91%), which did not melt prior to decomposition.

For C $_{91}$ H $_{58}$ O $_7$, Calcd. (%): C, 87.27; H, 4.37.

Found (%): C, 87.37; H, 4.60.

IR (v, cm^{-1}): 1738, 1780 (C=O of naphthalic anhydride), 1658 (C=O of benzophenone group), 1592 (C=C of aromatic ring).

^{1}H NMR (CDCl$_3$, δ, ppm): 6.50-6.60 (m, 8H), 6.60-6.70 (m, 4H), 6.70-6.80 (m, 4H), 6.80-6.98 (m, 28H), 7.03-7.14 (m, 4H), 7.40-7.48 (m, 2H), 7.60-7.66 (m, 2H), 8.22 (d, 2H, J = 8.1 Hz), 8.32 (d, 2H, J=8.1 Hz), 8.42 (d, 2H, J = 8.1 Hz).

^{13}C NMR (CDCl$_3$, δ, ppm): 116.59, 118.27, 125.56, 125.94, 126.05, 126.26, 126.78, 127.78, 128.09, 128.34, 128.44, 128.53, 129.35, 129.76, 129.87,130.49, 130.67, 130.68, 130.97, 131.58, 131.82, 132.26, 134.37, 134.38, 134.59, 135.58, 136.36, 138.87, 138.94, 139.01, 139.16, 139.45, 139.68, 140.24, 140.48, 140.71, 140.81, 140.95, 141.09, 141.62, 144.22, 144.51, 147.47, 147.54 (Ar), 160.33 (-CO-BNA), 160.53 (-CO-BNA), 196.09, 196.14, 196.19 (-CO-benzophenone).

2.2. Polymers

2.2.1. Synthesis of phenylated polyimides VA-VG

A three-necked flask equipped with a reflux condenser, a stirrer, and an inlet for feeding argon was charged with dianhydride **II** (0.4654 g, 0.5 mmol), p-phenylenediamine **IVA** (0.0541 g, 0.5 mmol), and a solution of benzoic acid (0.1212 g) in m-cresol (10 ml). The latter was used as a catalyst. After the reaction mixture was stirred for 5 h at 180°C, it was cooled and chloroform (2 ml) was added. The mixture was slowly poured into methanol. The precipitate was washed several times with methanol and dried at 150°C under vacuum for 20 h. The **VB-VG** phenylated PIs were prepared in a similar manner.

2.2.2. Synthesis of phenylated poly(naphthalimide)s VIA-VIG

A stirred mixture of equimolar amounts of dianhydride **III** (0.6678 g, 0.5 mmol) and p-phenylenediamine **IVA** (0.0541 g, 0.5 mmol), and p-chlorophenol (1 g) was heated at 80°C in the presence of benzoic acid (0.09 g) and benzimidazole (0.09 g) in a flow of argon over a 2-h period. After the temperature of the reaction mixture was raised to 140°C, it was stirred for another 5 h and then was slowly poured into methanol (100 ml). The polymer obtained was filtered off, washed several times with methanol, and dried under vacuum at 150°C for two days.

The **VIB-VIG** phenylated PNIs were prepared in a similar manner.

3. RESULTS AND DISCUSSION

3.1. Monomers synthesis

Within the framework of this study, we have synthesized new phenyl-substituted monomers, 4,4'-bis(2,3,6-triphenyl-4,5-dicarboxyphenyl)benzophenone dianhydride (**II**) and 4,4'-bis[2,3,5,6-tetraphenyl(4,5-dicarboxy-naphth-1-yl)phenyl]benzophenone dianhydride (**III**) (Scheme 1).

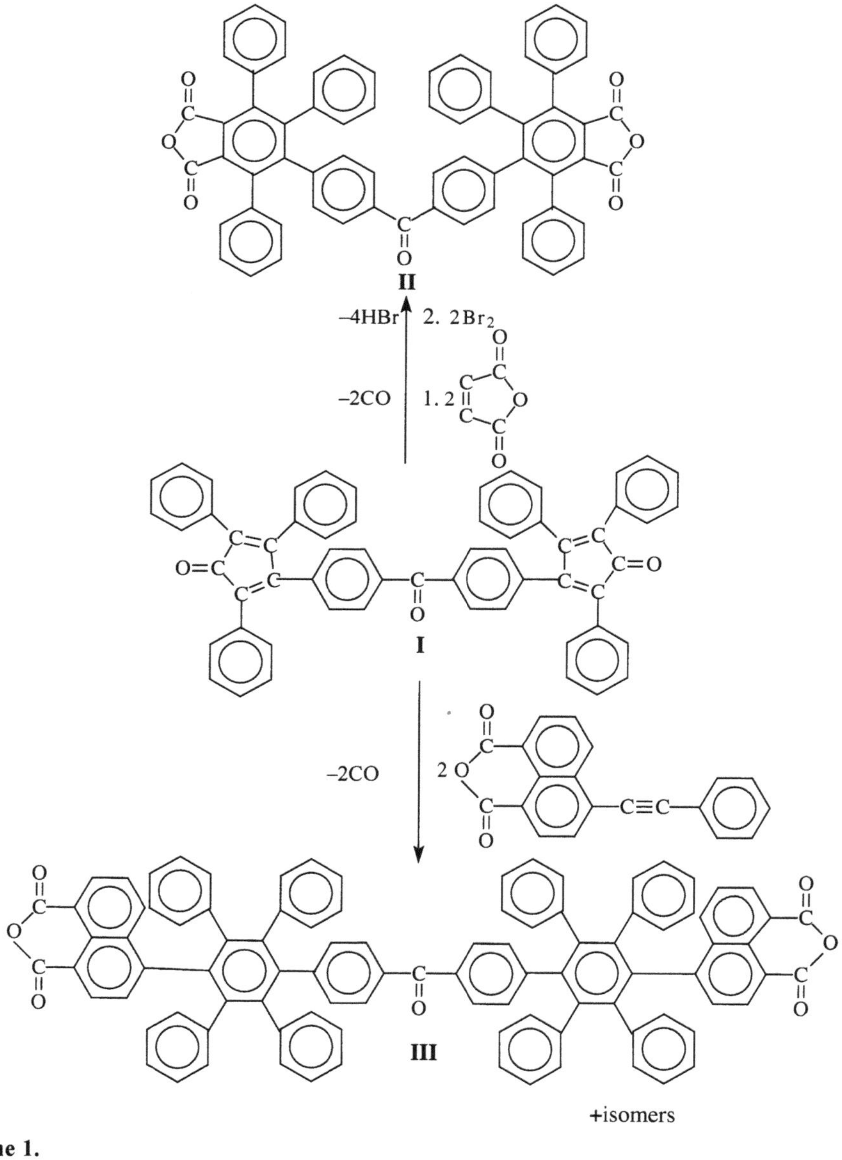

Scheme 1.

The reaction of 4,4'-bis(2,4,5-triphenylcyclopentadien-1-on-3-yl)benzophenone (**I**) [22] with a twofold molar amount of maleic anhydride in boiling bromobenzene, followed by dehydrogenation with bromine, yielded dianhydride **II**. This white crystalline compound showed good solubility in conventional organic solvents – N,N-dimethylformamide (DMF), dimethylsulfoxide (DMSO), tetrahydrofuran (THF), chloroform, toluene, etc.

The structure of compound **II** was confirmed by elemental analysis, ^{1}H and ^{13}C NMR, and FTIR spectroscopies. The above reaction was monitored by IR spectroscopy, in particular, by the disappearance of the intense band at 1709 cm^{-1} corresponding to the carbonyl of the cyclopentadienone moiety.

The IR spectrum of compound **II** shows absorption bands at 1843 and 1779 cm^{-1}, which are characteristic of the stretching vibrations of the carbonyl of phthalic anhydride, along with an intense absorption band at 1663 cm^{-1}, which corresponds to the stretching vibrations of the carbonyl of the benzophenone moiety.

The ^{1}H NMR spectrum of compound **II** exhibits a signal at $\delta = 6.85$ ppm (d, 4H, $J = 8.1$ Hz) and four multiplets corresponding to the other aromatic protons, while the ^{13}C NMR spectrum shows signals in the region of $\delta = 195.28$ ppm, characteristic of the carbonyl of the benzophenone moiety, and an intense signal at $\delta = 161.21$ ppm corresponding to the four almost equivalent carbons of the dianhydride **II** carbonyl groups.

NMR spectra do not provide information about the spatial arrangement of phenyl rings. Moreover, X-ray diffraction data were unavailable for compound **II**. Therefore, we carried out a series of calculations to gain insight in their special arrangement. In this work we made no attempt to calculate all possible conformations of compound **II**. The geometry of one of these conformations was estimated by the quantum semiempirical method PM3. The calculations showed that the benzene rings attached to the phthalic anhydride moiety were not in the same plane as the aromatic ring of the anhydride. The corresponding torsional angles that characterize the rotation angles of benzene rings relative to the phthalic anhydride together with the most typical bond lengths are summarized in Table 1. (Note the calculations showed that the distances between the H at C(19) and the H at C(90) (1.718 Å), and between the H at C(18) and the H at C(67) (2.645 Å) were very small lending support to the statement that compound **II** was completely noncoplanar.) This is an additional explanation for the high solubility of compound **II**.

The reaction of compound **I** with a twofold molar amount of 4-(phenylethynyl)naphthalic anhydride [23] by the Diels-Alder reaction in boiling 1,2,4-trichlorobenzene produced the new bis(naphthalic anhydride) – 4,4'-bis[2,3,5,6-tetraphenyl(4,5-dicarboxynaphth-1-yl)phenyl]benzophenone dianhydride (**III**) with a yield of 91%. The product was isolated as a pale-yellow amorphous substance showing good solubility in conventional organic solvents. The reaction was monitored by Raman spectroscopy by measuring the intensity of the band at 1709 cm^{-1} corresponding to the carbonyl of the cyclopentadienone moiety.

Table 1.
Torsion angles (φ) and some characteristic bonds of compound **II**

Angle	φ, deg	Bond	Bond length, Å
C(55)C(44)C(34)C(87)	87.7	C(34)C(44)	1.476
C(54)C(53)C(61)C(63)	−56.5	C(53)C(61)	1.494
C(65)C(63)C(61)C(55)	−56.5	C(61)C(63)	1.494
C(1)C(32)C(68)C(86)	87.8	C(68)C(32)	1.476
C(13)C(11)C(2)C(4)	−88.6	C(2)C(11)	1.472
C(3)C(99)C(100)C(101)	89.8	C(99)C(100)	1.472
C(99)C(1)C(1)C(10)C(17)	−80.8	C(10)C(1)	1.476
C(93)C(88)C(87)C(35)	−80.8	C(87)C(88)	1.476
C(56)C(45)C(36)C(35)	−89.5	C(35)C(45)	1.472
C(37)C(33)C(43)C(47)	90.9	C(43)C(33)	1.472

The structure of compound **III** was confirmed by elemental analysis, ^{1}H and ^{13}C NMR and FTIR spectroscopies. The IR spectrum of compound **III** exhibits an intense absorption at 1658 cm^{-1} due to the stretching vibrations of the CO group of the benzophenone moiety and intense bands at 1738 and 1780 cm^{-1}, which were assigned to the carbonyl groups of the anhydride **III**.

The ^{1}H NMR spectrum of compound **III** contains broad signals between 6.5 and 8.5 ppm (broadening may be related to the presence of various BNA isomers). This suggestion was unambiguously supported by ^{13}C NMR spectroscopy. The ^{13}C NMR spectrum of compound **III** features low-field signals at 160.33 and 160.53 ppm corresponding to two unequivalent C atoms of the anhydride group. Three signals at 196.19, 196.14, and 196.09 ppm, which were attributed to the carbon of the CO group of the benzophenone moiety (the ratio of integral intensities is 1 : 2 : 1) were observed in the region of $\delta \approx 200$ ppm. The product appears to be a mixture of three isomers with a ratio of A'-CO-A' : A'-CO-B' : B'-CO-B' = 1 : 2 : 1 (where A' and B' are the fragments of formulas)

A'   B'

However, it should be mentioned that these isomers cannot be distinguished based on the ^{1}H and ^{13}C NMR spectroscopies data. We made no attempts to separate the isomers of compound **III**, because they showed comparable solubility in organic solvents. In further experiments, a mixture of isomers was employed to prepare phenylated poly(naphthalimide)s (PNIs). The conformational characteristics of compound **III** were estimated by the quantum semiempirical method PM3. As evidenced by the theoretical calculations, torsional angles between the central phenyl ring and side phenyl groups are on average 60°, suggesting that compound **III** is completely noncoplanar.

3.2. Polymers synthesis

The reactions of the dianhydrides **II** and **III** with equimolar amounts of various aromatic diamines (**IVA–IVG**) produced a series of new phenylated PIs (**V**) and PNIs (**VI**).

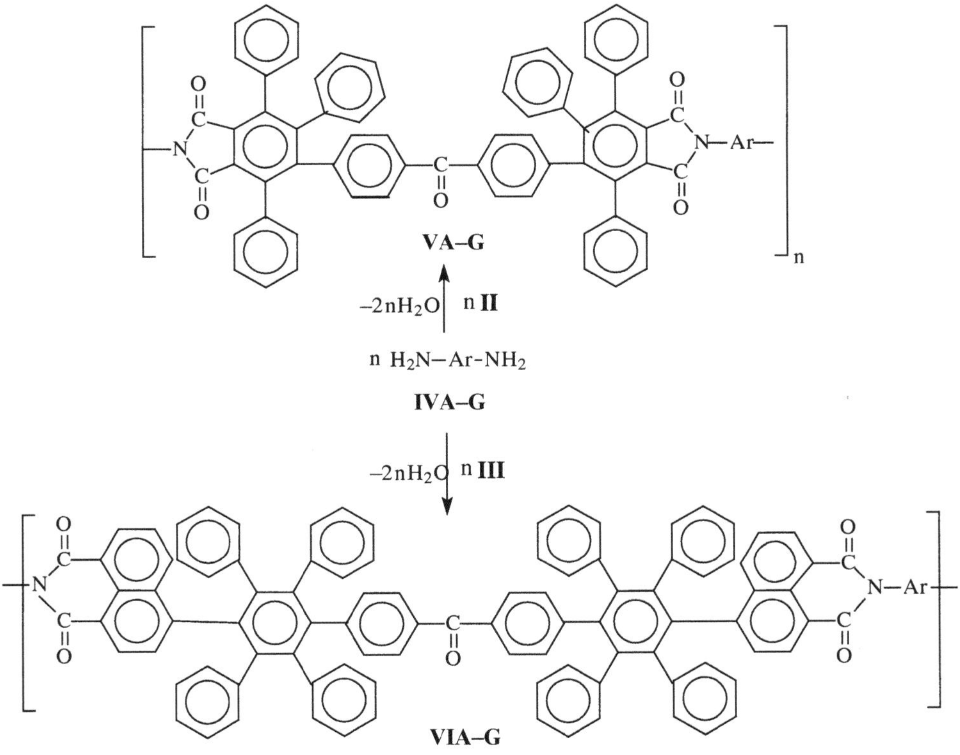

Scheme 2.

where Ar =

(A)   (B)   (C)

(D)   (E)   (F)

(G)

The reaction of **II** with aromatic diamines **IVA-IVG** was carried out in m-cresol using benzoic acid as catalyst. Polycyclocondensation was carried out at 180°C for 5 h. All the reactions proceeded under homogeneous conditions and gave rise to the formation of the target phenylated PIs with yields close to the theoretical values. The polymer viscosities ranged from 0.66 to 2.53 dl/g.

The structures of phenylated PIs were established by FTIR and ^{1}H and ^{13}C NMR spectroscopies. The formation of the **VA-VG** polymers was confirmed by the appearance of absorption bands at 1780 and 1730 cm^{-1}, which are characteristic of the CO groups of imides [24]. All of the ^{13}C NMR spectra of phenylated PIs exhibit an intense signal at 165 ppm, which corresponds to two equivalent carbons of the carbonyl of the phthalimide group, and a signal at –200 ppm, which is characteristic of the C=O group of the benzophenone moiety.

The X-ray diffractograms of the synthesized PIs show that they are amorphous. Possibly, this property of PIs is responsible for their improved solubility. Most phenylated PIs show good solubility in N-methylpyrrolidone (N-MP), DMF, and DMSO. They also dissolve rapidly at room temperature in m-cresol, THF, and chloroform. Upon heating, they show solubility even in pyridine and toluene. It is pertinent to note that the polyimides **VA** and **VB** containing rigid p-phenylene and p,p′-diphenylene moieties, respectively, are easily soluble in the majority of polar solvents. It appears that the high solubility of these polymers is due to their noncoplanar structure and to the presence of side phenyl groups in the polymer chains as well.

The thermal properties of the **VA-VG** polymers were determined by thermal mechanical analysis (TMA) and thermal gravimetric analysis (TGA). The results are presented in Table 2. As is seen, the T_g values of these polymers range from 345 to 395°C. As might be expected, the **VA** and **VB** polyimides containing rigid p-phenylene and p,p′-diphenylene moieties have the highest T_g values. The lower T_g values of polymers **VC** and **VE-VG** are apparently due to the introduction of bulky side trifluoromethyl groups and flexible bonds, respectively, into the polymer chain.

Table 2.
Thermal characteristics of polyimides **VA-VG**

Polymer	T_g, °C	$T_{10\%}$, °C (TGA)		T_{dec}-T_g, °C	
		air	argon	air	argon
A	390	600	665	210	275
B	395	610	665	215	270
C	365	625	675	260	310
D	385	605	655	220	270
E	360	580	645	220	285
F	345	565	610	220	265
G	350	575	610	225	260

Table 3.
Mechanical and dielectric characteristics of polyimides **VA-VG**

Polymer	Tensile properties of films (25°C)			ε' At humidity	
	σ, MPa	E, GPa	ε, %	0%	50%
A	84.3	2.85	7.5	2.80	3.25
B	86.6	2.50	8.0	2.80	3.14
C	85.4	3.03	3.6	2.75	2.95
D	81.9	2.80	7.0	3.10	3.44
E	78.3	2.28	6.0	2.95	3.25
F	78.3	2.28	6.0	3.06	3.32
G	68.0	2.50	4.0	2.72	2.89

All phenylated PIs exhibit an excellent thermal stability (Table 2): temperatures corresponding to the 10% weight loss under argon vary from 610 to 675°C. The corresponding values in air are lower by 50-65°C and lie in the range from 565 to 625°C. Wide ranges between the decomposition and glass transition temperatures (T_{dec}-T_g) of phenylated PIs indicate that, in principle, these polymers may be processed by compression molding.

All **VA-VG** polymers form films (Table 3). The tensile strength σ of these films ranges from 68.0 to 86.6 MPa, elastic modulus E is between 2.28 and 3.03 GPa, and elongation at break ε is in the range from 4 to 8%. As can be seen in Table 3, the elastic moduli of all the tested polymers are high and, for some polymers, they are comparable to those of Kapton [25] and Ultem (2.96 GPa) [26]. The tensile strength of phenylated polymers is lower than that of Kapton (172 MPa) and of Ultem (105 MPa) [26]. However, for the majority of the synthesized polymers, relative elongation at break ε is lower than that for commercial polyimide materials (60-70%).

Table 4.
Maxima of absorption and fluorescence spectra of phenylated polymers **VA-VG**

Polymer	λ^a_{max}, nm	λ^f_{max}, nm
A	390	490
B	392	550
C	380	500
D	410	560
E	395	500
F	389	500, 540
G	384	480

Table 3 also shows the dielectric constant, ε', of the phenylated PIs. These values vary from 2.72 to 3.10 and 2.89 to 3.44 at relative humidities of 0 and 50%, respectively, with an electric field frequency of 1 kHz. Phenylated PIs **VC** and **VG** containing trifluoromethyl and hexafluoroisopropylidene groups have the smallest dielectric constants (2.75 and 2.72, respectively, at the zero humidity). At a humidity of 50%, the dielectric constant of these polymers varies from 2.89 to 3.44; i.e., these values are 0.3 higher than those of the corresponding phenylated PIs measured at zero humidity. The reduced moisture absorption of fluoro-containing phenylated PIs ensures the long-term stability of their dielectric parameters.

Table 4 shows some photophysical characteristics of the phenylated PIs examined (for the corresponding polymer films under similar experimental conditions). For phenylated PIs, the maxima in the absorption spectra, λ^a_{max} appear in the 380-410 nm range. All the studied polymers fluoresce in the red region with maxima at $\lambda^f_{max} = 480\text{-}560$ nm. The fluorescence spectra of **VA, VB, VC, VE** and **VG** exhibit similar profiles with maxima at 490 nm. Unlike these polymers, the polymers **VD** and **VF** fluoresce strongly. The **VF** polyimide shows two maxima at 500 and 540 nm, whereas the λ^f_{max} of the **VD** polymer (560 nm) is shifted by 20 nm to longer wavelengths.

Phenylated PNIs were synthesized by high-temperature polycyclocondensation in p-chlorophenol via the reaction of compound **III** with the equimolar amounts of various aromatic diamines **IVA-IVG** (Scheme 2). All the reactions proceeded under homogeneous conditions. This enabled us to synthesize polymers with a reduced viscosity of 0.52-0.78 dl/g. The yields of the phenylated PNIs were close to the theoretical value.

The structures of phenylated PNIs were confirmed by FTIR, Raman, and ^{1}H and ^{13}C NMR spectroscopies. Carbonyl peaks were observed in the IR spectra of phenylated PNIs at 1713 and 1673 cm[-1]. The ^{13}C NMR spectra of **VIA-VIG** polymers exhibited two signals at −164.0 and −164.2 ppm, which were assigned to two nonequivalent carbons of naphthalimide carbonyl and a triplet at 200 ppm, corresponding to the carbonyl of the benzophenone moiety of the three isomers.

X-ray diffractograms showed that the phenylated PNIs were fully amorphous. This morphology probably contributes to their improved solubility. All the polymers were soluble in aprotic solvents – NMP, DMF, DMAA, DMSO, as well as in m-cresol and THF. When heated, the polymers were even soluble in hot chloroform and toluene.

The solubility of these polymers depended on the structure of the aromatic diamines used. Polymers **VIC**, **VIE-VIG** containing ether, methylene, hexafluoroisopropylidene, and bulky trifluoromethyl groups rapidly dissolved in organic solvents and displayed better solubility compared to the phenylated PNIs **VIA**, **VIB** with more rigid chains. Polymer **VID** was insoluble in chloroform and toluene possibly due to the presence of rigid and polar benzimidazole groups, which strongly enhance intermolecular dipole-dipole interaction. Finally, the solubility of the phenylated PNIs **VIA-VIG** was appreciably higher compared to the PNIs based on 1,4,5,8-naphthalenetetracarboxylic dianhydride [27-30]. The significant increase in solubility can be explained by the presence of various isomeric structures and to the large amount of phenyl substituents.

Polymers **VIA-VIG** displayed glass transition temperatures ranging from 340 to 400°C. These values decreased in the sequence **VIB>VIDV>IC>VIG>VIA>VIE>VIF**. As might be expected, a higher glass transition temperature was observed for the rigid-chain phenylated PNI **VIB** containing p-diphenylene moieties. For polymer **VID**, the high glass transition temperature was apparently related to the polarity and rigidity of the benzimidazole moiety. Contrary to expectations, polymer **VIA** containing p-phenylene moieties exhibited a lower T_g than polymer **VIB**. This may be due to the high percentage of side phenylene rings (per phenylene group of the backbone) in the repeat unit. In fact, polymers **VIA-VIG** have lower glass transition temperatures than those of nonphenylated PNIs [7, 31].

Phenylated PNIs displayed excellent thermal stability (Table 5). Dynamic TGA showed that the temperatures corresponding to 10% weight loss in air and argon were between 580 and 615, and 625 and 680°C, respectively. The wide ranges between the decomposition temperatures and the glass transition temperatures (T_{dec}-T_g) of the phenylated PNIs indicate that these polymers should be capable of being processed by compression molding.

Table 6 lists the mechanical properties of films based on phenylated PNIs. The tensile strength of these films ranged from 66.1 to 83.0 MPa, the elastic modulus was between 1.55 and 2.55 GPa, and the elongation at break ranged from 3 to 9%. It should be pointed out that the film mechanical properties are comparable to those based on the PNIs described in [32].

The dielectric constants, ε', of the phenylated PNIs were measured at relative humidies of 0 and 50% and an electric field frequency of 1 kHz (Table 6). When the relative humidity was 0%, the values of ε' for the **VIA-VIG** polymers varied from 2.85 to 3.00. These results are rather close to the calculated values (Table 6). The low values of ε' for the phenylated PNIs compared to the values reported for

Table 5.
Thermal characteristics of polymers **VIA-VIG**

Polymer	T_g, °C	T_{dec}-T_g		$T_{10\%}$, °C (TGA)	
		air	argon	air	argon
A	375	238	305	615	680
B	400	190	265	590	660
C	385	220	275	605	660
D	390	225	240	615	630
E	370	220	300	590	670
F	340	240	305	580	645
G	380	215	245	595	625

Table 6.
Mechanical and electrical characteristics of polymers **VIA-VIG**

Polymer	Tensile properties of films, 25°C			ε' Calc.	At humidity	
	σ, MPa	ε, %	E, GPa		0%	50%
A	79.5	6.0	2.55	2.87	2.95	3.17
B	83.0	3.0	1.80	2.86	2.89	3.19
C	82.0	9.0	1.55	2.81	2.85	2.91
D	66.1	6.0	2.17	2.97	3.0	3.25
E	69.0	4.0	1.76	2.88	3.00	3.28
F	70.0	8.0	1.60	2.86	2.91	3.30
G	65.0	4.5	1.70	2.82	2.80	2.95

common polyimides (3.1-3.5) [14] are related to the considerable free volume in phenylated PNIs. At a humidity of 50%, the dielectric constants of the latter polymers are higher than those measured at the zero humidity (by 14%) and are in the range of 2.91 to 3.25. This is apparently due to the moisture absorption of polymers **VIA-VIG**. At zero relative humidity, PNIs **VIC** and **VIG** containing trifluoromethyl and hexafluoroisopropylidene groups have low values of $\varepsilon' = 2.85$ and 2.80, respectively. While at 50% humidity, these values increase only slightly. It is evident that the introduction of fluorine into phenylated PNIs efficiently stabilizes the value of ε'. It is worth noting that the stability of dielectric constant is an important property of interlayer insulators in microelectronics.

Photoabsorption and fluorescent characteristics of the phenylated PNIs were determined using polymer films under similar experimental conditions. The polymers of this series show an intense fluorescence in solution (blue) and in film form (yellow). Table 7 lists the photoabsorption and fluorescence properties of

Table 7.
Photoabsorption and fluorescent characteristics of phenylated PNIs **VIA-VIG**

Polymer	λ^a_{max}, nm	λ^f_{max}, nm	Relative quantum yield
A	408	550	6.1
B	400	540	3.8
C	410	530	5.6
D*	431	580	1.0
E	415	510, 548	7.3
F	421	510, 550	7.3
G	399	510	4.1

* The quantum yield of polymer **VID** is taken to be 1.0

VIA-VIG polymers. Their UV absorption maxima λ^a_{max} are in the 399-431 nm range (films). All the phenylated PNIs display similar fluorescent profiles with λ^f_{max} in the region of 510-580 nm (a yellow emission) at the excitation wavelength of 400 nm. Relative quantum yields of these polymers vary from 1 to 7.3 (the quantum yield of **VID** is taken as unity). In the case of polymers **VIE** and **VIF**, the fluorescence is 7.3 times stronger than that for polymer **VID** (it appears that the regular structure of a polymer chain plays an important role in fluorescence).

4. CONCLUSION

New highly phenylated polyimides and poly(naphthalimide)s were prepared via the reaction of aromatic diamines with previously undescribed phenylated aromatic tetracarboxylic acid dianhydrides. All polyimides prepared showed excellent solubility in organic solvents combined with promising thermal, mechanical, electrical and optical properties.

Acknowledgement

This work was supported, in part, by the Civilian Research and Development Foundation (USA) under grant no. RC2-2203.

REFERENCES

1. N.A. Adrova, M.I. Bessonov, L.A. Laius and A.P. Rudakov, *Poliimidy – noviy klass termostoikikh polimerov (Polyimides – A New Class of Thermally Stable Polymers)*, p. 211, Nauka, Leningrad (1968).
2. M.I. Bessonov, M.M. Koton, V.V. Kudryavtsev and L.A. Laius, *Poliimidy – klass termostoikikh polimerov (Polyimides – A Class of Thermally Stable Polymers)*, p. 238, Nauka, Leningrad (1983).

3. C.E. Sroog, J. Polym. Sci., Macromol. Rev., **11**, 161 (1976).
4. C.E. Sroog, Prog. Polym. Sci., **16**, 561 (1991).
5. D. Wilson, H.D. Stenzenberger and P.M. Hergenrother (Eds.), *Polyimides*, Chapman and Hall (1990).
6. J. de Abajo, in: *Handbook of Polymer Synthesis,* H.R. Kricheldorf (Ed.), pt. B. p. 941, Marcel Dekker, New York (1992).
7. A.L. Rusanov, Russ. Chem. Rev., **61**, 815 (1992).
8. A.L. Rusanov, Adv. Polym. Sci., **111**, 116 (1994).
9. A.L. Rusanov, L.B. Elshina, E.G. Bulycheva and K. Müllen, *Polym. Sci., Ser. A*, **41**, 2 (1999).
10. F.W. Harris and L.H. Lanier, in: *Structure – Solubility Relations in Polymers,* F.W. Harris and R.B. Seymour (Eds.), p. 183, Academic Press, New York (1977).
11. A.K.St Clair, T.L.St Clair and E.N. Smith, in: *Structure – Solubility Relations in Polymers,* F.W. Harris and R.B. Seymour (Eds.), p. 199, Academic Press, New York (1977).
12. S.V. Vinogradova, Ya.S. Vygodskii, V.V. Korshak and T.N. Spirina, Acta Polymerica, **30**, 3 (1979).
13. S.J. Huang and A.E. Hoyt, Trends Polym. Sci., **3**, 262 (1995).
14. J. de Abajo and J.G. de la Campa, Adv. Polym. Sci., **140**, 23 (1998).
15. T. Takekoshi, Adv. Polym. Sci., **94**, 1 (1990).
16. F.W. Harris, S.O. Norris, L.H. Lanier and W.A. Feld, Am. Chem. Soc., Div. Org. Coat. Plastics Preprint, **33(1)**, 160 (1973).
17. F.W. Harris and S.O. Norris, J. Polym. Sci. Polym. Chem. Ed., **11**, 2143 (1973).
18. F.W. Harris, W.A. Feld and L.H. Lanier, Appl. Polym. Symp., **26**, 421 (1975).
19. F.W. Harris, W.A. Feld and L.H. Lanier, J. Polym. Sci. Polym. Chem. Ed., **13**, 283 (1975).
20. F.W. Harris, S.O. Norris, L.H. Lanier, B.A. Reinhardt, R.D. Case, S. Varaprath, S.M. Padaki, M. Torres and W.A. Feld, in: *Polyimides: Synthesis, Characterization and Applications*, Vol. 1, K.L. Mittal (Ed.), pp. 3-14, Plenum Press, New York (1984).
21. W.G. Kirn and A.S. Hay, Macromolecules, **26**, 5275 (1993).
22. A.L. Rusanov, M.L. Keshtov, S.V. Keshtova, P.V. Petrovskii, A.N. Shchegolikhin, A.A. Kirillov and V.V. Kireev, Russ. Chem. Bull., **47**, 325 (1998).
23. I.A. Khotina and A.L. Rusanov, Russ. Chem. Bull., **44**, 514 (1995).
24. T. Matsuda, Y. Hasuda, Sh.N. Shi and N. Yamada, Macromolecules, **24**, 5001 (1991).
25. C.E. Sroog, A.L. Endrey, S.V. Abramo, C.E. Berr, W.M. Edwards and K.L. Oliver, J. Polym. Sci. Part A: **3**, 1373 (1965).
26. R.O. Johnson and H.S. Burhils, J. Polym. Sci. Polym. Symp., **70**, 129 (1983).
27. A.L. Rusanov, A.M. Berlin, S.Kh. Fidler, G.S. Mironov, Yu.A. Moskvichev, G.V. Kolobov and V.V. Korshak, Vysokomol. Soyed., Ser. A, **23**, 1586 (1981).
28. U. Gaik, B. Kowalski and Z.J. Yedlinski, Izv. AN Kaz. SSR, Ser. Khim., No.5, 19 (1981).
29. Z.J. Yedlinski, U. Gaik and B. Kowalski, Makromol Chem., **183**, 1615 (1982).
30. T.V. Kravchenko, T.I. Dvalishvili, T.A. Romanova and G.D. Tkachev, Vysokomol. Soyed. **B 24**, No.11, 852 (1982).
31. G.A. Loughran and F.E. Arnold, Polym. Prepr. (Am. Chem. Soc., Div. Polym. Chem.)., **18**, No.1, 831 (1977).
32. H. Ghassemi and A.S. Hay, Macromolecules, **26**, 5824 (1993).
33. F.W. Mercer, Polym. Prepr. (Am. Chem. Soc., Div. Polym. Chem.), **32**, No. 2, 18 (1991).
34. A.L. Rusanov, T.A. Stadnik and K. Müllen, Russ. Chem. Rev., **68**, 760 (1999).

Polyimides and Other High Temperature Polymers, Vol. 2, pp. 113–136
Ed. K.L. Mittal
© VSP 2003

New photoalignable polyimides and their ability to control liquid-crystal alignment

MOONHOR REE,* SEUNG WOO LEE, SANG IL KIM, WOOYOUNG CHOI
and BYEONGDU LEE

Department of Chemistry, Center for Integrated Molecular Systems, BK21 Program, and Polymer Research Institute, Division of Molecular and Life Sciences, Pohang University of Science and Technology, San 31, Hyoja-dong, Nam-gu, Pohang 790-784, The Republic of Korea

Abstract—New soluble, photosensitive polyimides containing cinnamoyl and coumarin moieties as side groups were synthesized. Cinnamoyl group can undergo both [2+2] photodimerization and *trans-cis* photoisomerization but coumarin group undergoes only [2+2] photodimerization. Photo-induced alignment behaviors of the polyimide chains were investigated by irradiating with linearly polarized ultraviolet light (LPUVL). Both polymers in film forms gave a maximum dichroic ratio and optical retardation at 0.25-0.50 J/cm^2 exposure dose, indicating that the photoreaction induced polymer chains to orient preferentially. Such oriented polymers in a film form were found to induce nematic liquid-crystals (LCs) to align along a direction of 97-99° with respect to the electric vector of LPUVL, regardless of the kind of photosensitive side group. Such LC alignment behavior was observed even on films treated by multiple exposures with changing electric vector of LPUVL. In these cases, the director of LC alignment was governed mainly by the first exposure rather than the subsequent exposures with changing electric vector. The results suggest that the homogeneous, uni-axial alignment of LCs on the polymer film containing cinnamoyl side group is induced mainly by [2+2] photodimerization rather than *trans-cis* photoisomerization. As another clue for the [2+2] photodimerization, the formation of cyclobutyl group was spectrocopically detected from the films exposed to UV light. In addition, LC alignment on the films was found to be controlled mainly by the rubbing process, regardless of the sequence of LPUVL exposure and rubbing in the surface treatment process.

Keywords: Photoreactive polymer; alignment layer; ultraviolet light irradiation; photodimerization; photoisomerization; rubbing; rubbing density; molecular reorientation; liquid crystal (LC); LC cell; LC alignment; LC pretilt.

1. INTRODUCTION

Rubbing with fabrics is widely used in the liquid-crystal display (LCD) industry because of its high effectiveness in treating the surfaces of polymeric thin films to control the alignment of liquid-crystal (LC) molecules [1-8]. However, the proc-

---

*To whom all correspondence should be addressed. Phone: 82-54-279-2120,
Fax: 82-54-279-3399, E-mail: ree@postech.edu

ess has some shortcomings, such as dust generation, electrostatic problems, and poor control of rubbing strength and uniformity. In order to overcome the shortcomings in the rubbing process, several photoinduced LC alignment concepts using linearly polarized ultraviolet light (LPUVL) irradiation have recently been proposed [9-18]. These approaches have attracted considerable attention from both academic and industrial fields because of the capability of the rubbing-free production of LC aligning films. Nevertheless, these approaches still are not workable for the mass production of LCDs because of a number of unsolved problems, such as low thermal stability, low anchoring energy, low pretilt angle, limited processibility with ultraviolet (UV) light exposure, and unavailability of proper materials. Thus, there still remains a big challenge to deliver high performance materials suitable for processing LC aligning films with eliminating the rubbing step. A representative class of photoalignment materials is poly(vinyl cinnamate) (PVCi) and its derivatives [9-19]. However, such polymers have relatively very low glass transition temperature (T_g). Because of the low T_g, even the polymer chains in the film are oriented preferentially by exposure to LPUVL, but they are still mobile so that such chain orientation may not be stable enough towards environmental influences such as temperature, aging time, and so on. Furthermore, their photoalignment mechanism has not been fully understood yet.

In this study, new photosensitive polyimides with cinnamate and coumarin side groups were synthesized, which are stable both thermally and dimensionally because of their high T_g. Their photoreactivity and photoalignment characteristics were investigated, and UV-exposure and rubbing processibility were examined. In addition, the alignment behavior of LC molecules was investigated on films treated by various LPUVL exposures, rubbing, and their combinations.

2. EXPERIMENTAL

2.1. Synthesis

The polyimide containing hydroxyl side groups, 6F-HAB was synthesized from polycondensation of purified 2,2'-bis-(3,4-dicarboxyphenyl)hexafluoropropane dianhydride (6F) (Chriskev Company, USA) and 3,3'-hydroxy-4,4'-diaminobiphenyl (HAB) (Chriskev Co.) as follows. Equivalent moles of 6F and HAB monomers were dissolved together with two equivalent moles of isoquinoline catalyst in dry N-methyl-2-pyrrolidone (NMP). The solution was gently heated at 70°C for 2 h and then refluxed for 5 h. Thereafter, the reaction solution was poured into a mixture of methanol and water (6:4 volume ratio) with vigorous stirring, giving 6F-HAB polyimide in the form of precipitated powder. The precipitated powder was filtered and dried. The synthesized polyimide was dissolved in dimethyl-d_6 sulfoxide (DMSO-d_6), and characterized using a proton nuclear magnetic resonance (^{1}H NMR) spectrometer (Bruker Aspect 300 MHz). Its molecular weight was estimated using a gel permeation chromatography (GPC) system calibrated with polystyrene standards. In the measurement, a flow rate of 1.0 mL/min was employed and tetrahydrofuran (THF) was used as the eluent.

A cinnamate containing polyimide (6F-HAB-CI) was synthesized by the reaction of 6F-HAB polymer and cinnamoyl chloride. Four equivalent moles of cinnamic acid were dissolved in excess thionyl chloride under a nitrogen atmosphere. After heating this reaction solution at 80°C for 3 h, the residual thionyl chloride was removed under a reduced pressure. The reaction product was dissolved in dry THF, followed by adding one equivalent mole of 6F-HAB polymer and pyridine in excess. After stirring for 3 h at room temperature, the solution was poured into methanol under vigorous stirring, leading to precipitated polymer product. The precipitated polymer powder was filtered and dried, giving 6F-HAB-CI polymer. In addition, a polyimide containing coumarin side group was synthesized by the Mitsunobu reaction [20, 21] of 6F-HAB polyimide and 7-(2-hydroxyethoxy)-4-methylcoumarin in our laboratory as follows. 6F-HAB polymer, 7-(2-hydroxyethoxy)-4-methylcoumarin, and triphenyl phosphine (1:3:3 mole ratio) were dissolved in dry THF under a nitrogen atmosphere. Three equivalent moles of diisopropyl diazocarboxylate were slowly added into the solution. After stirring for 1 h at room temperature, the reaction solution was poured into methanol under vigorous stirring. Then, the polymer product was precipitated in powder form. The 6F-HAB-ETCOU polymer powder was filtered and dried. The obtained 6F-HAB-CI and 6F-HAB-ETCOU were identified in DMSO-d_6 by ^{1}H NMR spectroscopy.

2.2. Measurements

6F-HAB-CI polymer was dissolved in cyclohexanone while 6F-HAB-ETCOU polymer was dissolved in a mixture of NMP and cyclohexanone (1:1 volume ratio). The obtained polymer solutions were spin-cast on glass substrates and dried for 2 h in a vacuum oven at 100°C. The polyimide films were peeled off from the glass substrates and used for property measurements. T_g was measured over 25-400°C using a Seiko differential scanning calorimeter. During the measurements, dry nitrogen gas was used to purge with a flow rate of 80 cc/min and a ramping rate of 10.0°C/min was employed. In each run, a sample of about 5 mg was used. The T_g was taken as the onset temperature of glass transition in the thermogram. Thermal stability was measured over 50-800°C using a Perkin-Elmer thermogravimeter (Model TGA7). During the measurements, dry nitrogen gas was used to purge at a flow rate of 100 cc/min and a ramping rate of 5.0°C/min was employed.

The polymer solutions were spin-cast onto quartz slides or Si substrate and dried for 2 h in a vacuum oven at 100°C, giving thin films of ca. 100 nm thickness. For photoreactivity measurements, the films were irradiated with varying exposure dose by UV light. Here, a high-pressure Hg lamp system of 1.0 kW (ALTECH, South Korea), was employed as the UV light source together with an optical filter (Milles Griot, USA) which transmits a band beam of 260-380 nm

wavelength. The optically filtered UV light intensity was 50 mW/cm^2. The expo-
sure dose was measured using an International Light photometer (Model IL-1350,
International Light Inc., USA) with a sensor (Model SED-240, International Light
Inc., USA). For the irradiated films, UV spectra were recorded using an HP 8452
Hewlett-Packard spectrometer and FT-IR spectra were measured by an ATI
Mattson FTIR spectrometer.

Photoalignment behaviors of the polymers thin films were investigated as follows. The films were irradiated by LPUVL at varying exposure dose. Here, a linear dichroic polarizer (Oriel, USA) was used. The filtered LPUVL intensity was
10 mW/cm^2. For the irradiated films, optical retardation was measured using a
plane polariscope system equipped with a He-Ne laser source, a pair of polarizer
and analyzer, a photodiode detector, and a sample stage, which was made in our
laboratory. Polarized UV spectra were measured using a Hewlett-Packard spectrometer equipped with a linear dichroic polarizer. For each polymer system, a series of films deposited onto glass slides were irradiated by the filtered LPUVL
with varying exposure dose, according to the various exposure protocols shown in
Table 1. Each exposure was performed at either $\theta = 0°$ or $\theta = 90°$: here, θ is defined as the angle between the longer axis of the glass slide and the electric vector
of LPUVL. In addition, another series of films were subjected to rubbing and UV-
exposure in accordance with the various treatment protocols given in Table 1. The
rubbing process was conducted with a rubbing density (L/l) of 130 using a laboratory rubbing machine (Wande Company, USA) with a roller covered by a rayon
cloth (Yoshikawa YA-20-R). Here, the rubbing density (L/l) was determined using the following equation [22, 23]:

$$\frac{L}{l} = N \ (\frac{2\pi \ r \ n}{60 \ v} - 1)   \tag{1}$$

where L is the total length of the rubbing cloth which contacts a certain point of
the polymer film (mm), l is the contact length of circumference of rubbing roller
(mm), N is the cumulative number of rubbings, n and r are the speed (rpm) and
the radius (cm) of rubbing roller, and v is the velocity (cm/sec) of the substrate
stage.

Using these polymer films, LC cells were assembled and then filled with 4-
pentyl-4-biphenylcarbonitrile (5CB) containing 1.0 wt% of a dichroic dye (Disperse Blue 1) by capillary technique. Each LC cell was prepared as follows. Two
pieces of polymer film covered glass slides, which were treated at chosen treatment conditions, were assembled together anti-parallel along the longer axis of
the glass slide. The gap in the LC cells was 50 μm. For the LC cells, the alignment of LC molecules was determined by measuring the absorbance of a linearly
polarized He-Ne laser beam (632.8 nm wavelength) as a function of rotational angle φ. Here, φ is defined as the angle between the polarization of He-Ne laser
beam and the longer axis of the glass slide used in the LC cell.

Table 1.
Surface treatment protocols for photosensitive polyimide films adhered to glass slides

Photosensitive polymer film		Energy per exposure[a] (J/cm^2)	Rubbing density	Sequential treatment protocol
6F-HAB-CI (sample i.d.)	6F-HAB-ETCOU (sample i.d.)			
CI-0.25-#1	CU-0.25-#1	0.25		0°(θ)-exposure[b]
CI-0.25-#2	CU-0.25-#2	0.25		0°(θ)-exposure, 90°(θ)-exposure[c]
CI-0.25-#3	CU-0.25-#3	0.25		0°, 90°, 0°
CI-0.25-#4	CU-0.25-#4	0.25		0°, 90°, 0°, 90°
CI-0.25-#5	CU-0.25-#5	0.25		0°, 90°, 0°, 90°, 0°
CI-0.25-#6	CU-0.25-#6	0.25		0°, 90°, 0°, 90°, 0°, 90°
CI-0.25-#7	CU-0.25-#7	0.25		0°, 90°, 0°, 90°, 0°, 90°, 0°
CI-0.25-#8	CU-0.25-#8	0.25		0°, 90°, 0°, 90°, 0°, 90°, 0°, 90°
CI-0.50-#1	CU-0.50-#1	0.50		0°
CI-0.50-#2	CU-0.50-#2	0.50		0°, 90°
CI-0.50-#3	CU-0.50-#3	0.50		0°, 90°, 0°
CI-0.50-#4	CU-0.50-#4	0.50		0°, 90°, 0°, 90°
CI-1.50-#1	CU-1.50-#1	1.50		0°
CI-1.50-#2	CU-1.50-#2	1.50		0°, 90°
CI-RP	CU-RP	1.50	130	90°(θ)-rubbing,[d] 90°(θ)-exposure
CI-PR	CU-PR	1.50	130	90°(θ)-exposure, 90°(θ)-rubbing

[a]A linearly polarized UV light with 260-380 nm wavelength was used.

[b]Polymer film was exposed at θ = 0° to a linearly polarized UV light: θ is defined as the angle between the longer axis of the glass slide and the electric vector of the linearly polarized actinic UV light.

[c]Polymer film was exposed at θ = 90° to a linearly polarized UV light: θ is defined as the angle between the longer axis of the glass slide and the electric vector of the linearly polarized actinic UV light.

[d]Polymer film was rubbed at θ = 90°: θ is defined as the angle between the longer axis of the glass slide and the rubbing direction.

3. RESULTS AND DISCUSSION

3.1. Synthesis and thermal properties

Figure 1 shows two photosensitive polyimides synthesized in this study: 6F-HAB-CI and 6F-HAB-ETCOU. Each polymer was synthesized in two major steps, i.e., synthesis of soluble 6F-HAB polyimide and incorporation of side

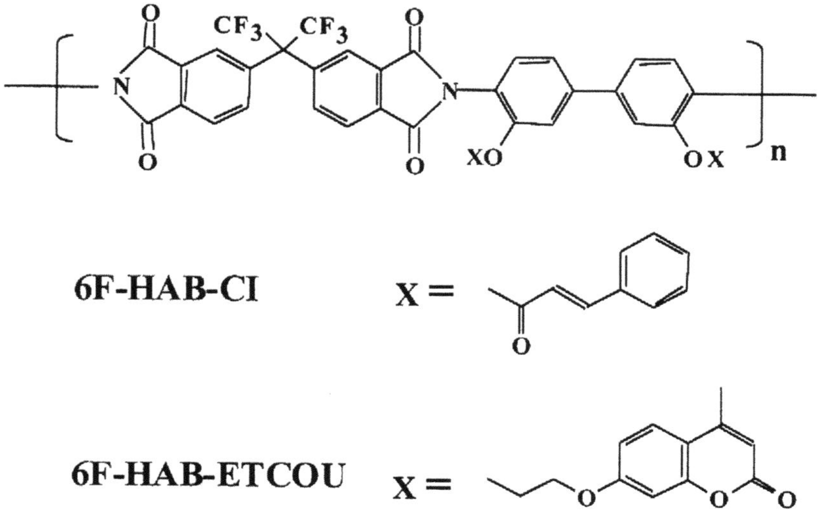

Figure 1. Chemical structures of photosensitive polyimides synthesized in this study.

groups into the polymer. Soluble 6F-HAB polyimide was synthesized directly from the polycondensation of the respective monomers using isoquinoline as a catalyst. The obtained polymer was characterized by ^{1}H NMR spectroscopy. In the ^{1}H NMR spectrum in Figure 2a, the proton peak of the hydroxyl side groups appears at 10.0 ppm while the protons of all the aromatic rings on the polymer backbone appear over the range of 6.9-8.3 ppm. The amino protons, which can originate from possible residues of partially imidized 6F-HAB poly(amic acid), are not detected. The polymer was measured to be 53,400 $\overline{M_w}$ and 1.87 polydispersity by the GPC analysis calibrated with polystyrene standards. In conclusion, a soluble 6F-HAB polyimide was successfully synthesized with a reasonably high molecular weight.

To the soluble 6F-HAB polyimide, cinnamoyl (CI) and 7-ethoxy-4-methylcoumarin (ETCOU) side groups were incorporated giving, respectively, 6F-HAB-CI and 6F-HAB-ETCOU. ^{1}H NMR spectra measured from the polymers are presented in Figures 2b and 2c. Neither polymer shows the proton peak originating from the hydroxyl groups of 6F-HAB polyimide. This indicates that the 6F-HAB-CI and 6F-HAB-ETCOU polymers were successfully prepared from the 6F-HAB polyimide. All the polymers obtained always formed good quality thin films through a conventional solution spin-casting and subsequent drying process.

The glass transition temperature T_g and the degradation temperature T_d were, respectively, 181°C and 340°C for 6F-HAB-CI polymer and 132°C and 300°C for 6F-HAB-ETCOU polymer. For 6F-HAB polymer, the T_d was 440°C but the T_g could not be detected in the range 50-400°C. Overall, both T_g and T_d of 6F-HAB polymer are lowered by incorporating photosensitive side groups.

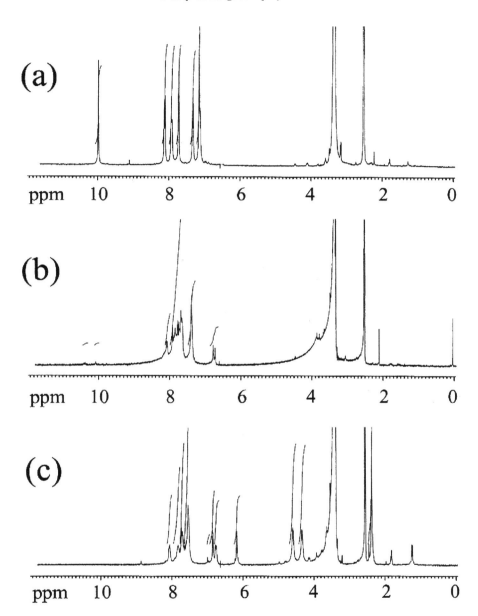

Figure 2. ^{1}H NMR spectra of the synthesized polymers: (a), 6F-HAB; (b), 6F-HAB-CI; (c), 6F-HAB-ETCOU.

3.2. Photoreactivity and photoalignment

Figure 3 shows UV absorption spectra of the photosensitive polymer films irradiated with the unpolarized UV light of 260-380 nm wavelength at varying exposure dose. 6F-HAB-CI polymer exhibits an absorption maximum at 278 nm

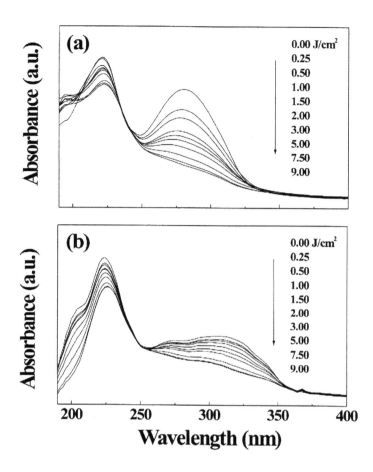

Figure 3. UV spectra measured from photosensitive polymer films exposed to an unpolarized UV light (260-380 nm wavelength) with varying exposure dose: (a), 6F-HAB-CI; (b), 6F-HAB-ETCOU.

($= \lambda_{max}$). This UV absorption originates from the photosensitive cinnamate side group in the polymer. The absorption peak intensity at $\lambda_{max} = 278$ nm drops drastically with increasing exposure dose up to 7.5 J/cm^2 and then stays almost constant with further increasing exposure dose. On the other hand, 6F-HAB-ETCOU polymer reveals a maximum absorption at 306 nm, which results from the photoreaction of coumarin side group. This peak is relatively weaker in intensity than that of 6F-HAB-CI polymer. The intensity of the absorption peak at $\lambda_{max} = 306$ nm drops gradually with increasing exposure dose and then levels off at 7.5 J/cm^2.

These results suggest that the cinnamate side group has a relatively higher extinction coefficient at $\lambda_{max} = 278$ nm than that of the coumarin side group at

λ_{max} = 306 nm. And, for the maximum absorption peak, its intensity drop due to the UV exposure is more significant in the cinnamate group containing polymer than in the coumarin group containing polymer. Therefore, 6F-HAB-CI polymer may have a relatively higher photoreactivity than 6F-HAB-ETCOU polymer. For both polymers the variations in the maximum absorption peaks due to the UV exposure remained constant for an exposure dose of >7.5 J/cm^2. These results suggest that the photoreaction possible in the film is almost completed for ca. 7.5 J/cm^2 exposure dose, regardless of the photosensitive side group.

The polymer films exposed to UV light were further examined by IR spectroscopy. The most characteristic IR vibrations of 6F-HAB-CI are shown in Figure 4a and listed in Table 2. Here, the vibrational peaks were assigned based on the results reported previously from some cinnamoyl derivatives [24-31]. As shown in Figure 4a, the vibration peaks (1633 and 981 cm^{-1}) of vinylene linkage in the cinnamoyl group are weakened with increasing UV-exposure dose. In general, [2+2] photodimerization consumes the vinylene linkages, whereas the *cis*-form [27, 28] of the vinylene linkage is expected to have a lower extinction coefficient than the *trans*-form in the IR absorption. Thus, such intensity drops might result from [2+2] photodimerization as well as *trans-cis* photoisomerization occurring possibly by UV-exposure. However, some supportive clues for the photodimerization were detected as follows. First, the conjugated ester carbonyl (C=O) appears at 1737 cm^{-1}, but becomes weak and broad as UV-exposure dose increases. Such peak weakening is attributed to a decrease in the population of the conjugated ester carbonyl by the photodimerization, while the peak broadening results from generation of non-conjugated ester carbonyl. Second, new IR peaks appear at 1469 and 1461 cm^{-1}, which correspond to the scissoring motion of cyclobutane ring which is a product of the photodimerization [32]. So it can be concluded that the [2+2] photodimerization takes place preferentially in the film by UV-exposure. In contrast, the IR spectroscopic study did not provide definite clues about the possibility of photoisomerization in the film.

Table 2.
Characteristic IR bands of photosensitive polyimide films

Wavenumber (cm^{-1})		Assignments
6F-HAB-CI	6F-HAB-ETCOU	
1788	1785	imide C=O symmetric stretching
1732	1725	imide C=O asymmetric stretching
1737	–	ester C=O stretching
–	1735	pyrone C=O stretching
1633	1613	vinylene C=C stretching
1376	1378	imide C-N-C stretching
1255	1255	CF$_3$ stretching
981	984	vinylene C-H out-of-plane bending

M. Ree et al.

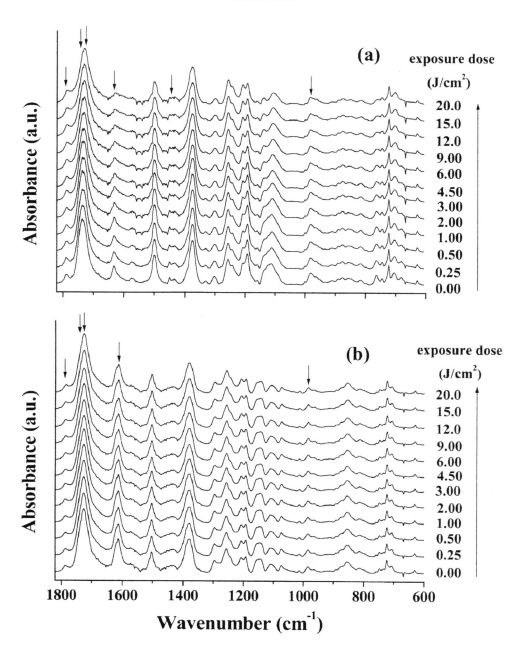

Figure 4. FT-IR spectra measured from photosensitive polymer films exposed to an unpolarized UV light (260-380 nm wavelength) with varying exposure dose: (a), 6F-HAB-CI; (b), 6F-HAB-ETCOU.

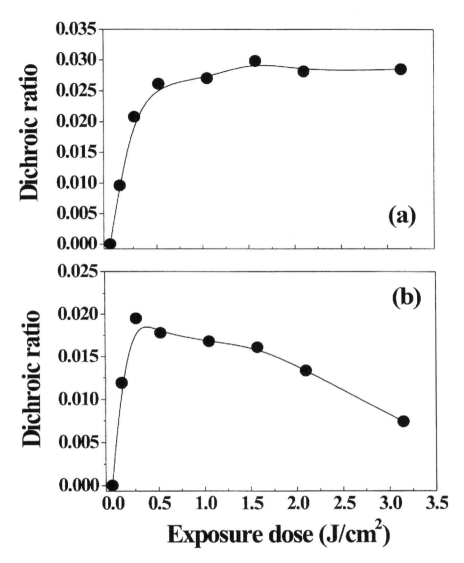

Figure 5. Dichroic ratios measured from photosensitive polymer films exposed to a linearly polarized UV light (260-380 nm wavelength) with varying exposure dose: (a), 6F-HAB-CI; (b), 6F-HAB-ETCOU.

Variations of some vibrational peaks due to UV-exposure were also found in 6F-HAB-ETCOU polymer films (see Figure 4b). By increasing the UV-exposure dose, both stretching vibration of C=C and out-of-plane bending vibration of C-H in the vinylene linkage of coumarin group decreased in intensity. In addition, the stretching vibration of C=O in the pyrone, which appears at 1735 cm^{-1} as a shoulder, is also decreased and shifted to the high frequency region with increasing

UV-exposure dose. These results are due mainly to photodimerization which takes place on UV-exposure, because coumarin group is known to undergo only [2+2] photodimerization because of its fused-ring structure [27, 33-37].

For the polymer films exposed to the linearly polarized UV light, the dichroic ratio was monitored as a function of UV-exposure dose. The dichroic ratio was determined from the following equation:

$$\text{Dichroic ratio} = (A_\perp - A_{//})/(A_\perp + A_{//}) \qquad (2)$$

where $A_\perp$ and $A_{//}$ are the absorbances of the polymer film at λ_{max} measured on exposure to a linearly polarized UV light perpendicular to and parallel with the electric vector of LPUVL used in the exposure, respectively. The results are presented in Figure 5a.

All 6F-HAB-CI films exposed to LPUVL always revealed a positive dichroic ratio. This indicates that in the photoreaction the cinnamate side groups in the direction parallel to the electric vector of LPUVL are consumed more rapidly than those in the direction perpendicular to the electric vector of LPUVL. The dichroic ratio increases sharply with increasing exposure dose, reaching 0.026 at 0.5 J/cm². Thereafter, the dichroic ratio levels off and stays almost constant. All 6F-HAB-ETCOU films exposed to LPUVL also show a positive dichroic ratio. This suggests that the coumarin side groups undergo a similar selective photoreaction as cinnamate groups do. The dichroic ratio increases very sharply with increasing exposure dose and then reaches 0.0195 at 0.25 J/cm². Thereafter, the dichroic ratio, however, decreases gradually with further increase of exposure dose (see Figure 5b).

The measured optical retardations are plotted in Figure 6 as a function of exposure dose. For 6F-HAB-CI film, the optical retardation initially increases sharply with increasing exposure dose, reaching 1.5 nm at 0.5 J/cm², and it increases only slightly with further increase of exposure dose. In the case of 6F-HAB-ETCOU film, the optical retardation also initially increases sharply with increasing exposure dose and reaches a maximum value of 1.2 nm at 0.25 J/cm². Thereafter, the retardation declines slowly with further increase of exposure dose.

All UV-exposed polymer films always exhibit a positive retardation, which is attributed to an optical anisotropy (i.e., in-plane birefringence) caused by exposure to LPUVL. Both polymer systems are positive birefringent, i.e., the polarization along the chain axis is relatively larger than that along the normal to the chain axis. Thus, the measured optical retardations might result from the reorientation of polymer chains along a preferential direction in the film, which is induced by the directionally selective photoreaction occurring during exposure to LPUVL.

In comparison, in the 6F-HAB-ETCOU film both the dichroic ratio and optical retardation declined with increasing exposure dose after reaching their maximum values, whereas 6F-HAB-CI film does not show such declinations (see Figures 5 and 6). Such declining behaviors in the 6F-HAB-ETCOU film may be attributed to the relatively high chain mobility in the polymer. The polymer exhibits the T_g

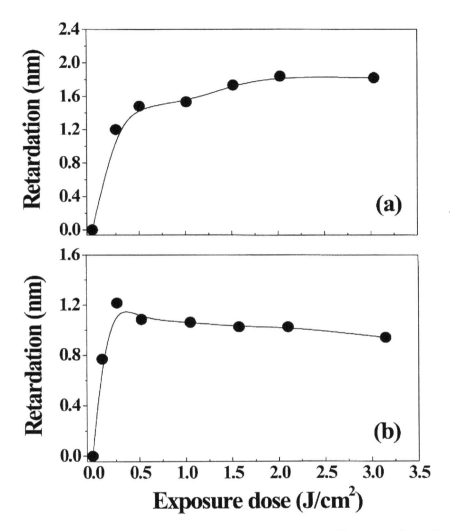

Figure 6. Optical retardations measured from photosensitive polymer films exposed to a linearly polarized UV light (260-380 nm wavelength) with varying exposure dose: (a), 6F-HAB-CI; (b), 6F-HAB-ETCOU.

at 132°C which is 50°C lower than that of the 6F-HAB-CI polymer. Both polymers have the same polymer backbone except for the different side groups. The coumarin moiety is relatively bulkier than the cinnamoyl group. In general, a bulkier group in a polymer gives higher T_g. Such low T_g in 6F-HAB-ETCOU polymer might result from a relatively high flexibility of the spacer in the ETCOU side group. The flexible spacer causes a relatively high mobility to the coumarin side group leading, in turn, to an increase in the polymer chain mobility. On the other hand, UV-exposure is generally known to generate thermal heat in the polymer film: longer UV-exposure time generates more thermal heat in the film. Thus, a high exposure dose, which needs relatively long exposure time, gives a chance as well as a thermal energy to the polymer chains to mobilize. The mobile polymers chains allow the coumarin side groups to reorient favorably to LPUVL to some extent and involve in the photoreaction, even though they are initially in the direction perpendicular to the electric vector of LPUVL, leading to a declination in both the dichroic ratio and optical retardation. Such declination in the optical retardation is always observed for films of polyvinyl derivatives with cinnamoyl moiety which also exhibit relatively low T_gs because of their flexible polymer backbone [9-11, 17, 18].

3.3. LC alignment on films subjected to UV exposure

Figure 7 shows polar diagrams of LC cells fabricated with 6F-HAB-CI films subjected to LPUVL of 0.25 J/cm^2 exposure dose through 0°(θ)- and 90°(θ)-exposure in a sequential manner. The first 0°(θ)-exposure generates a very weak anisotropy in the polar absorbance diagram. The alignment of LC molecules is very weak along the direction of $\varphi = 150° = 330°$. The subsequent 90°-exposure enhances slightly the development of anisotropy in the polar diagram, but changes the direction of LC alignment to $\varphi = 176° = 356°$. The LC alignment is changed again to $\varphi = 135° = 315°$ by the subsequent 0°(θ)-exposure. The LC alignment is moved again to the direction of $\varphi = 173° = 353°$ by the subsequent 90°(θ)-exposure. Further 0°(θ)-exposure turns again the direction of LC alignment to $\varphi = 102° = 282°$. In this case, the anisotropy in the polar diagram is very distinctive, indicating that a relatively high degree of LC alignment had taken place along the direction of $\varphi = 102° = 282°$. Thereafter, the direction of LC alignment varies less sensitively with further sequential 0°(θ)- and 90°(θ)-exposures. Finally, the principal director of LC alignment is settled along the direction of $\varphi = 99° = 279°$, regardless of the history of the sequential 0°(θ)- and 90°(θ)-exposures.

The initially observed large variations in the director of LC alignment may be attributed to the relatively low energy dose per UV-exposure. The exposure dose of 0.25 J/cm^2 seems to be too low to induce the photoreaction of cinnamoyl side groups in a high yield. In fact, 0.25 J/cm^2 is just half of the minimum exposure dose (0.50 J/cm^2) needed for the photoreaction of cinnamoyl moieties in the film to give maximum dichroic ratio and optical retardation as aforementioned. Thus, each exposure causes the directional photoreaction of cinnamoyl side groups to a relatively low extent, leading to a degree of reorientation of polymer chains and side groups in the film. This situation is directly reflected in the induction of the alignment of LC molecules in the cell, consequently causing a strong dependency of LC alignment on the history of the sequential 0°(θ)- and 90°(θ)-exposures. This dependency is large at the early stage of the sequential 0°(θ)- and 90°(θ)-exposures because the number of unreacted cinnamoyl moieties on the surface is relatively high. Then, the dependency becomes small with increasing number of

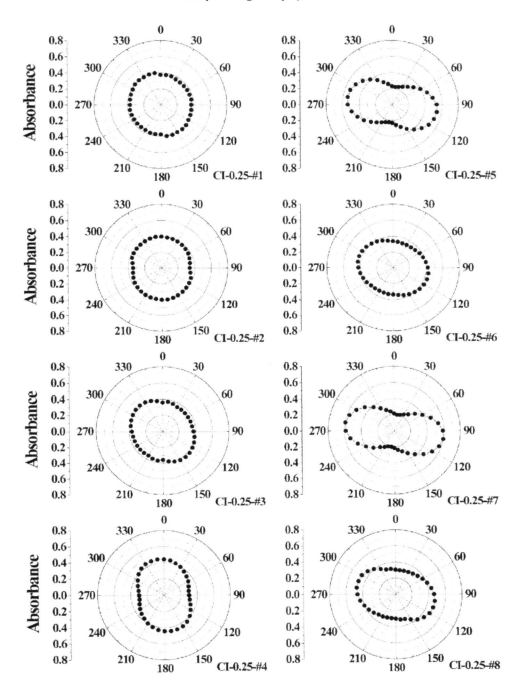

Figure 7. Polar diagrams of absorbances measured from LC cells fabricated with 6F-HAB-CI films which were treated by 0°(θ)- and 90°(θ)-exposure protocols in Table 1: here, θ is defined as the angle between the longer axis of the glass slide and the electric vector of the linearly polarized actinic UV light (260-380 nm wavelength). An energy of 0.25 J/cm^2 per exposure was employed.

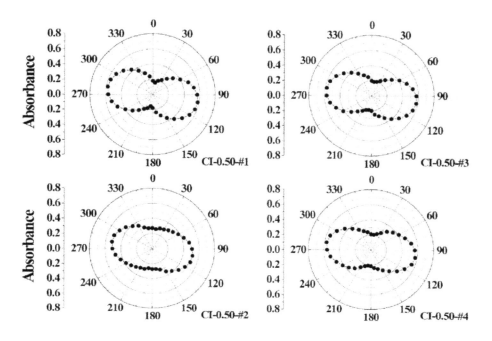

Figure 8. Polar diagrams of absorbances measured from LC cells fabricated with 6F-HAB-CI films which were treated by 0°(θ)- and 90°(θ)-exposure protocols in Table 1: here, θ is defined as the angle between the longer axis of the glass slide and the electric vector of the linearly polarized actinic UV light (260-380 nm wavelength). An energy of 0.50 J/cm^2 per exposure was employed.

sequential exposures because of large consumption of unreacted cinnamoyl moieties. However, the polar diagrams obtained for the CI-0.25#5, CI-0.25#6, CI-0.25#7 and CI-0.25#8 cells indicate that the photoalignment of polymer chains generated by the first UV-exposure still governs strongly the LC alignment even on the film sequentially exposured more than two times.

The sequential 0°(θ)- and 90°(θ)-exposures were conducted further with a high exposure dose of 0.50-1.50 J/cm^2. Figure 8 shows the polar diagrams measured for the LC cells fabricated with polymer films exposed sequentially to a polarized UV light of 0.50 J/cm^2 exposure dose. The first 0°(θ)-exposure causes the LC alignment along the direction of φ = 100° = 280°. For the subsequent 90°(θ)-exposure, the LC alignment is disturbed to some extent, but the principal director of the LC alignment apparently is not changed. The slight disturbance in the LC alignment is healed completely by the subsequent 0°(θ)-exposure. Thereafter, the LC alignment apparently is not sensitive to further 90°(θ)-exposure.

Figure 9 presents the polar diagrams for the LC cells prepared with polymer films subjected to a linearly polarized UV light with an exposure dose of 1.50 J/cm^2 by sequential 0°(θ)- and 90°(θ)-exposures. The first 0°(θ)-exposure induces

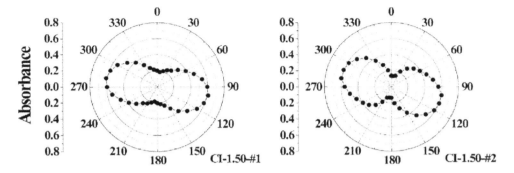

Figure 9. Polar diagrams of absorbances measured from LC cells fabricated with 6F-HAB-CI films which were treated by 0°(θ)- and 90°(θ)-exposure protocols in Table 1: here, θ is defined as the angle between the longer axis of the glass slide and the electric vector of the linearly polarized actinic UV light (260-380 nm wavelength). An energy of 1.50 J/cm^2 per exposure was employed.

LC molecules to orient along the direction of φ = 99° = 279°, but the LC alignment is influenced only slightly by the subsequent 90°(θ)-exposure, i.e., the direction of LC alignment is turned very slightly to the axis of φ = 100° = 279°. Instead, the LC alignment along the preferential direction is enhanced.

Similar effects of the sequential 0°(θ)- and 90°(θ)-exposures on the LC alignment were observed for the LC cells fabricated with 6F-HAB-ETCOU polymer films. The results are illustrated in Figures 10, 11, and 12.

These results provide us several informations as follows. First, 5CB LC molecules are aligned uniaxially on the 6F-HAB-CI and 6F-HAB-ETCOU films prepared by single exposure to LPUVL. This alignment behavior is highly distinctive on the polymer films subjected to a high exposure dose. The principal director in the LC alignment induced by the photoaligned film is along the direction with an angle of 97-99°, rather than 90°, to the electric vector of LPUVL, depending on the photosensitive side group. These LC alignments are somewhat different from those on the photoaligned polyvinyl derivatives containing cinnamoyl and coumarin moieties in which the alignment angle is perpendicular to the electric vector of LPUVL [9-11, 17, 18].

Second, on the polymer films subjected to multiple LPUVL exposures, the alignment of LC molecules is found to depend upon the history of changing electric vector of LPUVL. In particular, the direction of LC alignment is varied sensitively with changing electric vector of LPUVL in the multiple exposures when an energy dose per exposure is low. However, such dependency apparently is small for an energy dose per UV-exposure greater than 0.50 J/cm^2 at which both dichroic ratio and optical retardation level off or reach maxima in their variations with exposure dose (see Figures 5 and 6). On the film treated by multiple exposures with changing electric vector of LPUVL of ≥0.50 J/cm^2 energy dose, LC molecules have a strong tendency to align mainly along the direction

M. Ree et al.

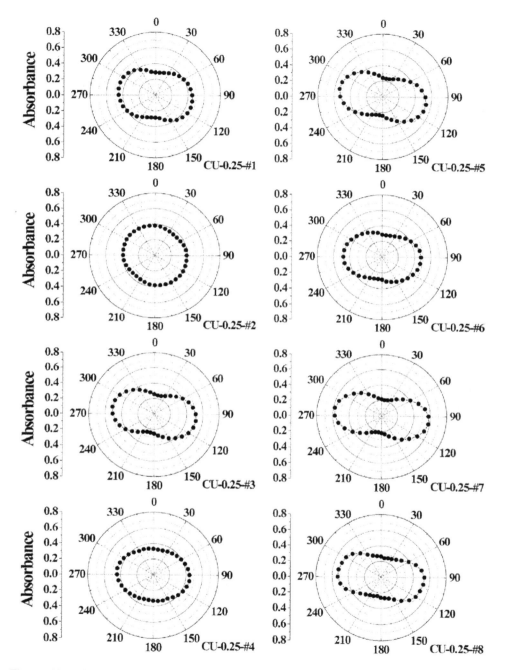

Figure 10. Polar diagrams of absorbances measured from LC cells fabricated with 6F-HAB-ETCOU films which were treated by 0°(θ)- and 90°(θ)-exposure protocols in Table 1: here, θ is defined as the angle between the longer axis of the glass slide and the electric vector of the linearly polarized actinic UV light (260-380 nm wavelength). An energy of 0.25 J/cm² per exposure was employed.

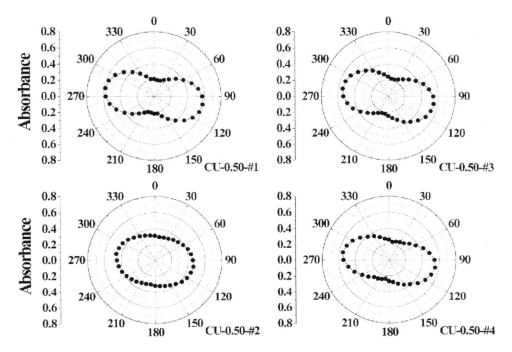

Figure 11. Polar diagrams of absorbances measured from LC cells fabricated with 6F-HAB-ETCOU films which were treated by 0°(θ)- and 90°(θ)-exposure protocols in Table 1. Here, θ is defined as the angle between the longer axis of the glass slide and the electric vector of the linearly polarized actinic UV light (260-380 nm wavelength). 0.50 J/cm² per exposure was employed.

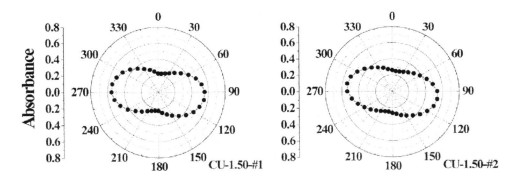

Figure 12. Polar diagrams of absorbances measured from LC cells fabricated with 6F-HAB-ETCOU films which were treated by 0°(θ)- and 90°(θ)-exposure protocols in Table 1. Here, θ is defined as the angle between the longer axis of the glass slide and the electric vector of the linearly polarized actinic UV light (260-380 nm wavelength). 1.50 J/cm² per exposure was employed.

associated with the orientation of polymer chains and reacted photosensitive side groups generated by the first exposure. This tendency becomes strong with increasing exposure dose, regardless of the photosensitive side group.

Finally, the coumarin moiety cannot undergo photoisomerization because of its fused ring (see the chemical structure in Figure 1). Therefore, the alignment of LC molecules induced on the photoaligned 6F-HAB-ETCOU film should be attributed to the directionally selective photodimerization of coumarin side groups by exposing to LPUVL. In contrast to the coumarin moiety, the cinnamoyl moiety can undergo both photodimerization and photoisomerization. If the LC alignment on the 6F-HAB-CI film is induced by the directionally selective photoisomerization of cinnamoyl side groups occurring by exposing to LPUVL, the direction of the LC alignment should vary sensitively on sequential $0°(\theta)$- and $90°(\theta)$-exposures of LPUVL, regardless of the exposure dose. However, such variations in the LC alignment are not detected for >0.50 J/cm^2 exposure dose. Furthermore, the alignment behavior of LC molecules on the UV-exposed 6F-HAB-CI films is very similar to that on the UV-irradiated 6F-HAB-ETCOU films containing coumarin moieties which cannot undergo photoisomerization, as described earlier. These facts suggest that the LC alignment induced on the 6F-HAB-CI film cannot be attributed to photoisomerization of cinnamoyl side groups. Furthermore, the photoisomerization perhaps does not occur for the cinnamoyl side groups being exposed to LPUVL. In conclusion, the directionally selective photodimerization of cinnamoyl side groups plays a major role in inducing the alignment of LC molecules on the surface of 6F-HAB-CI film as observed for the LC cell fabricated with 6F-HAB-ETCOU films.

3.4. LC alignment on films treated by both UV exposure and rubbing

The CI-RP in Figure 13 shows the polar diagrams measured from the LC cell fabricated with a 6F-HAB-CI film which was treated by $90°(\theta)$-rubbing and subsequent $90°(\theta)$-exposure. Here, θ is the angle between the longer axis of the polymer film adhered to glass slide and either the electric vector of LPUVL or the rubbing direction. The rubbing treatment generates microgrooves in the film along the rubbing direction, thus inducing LC molecules to align along the rubbing direction (namely, the direction of $\varphi = 90° = 270°$ in the polar diagram), whereas the $90°(\theta)$-exposure has the ability to induce LC molecules to align along the direction perpendicular to the rubbing direction. However, the aligned LC molecules in the cell exhibit a principal director along the direction of $\varphi = 90° = 270°$ (namely, rubbing direction). In comparison, the degree of LC alignment in this cell is higher than that in the cell prepared with a polymer film subjected to a $0°(\theta)$-exposure with an energy of 1.50 J/cm^2 (see the CI-1.50-#1 in Figure 9).

For the LC cell fabricated with a 6F-HAB-CI film treated by $90°(\theta)$-exposure and followed by $90°(\theta)$-rubbing, LC molecules also are aligned preferentially along the rubbing direction (see the CI-PR in Figure 13). The degree of LC alignment in the cell is high, compared to that on the film treated by 90°(θ)-rubbing and subsequent 90°(θ)-exposure (see Figure 13).

As seen in Figure 14, similar polar diagrams were obtained for the LC cells fabricated with 6F-HAB-ETCOU films that were treated in the same manner as the 6F-HAB-CI films. However, the LC alignment in the cell apparently varies only slightly with the sequence of 90°(θ)-exposure and 90°(θ)-rubbing in the film treatment process. Further, the degrees of LC alignment in these cells are very similar to that of the cell fabricated with a polymer film subjected to 0°(θ)-

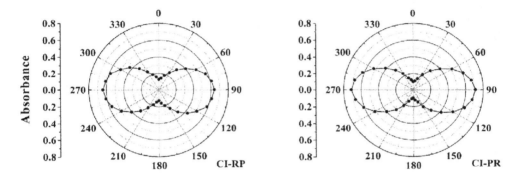

Figure 13. Polar diagrams of absorbances measured from LC cells fabricated with 6F-HAB-CI films which were treated by 90°(θ)-rubbing and 90°(θ)-exposure: CI-RP, rubbed and subsequently UV-exposed; CI-PR, UV-exposed and subsequently rubbed. Here, θ is defined as the angle between the longer axis of the glass slide and either the rubbing direction or the electric vector of the linearly polarized actinic UV light (260-380 nm wavelength). A rubbing density of 130 and an exposure dose of 1.50 J/cm^2 were employed.

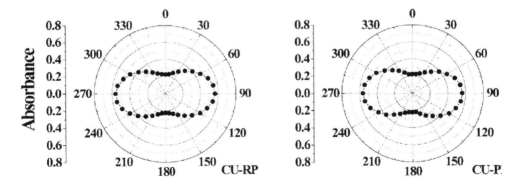

Figure 14. Polar diagrams of absorbances measured from LC cells fabricated with 6F-HAB-ETCOU films that were treated by 90°(θ)-rubbing and 90°(θ)-exposure: CU-RP, rubbed and subsequently UV-exposed; CU-PR, UV-exposed and subsequently rubbed. Here, θ is defined as the angle between the longer axis of the glass slide and either the rubbing direction or the electric vector of the linearly polarized actinic UV light (260-380 nm wavelength). A rubbing density of 130 and an exposure dose of 1.50 J/cm^2 were employed.

exposure with an energy of 1.50 J/cm^2 (see the CU-1.50-#1 in Figure 12).

These results provide us some information as follows. First, for the polymer films treated by 90°(θ)-exposure and subsequent 90°(θ)-rubbing, the LPUVL exposure process does not seem to make any contribution to the LC alignment on the film surface. This appears distinctively for the LC cell fabricated with 6F-HAB-CI film. This might result from the preferential reorientation of the polymer chains and side groups, including the photoaligned polymer chains and side groups, along the rubbing direction through the 90°(θ)-rubbing process. That is, some fractions of polymer chains and side groups are oriented first along the direction with an angle of 97-99° with respect to the polarization of LPUVL through the 90°(θ)-exposure process but they reorient along the rubbing direction by the subsequent 90°(θ)-rubbing process. Conclusively, the LC alignment is controlled mainly by the surface texture generated through the 90°(θ)-rubbing process.

Second, for the polymer film treated by 90°(θ)-rubbing and subsequent 90°(θ)-exposure, the surface textures created by both processes may involve in a competitive manner to induce the alignment of LC molecules on the film surface. Perhaps, the microgrooves generated by the rubbing process are well maintained through the subsequent 90°(θ)-exposure of linearly polarized UV light. However, the photosensitive side groups lain parallel to the electric vector of the polarized actinic UV light undergo photoreaction selectively by the 90°(θ)-exposure process, regardless of whether they are localized on the microgrooved or non-microgrooved surfaces of the film. Consequently, the 90°(θ)-exposure reorients the polymer chains and side groups of the rubbed film surface in part towards the direction with an angle of 97-99° with respect to the rubbing direction. In comparison, the fractions of the photoaligned polymer chains and side groups may be much smaller than those of the non-photoaligned polymer chains and side groups, because of the directionally selective photoreaction to the polarized light and its relatively low quantum yield. Considering these factors (namely, the directionally maintained microgrooves and the relatively high fraction of non-photoaligned polymer chains including side groups), LC molecules on the film should be aligned preferentially along the rubbing direction rather than the photoalignment direction.

Finally, mechanical rubbing process generally generates a relatively high roughness as well as microgrooves along the rubbing direction on polymer films. Of course, the surface roughness and the dimension of the directional microgrooves are dependent upon the rubbing density as well as the kind of velvet cloth. On the contrary, LPUVL exposure process generates relatively smooth surface in the polymer film. Considering these surface textures, the measured polar diagrams indicate that the relatively high surface roughness consisting of directionally aligned microgrooves plays a major role in the alignment of LC molecules in the cells, as compared to the photoaligned polymer chains and their related surface characteristics.

4. CONCLUSION

Soluble 6F-HAB polymer and its derivatives containing photosensitive cinnamoyl and coumarin side groups were successfully synthesized with reasonably high molecular weights. The polymers gave good quality films through conventional spin-casting and drying process. Thermal properties were determined. T_g and T_d were 181°C and 340°C for the 6F-HAB-CI polymer and 132°C and 300°C for the 6F-HAB-ETCOU polymer, respectively.

The photoreaction in the film is almost completed at ca. 7.5 J/cm^2 exposure dose, regardless of the photosensitive side group when the film is exposed to an unpolarized UV light of 260-380 nm wavelength. However, both dichroic ratio and optical retardation of the film due to a linearly polarized UV light reach their maximum values at only 0.25-0.50 J/cm^2 exposure dose. These are evidences that the polymer chains including photosensitive side groups are aligned along a preferential direction via the directionally selective photoreaction by exposing to LPUVL.

For both photosensitive polymers, the photoaligned polymer chains and side groups are evidently highly involve in inducing the alignemnt of LC molecules on the surface. The director of the LC alignment is along the direction with an angle of 97-99° with respect to the polarization of UV light. The LC alignment in the film is mainly induced by the photoaligned polymer chains and side groups made by the first LPUVL exposure even though the film is subsequently subjected to multiple exposures with changing electric vector of LPUVL. This behavior becomes significant with increasing exposure dose in the UV treatment. Further, the formation of cyclobutyl moiety, which is the product of the [2+2] photodimerization of cinnamoyl side group, was spectroscopically detected for 6F-HAB-CI films exposed to UV light. These results indicate that for both polymers, the photoalignments of polymer chains and side groups are caused by the [2+2] photodimerization of the photosensitive side groups, and the photoaligned polymer chains and side groups on the film induce uniaxial alignment of LC molecules on the surface. Even if the *trans-cis* photoisomerization occurs in the 6F-HAB-CI film, its fraction will be relatively small.

In addition, the LC alignment behavior was investigated on films treated by both UV-exposure and rubbing. The LC alignment is observed to be governed mainly by the rubbing process, regardless of the sequence of LPUVL exposure and rubbing. The LPUVL exposure contributes negligibly to the LC alignment. These results are explained in terms of the surface topography including roughness, microgrooves, rubbing-induced polymer alignment, and photodimerization-induced polymer alignment.

Acknowledgements

This study was supported, in part, by the Center for Integrated Molecular Systems (KOSEF), by the Ministry of Education (New Materials Program), and by the

M. Ree et al.

Ministry of Industry & Energy and the Ministry of Science & Technology (Electronic Display Industrial Research Association – G7 Project Program).

REFERENCES

1. W.C. O'Mara, *Liquid Crystal Flat Panel Displays*, Van Nostrand Reinhold, New York (1993).
2. E. Tannas, Jr., W.E. Glenn and J.W. Doane (Eds.), *Flat-Panel Display Technologies,* Noyes, Park Ridge, N.J. (1995).
3. S.I. Kim, M. Ree, T.J. Shin and J.C. Jung, *J. Polym. Sci.: Part A: Polym. Chem.*, **37**, 2909 (1999).
4. M. Ree, S.I. Kim, S.M. Pyo, T.J. Shin, H.K. Park and J.C. Jung, *Macromolecular Symp.*, **142**, 73 (1999).
5. E.S. Lee, P. Vetter, T. Miyashita and T. Uchida, *Jpn. J. Appl. Phys.*, **32**, L1339 (1993).
6. H. Kikuchi, J.A. Logan and D.Y. Yoon, *J. Appl. Phys.*, **79**, 6811 (1996).
7. K.-W. Lee, S.-H. Paek, A. Lien, C. During and H. Fukuro, *Macromolecules*, **29**, 8894 (1996).
8. D.-S. Seo, K. Araya, N. Yoshida, M. Nishikawa, Y. Yabe and S. Kobayashi, *Jpn. J. Appl. Phys.*, **34**, L503 (1995).
9. M. Schadt, K. Schmitt, V. Kozinkov and V. Chigrinov, *Jpn. J. Appl. Phys.*, **31**, 2115 (1992).
10. M. Schadt, M. Seiberle, A. Schuster and S.M. Kelly, *Jpn. J. Appl. Phys.*, **34**, L764 (1995).
11. K.Y. Han, B.H. Chae, S.H. Yu, J.K. Song, J.G. Park and D.Y. Kim, *SID 97 DIGEST*, **28**, 707 (1997).
12. S.I. Kim and M. Ree, *Proc. 1st Korea Liquid Crystal Conf.*, **1**, 55 (1998).
13. M. Schadt, H. Seiberle and A. Schuster, *Nature*, **381**, 212 (1996).
14. K.Y. Han, B.H. Chae, S.H. Yu, J.K. Song, J.G. Park and D.Y. Kim, *AM-LCD'96/IDW*, 403 (1996).
15. J.K. Song, K.Y. Han and V.G. Chigrinov, *AM-LCD'96/IDW*, 407 (1996).
16. Y.-K. Jang, H.-S. Yu, S.H. Yu, J.K. Song, B.H. Chae and K.Y. Han, *SID 97 DIGEST*, **28**, 703 (1997).
17. Y. Iimura, S. Kobayashi, T. Hashimoto, T. Sugiyama and K. Katoh, *HEICE Trans. Electron.*, **E39(8)**, 1040 (1996).
18. K. Ichimura, Y. Akita, H. Akiyama, K. Kudo and Y. Hayashi, *Macromolecules*, **30**, 903 (1997).
19. L.M. Minsk, J.G. Smith, W.P. van Deusen and J.F. Wright, *J. Appl. Polym. Sci.*, **2**, 302 (1958).
20. O. Mitsunobu, *Synthesis*, 1 (*Jan.* 1981).
21. T.-A. Chen, A.K.-Y. Jen and Y. Cai, *J. Am. Chem. Soc.*, **117**, 7295 (1995).
22. S.I. Kim, S.M. Pyo, M. Ree, M. Park and Y. Kim, *Mol. Cryst. Liq. Cryst.*, **316**, 209 (1998).
23. T. Uchida, M. Hirano and H. Sakai, *Liquid Crystals*, **5**, 1127 (1989).
24. G.M. Schmidt, *J. Chem. Soc.*, 2014 (1964).
25. M.D. Cohen and G.M. Schmidt, *J. Chem. Soc.*, 1996 (1964).
26. M. Ghosh, S. Chakrabarti and T.N. Misra, *J. Raman Spectro.*, **29**, 807 (1998).
27. S. Perny and P.L. Barny, *Liquid Crystals*, **27**, 329 (2000).
28. A.H. Ali and S.V. Srinivasan, *Polym. Int.*, **43**, 310 (1997).
29. M. Yamamoto, N. Furuyama and K. Itoh, *J. Phys. Chem.*, **100**, 18483 (1996).
30. A. Saito, T. Yajima, M. Yamamoto and K. Itoh, *Langmuir*, **11**, 1277 (1995).
31. R.F. Bryan and D.P. Freyberg, *J. Chem. Soc., Perkin Trans.*, **2**, 1835 (1975).
32. H.E. Ulery and J.R. McClenon, *Tetrahedron*, **19**, 749 (1963).
33. M. Obi, S. Morino and K. Ichimura, *Chem. Mater.*, **11**, 656 (1999).
34. Y. Chen and J.D. Wu, *J. Polym. Sci.: Polym. Chem.*, **32**, 1867 (1994).
35. Y. Chen and R.T. Hong, *J. Polym. Sci.: Polym. Chem.*, **35**, 2999 (1997).
36. U. Ghosh and T.N. Misra, *J. Polym. Sci.: Polym. Chem.*, **25**, 215 (1987).
37. M. Ghosh, S. Chakrabarti and T.N. Misra, *J. Phys. Chem. Solids*, **57**, 1891 (1996).

Polyimides and Other High Temperature Polymers, Vol. 2, pp. 137–154
Ed. K.L. Mittal
© VSP 2003

Proton conducting polyimides from novel sulfonated diamines

JIANHUA FANG, XIAOXIA GUO, TATSUYA WATARI,
KAZUHIRO TANAKA, HIDETOSHI KITA and KEN-ICHI OKAMOTO[*]

*Department of Advanced Materials Science & Engineering, Faculty of Engineering,
Yamaguchi University, 2-16-1 Tokiwadai, Ube, Yamaguchi 755-8611, Japan*

Abstract—Three kinds of novel sulfonated diamine monomers, 4,4'-diaminodiphenyl ether-2,2'-disulfonic acid (ODADS), 9,9-bis(4-aminophenyl)fluorene-2,7-disulfonic acid (BAPFDS), and 4,4'-bis(4-aminophenoxy)biphenyl-3,3'-disulfonic acid (BAPBDS), were successfully synthesized by direct sulfonation of the corresponding parent diamines, 4,4'-diaminodiphenyl ether (conventional name: 4,4'-oxydianiline (ODA)), 9,9-bis(4-aminophenyl)fluorene (BAPF), and 4,4'-bis(4-aminophenoxy)biphenyl (BAPB), respectively, using fuming sulfuric acid as the sulfonating reagent. Various sulfonated polyimides were prepared from 1,4,5,8-naphthalenetetracarboxylic dianhydride (NTDA), ODADS, BAPFDS, BAPBDS, and common non-sulfonated diamines. Proton conductivities of the resulting sulfonated polyimide membranes were measured as functions of relative humidity and temperature. With increasing relative humidity (at a given temperature) or temperature (in water) the proton conductivities increased. The proton conductivity was mainly determined by the ion exchange capacity (IEC), i.e., the higher the IEC, the larger the conductivity. All the sulfonated polyimide membranes displayed much better stability toward water than those derived from the widely used sulfonated diamine 2,2'-benzidinedisulfonic acid (BDSA) with similar IEC, and among them NTDA-BAPBDS exhibited the best water stability. This is because of the flexible structure and/or the high basicity of the sulfonated diamine moieties of these polyimides in comparison with that of BDSA-based ones. Fenton's reagent test revealed that these polyimide membranes also had fairly good stability towards oxidation. Polyimide membranes with good water stability as well as high proton conductivity were developed. NTDA-BAPBDS polyimide membrane, for example, did not lose mechanical properties after being soaked in water at 80°C for more than 1000 h, while its proton conductivity was still at high level (comparable to that of Nafion 117).

Keywords: Polyimide; sulfonated diamine; synthesis; membrane; proton conductivity; water stability.

1. INTRODUCTION

Aromatic polyimides, known for their excellent thermal stability, high mechanical strength, good film forming ability, and superior chemical resistance, have found

---

[*]To whom all correspondence should be addressed. Phone: 81-836-85-9660,
Fax: 81-836-85-9601, E-mail: okamotok@yamaguchi-u.ac.jp

wide applications in many industrial fields [1, 2]. Recently sulfonated six-membered ring polyimides have been identified to be promising proton conducting membrane materials for fuel cell application [3-8]. Mercier and coworkers first synthesized a series of sulfonated copolyimides from naphthalene-1,4,5,8-tetracaboxylic dianhydride (NTDA), 2,2'-bendizine sulfonic acid (BDSA, a commercially available sulfonated diamine), and common non-sulfonated diamine monomers [3-6]. These sulfonated polyimide membranes were tested in a fuel cell system which showed fairly good performance. However, the proton conductivity of these membranes was rather low ($< 10^{-2}$ S·cm^{-1}, in water) due to the low ion exchange capacity (IEC) which is essential for maintaining hydrolysis stability of the membranes. Here, IEC refers to the equivalent of exchangeable ions (cations, in case of sulfonated polymers) of unit weight of polymer. The fairly short lifetime (maximum: 3000 h) compared with the fuel cell with Nafion membrane is another problem. Litt's group has also employed BDSA as the sulfonated diamine monomer for preparation of various random and sequenced copolyimides [7, 8]. They reported that some copolyimide membranes derived from bulky or angled diamine comonomers displayed higher conductivities than Nafion 117. However, the water stability of their membranes is still a problem.

Proton conductivity and membrane stability are two important factors which greatly affect the performance of a fuel cell system. The enhancement of proton conductivity can be achieved by increasing the IEC. However, too high IEC generally leads to high degree of swelling or even dissolution in water of the membranes. To maintain membrane stability, the IEC should be controlled at a relatively low level. On the other hand, it is well known that the chemical structure of polymers has considerable effect on the properties of the membranes. However, little information on this aspect can be found in the literature. To develop polyimide membranes with high proton conductivity and good water stability, it is essential to systematically study the "structure-property" relationships of the sulfonated polyimides. Besides BDSA, the only other commercially available sulfonated diamines are 2,4-diaminobenzenesulfonic acid (DABS) and 5,5'-dimethylbenzidine-2,2'-disulfonic acid (DMBDS). DABS-based polyimides generally have poor solubility and poor mechanical properties, and, therefore, DABS is seldom used. DMBDS-based polyimides are expected to have similar properties to BDSA-based ones due to their similar structure. Therefore, the development of novel sulfonated diamine monomers is strongly desired. In this paper, we report on our recent progress on the development of novel sulfonated diamines and related polyimides and summarize the work in order to make a contribution to the understanding of the relationships between polyimide structure and the proton conductivity and water stability.

2. EXPERIMENTAL SECTION

2.1. Materials

NTDA, 4,4'-diaminodiphenyl ether (ODA), 4,4'-bis(4-aminophenoxy)biphenyl (BAPB), 9,9-bis(4-aminophenyl)fluorene (BAPF), 2,2-bis(4-aminophenyl)-hexafluoropropane (BAPHF), and 4,4'-bis[4-(3-aminophenoxy)phenyl]sulfone (BAPPS) were purchased from Tokyo Kasei Co. NTDA and ODA were purified by vacuum sublimation prior to use. BAPB, BAPF, and BAPHF were used as-received. BAPPS was recrystallized from ethanol prior to use. BDSA, triethyl-amine (Et$_3$N), m-cresol, concentrated sulfuric acid (95%), and fuming sulfuric acid (SO$_3$, 60%) were purchased from Wako Chemical Co. BDSA was soaked in boiling de-ionized water for several hours, filtered while hot, washed thoroughly with boiling water followed by drying at 90°C in vacuum prior to use. Et$_3$N was distilled and dried with molecular sieve 4 A prior to use. m-Cresol was used as received.

2.2. Synthesis of sulfonated diamines

4,4'-diaminodiphenyl ether-2,2'-disulfonic acid (ODADS) was synthesized according to a previously reported method [9]. To a 100 ml 3-neck flask equipped with a mechanical stirring device was charged 2.00 g (10.0 mmol) of 4,4'-diaminodiphenyl ether (ODA). The flask was cooled in an ice-bath, and then 1.7 mL of concentrated (95%) sulfuric acid was slowly added with stirring. After ODA was completely dissolved, 3.5 mL of fuming (SO$_3$ 60%) sulfuric acid was slowly added to the flask. The reaction mixture was stirred at 0°C for 2 h and then slowly heated to 80°C and kept at this temperature for additional 2 h. After cooling to room temperature, the slurry solution mixture was carefully poured into 20 g of crushed ice. The resulting white precipitate was filtered off, and then re-dissolved in a sodium hydroxide solution. The basic solution was filtered and the filtrate was acidified with concentrated hydrochloric acid. The white solid was filtered off, washed with water and methanol successively, and dried at 80°C in vacuum. 3.05 g of white product was obtained (yield: 85%), mp: 265°C.

The above synthesis procedure was followed for the synthesis of 9,9-bis(4-aminophenyl)fluorene-2,7-disulfonic acid (BAPFDS) except that the sulfonation temperature was 60°C as reported in a previous paper [10]. Reddish solid product was obtained. Yield: 83%. m.p. (DSC): 273°C.

4,4'-Bis(4-aminophenoxy)biphenyl-3,3'-disulfonic acid (BAPBDS) was synthesized similarly to BAPFDS. White solid product was obtained. Yield: 93%. m.p. (DSC): 271°C.

2.3. Polymerization

Sulfonated polyimides were prepared by a one-step method in m-cresol. The synthesis procedures for the homopolyimides are described as follows using NTDA-ODADS as an example.

0.540 g (1.5 mmol) of ODADS, 5.0 mL of m-cresol, and 0.36 g (3.6 mmol) of triethylamine were successively added to a 100 ml completely dried 4-neck flask under nitrogen flow with stirring. After ODADS was completely dissolved, 0.402 g of NTDA (1.5 mmol) and 0.26 g (2.13 mmol) of benzoic acid were added. The mixture was stirred at room temperature for a few minutes, and then heated at 80°C for 4 h and 180°C for 20 h. After cooling to about 100°C, additional 10-20 mL of m-cresol was added to dilute the highly viscous solution, which was then poured into 100 mL of acetone. The fiber-like precipitate was filtered off, washed with acetone, and dried in vacuum.

The various random copolyimides were prepared in a similar way to that for the homopolyimides. A typical procedure is described as follows using NTDA-ODADS/ODA(1/1) as an example.

0.360 g (1.0 mmol) of ODADS, 6.0 mL of m-cresol, and 0.24 g (2.4 mmol) of triethylamine were successively added to a 100 ml completely dried 4-neck flask under nitrogen flow with stirring. After ODADS was completely dissolved, 0.200 g (1.0 mmol) of ODA, 0.536 g (2.0 mmol) of NTDA, and 0.34 g (2.8 mmol) of benzoic acid were added. The mixture was stirred at room temperature for a few minutes and then heated at 80°C for 4 h and 180°C for 20 h. After cooling to around 100°C, additional 10-20 mL of m-cresol was added to dilute the highly viscous solution, which was then poured into 100 mL of acetone. The fiber-like precipitate was filtered off, washed with acetone, and dried in vacuum.

2.4. Film formation and proton-exchange

Polyimide films (in triethylammonium salt form) were prepared by the conventional solution cast method. NTDA-BAPBDS, NTDA-ODADS/BAPB(1/1), NTDA-BDSA/ODA(1/1), and NTDA-ODADS/ODA(1/1) copolyimide films were prepared by casting their m-cresol solutions onto glass plates or dishes and dried at 120°C for 10 h. Other polyimide films were prepared by casting their DMSO solutions at 80°C for 10 h. The as-cast films were soaked in methanol at 60°C for 1 h to remove the residual solvent, and then the proton exchange was performed by immersing the films into 1.0 N hydrochloric acid at room temperature for 5-10 h. The films in proton form were thoroughly washed with de-ionized water and then dried in vacuum at 150°C for 20 h. The thickness of the films was in the range of 20-40 μm.

2.5. Measurements

Infrared (IR) spectra were recorded on a Horiba FT-200 spectrometer using KBr pellets. ¹H NMR spectra were recorded on a JEOL EX270 (270 MHz) instrument.

Differential scanning calorimetry (DSC) was performed with a Seiko DSC-5200 at a heating rate of 10°C/min. Thermogravimetry-mass spectrometry (TG-MS) was conducted with a JEOL MS-TG/DTA220 instrument in helium (flow rate: 100 cm^3/min) at a heating rate of 5°C/min. Gel permeation chromatography (GPC) was performed with a HLC-8020 apparatus (column: Shodex KD-80M). NMP was used as the eluent at a flow rate of 1.0 mL/min. Polymer solutions containing 0.05 M LiCl and 0.05 M phosphoric acid were filtered through a 0.5 μm PTFE filter prior to injecting into the column. Molecular weights were calculated against poly(ethylene oxide) standards.

Proton conductivity was measured by an ac impedance method with two black platinum electrodes using Hioki 3552 Hitester instrument over the frequency range from 100 Hz to 100 kHz.

Water sorption experiments were carried out by immersing three sheets of film (20-30 mg per sheet) of the polyimide into water at 80°C for 5 h. Then the films were taken out, wiped with tissue paper, and quickly weighted on a microbalance. Water uptake of the films, S, was calculated from

$$S = (W_s - W_d)/W_d \times 100 \ (\%) \tag{1}$$

where W_d and W_s are the weights of the dry and corresponding water-swollen film sheets, respectively. The average value of three sheets of film was used as the water uptake of the polyimide film.

3. RESULTS AND DISCUSSION

3.1. Monomer synthesis and characterization

All the sulfonated diamines, ODADS, BAPFDS, and BAPBDS, were prepared by direct sulfonation of the corresponding parent diamines, ODA, BAPF, and BAPB, respectively. Primarily the parent diamines reacted with concentrated sulfuric acid to form the sulfuric acid salt of the diamines. SO$_3$ in fuming sulfuric acid was the real sulfonating agent, and the sulfonation reaction was carried out at different temperatures depending on the starting diamine. In the case of ODA, the appropriate sulfonation temperature was 80°C, and the sulfonation reaction could not occur below this temperature. This is because the protonated amino group is a strong electron-withdrawing group which deactivates the phenyl rings. For the same reason, the sulfonation reaction mainly occurred in the meta-position of the amino groups of ODA. The ether bond in the para-position of amino group also supported such a meta-position substitution.

For BAPF, the sulfonation reaction could readily be carried out at 60°C. The two phenyl-rings to which the two amino groups are attached are deactivated due to the strong electron withdrawing effect of the protonated amino groups, whereas the central fluorenylidene ring is highly reactive. As a result, the sulfonation reaction occurred in 2,7-positions of the fluorenylidene ring (these two positions were more reactive than the others), and the reaction temperature was lower than that for ODA. Just like the case of BAPF, BAPB could also readily undergo sulfonation reaction at 60°C. At this temperature the sulfonation reaction selectively occurred at the two central phenyl rings due to their high reactivity.

The chemical structures of all the resulting sulfonated diamines were characterized by IR and ^{1}H NMR spectroscopies.

3.2. Preparation and properties of sulfonated polyimides

As shown in Scheme 1, polymerization of NTDA and the sulfonated diamines (ODADS, BAPFDS, BAPBDS, and BDSA) was carried out by "one-step" method in m-cresol in the presence of triethylamine (Et$_3$N) and benzoic acid. Et$_3$N was used to liberate the protonated amino groups for polymerization with NTDA, and benzoic acid functioned as catalyst. This is a literature method which has been employed for preparation of a series of BDSA-based polyimides [6]. Random copolymerization of NTDA, the sulfonated diamines, and non-sulfonated diamines was also carried out using this method. For comparison purpose, BDSA-based polyimides were also prepared. The as-synthesized polyimides were in the triethylammonium sulfonate form and were converted to the proton form by treating with 1.0 N hydrochloric acid at room temperature. The completion of proton exchange was confirmed by the disappearance of the peaks corresponding to the protons of triethylamine in the ^{1}H NMR spectra of the polyimides (for NTDA-BAPBDS, IR spectroscopy was used instead of ^{1}H NMR spectroscopy because it was insoluble in DMSO-d$_6$ or other solvents). The IR spectra of the sulfonated polyimides were recorded. For NTDA-ODADS, the strong absorption bands around 1717 cm^{-1} and 1671 cm^{-1} are assigned to the stretching vibration of carbonyl groups of the imido rings. The broad band around 1255 cm^{-1} and the band around 1088 cm^{-1} correspond to the stretching vibration of sulfonic acid groups.

Solubility behaviors of the sulfonated polyimides (in triethylammonium salt form) are shown in Table 1. All the polyimides except NTDA-BDSA/BAPB(1/1) are soluble in m-cresol but insoluble in common dipolar aprotic solvents such as 1-methyl-2-pyrrolidinone (NMP) and N,N-dimethylacetamide (DMAc) which are good solvents for many five-membered ring (non-sulfonated) polyimides. BAPFDS-based polyimides are still soluble in DMSO besides in m-cresol; whereas some ODADS- or BDSA-based polyimides are insoluble in DMSO, indicating a better solubility of the former type of polyimides. Proton exchange generally led to significant improvement in solubility of the polyimides. For example, most of the sulfonated polyimides in proton form could be dissolved in NMP by slight heating. NTDA-BAPBDS is an exception, as no large difference in solubility of this polyimide was observed before and after proton exchange.

Thermal stability of the sulfonated polyimides (in proton form) was investigated by TG-MS measurements. For NTDA-ODADS, the weight loss starting from 275°C is due to the decomposition of sulfonic acid groups judging from the evolution of sulfur monoxide and sulfur dioxide, indicating fairly good thermal

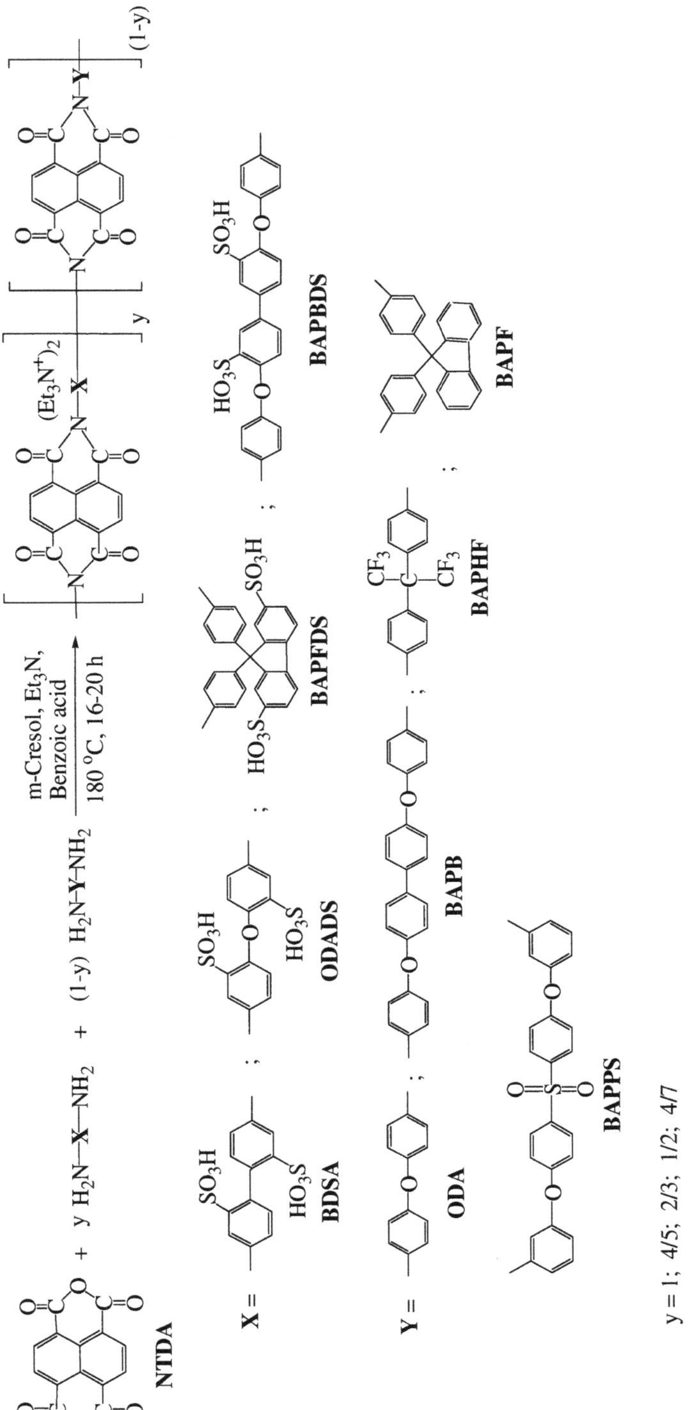
NTDA
+ y H2N—X—NH2 + (1-y) H2N-Y-NH2
m-Cresol, Et3N, Benzoic acid
180 °C, 16-20 h
(Et3N+)2
y
(1-y)
X =
BDSA
ODADS
BAPFDS
BAPBDS
Y =
ODA
BAPB
BAPHF
BAPF
BAPPS
y = 1; 4/5; 2/3; 1/2; 4/7

Scheme 1.

Table 1.
Solubility behaviors of various sulfonated polyimides in their triethylammonium salt forms

Polyimide	m-Cresol	DMSO	NMP
NTDA-ODADS	+	+	–
NTDA-ODADS/ODA(1/1)	+	–	–
NTDA-ODADS/BAPB(1/1)	+	–	–
NTDA-ODADS/BAPF(1/1)	+	+	–
NTDA-ODADS/BAPHF(1/1)	+	+	–
NTDA-BAPFDS	+	+	–
NTDA-BAPFDS/ODA(4/1)	+	+	–
NTDA-BAPFDS/ODA(2/1)	+	+	–
NTDA-BAPFDS/ODA(1/1)	+	+	–
NTDA-BAPFDS/BAPB(4/3)	+	+	–
NTDA-BAPFDS/BAPPS(2/1)	+	+	–
NTDA-BAPBDS	+	–	–
NTDA-BDSA	+	+	–
NTDA-BDSA/ODA(1/1)	+	–	–
NTDA-BDSA/BAPB(1/1)	+ –	–	–
NTDA-BDSA/BAPF(1/1)	+	+	–
NTDA-BDSA/BAPHF(1/1)	+	+	–

"+": soluble, "+ –": partially soluble, "–": insoluble.

Table 2.
Molecular weights and molecular weight distributions of proton exchanged polyimides

Polyimide	Mn	Mw	Mw/Mn
NTDA-ODADS	18000	53000	2.9
NTDA-ODADS/ODA(1/1)	16000	72000	4.5
NTDA-BAPFDS	16000	32000	2.0
NTDA-BAPFDS/ODA(2/1)	17000	39000	2.3
NTDA-BDSA	15000	34000	2.3
NTDA-BDSA/ODA(1/1)	14000	47000	3.4

stability of this polyimide. The sulfonic acid group of NTDA-BAPFDS started to decompose around 270°C.

Table 2 lists the molecular weights and molecular weight distributions of some of the prepared polyimides (in proton form) determined by gel permeation chromatography (GPC). The number-averaged molecular weights (Mn) of these sulfonated polyimides are roughly close to each other. However, the molecular weight distributions are different. NTDA-ODADS/ODA(1/1) showed fairly broad molecular weight distribution in comparison with others.

3.2.1. Proton conductivity

The proton conductivity of the sulfonated polyimide membranes was measured as a function of relative humidity at 50°C as shown in Fig. 1. The conductivity of Nafion 117 has been widely investigated by several groups but the deviation between the results is rather large [11, 12]. The data of Nafion 117 in Fig. 1, which were obtained in this study, are in consistent with those reported by Miyatake *et al.* [11]. The IEC values and the conductivities of the membranes at 80% relative humidity and 50°C are listed in Table 3. It can be seen that the proton conductivities of the polyimide membranes are strongly dependent on both the IEC as well as the relative humidity. For the same sulfonated diamine-based polyimide membranes, the larger IEC resulted in higher proton conductivity. With increasing relative humidity the proton conductivity increased sigmoidally for all the polyimide membranes, which is a typical behavior as has been observed for Nafion 117 and many other sulfonated polymer membranes.

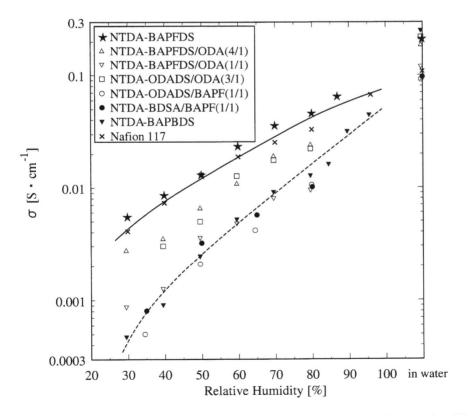

Figure 1. Proton conductivities of various sulfonated membranes as a function of relative humidity at 50°C.

As can be seen that in the relative humidity range from 30 to 95% no polyimide membrane displayed better proton conducting performance than Nafion 117 despite much larger IEC values of the sulfonated polyimide membranes. This is probably because the sulfonated polyimide membranes have homogeneous structure (no significant microphase separation), whereas Nafion has unique ion-rich channels (clusters) which are favorable for proton transport. Nevertheless, it should be noted that the proton conductivities of all the present polyimide membranes are much larger (over one order) than those reported in the literature [6]. Among the prepared polyimides, NTDA-BAPFDS displayed the highest proton conductivity which was comparable to that of Nafion 117.

The sulfonated polyimides displayed much larger conductivities for the membranes equilibrated in liquid water than those at the saturated water vapor because of the much more swollen state of the former, whereas the corresponding increase in the conductivity for Nafion 117 was rather small. As a result, some of the sulfonated polyimides displayed higher proton conductivities in liquid water than Nafion 117.

Fig. 2 shows the temperature dependence of proton conductivity of the polyimide membranes in water. The conductivity of Nafion 117 shown in this figure was determined in our laboratory. For all the membranes the proton conductivity increased with increasing temperature. In addition, unlike the case of the humidity dependence shown in Fig. 1, in water all the polyimide membranes displayed similar or higher proton conductivities than Nafion 117, indicating good proton conducting performance of these membranes. This is probably because in water the polyimide membranes are in a hydrated state and the transport of protons becomes quite easy, making the effect of channels (for facilitation of proton transport) less important.

3.2.2. Water uptake and membrane stability

Table 3 lists water uptake and water stability of the polyimide membranes at 80 or 50°C. It can be seen that, for the same sulfonated diamine-based polyimides, the membranes with higher IEC values due to the higher sulfonic content tended to have larger water uptake. The stability test toward water of the polyimide membranes was performed by immersing the membranes into distilled water at a given temperature (50°C or 80°C) and characterized by the time elapsed until the hydrated membranes lost mechanical properties. The criterion for the judgment of the loss of mechanical properties was that the membrane broke when lightly bent. As can be seen from Table 3, NTDA-BDSA homopolyimide membrane displayed the poorest stability among the prepared membranes. It completely dissolved in water at 50°C within a few seconds. NTDA-ODADS, however, could maintain mechanical strength after being soaked in water for 10 minutes at the same temperature, and it took more than 2 h to completely dissolve in water. This indicates that NTDA-ODADS is much more stable toward water than NTDA-BDSA despite their similar IEC values. Another kind of homopolyimide, NTDA-BAPFDS, was insoluble in water at 50°C and could maintain mechanical strength even after being soaked in water at this temperature for 50 h indicating significantly improved water stability in comparison with NTDA-ODADS. However, this polyimide could still dissolve in water at high temperature (80°C) because the IEC value was still at high level. In contrast, NTDA-BAPBDS polyimide membrane had very good water stability despite its high IEC (close to that of NTDA-BAPFDS). It could maintain mechanical properties even after being soaked in water at 80°C for more than 1000 h.

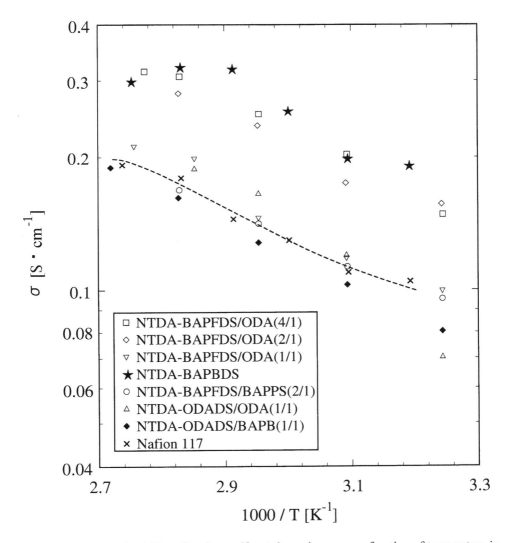

Figure 2. Proton conductivities of various sulfonated membranes as a function of temperature in water.

Table 3.
Thickness, water uptake, and water stability of various sulfonated polyimide membranes

Membrane	Thickness [μm]	IEC [meq g⁻¹]	Water uptake[a] [%w/w]	σ[a] [S cm⁻¹]	Membrane stability test		
					T [°C]	Time	Stability[b]
NTDA-ODADS	22	3.37	—	0.039	50	10 minutes	O
					50	2 h	Dissolved
NTDA-ODADSA/OD(3/1)	44	2.70	113[c]	0.022	80	5 h	x
					80	15 h	Dissolved
NTDA-ODADS/ODA(1/1)	34	1.95	87	0.017	80	25 h	O
NTDA-ODADS/BAPB(1/1)	36	1.68	57	0.011	80	200 h	O
NTDA-ODADS/BAPHF(1/1)	31	1.73	72	0.012	80	11 h	x
NTDA-ODADS/BAPF(1/1)	37	1.71	69	0.011	80	13 h	x
NTDA-BAPFDS	40	2.70	122	0.045	50	50 h	O
					80	5 h	Dissolved
NTDA-BAPFDS/ODA(4/1)	15	2.36	100	0.024	80	6 h	x
NTDA-BAPFDS/ODA(2/1)	28	2.09	76	0.020	80	20 h	O
NTDA-BAPFDS/ODA(1/1)	30	1.71	78	0.0094	80	26 h	O
NTDA-BAPFDS/BAPB(4/3)	30	1.68	56	NM	80	27 h	O
NTDA-BAPFDS/BAPPS(2/1)	25	1.87	60	NM	80	20 h	O
NTDA-BAPBDS	20	2.63	107	0.020	80	1000 h	O
NTDA-BDSA	20	3.47	—	0.038[d]	50	A few seconds	Dissolved
NTDA-BDSA/ODA(1/1)	34	1.98	79	0.013	80	5.5 h	x
NTDA-BDSA/BAPHF(1/1)	28	1.75	61	0.012	80	6 h	x
NTDA-BDSA/BAPF(1/1)	27	1.73	63	0.010	80	6 h	x

[a]Water uptakes in water at 80°C; Proton conductivity at 80% relative humidity and 50°C.
[b]O: mechanical strength was maintained; x: somewhat brittle.
[c]measured at 50°C.
[d]From ref. 8 at 75% relative humidity and 25°C.
—: could not be measured. NM: not measured.

In order to obtain more information on the relationship between the chemical structure of the sulfonated polyimides and their water stability, various copolyimides were prepared and their water stability was measured. As shown in Table 3, decreasing the sulfonation degree by incorporation of non-sulfonated diamines to the polyimide structure led to a significant improvement in water stability of the membranes. All the copolyimide membranes were insoluble at elevated temperature (e.g. 80°C). However, there is still a large difference in water stability between these copolyimide membranes. NTDA-BDSA/ODA(1/1), for example, became somewhat brittle (the membrane broke when lightly bent) after being soaked in water at 80°C for 5.5 h; whereas NTDA-ODADS/ODA(1/1) membrane remained very tough. It did not become brittle at all even after it was soaked in water at the same temperature for 25 h. This clearly indicates that ODADS-based copolyimide membranes are much more stable than the corresponding BDSA-based ones despite their similar IEC values. Usually water stability of polymer membranes is strongly dependent on the water uptake which is mainly determined by the IEC. Membranes with lower water uptake should have better water stability. However, in the present case, the difference in water stability between NTDA-ODADS/ODA(1/1) and NTDA-BDSA/ODA(1/1) is not because of the difference in water uptake as the former displayed even slightly larger water uptake than the latter. A comparison of the chemical structures revealed that these two kinds of copolyimides had quite different chain flexibilities. BDSA moiety is highly rigid because the two phenyl rings cannot rotate along the axis due to the steric effect of the two sulfonic groups. In contrast, ODADS is fairly flexible due to the existence of a flexible linkage, ether bond (Fig. 3), and therefore the backbone of NTDA-ODADS/ODA(1/1) should be much more flexible than that of NTDA-BDSA/ODA(1/1). A flexible chain can undergo relaxation more easily than a rigid one, and this is likely the main reason that NTDA-ODADS/ODA(1/1) displayed much better water stability than NTDA-BDSA/ODA(1/1). This effect has also been observed for non-sulfonated five-membered ring polyimides, i.e., the rod-like polyimides showed much poorer hydrolysis stability than the flexible ones [13]. The chain flexibility of sulfonated polyimide is also affected by the non-sulfonated diamine moieties. NTDA-ODADS/BAPF(1/1) and NTDA-ODADS/BAPHF(1/1) copolyimides have less flexible structure than NTDA-ODADS/ODA(1/1), and their water stability is much poorer than that of the latter despite their lower IEC values and lower water uptake. This indicates that water stability of the membranes not only depends on the flexibility of the sulfonated diamine moiety but is also affected by the flexibility of the non-sulfonated diamine moiety, i.e., water stability is determined by the flexibility of the whole polymer chain. NTDA-BDSA/BAPF(1/1) and NTDA-BDSA/BAPHF (1/1) have more rigid structure and, therefore, even poorer water stability than the corresponding ODADS-based ones. However, it should be noted that for BDSA-based copolyimides (i.e., "rigid"-type) the replacement of the non-sulfonated diamine moieties did not cause a large change in water stability of the membranes. NTDA-BDSA/ODA(1/1), for example, displayed quite similar water stability to that of

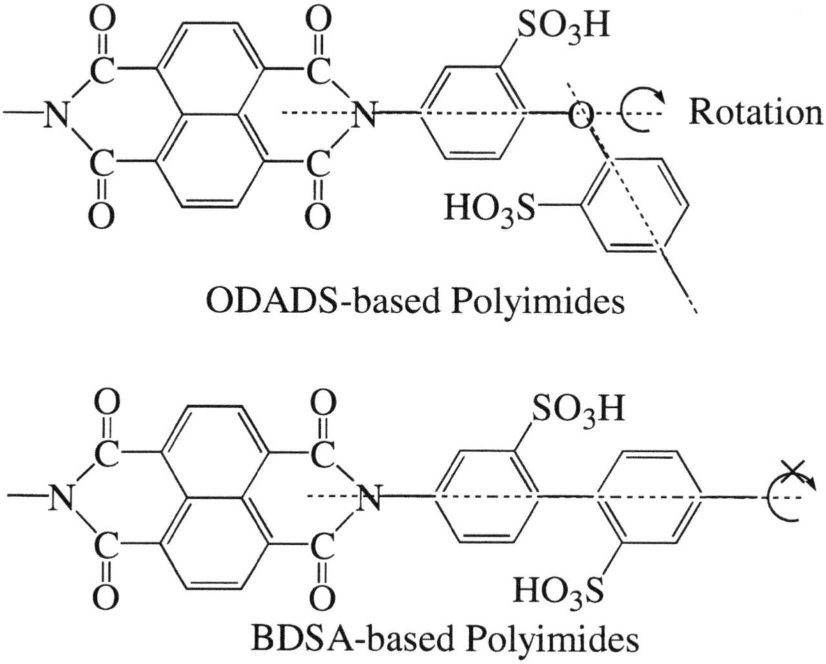

ODADS-based Polyimides

BDSA-based Polyimides

Figure 3. Schematic diagrams showing different intrasegmental mobilities for ODADS and BDSA-based polyimides.

NTDA-BDSA/BAPF(1/1) and NTDA-BDSA/BAPHF (1/1) despite a less rigid structure of the former; i.e., the favorable effect of the less rigid structure of NTDA-BDSA/ODA(1/1) on water stability was offset by the unfavorable effect of the higher water uptake. In contrast, for the "flexible"-type copolyimides, the reduction of the IEC (and therefore the water uptake) by changing the non-sulfonated diamine moiety led to great improvement in water stability of the membranes. NTDA-ODADS/BAPB(1/1) displayed the best water stability among ODADS-based polyimide membranes because of its highly flexible structure and the lowest IEC.

On the other hand, although BAPFDS is a typical rigid and bulky diamine monomer, the stability of BAPFDS-based copolyimide membranes is not so poor as expected. In fact, BAPFDS-based copolyimide membranes displayed much better water stability than BDSA-based ones with the same non-sulfonated dia-mine moiety (ODA) and similar IEC. As shown in Table 3, NTDA-BAPFDS /ODA(2/1) had much better water stability than NTDA-BDSA/ODA(1/1) despite slightly larger IEC of the former. NTDA-BAPFDS/ODA(1/1) and NTDA-ODADS/BAPF(1/1) had the same IEC and chain flexibility, and the difference in structures between the two kinds of copolyimides is only that the sulfonic acid groups are attached to different diamine moieties. However, NTDA-BAPFDS

/ODA(1/1) displayed much better water stability than NTDA-ODADS /BAPF(1/1). This clearly indicates that the stability of polyimide membranes not only depends on the flexibility of polymer chains and the IEC but also depends on other factor(s). A common structural feature between ODADS and BDSA is that the sulfonic acid groups are directly attached to the phenyl rings to which the amino groups are attached, and here they are noted as "type 1" sulfonated diamines. Unlike the case of ODADS and BDSA, the sulfonic acid groups of BAPFDS are attached to the bridged phenyl rings, and it is noted as "type 2" sulfonated diamine (Scheme 2). Because the sulfonic acid group is a strong electron-withdrawing group, the electron density of the phenyl rings to which the amino groups are attached should be larger for BAPFDS than for ODADS or BDSA, i.e., BAPFDS is more basic than ODADS or BDSA. It is well known that aromatic diamines with higher basicity are generally more reactive with dianhydrides than those with lower basicity [14]. Since hydrolysis is the reverse reaction of polymerization, polyimides derived from more basic diamines should have higher hydrolysis stability. As a result, the high basicity of BAPFDS is favorable for maintaining the stability of the imido rings, which offsets the unfavorable effect on water stability due to its rigid structure, and this might be the reason that BAPFDS-based polyimide membranes had much better water stability than BDSA-based ones with similar IEC. BAPBDS is a "type 2" and highly flexible sulfonated diamine, and therefore NTDA-BAPBDS polyimide membrane displayed very high water stability in comparison to others.

It should be noted that in the literature [6, 15] the criterion for the judgment of the loss or maintaining mechanical properties of membranes is described by the statements, "the membranes become highly brittle" or "keep their form", which are not so strict as that proposed in this paper. Using the literature description, the time for characterization of the stability of the copolyimide membranes should be much longer than these listed in Table 3. NTDA-ODADS/ODA(1/1) membrane, for example, did not break into pieces (i.e., kept its form) after being soaked in water at 80°C for more than 200 h even by vigorously shaking the bottle where the membrane and distilled water were charged.

Membrane stability to oxidation was also investigated. The copolyimide membranes (the size of each sheet: 0.5×1.0 cm^2) were soaked in Fenton's reagent (30 ppm FeSO$_4$ in 30% H$_2$O$_2$, see ref. 15) at room temperature. Oxidative stability of the membranes was characterized by the time elapsed when the membranes started to become slightly brittle (the membranes broke when bent) or started to dissolve in the solution. As shown in Table 4, the sulfonated copolyimide membranes displayed fairly good stability to oxidation, which is much better than that of the sulfonated arylene ether/fluorinated alkane copolymers reported in the literature [15]. In addition, the oxidative stability of NTDA-BAPBDS polyimide membrane is relatively poorer than that of others.

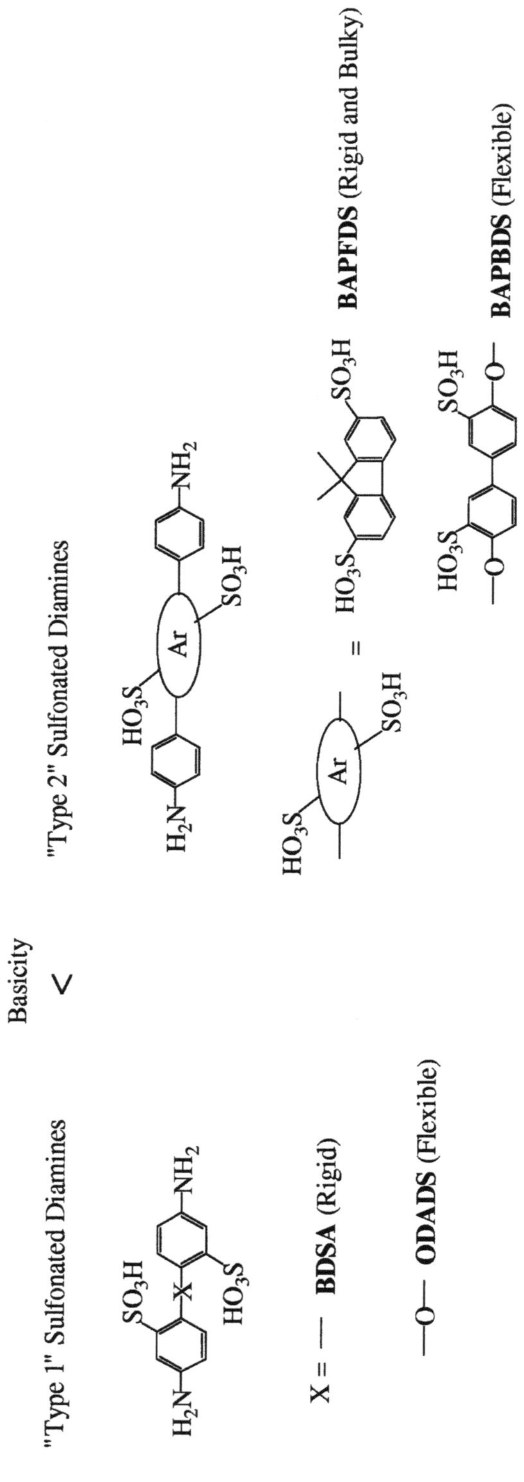

"Type 1" Sulfonated Diamines

"Type 2" Sulfonated Diamines

Basicity

<

X = — **BDSA** (Rigid)

—O— **ODADS** (Flexible)

BAPFDS (Rigid and Bulky)

BAPBDS (Flexible)

Scheme 2.

Table 4.
Results of Fenton's reagent test of various sulfonated polyimide membranes at room temperature

Membrane	Thickness [μm]	τ_1* [h]	τ_2* [h]
NTDA-ODADSA/ODA(1/1)	29	20	24
NTDA-ODADS/BAPB(1/1)	37	29	32
NTDA-ODADS/BAPF(1/1)	40	29	32
NTDA-ODADS/BAPHF(1/1)	26	24	29
NTDA-BAPFDS/ODA(4/1)	58	17	21
NTDA-BAPFDS/ODA(1/1)	23	18	22
NTDA-BAPFDS/BAPB(4/3)	31	22	26
NTDA-BAPFDS/BAPPS(2/1)	26	21	26
NTDA-BAPBDS	19	13	25
NTDA-BDSA/ODA(1/1)	21	13	20
NTDA-BDSA/BAPF(1/1)	34	23	26
NTDA-BDSA/BAPHF(1/1)	25	18	20

*τ_1 and τ_2 refer to the time elapsed when the membranes became slightly brittle and started to dissolve in the solution, respectively.

4. CONCLUSIONS

1. Three kinds of sulfonated diamine monomers ODADS, BAPFDS, and BAPBDS were successfully synthesized by direct sulfonation of the parent diamines. Various sulfonated polyimides were prepared from NTDA, ODADS, BAPFDS, BAPBDS, and non-sulfonated diamines.

2. At relative humidities below 100%, the proton conductivities of the sulfonated polyimide membranes are generally lower than that of Nafion 117 in water, most of the sulfonated polyimides showed similar or higher proton conductivities than Nafion 117.

3. Water stability of the sulfonated polyimide membranes is greatly affected by the flexibility of polymer chains, i.e., a flexible chain structure results in good water stability. ODADS-based polyimide membranes displayed similar proton conductivities but much better water stability than the corresponding BDSA-based ones because of the more flexible structure of the former.

4. Besides the flexibility and the IEC, the basicity of the sulfonated diamine moieties also has a great effect on water stability of the polyimide membranes. BAPFDS-based polyimide membranes displayed much better water stability than the corresponding BDSA-based ones with similar IEC because for the former the unfavorable effect due to the rigid structure was offset by the favorable effect due to the high basicity of BAPFDS.

5. NTDA-BAPBDS polyimide membrane displayed very good water stability in comparison to others because of the highly flexible structure and the high basicity of BAPBDS.

REFERENCES

1. M.K. Ghosh and K.L. Mittal (Eds.), *Polyimides: Fundamentals and Applications*, Marcel Dekker, New York (1996).
2. K.L. Mittal (Ed.), *Polyimides and Other High Temperature Polymers: Synthesis, Characterization and Applications*, Vol. 1, VSP, Utrecht (2001).
3. S. Faure, R. Mercier, P. Aldebert, M. Pineri and B. Sillion, *French Pat.* 9605707 (1996).
4. S. Faure, N. Cornet, G. Gebel, R. Mercier, M. Pineri and B. Sillion, *Proceedings of Second International Symposium on New Materials for Fuel Cell and Modern Battery Systems*, O. Savadogo and P.R. Roberge (Eds.), held at Montréal, Canada, July 6-10, p. 818 (1997).
5. E. Vallejo, G. Pourcelly, C. Gavach, R. Mercier and M. Pineri, *J. Membrane Sci.* **160**, 127 (1999).
6. C. Genies, R. Mercier, B. Sillion, N. Cornet, G. Gebel and M. Pineri, *Polymer* **42**, 359 (2001).
7. Y. Zhang, M. Litt, R.F. Savinell and J.S. Wainright, *Polym. Prepr.(Am. Chem. Soc., Div. Polym. Chem.)* **40(2)**, 480 (1999).
8. H. Kim and M. Litt, *Polym. Prepr. (Am. Chem. Soc., Div. Polym. Chem.)* **42(2)**, 486 (2001).
9. J. Fang, X. Guo, S. Harada, K. Tanaka, H. Kita and K. Okamoto, *Macromolecules* **35**, 9022 (2002).
10. X. Guo, J. Fang, T. Watari, K. Tanaka, H. Kita and K. Okamoto, *Macromolecules*, **35**, 6707 (2002).
11. N. Miyatake, J. Wainright and R. Savinell, *J. Electrochem. Soc.* **148**, A898 (2001).
12. Y. Sone, P. Ekdunge and D. Simonsson, *J. Electrochem. Soc.* **143**, 1254 (1996).
13. C.E. Sroog, *Prog. Polym. Sci.* **16**, 561 (1991).
14. T. Takekoshi, in *Polyimides: Fundamentals and Applications,* M.K. Ghosh and K.L. Mittal (Eds.), p. 7, Marcel Dekker, New York (1996).
15. K. Miyatake, K. Oyaizu, E. Tsuchida and A.S. Hay, *Macromolecules*, **34**, 2065 (2001).

Polyimides and Other High Temperature Polymers, Vol. 2, pp. 155–163
Ed. K.L. Mittal
© VSP 2003

High-modulus poly(p-phenylenepyromellitimide) films obtained using a novel gel-drawing technique

JIRO SADANOBU*

Polymer Research Institute, Teijin Limited, 2-1 Hinode-cho, Iwakuni Yamaguchi 740-8411, Japan

Abstract—A new concept for the processing and fabrication of rigid-rod polyimide, poly(p-phenylenepyromellitimide), (PPPI) is proposed. In this approach, the highly swollen polyisoimide gel-film was used as a precursor and was biaxially stretched using gel-drawing technique. The gel-drawn polyisoimide film was converted to a polyimide film by heat treatment at above 350°C. The gel-drawn PPPI film possesses a tensile strength of 500 MPa and a Young's modulus of 20 GPa.

Keywords: Poly(p-phenylenepyromellitimide); polyimides; polyisoimides; gel-drawing; rigid-rod polyimides.

1. INTRODUCTION

There has been an increasing demand for high-performance organic films with dimensionally, thermally and mechanically improved properties from the viewpoint of emerging electronics applications [1]. Current trend of downsizing the electronics components, such as printed circuits, IC carrier tapes, chip scale packagings, magnetic recording tapes and so on, additionally requires the thickness to be smaller than 10 μm but still maintaining good processability. For conventional organic films such a low thickness range has not been reached because reducing the thickness causes difficulties in material handling. To remedy the situation, development of high-modulus organic thin films is highly promising. Although conventional polyimide film has been widely used in the field of electronics, its Young's modulus is limited to 10 GPa, which is not enough for thin film applications. In the last decade some Aramid films with enhanced Young's modulus have been developed, however, the low hygroscopic stability limits their applications in electronics [2]. Therefore, the development of high performance organic films with high Young's modulus is still a challenge.

Poly(p-phenylenepyromellitimide), (PPPI) is the simplest form of crystalline polyimides, depicted in Figure 1, prepared from pyromellitic anhydride and p-

---

*Phone: +81-6-6268-2611, Fax: +81-6-6268-2633, E-mail: j.sadanobu@teijin.co.jp

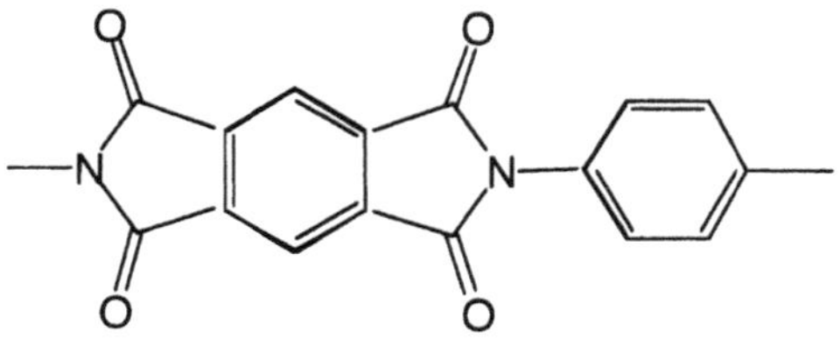

Figure 1. Repeat unit of PPPI.

phenylene diamine. It has fully rigid rod-like conformation with an estimated theoretical Young's modulus of 505 GPa [3] that exceeds any of current commercial polymers [4]. Although this well-known polymer should be a candidate for a high modulus film, it has never been employed for industrial usages since the materials fabricated from PPPI are too brittle to handle [5]. There have been many efforts to improve the mechanical properties of PPPI, however almost all efforts resulted in failure. Some elaborated works on uniaxially oriented fibers and films of PPPI succeeded in attaining relatively high Young's modulus, however could never overcome the brittleness [6, 7].

In this paper we report the first successful formation of high modulus PPPI thin films. This work is based on a novel gel-drawing technique for specially designed precursor gel-film.

2. CONCEPT

One of the causes for the brittleness of PPPI is hydrolytic instability of the precursor poly(amic acid) that is easily degraded by ambient moisture and even by water generated on thermal imidization. Another is the coarse grained structure developed due to intrinsic high-crystallinity of PPPI, that may induce stress concentration on application of force. To overcome these problems we came up with a new concept of film fabrication, i.e., *the gel-drawing of precursor polyisoimide gel film*. Polyisoimide is an isomer of polyimide with semi-flexible backbone structure [8-10]. In this approach, the highly swollen polyisoimide gel-film was used as precursor and was biaxially stretched using gel-drawing technique.

The flow chart for the new film formation procedure is illustrated in Figure 2. Since the precursor polyisoimide for PPPI was not soluble in any organic solvent, the solution of poly(amic acid) in N-methyl-2-pyrrolidone (NMP) was used for film casting. The cast film of poly(amic acid) was immersed into the solution of N,N'-dicyclohexyl carbodiimide (DCC) in NMP and was converted to polyisoimide gel-film. The reaction scheme for isoimidization is shown in Figure 2. Polyisoimide gel-film is subjected to a simultaneous biaxial gel-drawing at room temperature. After the gel-drawing, polyisoimide is thermally converted to polyimide with no generation of water on heating [9]. Finally, the heat-treatment at above 350°C provides a highly oriented PPPI film with a high degree of crystallization.

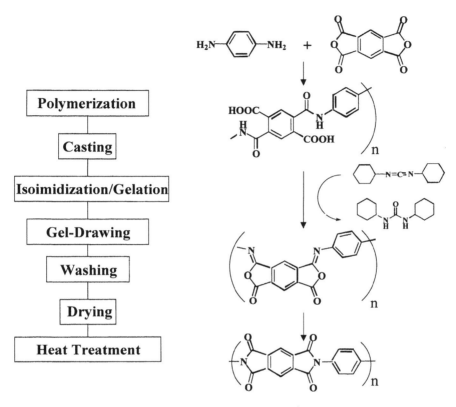

Figure 2. PPPI gel-drawing process and reaction scheme.

3. PROCEDURE

3.1. Polymerization

Pyromellitic anhydride (PMA) was added to an NMP solution of p-phenylenedi-amine (PPD) under N_2 at 0°C with stirring. The mole ratio of PMA/PPD was 1.0/1.0 and the solid content in the reaction mixture was 6 wt%. After solution was stirred for three hours, further reaction was continued for two hours at room temperature. The resultant solution of poly(amic acid) was directly used for film casting. The inherent viscosity of the poly(amic acid) was typically 4.5 dl/g (0.5 g/dl in NMP at 30°C).

3.2. Gelation

The poly(amic acid) solution film was cast on a glass plate using a knife blade, followed by immersing into a gelation bath containing 15 wt% solution of DCC in NMP at room temperature. DCC diffused into the cast film from the solution in the gelation-bath and the dehydration reaction of poly(amic acid) occurred. The

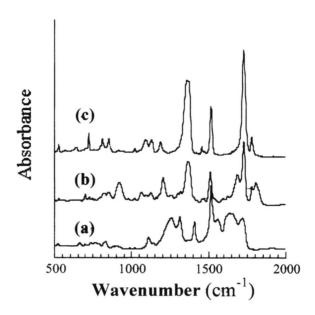

Figure 3. FT-IR spectra of (a) as-cast film, (b) gel-film and (c) gel-drawn film after heat treatment at 350°C; spectra of as-cast film and gel-film were recorded after removing the solvent by extraction.

cast film immediately turned reddish and formed a highly swollen gel. The reddish color of the gel-film indicates conversion of poly(amic acid) to polyisoimide. The FT-IR spectrum of gel-film shows clear isoimide bands at 910 cm^{-1} and 1803 cm^{-1} without bands of poly(amic acid) (see Figure 3) [10, 11]. The control of the degree of swelling is a key issue for determining the stretchability of gel-film. Using RT or lower temperature for gelation minimizes the dimensional contraction during the dehydration reaction and attains a high degree of swelling for the resultant gel-film. After gelation was completed by carrying out the reaction at RT, typical polymer to solvent ratio was 1:18 by weight. After gelation the gel-film was stripped from the glass plate and washed with NMP to remove the unreacted DCC and the by-product N,N'-dicyclohexylurea.

3.3. Gel-drawing

The resultant highly swollen polyisoimide gel-film was self-supporting and had rubber-like ductility that enabled the gel-film to be stretched even at RT. The gel-film was simultaneously drawn in two perpendicular directions in dry air. (In this paper biaxial draw ratio of (n) x (n) is referred to as λ=n^2.) The typical maximum biaxial draw-ratio used at RT was 5.5. The deformation at gel-drawing was completely uniform without any necking. Stretching at RT keeps a high degree of swelling throughout gel-drawing process. Almost no conversion of polyisoimide in gel-film to polyimide was detected after stretching. The drawn film was washed with isopropanol to remove NMP.

3.4. Heat treatment

The drawn gel-film was dried in dehumidified air by fixing the ends of the film to a metal frame to prevent shrinkage. The dried film was subjected to high temperature heat treatment to convert polyisoimide to polyimide under the prescribed heating condition of 200°C/10 min, 250°C/8 min, 350°C/5 min and 350 to 450°C/5 min. In this report the heat treatment temperature (HTT) represents the final treatment temperature. Imidization gradually occurred above 200°C and completed at around 300°C. The conversion to polyimide was confirmed by FT-IR spectroscopy as shown in Figure 3(c). The final thickness of studied PPPI film was 8.0 μm.

4. HIGH-MODULUS PPPI FILMS

Figure 4 shows the dependence of tensile properties of heat-treated PPPI films on gel-draw ratio, where HTT was 450°C. Undrawn films are too brittle to carry out tensile test after heat treatment. Both tensile strength and modulus dramatically increase with increasing draw ratio; especially the improvement in tensile strength is quite dramatic. One can see that a tensile strength of 500 MPa and a Young's Modulus of 20 GPa are simultaneously realized at a draw ratio of 5.0. We can conclude that the gel-drawing technique overcomes the problem of intrinsic brittleness of PPPI polymer. Figure 5 shows dependence of Young's modulus on the heat treatment temperature. Young's modulus of PPPI almost saturated at 350°C, which corresponds to completion of conversion from polyisoimide to polyimide.

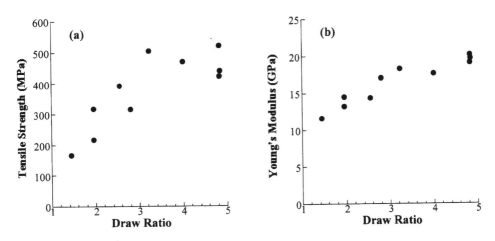

Figure 4. Dependence of (a) tensile strength and (b) Young's modulus for gel-drawn (λ=4.8) PPPI film on gel-draw ratio.

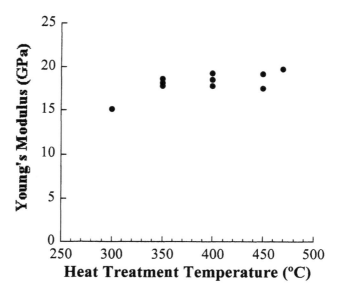

Figure 5. Dependence of Young's modulus for gel-drawn (λ=4.8) PPPI film on heat treatment temperature.

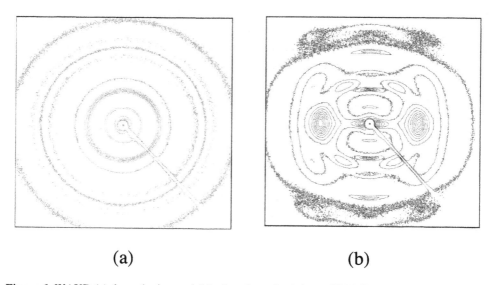

(a)   (b)

Figure 6. WAXD (a) through view and (b) edge view of gel-drawn PPPI film.

During heat treatment, crystallization of PPPI takes place to form higher order structure. Figure 6 illustrates WAXD pattern for gel-drawn PPPI film after heat treatment at 450°C. Sharp (00L) reflections (L=1-8) are observed in the through view (see Figure 6(a)). The strong diffraction intensity of the higher order reflection indicates a completeness of translational symmetry along the polymer fiber axis, which is associated with the rigid-rod configuration of PPPI chain [5, 11]. In edge view, the peaks from (00L) reflections are observed along the meridional axis (see Figure 6(b)), which evidences high planar orientation of the fiber axis of PPPI crystal in the film plane. On the other hand, only a broad peak is observed in the equatorial direction, which indicates restricted lateral crystal growth.

Figure 7 shows an SEM image of the cross section of the PPPI film. An elongated thin lamellar structure aligned parallel to the film surface is observed. The thickness of a lamella is roughly estimated to be 10 nm or less. This kind of lamella-like nanostructure has never been observed for conventional polyimide films and is unique to gel-drawn PPPI film.

We assume that dramatic improvement in tensile strength for gel-drawn PPPI film could be attributed to the following three aspects attained by the gel-drawing technique of the precursor polyisoimide gel-film.

(1) Minimal hydrolytic degradation preserves the high molecular weight after completion of imidization.

(2) Highly planar orientation and restricted lateral crystal growth induce the formation of lamella-like nanostructure.

(3) Controlled nanostructure results in suppressing stress concentrations on application of force.

Figure 7. SEM image of the cross section of gel-drawn PPPI film.

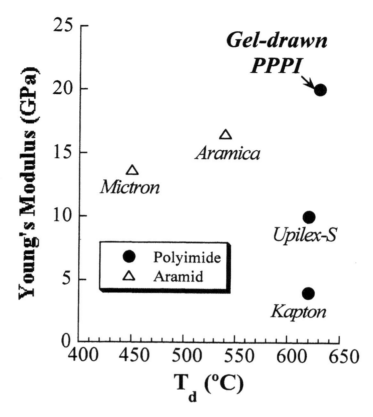

Figure 8. Comparison of gel-drawn PPPI film with commercial heat-resistant films.

The performance of gel-drawn PPPI film is compared with commercially available heat-resistant polymer films in Figure 8 from the viewpoint of thermal and mechanical properties. Here thermal degradation temperature (T_d) represents the temperature at which 5% weight loss has occurred in thermo-gravimetric analysis. One can see that the gel-drawn PPPI film possesses both the highest Young's Modulus and the highest T_d.

5. CONCLUSION

The PPPI was once discarded because of its intractable nature. Novel technique of gel-drawing overcomes the earlier problems. The breakthrough is based on choosing a special polyisoimide precursor and designing highly swollen gel structure. Gel-drawing induces highly planar orientation of polymer chains with restricted lateral crystal growth, which results in much improved tensile strength and high Young's modulus.

REFERENCES

1. 'Japan Jisso Technology Roadmap 2001', Japan Electronics & Information Technology Industries Association, Tokyo (2001).
2. T. Fujita, T. Fujiwara, E. Sato and K. Nagasawa, Polym. Eng. Sci., **29**, 1237 (1989).
3. K. Tashiro and M. Kobayashi, Sen-i Gakkaishi, **43**, 78 (1987).
4. K. Tashiro, M. Kobayashi and H. Tadokoro, Macromolecules, **10**, 413 (1977).
5. L.N. Korzhavin, N.R. Prokopchuk, Yu.G. Baklagina, F.S. Florinskii, N.V. Yefanova, A.M. Dubnova, S.Ya. Frenkel and N.M. Koton, Vysokomol. Soedin., **A18**(6), 1235 (1976).
6. M. Kakimoto, H. Orikabe and Y. Imai, Polymer Preprints, **34** (1), 746 (1993).
7. A. Masuda, S. Kotobuki, S. Nakamura, S. Oshida, M. Kochi and R. Yokota, Kobunshi Ronbunshu, **56** (5), 282 (1999).
8. J. Zurakowska-Orszagh, A. Orzeszko and T. Chreptowicz, Euro. Polym. J., **16**, 289 (1980).
9. A.K. Saini, C.M. Carlin and H.H. Patterson, J. Polym. Sci.: Part A: Polym. Chem., **31**, 2751 (1993).
10. J.S. Wallace, L-S. Tan and F.E. Arnold, Polymer, **31**, 2411 (1990).
11. J-H. Jou and P-T. Huang, Macromolecules, **24**, 3796 (1991).

Polyimides and Other High Temperature Polymers, Vol. 2, pp. 165–183
Ed. K.L. Mittal
© VSP 2003

Effect of structure on the thermal behaviour of bisitaconimide resins

INDRA K. VARMA*

Centre for Polymer Science and Engineering, Indian Institute of Technology, Delhi, Hauz Khas, New Delhi-110016, India

Abstract—Many addition polyimides such as bismaleimides and bisnadimides have been extensively investigated in the past. These resins are currently being used as matrix resins for advanced fibre-reinforced composite materials, which find applications in the aerospace, defence and electronics industries. As one of the thermoset materials, these polymers offer superior thermal oxidative stability and better retention of properties in hot wet environments than the conventional epoxy resins.

The potential of bisitaconimide resins which cure at lower temperatures than bismaleimide or bisnadimide resins and yield a cross-linked network with thermal stability comparable to other types of addition polyimides has not been fully explored. In this paper, recent studies on the synthesis and characterisation of novel itaconimide resins are described. The effect of structure on thermal behaviour is discussed.

Keywords: Bisitaconimide; biscitraconimide; isomerisation; thermosetting polyimides; char yield.

1. INTRODUCTION

Addition polyimides, such as 5-norbornene-2, 3-dicarboximide (nadimide), maleimide, or ethynyl-terminated imide resins, are a leading class of thermosetting polyimides. Since the early 1970's these resins have been extensively investigated as matrix resins for advanced fiber-reinforced composites, adhesives and in multilayer circuit boards in the electronic industry [1].

Although extensive studies have been reported on nadimide and maleimide resins, very few papers have been published on imide derivatives of itaconic anhydride [2]. This may be due to higher material cost as well as due to some side reactions during their synthesis (such as isomerisation to citraconimides). However, there are some advantages of bisitaconimides. Nadimides and maleimides are prepared from anhydrides based on petrochemicals whereas itaconic acid is obtained from easily renewable resources such as *corn starch* by fermentation processes using *Aspergillus itaconicus* or *Aspergillus terreus fungi* [3]. The dwindling petrochemical feedstocks have necessitated the need to focus attention on devel-

---

*Phone: +91-11-650-1425, Fax: +91-11-659-1421, E-mail: ikvarma@hotmail.com

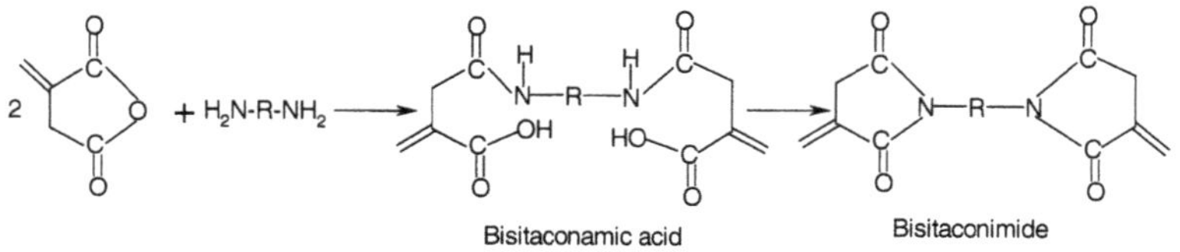

Scheme 1. Reaction sequence for the synthesis of bisitaconimides.

oping polymers based on annually renewable resources. Itaconic acid derivatives homopolymerise by an addition-type reaction more readily than 1,2-ethylene derivatives (e.g. maleimide or nadimide) resins. There is thus a need to investigate addition polyimides based on these monomers.

The main focus of present studies was to examine the effect of structure on curing characteristics and thermal behaviour of itaconimide resins. The synthesis of bisitaconimides was carried out in two steps according to the reaction Scheme 1.

Bisitaconimides of varying backbone structure and varying molecular weight were synthesized by using aromatic amines having ether, sulfone, phoshine oxide, amido group or fused aromatic ring structure. The effect of partial isomerisation of bisitaconimide to biscitraconimide on thermal behaviour was also investigated.

2. EXPERIMENTAL

2.1. Materials

Chloroform, toluene (Qualigens), phosphorus pentoxide (P_2O_5), N,N'-dimethyl formamide (Merck), itaconic acid (Aldrich), 4,4'-diaminodiphenyl sulfone (S_p), 3,3'-diaminodiphenyl sulfone (S_m), 4,4'-diaminodiphenyl ether (Fluka) (E), 1,3-bis(3-aminobenzamide)benzene (A) and 9,9-bis(4-aminophenyl)fluorene (Monsanto) (F) were used as received. Acetone (Qualigens) was kept over anhydrous potassium carbonate for 24 h and distilled before use. Tetrahydrofuran (THF) (Central Drug House, New Delhi) was refluxed over stannous chloride and then distilled before use. Acetic anhydride (British Drug House, London) was distilled before use and sodium acetate (Sarabhai Chemicals, Vadodara) was fused by heating and stored in a dessicator.

The amines 2,2'-bis[4-(4-aminophenoxy) phenyl] propane (P), 1,3-bis(4-aminophenoxy) benzene (R), 1,4-bis (4-aminophenoxy) benzene (H) [4], tris(3-aminophenyl)phosphine oxide (T) and bis(3-aminophenyl)methyl phosphine oxide (B) were prepared according to the procedure reported elsewhere [5, 6].

2.2. Synthesis of bisitaconimides

Itaconic anhydride was prepared from itaconic acid by using phosphorus pentoxide (P_2O_5) as a dehydrating agent and chloroform as solvent with yield = 80%, melting point = 69-70°C.

In a typical synthesis of bisitaconimides, 0.005 mole of diamine was dissolved in 20 mL acetone and itaconic anhydride (0.011 mol) was added slowly with stirring. The contents were heated at 60°C for 4-5 h. Cyclodehydration of itaconamic acid was carried out using acetic anhydride and fused sodium acetate.The contents were heated for 4-5 h at 60°C. The imides were precipitated in ice-cold water, filtered, washed successively with water, sodium bicarbonate solution, water, and dried in vacuum oven at 50-75°C. The purification was done by column chromatography using a silica column and chloroform as an eluent. Bisitaconimides having sulfone and phosphine oxide groups were prepared by refluxing anhydride and amine in glacial acetic acid.

2.3. Characterisation

Structural characterisation was done by recording the IR spectra in a film form or in KBr pellets using BioRad Digilab FTS-40 FT-IR spectrometer.

^{1}H-NMR spectra were recorded on a Bruker AC 300 MHz spectrometer using DMSO-d_6 / CDCl$_3$ as solvents and tetramethylsilane (TMS) as an internal standard.

A DuPont 2100 thermal analyser having 910 DSC and 951 TG modules was used for thermal characterisation. DSC scans were recorded in static air atmosphere using 7 ± 2.5 mg of sample in a temperature range of 50-350°C. The characteristic melting temperature (T_m), onset temperature of curing (T_1), exothermic peak temperature (T_{exo}) and temperature of completion of exotherm (T_2) were recorded. TG studies were carried out in nitrogen atmosphere (flow rate 60 cm^3/min) at a heating rate of 20°C/min (sample weight 10 ± 2 mg).

3. RESULTS AND DISCUSSION

3.1. Isomerisation of bisitaconimides

In order to investigate the isomerisation of bisitaconimides to biscitraconimides [7], the synthesis was carried out in solvents of different polarities. In Table 1 the details of the reaction conditions and sample designations are given.

Table 1.
Preparation of bisitaconamic acids and bisitaconimides: Details of reaction conditions and sample designation

	Bisitaconamic acid		*Bisitaconimide*	
	Itaconic anhydride $= 2.24$ g		Bisitaconamic acid $= 3.5$ g	
	4,4'-diaminodiphenyl ether $= 2$ g		Acetic anhydride $= 1.5$ mL	
	Temperature $= 60$°C		Anhyd. sodium acetate $= 1$ g	
	Solvent $= 40$ mL		Solvent $= 20$ mL	
S. No	Bisitaconamic acid		Bisitaconimide	
	Solvent	Sample designation	Solvent	Sample designation
1	Tetrahydrofuran	IAF	Tetrahydrofuran	IEF
2	Acetone	IAA	Acetone	IEA
3	Chloroform	IAC	Chloroform	IEC
			Toluene	IET
			Acetic anhydride	IEN

3.1.1. Characterisation of bisitaconamic acids

Bisitaconamic acids were obtained in 80% yield. These monomers were white in colour and soluble in polar solvents like N,N-dimethylformamide, dimethyl sulfoxide and N,N-dimethylacetamide.

In the FT-IR spectra of bisitaconamic acids, multiple stretching vibrations due to –OH (carboxylic group) and –NH– (amide group) were observed in the region 2630-3500 cm^{-1}. A prominent and sharp absorption peak was observed at 3295 ± 2 cm^{-1} in all the samples (Fig. 1) while absorptions at 3215 and 3138 cm^{-1} appeared as shoulders. The >C=C< stretch was observed at 1631 ± 1 cm^{-1} in all the bisitaconamic acids, while the >C=O stretches of carboxylic acid and amide group were present at 1683 ± 1 cm^{-1} and 1655 cm^{-1}, respectively.

In the ^{1}H-NMR spectra, the vinylidene protons of bisitaconamic acids, being magnetically non-equivalent, were observed as singlets at 5.7 and 6.1 ppm. The methylene protons merged with the dimethyl sulfoxide protons (solvent) at 3.4 ppm. The –COOH proton was observed at 10.0 ppm. A complex pattern was observed in the region 7-8 ppm due to aromatic protons. The isomerisation of bisitaconamic acid to biscitraconamic acid is indicated by the appearance of an additional proton resonance signal due to methyl protons at 2.06 ± 0.03 ppm. The intensity of this peak depended on the type of solvent and reaction conditions employed.

In the TG traces of bisitaconamic acids, a two-step decomposition was observed (Fig. 2). The TG traces were characterised by determining T_i = initial decomposition temperature, T_{max} = temperature of maximum rate of weight loss, T_f = final decomposition temperature and char yield (Y_c) at 800°C. The T_i and T_f were obtained by extrapolation. The initial weight loss was observed in the temperature range of 180-240°C. This may be due to cyclodehydration of amic acids (Table 2). The weight loss due to such reaction should be around 8.5%. However, slightly higher weight loss was observed in all samples ($10 \pm 1\%$). This may be attributed to the loss of strongly bonded water from the samples. Major weight loss was observed above 400°C, and was due to scission of the backbone.

Table 2.
TG results of bisitaconamic acids

Samples	Decomposition temperatures (°C)			Weight loss (%)	Y_c (%) at 800°C
	T_i	T_{max}	T_f		
IEC	209	231	239	9.6	
	410	456	519	58.9	27.4
IEA	181	208	236	11.0	
	434	466	518	55.4	–
IEF	198	207	216	10.7	
	434	467	524	58.7	28.1

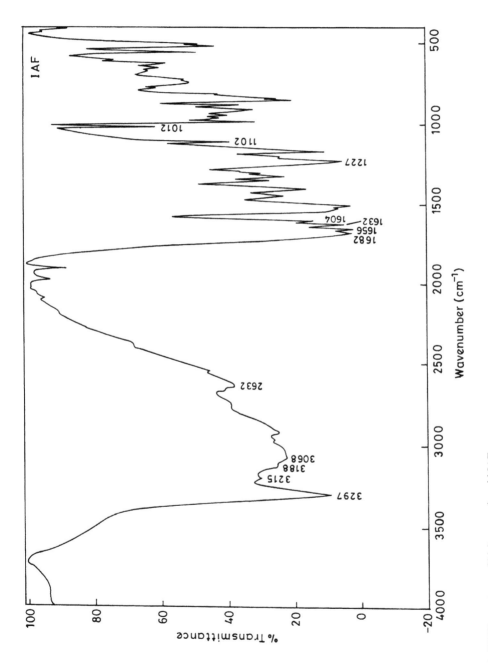

Figure 1. FT-IR spectrum of bisitaconamic acid IAF.

I.K. Varma

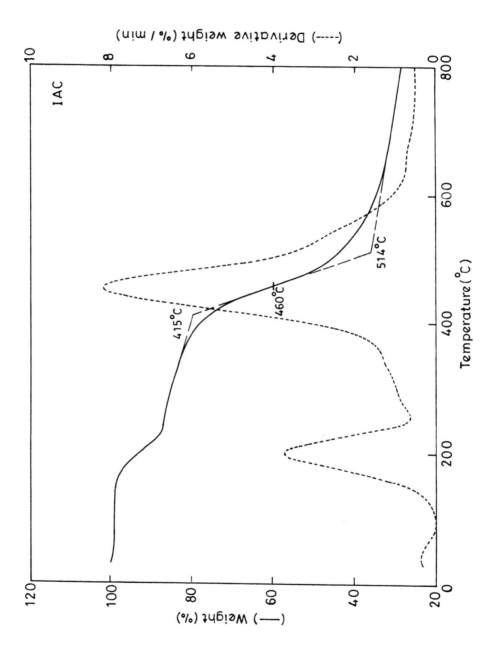

Figure 2. TG trace of bisitaconamic acid IAC.

3.1.2. Characterisation of bisitaconimides

Bisitaconimides (60-70% yield) obtained by cyclodehydration of bisitaconamic acids were yellow in colour and were soluble in low boling solvents like acetone and chloroform.

In the FTIR spectra of bisitaconimides, the characteristic absorption peaks due to imide groups were observed at 1775 ± 8 cm^{-1} and 1710 ± 10 cm^{-1}. The >C=C< stretch of bisitaconimides was observed at 1660 ± 5 cm^{-1}. Isomerisation of itaconimide to citraconimide was indicated in sample IET where an additional absorption peak was observed at 1643 cm^{-1} due to the >C=C< stretch in a 5 membered ring (Fig. 3).

In the ^{1}H-NMR spectra of bisitaconimides (Fig. 4) vinylidene protons were observed as singlets at 5.6 ± 0.1 ppm and 6.4 ± 0.1 ppm. The methylene protons appeared at 3.4 ± 0.1 ppm. Aromatic proton resonance signals were observed at 7-8 ppm. An additional proton resonance signal of varying intensity was present at 2.06 ± 0.05 ppm. This is due to $-CH_3$ protons of citraconimide, thereby indicating isomerisation of bisitaconimide to biscitraconimide.

The intensity of the 6.4 ppm signal was also higher compared to 5.6 ppm due to the presence of vinylidene proton in citraconimide. The content of biscitraconimide in these products was estimated from the intensity ratio of proton resonance signals at 2.06 and 5.6 ppm, and was found to be highest in IET (77%) and lowest in IEF (7.2%). The extent of isomerisation was comparable in IEA (29.5%), IEC (24%) and IEN (24%).

The DSC scans (Fig. 5) of IET, IEC and IEF showed sharp melting endotherms at 191°C, 201°C and 208°C, respectively and the curing exotherms were immediately after melting. The melting endotherms in IEA and IEN were at 196 and 201°C, respectively. It may be concluded that as the % biscitraconimide increases the melting temperature decreases. The exothermic peak temperature was lowest in IET and increased as the % biscitraconimide decreased. This could be due to presence of electron donating methyl group in citraconimide that increases the reactivity of the double bond. Due to partial overlap of melting endotherm and curing exotherm it was difficult to calculate the heat of curing (ΔH). The extent of isomerisation was maximum in bisitaconimide prepared by refluxing in toluene and lowest when tetrahydrofuran was used as solvent.

The TG traces of bisitaconimide resins cured at 200°C showed one- or two-step decomposition (Fig. 6). The characteristic decomposition temperatures (T_i, T_{max} or T_f) and char yield at 800°C of bisitaconimides are summarized in Table 3. In case of IEC, IEN and IET a weight loss of 4-6% was observed in the temperature range 240-400°C. T_i value for the first step was lowest for IET, while for IEC and IEN the T_i values were 290°C and 272°C, respectively.

The citraconimide content was highest in IET; therefore, the poor thermal stability of this resin compared to other samples may be attributed to the methyl group. In all the imide resins, major weight loss due to scission of the main chain was observed at temperatures >440°C. The T_{max} values for all the imides were ob-

I.K. Varma

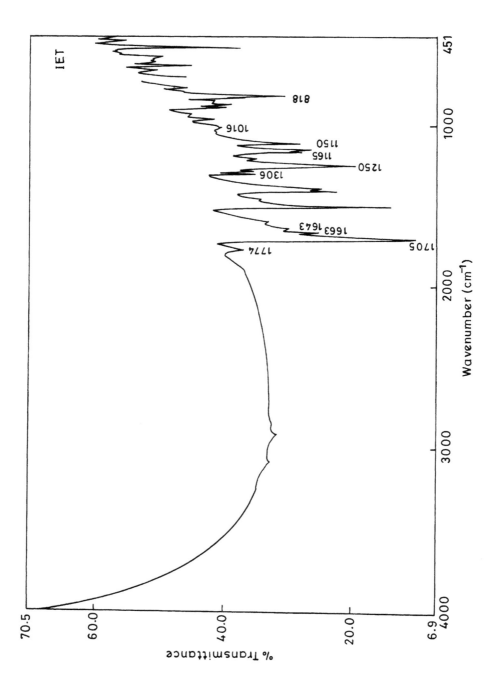

Figure 3. FT-IR spectrum of bisitaconimide acid IET.

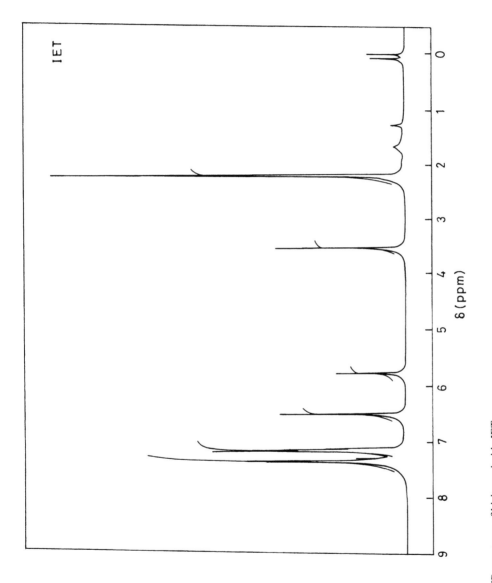

Figure 4. ^{1}H-NMR spectrum of bisitaconimide IET.

I.K. Varma

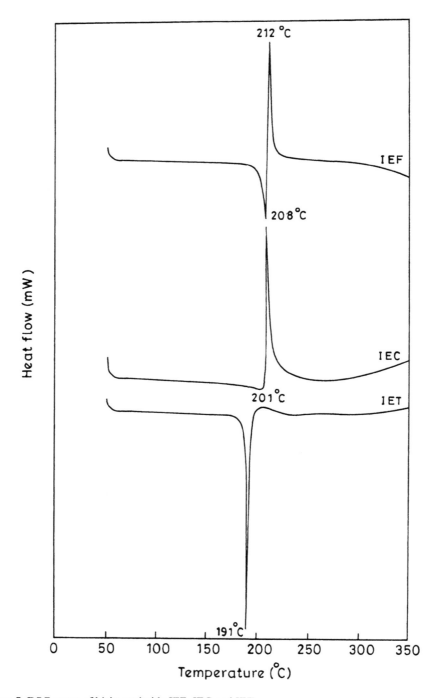

Figure 5. DSC scans of bisitaconimide IEF, IEC and IET.

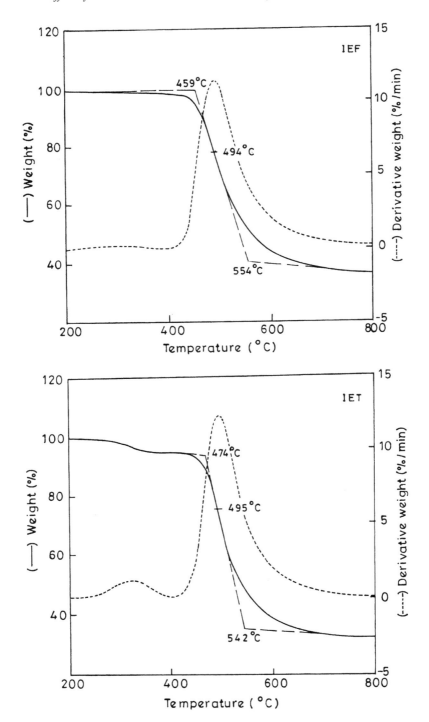

Figure 6. TG traces of bisitaconimide IEF and IET cured at 200°C for 2 h.

Table 3.
TG results of bisitaconimides (cured at 200°C for 2 h)

Samples	Decomposition temperatures (°C)			Weight loss (%)	Y_c (%) at 800°C
	T_i	T_{max}	T_f		
IEF	459	494	554	63.0	37
IEC	290	346	400	6.0	
	452	488	562	57.0	37
IEN	272	337	400	4.4	
	441	504	567	61.6	34
IEA	455	489	548	60.0	40
IET	240	321	382	5.7	
	474	495	542	62.3	32

served in the range of 488-504°C. Highest char yield of 40% was obtained in IEA (where acetone was used as a solvent for synthesis) and lowest (32%) in IET. On the basis of these studies it may be concluded that the presence of low content (< 30%) of citraconimides in bisitaconimides does not affect the thermal stability of cured resins.

3.2. Effect of structure on thermal behaviour of bisitaconimides

In order to study the effect of structure on thermal behaviour, bisitaconimide monomers having ether, phosphine oxide, sulphone, amide and fused aromatic rings were synthesized. Bisitaconimide oligomers of higher molecular weight were also investigated. Curing characteristics and thermal behaviour of these resins are described in the following text. A comparison with the state-of-the-art bismaleimides and bisnadimides is also made.

3.2.1. Thermal behaviour of itaconimides containing ether groups
The structures of itaconimides based on amines containing ether groups are given in Table 4. The molecular weight of these imides varied from 388 to 598 g/mol.

Characterization of the resins was done by IR and ^{1}H-NMR spectroscopies. The IR spectrum of IE resin is given in Fig 7. The presence of citraconimide (<30%) was indicated in these samples. The curing exotherm depended on the structure of the resin and was highest in IP and lowest in IE resin. An increase in molecular weight of bisitaconimide leads to an increase in T_{exo} values (Table 5).

The melting points of bisnadimides of corresponding structures were significantly higher than the bisitaconimides [8-10].

A two-step decomposition was observed in most of the isothermally cured resin samples. A weight loss of 2-3% was observed in the temperature range of 245-400°C while the major weight loss was observed at temperature greater than 425°C. The char yield of the resins at 800°C varied from 39-42%. Cured bisnadimides of similar structures had slightly higher char yields (Table 6).

Table 4.
Structures of itaconimides based on amines containing ether groups

S. No	Structure of amines	Designation of		Molecular weight (g/mol)
		amine	itaconimide	
1		E	IE	388
2		R	IR	480
3		H	IH	480
4		P	IP	598

Table 5.
Characteristic thermal transitions of bisitaconimides

Bisitaconimide	Temperatures (°C)			
	T_m	T_1	T_{exo}	T_2
IE	196 (264)	200	206	211
IR	163 (217)	182	223	283
IH	193 (295)	199	210	245
IP	83 (174)	200	257	311

Figures in parentheses indicate the melting points of corresponding bisnadimide resins [8]

Table 6.
Thermal behaviour of cured (at 200°C for 2 h) bisitaconimides

Bisitaconimide	Temperatures (°C)			Weight loss (%)	Char yield (%) at 800°C
	T_i	T_{max}	T_f		
IE	290	323	395	3	39
	448	482	542	58	(41)
IR	286	323	413	2	39
	445	490	541	59	(39)
IH	245	282	382	2	42
	427	454	525	56	(55)
IP	464	499	560	65	35
					(46)

Figures in parentheses indicate char yields of corresponding bisnadimides [8-10]

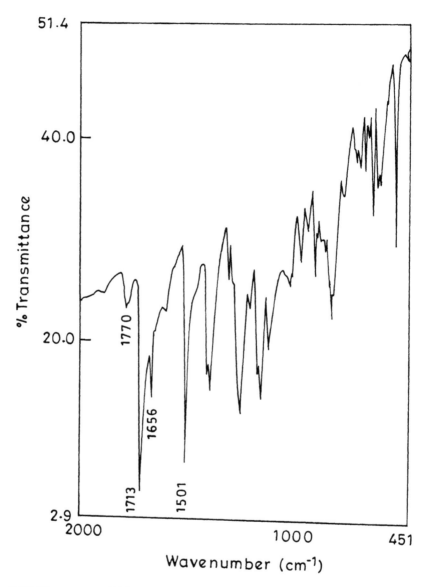

Figure 7. FT-IR spectrum of bisitaconimide IE.

3.2.2. Thermal behaviour of itaconimides based on amines containing phosphine oxide and sulfone groups

Itaconimides based on bis(3-aminophenyl) methyl phosphine oxide, tris(3- aminophenyl) phosphine oxide 3,3'-diaminodiphenyl sulfone and 4,4'-diaminophenyl sulfone were prepared for these studies. The structures of these bisitaconimides are given in Table 7.

In the IR spectra of these resins characteristics peaks of sulfone and phosphine oxide groups were present along with the absorption peaks of the imido group. Characteristic thermal transitions as determined by DSC are given in Table 8.

The char yield of the cured resin (Table 9) was highest in itaconimide having phosphine oxide and aromatic groups (IT). Presence of sulfone group at meta or para position did not affect the char yield.

Table 7.
Itaconimides based on amines containing phosphine oxide or sulfone group

S. No	Structures of amines	Designation of amine	itaconimide	Molecular weight (g/mol)
1		T	IT	605
2		B	IB	434
3		S_p	IS_p	436
4		S_m	IS_m	436

Table 8.
Characteristic thermal transitions of bisitaconimides

Bisitaconimide	Temperatures (°C)			
	T_m	T_1	T_{exo}	T_2
IT	–	180	210	282
IB	–	252	262	300
IS_p	107	190	220	301
IS_m	88	177	216	297

Table 9.
Thermal behaviour of bisitaconimides (cured at 200°C for 2 h)

Bisitaconimide	Temperatures (°C)			Char yield (%) (800°C)
	T_i	T_{max}	T_f	
IT	475	515	533	65.6, 60*, 64.5**
IB	398	448	493	55. 50*. 51**
IS_p	402	448	499	33, 44*
IS_m	392	454	526	33.0

   *   Char yield of corresponding bisnadimide resins
  ** Char yield of corresponding bismaleimide resins [11, 12]

3.2.3. Itaconimides containing amido or fused aromatic groups
Bisitaconimides of comparable molecular weights were prepared using amines having fused aromatic ring (IF) or amide group (IA) (Table 10).

The curing of IF resins occurred at higher temperature. Similar behaviour has been observed earlier in bismaleimide resins [13] (Table 11).

The char yield of cured resins was comparable (Table 12).

Table 10.
Itaconimides containing amide or fused aromatic group

S. No	Structures of amines	Designation of		Molecular weight (g/mol)
		amine	itaconimide	
1	H₂N—⟨⟩—CONH—⟨⟩—NHCO—⟨⟩—NH₂	A	IA	534
2	H₂N—⟨⟩⟨⟩—NH₂ (fused fluorene)	F	IF	536

Table 11.
Characteristic thermal transitions of bisitaconimides

Bisitaconimide	Temperatures (°C)			
	T_m	T_1	T_{exo}	T_2
IA	232	239	244	294
IF	–	293	306	320

Table 12.
Thermal behaviour of cured (at 200°C for 2 h) bisitaconimides

Bisitaconimide	Temperatures (°C)			Weight loss (%)	Char yield (%) (800°C)
	T_i	T_{max}	T_f		
IA	218	231	249	6.9	
	404	449	518	51.7	42
IF	268	314	332	4	
	482	516	562	55	41

3.2.4. Bisitaconimide oligomers

Bisitaconimde oligomers based on 4,4'-diaminodiphenyl ether/1,4-bis(4-amino-phenoxy) benzene and aromatic tetracarboxylic acid dianhydride were prepared according to reaction Scheme 2.

The synthesis was carried out by reacting 0.01 mol of appropriate amine with 0.01 mol of itaconic anhydride in 20 mL dimethyl formamide. After stirring at 60°C for 3 h, 0.005 mol of aromatic tetracarboxylic acid dianhydride was added and the reaction was continued for 3 h. Cyclodehydration of amic acid to imide was carried out by acetic anhydride and anhydrous sodium acetate (4 h). The imide resins were precipitated in ice-cold water, filtered, washed successively with water, sodium bicarbonate solution, water and dried in vacuum oven at 50-75°C. The structure and letter designation of these bisitaconimides are given in Scheme 3.

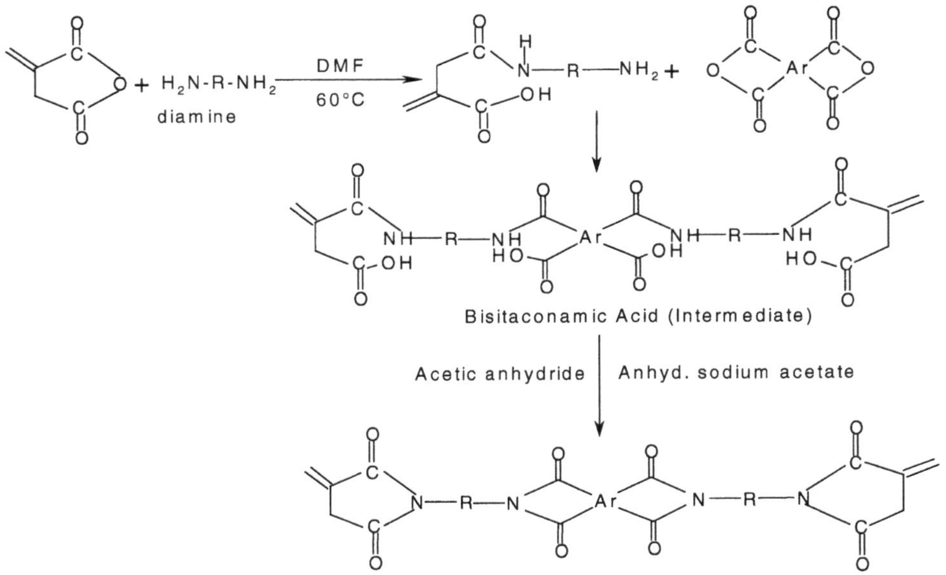

Scheme 2. Reaction scheme for the synthesis of bisitaconimide oligomers.

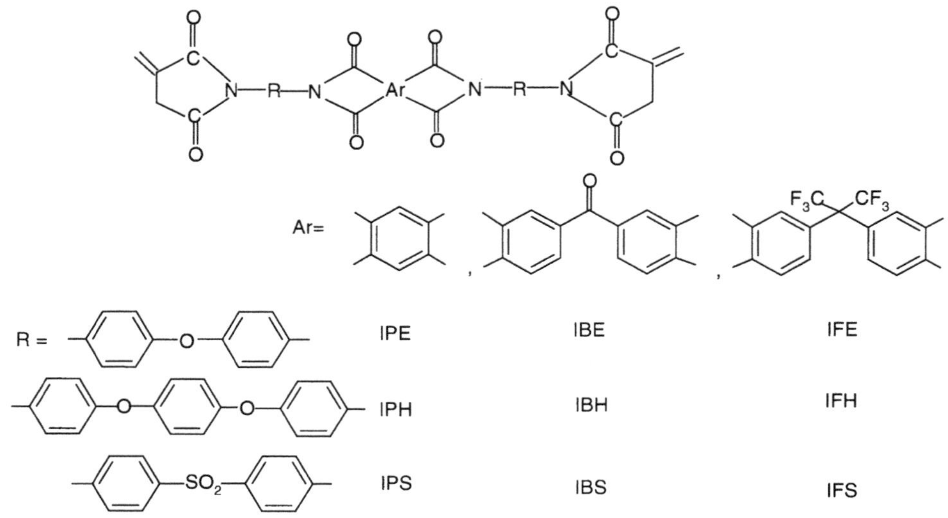

Ar=

IPE	IBE	IFE

R =

IPH	IBH	IFH
IPS	IBS	IFS

Scheme 3. Structure and letter designation of various bisitaconimide oligomers.

Table 13.
TG results of cured (at 220°C for 2 h) bisitaconimides

Bisitaconimides	Decomposition temperatures (°C)			Weight loss (%)	Y_c (%) at 800°C
	T_i	T_{max}	T_f		
IPE	326.8	385.0	449.4	16.43	
	449.4	571.3	624.2	35.23	46.1
IBE	348.6	415.1	490.0	14.30	
	489.9	600.8	649.1	34.68	50.1
IFE	411.2	476.4	530.3	18.48	
	530.3	570.7	627.0	18.32	
	627.0	651.6	713.5	12.43	50.8
IPH	372.1	431.5	500.0	19.54	
	500.0	573.4	623.7	35.33	45.1
IBH	355.8	49.0	485.4	16.32	
	485.4	588.7	667.4	35.55	48.1
IFH	404.5	451.6	516.8	18.20	
	516.8	570.8	604.5	14.73	
	604.5	627.0	683.1	14.97	51.6
IPS$_p$	251.7	285.3	316.8	6.87	
	316.8	382.0	500.0	27.67	
	500.0	546.0	660.7	12.58	52.8
IBS$_p$	274.0	292.1	319.1	2.41	
	319.1	427.0	532.6	27.90	
	532.6	570.7	651.7	15.20	53.9
IFS$_p$	380.3	439.2	511.2	26.85	
	511.2	559.2	623.6	26.80	46.5

Isothermal curing of these oligomers was carried out at 220°C for 2 h and relative thermal stability was evaluated by thermogravimetric analysis (Table 13). The char yield of these oligomers was higher than that of the corresponding monomeric bisitaconimides.

4. CONCLUSIONS

These studies indicate that bisitaconimide resins of varying thermal stability and curing characteristics can be prepared by structural modification. These resins cure at a significantly lower temperature than the corresponding bisnadimides and at a slightly lower temperature than bismaleimide resins. Our studies have shown that isomerisation of bisitaconimide to citraconimide is dependent on temperature. The effect of solvent polarity was only marginal. The presence of biscitraconimide (up to 30%) in bisitaconimide reduced the curing temperature without affecting the thermal stability. Cured resins are stable upto approximately 400°C and char yield at 800°C is slightly lower than bismaleimides or bisnadimides of similar structure.

REFERENCES

1. A. Nagai and A. Takahashi, in: *Polymeric Materials Encyclopedia*, Vol. 6, J.C. Salamone (Ed.), pp. 4013-4025. CRC Press, Boca Raton, FL (1996).
2. S.L. Hartford, S. Subramanian and J.A. Parker, J. Polym. Sci. Polym. Chem. Ed., **16**, 137 (1978).
3. D. Radiec and L. Gargallo, in: *Polymeric Materials Encyclopedia*, Vol. 8, J.C. Salamone (Ed.), pp. 6346-6350. CRC Press, Boca Raton, FL (1996).
4. N. Gupta and I.K. Varma, J. Appl. Polym. Sci., **68**, 1759 (1998).
5. I.K. Varma and U. Gupta, J. Macromol. Sci. Chem. Ed., **A-23**, 19 (1986).
6. I.K. Varma, G.M. Fohlen and J.A. Parker, J. Macromol. Sci. Chem. Ed., **A-19**, 209 (1983).
7. A. Solanki, V. Chaudhary and I.K. Varma, Polym. Int., **51**, 493 (2002).
8. R. Madan, R.C. Anand and I.K. Varma, J. Polym. Sci., Polym. Chem. Ed., **35**, 2917 (1997).
9. A. Mathur and I.K. Varma, Polymer, **33**, 4845 (1992); J. Appl. Polym. Sci., **46**, 1749 (1992); Angew Makromol. Chem., **206**, 53 (1993).
10. A. Solanki, V. Chaudhary and I.K. Varma, J. Thermal Anal. & Cal., **66**, 749 (2001).
11. I.K. Varma, Mater. Res. Innovation, **4**, 306 (2001).
12. I.K. Varma and V.B. Gupta, in: *Polymer Matrix Composites, A Comprehensive Composite Materials Publication*, R. Talreja and J.-A.E. Manson (Eds.), pp. 1-56. Elsevier Science Ltd., Oxford, U.K. (2001).
13. I.K. Varma, G.M. Fohlen and J.A. Parker. J. Polym. Sci. Polym. Chem. Ed., **20**, 283 (1982).

Polyimides and Other High Temperature Polymers, Vol. 2, pp. 185–204
Ed. K.L. Mittal
© VSP 2003

1-Amino-4,5-8-naphthalenetricarboxylic acid-1,8-lactam-4,5-imide-containing macrocycles: Synthesis, molecular modeling and polymerization

PATRICIA PONCE, LIOUDMILA FOMINA, PATRICIA GARCÍA and SERGUEI FOMINE*

Instituto de Investigaciones en Materiales, Universidad Nacional Autónoma de México, Apartado Postal 70-360, CU, Coyoacán, México DF 04510, México

Abstract—Novel macrocycles containing 1-amino-4,5-8-naphthalenetricarboxylic acid-1,8-lactam-4,5-imide and 1,4,5-8-naphthalenetetracarboxylic bisimide fragments have been synthesized by the high temperature pseudo high dilution acylation of the corresponding diols with isopthaloyl chloride, 4,4'- and 2,2'-dichlorocarbonyl biphenyls with up to 60% yield. An important side reaction that impedes cyclization was found to be the reaction of diol OH groups with HCl during the acylation. The ring strain in synthesized macrocycles and model cycles was estimated using the isodesmic reaction approach at B3LYP/6-311+G(d,p)//HF/3-21G level of theory. Lactamimide containing macrocycles were found to be more strained as compared to those containing bisimide. The ring opening polymerization (ROP) of synthesized macrocycles in the molten state shows that the driving force for this process is the strain release on ring opening. The ROP of lactamimide containing macrocycles was found to be an efficient way to obtain lactamimide containing polymers, otherwise difficult to synthesize.

Keywords: Macrocycles; ring opening polymerization; molecular modeling.

1. INTRODUCTION

The preparation of engineering thermoplastics via ROP is commercially attractive for a number of reasons. The ROP reaction would convert a low molecular weight, low viscosity cyclic precursor into a high molecular weight polymer without formation of any by-product. In case of intractable polymers, ROP offers a unique opportunity for the reactive processing of these polymers. The discovery by Brunelle and coworkers of the high yield synthesis and facile polymerization of cyclic polycarbonates [1] has sparked much interest in the macrocyclic monomer technique. In the past 15 years this area has been rapidly extended to other systems such as cyclic esters [2, 3], amides [4], aryl ethers [5-7], aramids [8] and

---

*To whom all correspondence should be addressed. Phone: +556224726, Fax: +556161201, E-mail: fomine@servidor.unam.mx

imides [9]. The most recent coverage of ROP of cyclic esters can be found in [10].

The present authors have recently reported the synthesis and characterization of polylactamimides, a novel class of polymers combining good thermostability, with luminescent and nonlinear optical properties [11-13]. However, in some cases intractable products were obtained. The macrocyclic monomer technique could be a new promising approach for the synthesis of polylactamimides via reactive processing of macrocyclic monomers. Recent theoretical study of ROP of lactamimide-containing macrocycles showed that this process was enthalpy driven [14] unlike macrocylcic polycarbonates where the polymerization proceeded due to entropy gain on ring opening [15]. However, Kricheldorf et al. recently discussed the formation of cyclic polyesters as a kinetically controlled reaction [16]. Therefore, the successful ROP of lactamide containing macrocycles is a challenging problem involving careful selection, design and synthesis of the corresponding macrocyclic systems showing ring strain to undergo ROP.

This paper describes the molecular design and synthesis of lactamimide-containing macrocycles, as well as their molecular modeling and polymerization.

2. EXPERIMENTAL PART

2.1. Materials

All reagents were used as received from Aldrich.

N,N'-Bis(6-hydroxyhexyl)-1,4,5-8-naphthalenetetracarboxylic bisimide (1) was prepared according to [11] with 91% yield, $T_m = 209°C$.

2

N,N'-Bis(6-hydroxyhexyl)-1-amino-4,5-8-naphthalenetricarboxylic acid-1,8-lactam-4,5-imide (**2**): A mixture of 3.018 g (0.0065 mol) of bisimide **1**, DMSO (45 ml), KOH (10.5 g (0.18 mol)) and methanol (60 ml) was stirred for 8 hrs under oxygen flow at 70°C. The reaction mixture was acidified by dilute HCl, the precipitate formed was filtered off and refluxed with a mixture of ethanol (100 ml) and conc. HCl (15 ml) for 2 hrs. The hot solution was filtered and poured into water. The precipitate formed was filtered off and dried in vacuum: yield 97%, mp 135°C. The use of oxygen during the ring contraction increased the yield from 57% according to [11] to 97% and made the the purification by column chromatography unnecessary. T_m= 118-120°C. ^{1}H NMR: (DMSO-d$_6$), 8.24 (d, 1H, J=7.2 Hz, H^6), 8.14 (d, 1H, J=7.5 Hz, H^3), 8.03 (d, 1H, J=7.2 Hz, H^7), 7.25 (d, 1H 7.5 Hz, H^2), 4.20 (s, 2H, OH), 3.90 (t, 2H, J=7.5 Hz, H^{14}), 3.80 (t, 2H J=6.6 Hz, H^{15}), 1.70-1.10 (m 16H, CH$_2$). ^{13}C NMR: 166.7 (C^{11}), 162.5, 162.4 (C^{12}, C^{13}), 144.0 (C^1), 134.1, 130.1, 125.0 (C^3, C^6, C^7), 129.0 (C^{10}), 125.9 (C^8), 123.5, 122.9 (C^4, C^5), 115.4 (C^9), 106.7 (C^2), 60.6, 60.5 (C^{16}, C^{17}), 32.4, 28.1, 27.8, 26.5, 26.1, 25.3, 25.1 (CH$_2$ aliph). Mass spectrum: M$^+$ = 438.

2.2. General procedure for macrocycles synthesis

The corresponding lactamimide or bisimide (4 g) and equimolar amount of acid dichloride dissolved in 40 ml of 1,2-dichlorobenzene (DCB) were added dropwise to 100 ml of DCB at 170°C for 16 hrs under nitrogen flow. Dichloride of 4,4'-biphenyl dicarboxylic acid was added as a solid due to its insolubility in DCB. Hot solution was filtered and the solvent was evaporated under vacuum The residue was purified by column chromatography on silica using CHCl$_3$ as eluent. The yields and melting points of synthesized macrocyles are listed in Table 1.

Dichlorides of 2,2'- and 4,4'-dicarboxybiphenyls needed for macrocyclization were prepared from the corresponding diacids by refluxing them with an excess of SOCl$_2$ in the presence of a few drops of DMF until the reaction mixture became clear The reaction was also monitored by IR. The excess SOCl$_2$ was removed in vacuum.

Table 1.
Yields, melting points and molecular weights of synthesized macrocycles

Macrocycle	mp (°C)	Yield (%)	M+$^{a)}$
BIISO	100	30	596
LIISO	134	26	568
LI2Bph	243	60	644
LI4Bph	245	21	644

a) Molecular ion

2.3. Polymerization

All macrocycles were tested in the molten state polymerization as follows: a macrocycle (150 mg) was heated at 260-270°C for C 2.5 hrs under a nitrogen flow. Soluble polymers were dissolved in chloroform and precipitated in methanol. Insoluble products were analyzed in their "as is" state.

2.4. Measurements

^{1}H-NMR and ^{13}C-NMR spectra were taken using a Varian spectrometer at 300 and 75.5 MHz, respectively, in DMSO-d_6 or CDCl$_3$ with tetramethylsilane (TMS) as an internal standard. FT-IR spectra were recorded using a Nicolet 510p spectrometer. Differential scanning calorimetry (DSC) and thermogravimetric analysis (TGA) were performed at a heating rate of 10°C/min under nitrogen with a Du-Pont 2100 instument. Mass spectra were recorded on a JEOL JMS-AX505HA spectrometer.

2.5. Computational details

All calculations were carried out using global minimum structures found with Monte Carlo conformational search algorithm implemented in Titan program with MMFF force field [17]. To estimate the macrocycle strain energy the isodesmic reaction technique was used [18]. All geometry optimizations in this method were run with Jaguar 4.1 program [19] at Hartree-Fock (HF) level using standard 3-21G basis set. Single point energies of HF/3-21G minimized structures were calculated with the Becke three parameter hybrid functional (B3) in combination with the Lee-Yang-Parr (LYP) [20, 21] correlation functional (LYP) using large 6-311+G(d,p) basis set. This model designated as B3LYP/6-311+G(d,p)//HF/3-21G shows a mean deviation of 3 kcal/mol for G2 molecule set [18].

3. RESULTS AND DISCUSSION

3.1. Synthesis of macrocycles

Scheme 1 shows the synthesis route to lactamimide containing cyclic esters. The preparation of lactamimide containing diols includes bisimide synthesis **1** from dianhydride 1,4:5,8-Naphthalenetetracarboxylic acid dianhydride and corresponding amino alcohol followed by ring contraction reaction to produce lactamimide containing diol **2**. A modified synthesis procedure involving the use of oxygen atmosphere during the ring contraction reaction allows drastically increased yield and purity of synthesized lactamimides compared to conventional techniques [11]. It has been shown recently [22] that the mechanism of the ring contraction reaction of six-membered bisimides involves the formation of a carboanion followed by its oxidation by air oxygen. It seems that oxygen favors the oxidation of this carboanion reducing the possibility of side reactions.

The pseudo high dilution technique has been used to synthesize well-defined macrocyclic lactamimide esters. The best reaction conditions for the macroacylation were found to be the high temperature acylation with no catalyst. The reaction was carried out in boiling DCB when equimolar amounts of diol as solid and corresponding diacid dichloride were slowly added to hot DCB. As can be seen from Table 1 this technique allows to obtain well-defined macrocycles with yields

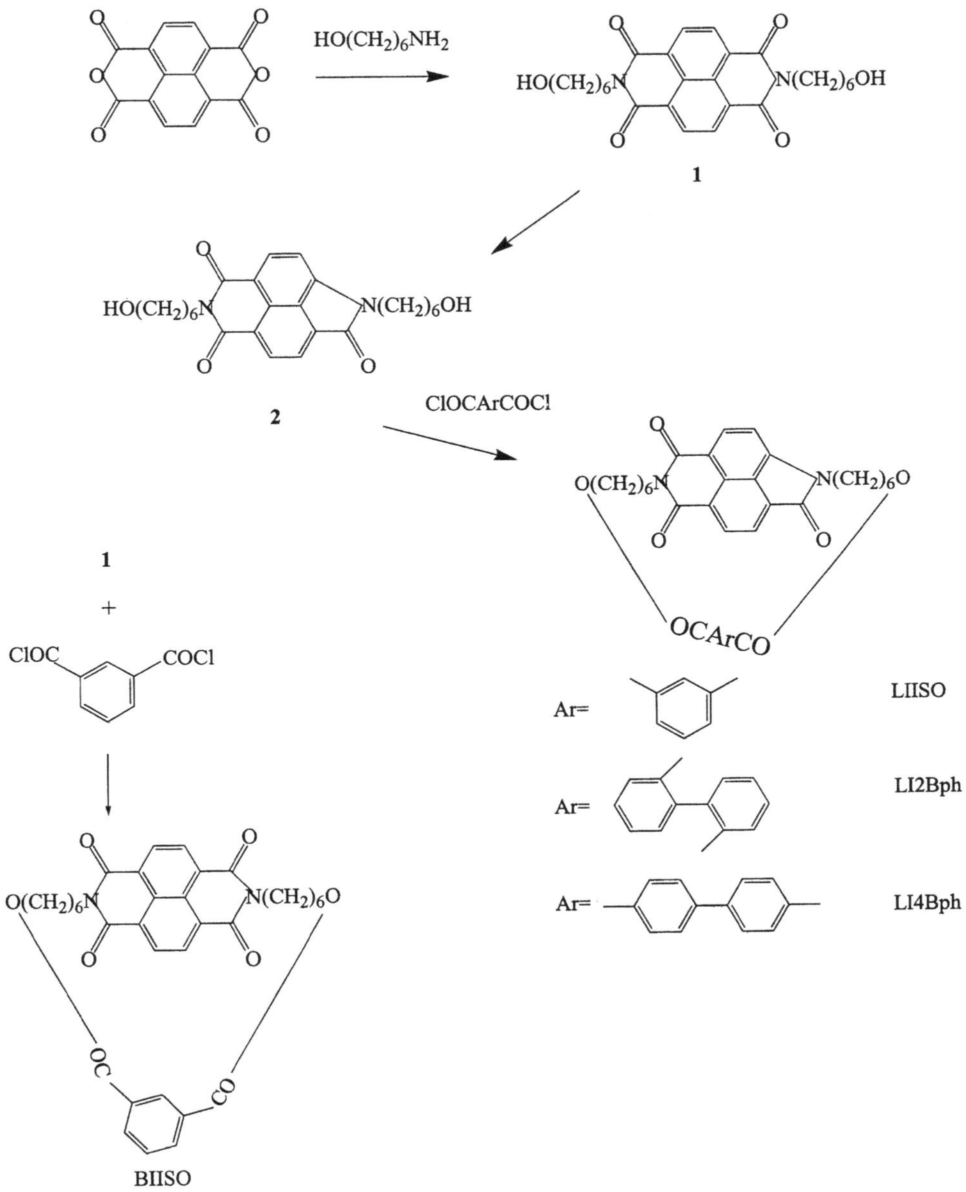

Scheme 1. Synthesis route to macrocycles.

up to 60%. To confirm the cyclic nature of synthesized products they were characterized by mass and NMR spectroscopies. Figs. 1, 2 and Figs. 3, 4 show mass and ^{1}H-NMR spectra of synthesized macrocycles, respectively. The most intense peak in the mass spectra of synthesized molecules is due to molecular ion as can

P. Ponce et al.

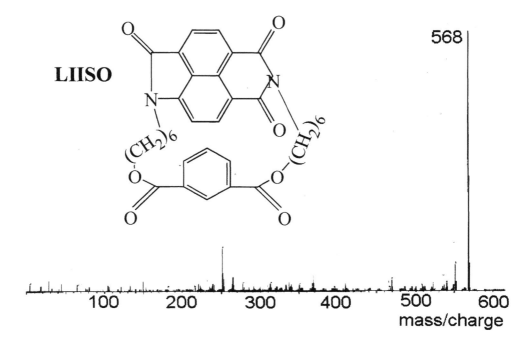

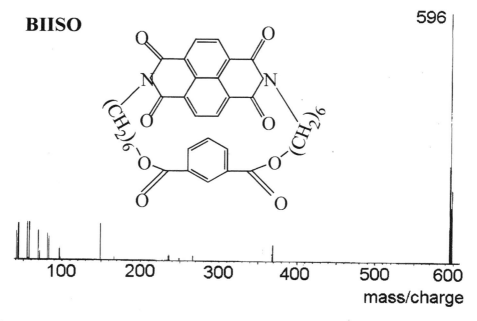

Figure 1. Mass spectra of **LIISO** and **BIISO** macrocycles.

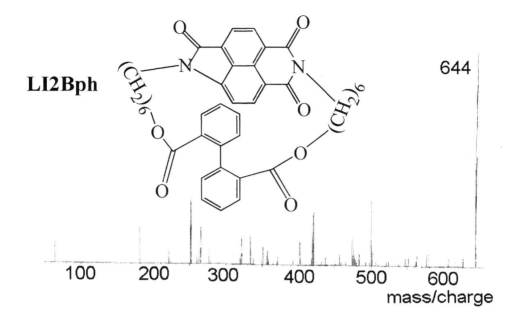

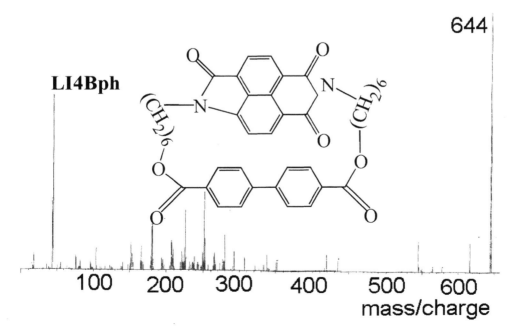

Figure 2. Mass spectra of **LI2Bph** and **LI4Bph** macrocycles.

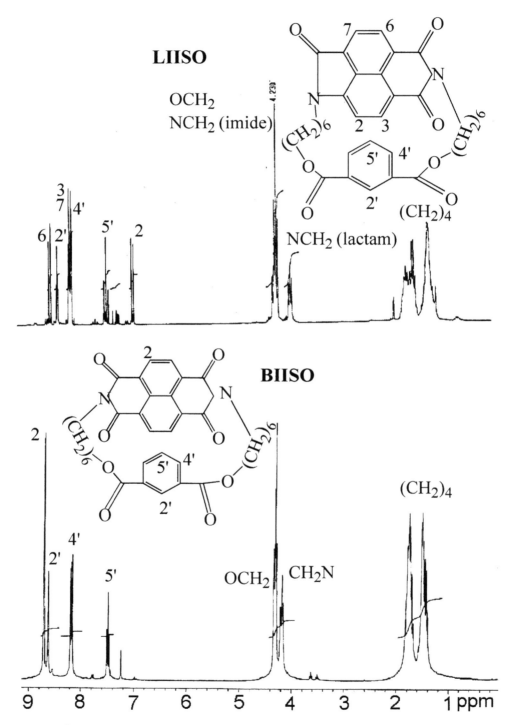

Figure 3. ^{1}H-NMR spectra of **LIISO** and **BIISO** macrocycles.

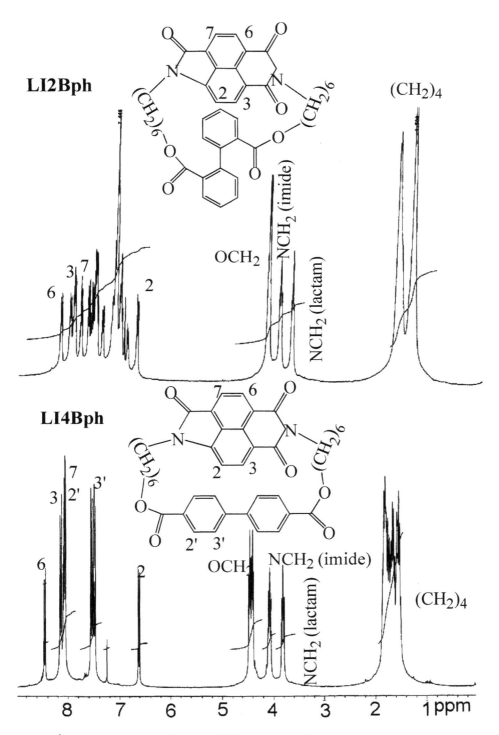

Figure 4. ^{1}H-NMR spectra of **LI2Bph** and **LI4Bph** macrocycles.

be seen from the comparison of the molecular weights of synthesized products
(Table 1) and the mass spectra obtained (Figs. 1, 2). Additional confirmation of
the molecular structure comes from the analysis of ¹H-NMR spectra (Figs. 3 and
4), where all signals can be assigned to the protons of the corresponding cyclic
structure. The yields of individual macrocycles (not a mixture of cycles) are high
(up to 60% in case of **LI2Bph** cycle). It seems that this phenomenon is due to favorable conformation adopted by a dimer on cyclization. Scheme 2 shows the cyclization reaction. The dimer formed by reaction of one equivalent of diol with

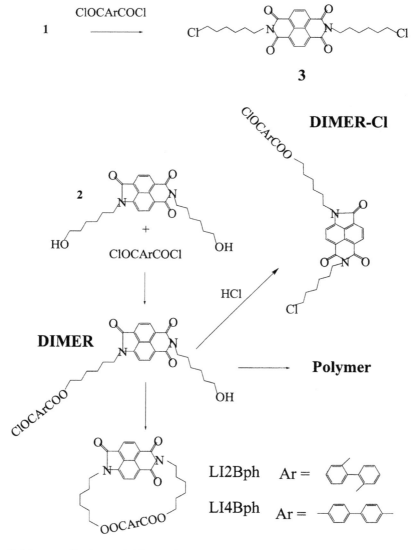

Scheme 2. Macrocyclization and side reactions.

one equivalent of acid dichloride can either form a cycle or produce a linear oligomer. The shorter the distance between the reactive sites of a dimer (chlorocarbonyl carbon and hydroxide oxygen) the lower the cyclization activation entropy which is of primary importance for the large cycle formation. Fig. 5 shows the lowest energy conformers found using the Monte-Carlo conformational search algorithm (within 1 kcal mol, which are the most populated ones at the reaction temperature of 170°C) for dimer precursors for **LI2Bph** and **LI4Bph**. As can be seen from Fig. 5 the C-O distances are significantly shorter for **LI2Bph** precursor compared to **LI4Bph**, especially in the lowest energy conformer. These results explain the high yield of **LI2Bph** cycle where the conformation of the lowest energy conformer is very favorable for the cycle formation.

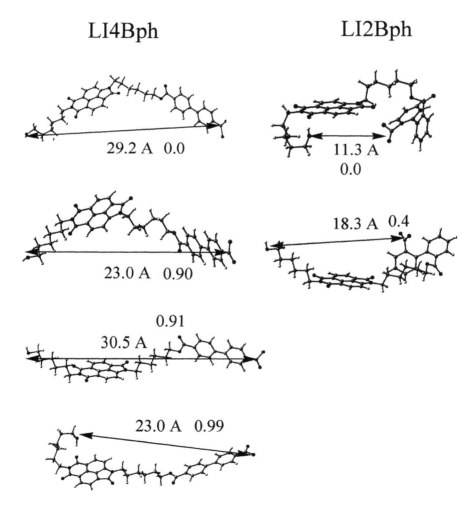

Figure 5. Lowest energy conformers of **LI4Bph** and **LI2Bph** precursors and their relative energies (kcal/mol) located by Monte-Carlo conformational search algorithm using MMFF force field.

The most important side reaction in the macrocyclization process is the polymer formation. Another important side reaction that was found to impede the cycle formation was OH replacement by Cl as shown in Scheme 2. Although we were not able to isolate the intermediate **DIMER-Cl**, the dichloride **3** was isolated with up to 25% yield and characterized by mass and NMR spectroscopies (Fig. 6). It seems that the substitution of OH by Cl and the formation of chlorinated prod-

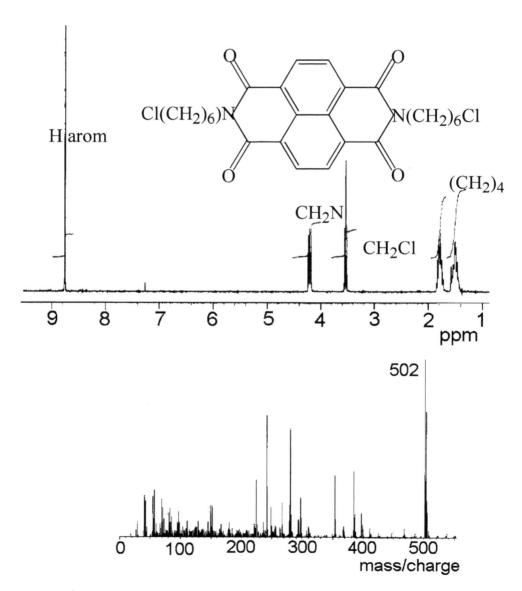

Figure 6. ^{1}H-NMR and mass spectra of N,N'-Bis(6-chlorohexyl)-1,4,5-8-naphthalenetetracarboxylic bisimide.

ucts occurs by the reaction of HCl, which is the by-product of acylation, with primary OH of diols or **DIMER** type dimers as shown in Scheme 2.

Apart from lactamimide containing diol **2**, the bisimide-containing diol **1** was used to prepare bisimide-containing macrocycles for comparison purpose. To our surprise, only one bisimide-containing macrocycle (**BIISO**) was isolated and in case of other acid dichlorides (2,2'- and 4,4'-dichlorocarbonyl biphenols) no cyclic products were isolated. The conformational search carried out for bisimide cycle precursors similar to that carried out for lactamimide-containing macrocycles showed that the low energy conformers of dimer precursors were even more favorable for cyclization in comparison to lactamimide containing dimers. We believe it is the very low solubility of bisimide-containing macrocycles that impedes the isolation of biphenyl-containing macrocycles. Thus, the only isolated bisimide-containing macrocycle (**BIISO**) is the one obtained based on isophthaloyl chloride which always gives much soluble products compared to rigid biphenyl fragment.

3.2. Molecular modelling of macrocycles

Taking advantage of having well-defined unique cyclic structures and not a mixture of cyclic oligomers it is possible to correlate their polymerizability with the ring strain of prepared macrocycles, since as it has been recently shown that the ROP of lactamimide-containing macrocycles is due to strain release and not entropy gain [14]. The isodesmic reaction approach has been adopted to estimate the ring strain [18]. Isodesmic reactions used for the ring strain calculations are shown in Scheme 3. The numbers of bond and atom types do not change during the reaction and, therefore, the energetic effect of this transformation represents the ring strain. This scheme has been tested by comparing the experimental ring opening enthalpy of oxirane and caprolactam with calculated ring strain energy estimated using isodesmic reaction scheme (Scheme 3). In case of oxirane and caprolactam the calculated strain energies are 24.9 and 5.1 kcal/mol, while the experimental ring opening enthalpies are 25.0 and 4.0 kcal/mol, respectively [23]. As seen the isodesmic reaction approach using B3LYP/6-311+G(d,p)//HF/3-21G level of theory allows one to estimate strain energies within 2 kcal/mol. Table 2 shows calculated strain energies of synthesized macrocycles as well as those of bisimide containing cycles that were not synthesized. The only cycles showing significant strain energy are lactamimide-containing cycles **LI2Bph** and **LI4Bph** where the strain energies are as high as 10 kcal/mol. It is interesting to note that the corresponding bisimide-containing macrocycles show very low strain energy of 3.4 and −1.3 kcal/mol, respectively. The lack of one carbonyl group in lactamimide causes strong difference in strain energies of corresponding macrocycles. When analyzing the optimized geometries of macrocycles studied (Fig. 7) one can see the origin of strain in **LI2Bph** and **LI4Bph** cycles. Benzene rings of biphenyl moieties are not planar in these cycles. This deviation in **LI2Bph** and

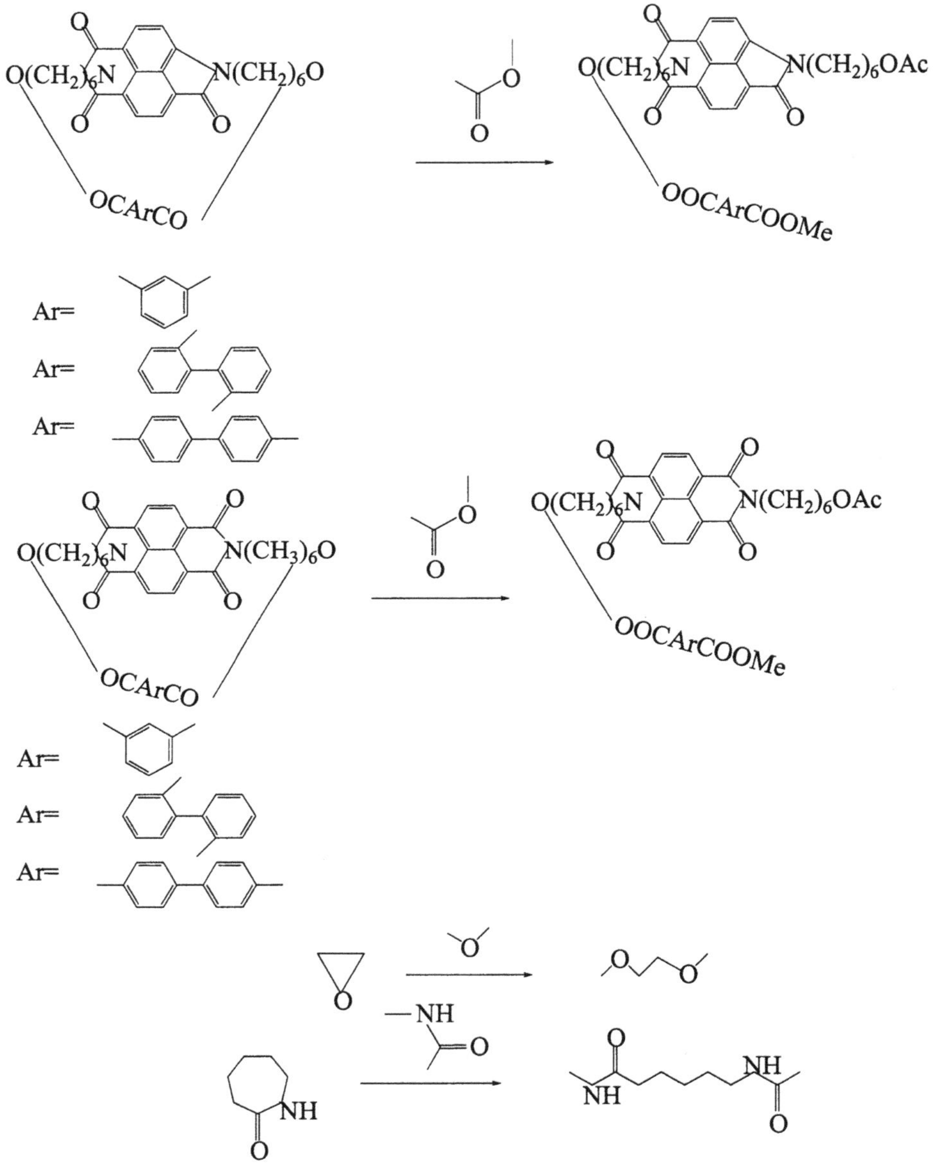

Scheme 3. Isodesmic reactions used for the estimation of ring strain.

LI4Bph macrocycles is 7-9°, while in corresponding bisimide containing cycles the out of plane deviations for benzene rings of biphenyl fragment are 5 and 1°, respectively. The direct correlation between the calculated strain energies and the corresponding geometrical changes in macrocycles studied indicates that the most important contribution to the strain energy of prepared cycles comes from angular

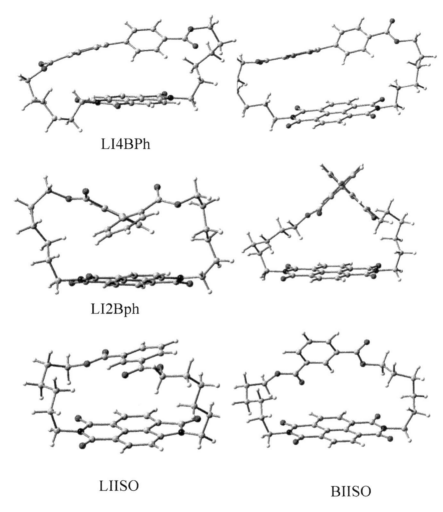

LI4BPh

LI2Bph

LIISO

BIISO

Figure 7. Lowest energy conformers of synthesized and model macrocycles optimized at HF/3-21G level of theory.

Table 2.
Strain energies (kcal/mol) calculated at B3LYP/6-311+G(d,p)//HF/3-21G level using isodesmic reaction scheme

Cycle	Strain energy
LIISO	−2.9
BIISO	−2.9
LI2Bph	9.9 (3.4)[a]
LI4Bph	10.1 (−1.3)[a]

a) Strain energy calculated for the corresponding bisimides containing macrocycles strain. Unlike biphenyl containing macrocycles the macrocycles **BIISO** and **LIISO** are not strained at all as shown from the calculations. Apparently, isophthaloyl moiety is not large enough to cause any significant strain in these molecules. It follows from the results of molecular modeling the macrocycles most active in ROP are **LI2Bph** and **LI4Bph** while **BIISO**, and **LIISO** are much less active in ROP.

3.3. ROP of macrocycles

The polymerization of synthesized macrocycles was carried out in the molten state at 260-270°C for 2.5 hrs under a nitrogen flow. Both biphenyl containing macrocycles (**LI2Bph** and **LI4Bph**) readily underwent ROP producing corresponding polymers (Scheme 4). Both **LI2Bph** and **LI4Bph** macrocycles produce polymers insoluble in common organic solvents which made it impossible to measure their molecular weights. However, both polymers (**PLI2Bph** and **PLI4Bph**) can be processed from the melt at 280°C into strong fibers proving the high molecular weight of the polymers obtained. The structure of the polymers can easily be confirmed by a comparison of monomer and polymer FT-IR spectra (Fig. 8). The FT-IR spectra are very much alike in the region of functional groups (above 1500 cm^{-1}) for both polymers and monomers differing only in the low frequency region due to conformational changes in polymers. Unlike **LI2Bph** and **LI4Bph** macrocycles, **LIISO** and **BIISO** produced soluble products on polymerization. In case of **LIISO** only low molecular weight soluble polymer with M_w of 4000 and polydispersity of 1.7 was obtained together with a large amount of unreacted monomer (60%). The ^{1}H-NMR and FT-IR spectra (Fig. 9) confirm the structure of the polymer.

It is noteworthy that **BIISO** did not polymerize at all under the experimental conditions. **BIISO** was almost quantitatively isolated after the polymerization. As can be seen that there is a direct correlation between the polymerizability of macrocycles with calculated ring strain showing that the driving force for ROP of synthesized macrocycles is the strain release and not an entropy gain during the ring opening.

The thermal properties of synthesized polymers are listed in Table 3. Only **PLI2Bph** is crystalline showing a melting endotherm above 200°C, while others are amorphous as follows from DSC data showing only T_g in the range of 75-130°C.

Low T_g of **BLIISO** is probably attributed to its low molecular weight, while the crystallinity of **LI4Bph** is due to rigid-rod character of 4,4'-diphenyl moiety in this polymer. Due to the presence of aliphatic spacers the thermostability of the polymers is not very impressive. **LI4Bph** shows the highest 10% weight loss temperature of 360°C (Table 3).

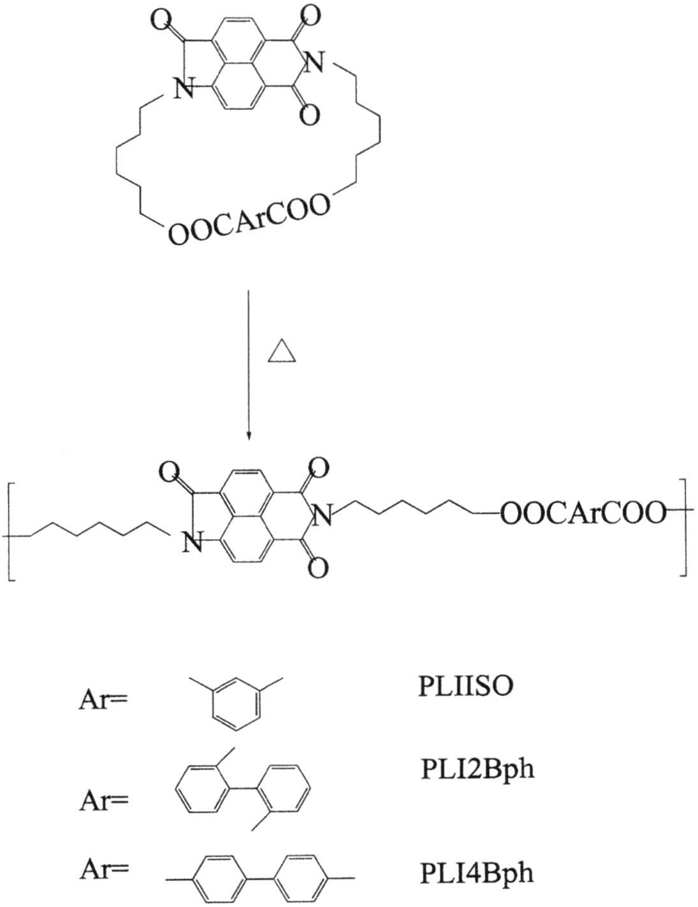

Ar=	(3-substituted benzene)	PLIISO
Ar=	(2,2'-dimethylbiphenyl)	PLI2Bph
Ar=	(4,4'-biphenyl)	PLI4Bph

Scheme 4. Ring opening polymerization of macrocycles.

Table 3.
Thermal properties of synthesized polymers

Polymer	T_m (°C)[a]	T_g (°C)[b]	T_{10} (°C)[c]
PLIISO	–	75	312.6
PLIISO	–	–	314.9
PLI2Bp	–	110	322.2
PLI2Bph	–	130	325
PLI4Bph	220	–	353.2
PLI4Bph	219	–	360

a) Melting point
b) Glass transition temperature
c) 10% weight loss temperature

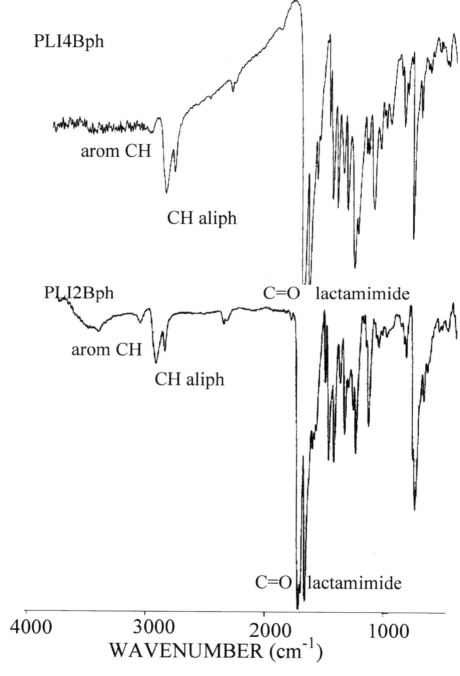

Figure 8. FT-IR spectra of **PLI4Bph** and **PLI2Bph**.

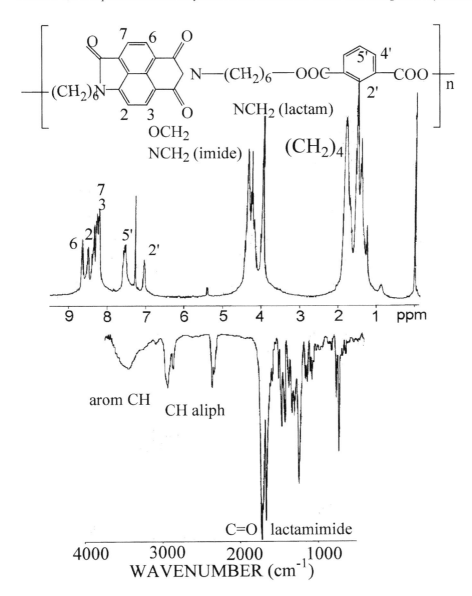

Figure 9. ^{1}H-NMR and FT-IR spectra of **PLIISO**.

4. CONCLUSIONS

Novel, well-defined lactamimide containing macrocycles were prepared by high temperature acylation of lactamimide containing diols with acid dichlorides under the pseudo high dilution conditions with up to 60% yield. The conformational analysis of low energy conformers corresponding to macrocycle precursors suggests that the cyclization is governed by entropy factor and not by enthalpy of cyclization.

On the other hand, both the experimental results and the molecular modelling data reveal that the driving force for ROP of the macrocycles prepared is the strain release on ring opening.

The molecular modelling is a powerful tool to design and select macrocyclic monomers.

The ROP of lactamimide containing macrocycles is an efficient way to obtain lactamimide containing polymers otherwise difficult to synthesize.

Acknowledgments

This work was supported by a grant from DGAPA under contract 102999. Thanks are also due to M.A. Canseco, C. Vazquez, L. Velasco and G. Cedillo for their assistance in thermal analysis, mass spectrometry and NMR spectroscopy measurements, respectively.

REFERENCES

1. D.J. Brunelle and T.G. Shannon, *Macromolecules* **24**, 3035 (1991).
2. H.Y. Juang, T.L. Chen and J.P. Xu, *Macromolecules* **30**, 2839 (1997).
3. H.Y. Juang, T.X. Liu, H.F. Zhang, T.I. Chen and Z.S. Mo, *Polymer* **37**, 3427 (1996).
4. P. Hubbard, W.J. Brittain, W.J. Simonsick and C.W. Ross, *Macromolecules* **29**, 8304 (1996).
5. K.P. Chan, Y. Wang and A.S. Hay, *Macromolecules* **28**, 653 (1995).
6. K.P. Chan, Y. Wang, A.S. Hay, X.I. Hronowsky and R.J. Cotter, *Macromolecules* **29**, 6705 (1996).
7. Y. Ding and A.S. Hay, *Macromolecules* **29**, 3090 (1996).
8. M. Memeger, L. Lazar, D. Ovenall and N.A. Leach, *Macromolecules* **26**, 3476 (1993).
9. T. Takekoshi and J.M. Terry, *U.S. Patent 5,362,845* (1994).
10. D.J. Brunelle, *Cyclic Polymers*, 2nd edition, Kluwer (2000).
11. S. Fomine, L. Fomina, V. Garcia and R. Gaviño, *Polymer* **39**, 6415 (1998).
12. S. Fomine, L. Fomina, R. Arreola and J.C. Alonso, *Polymer* **40**, 2051 (1999).
13. L. Fomina, S. Fomine, P. Ponce, T. Ogawa, L. Alexandrova and R. Gaviño, *Macromol. Chem., Phys.* **200**, 239 (1999).
14. P. Ponce, L. Fomina, E. Rivera and S. Fomine, *Macromolecular Theory Simulations* **8**, 20 (2000).
15. D.J. Brunelle, *Ring Opening Polymerization. Mechanism, Catalysis, Structure, Utility*, p. 10, Hanser, Munich (1993).
16. H.R. Kricheldorf, M. Rabenstein, M. Maskos and M. Schmidt, *Macromolecules*, **34**, 713 (2001).
17. Titan Version 1.0.5.
18. J. Foreman and A. Frisch, *Exploring Chemistry with Electronic Structure Methods*, 2nd edition, Gaussian, Inc., Pittsburgh, PA (1996).
19. Jaguar 4.1, Schrodinger, Inc., Portland, Oregon (2000).
20. A. Becke, *Phys. Rev.A* **38**, 3098 (1988).
21. C. Lee, W. Yang and R. Parr, *Phys. Rev.B* **37**, 785 (1988).
22. P. Ponce, L. Fomina and S. Fomine, *J. Phys. Org. Chem.* **14**, 657 (2001).
23. K.J. Ivin, in *Polymer Handbook* 2nd ed., Chapter II-8, Wiley, New York (1975).

Polyimides and Other High Temperature Polymers, Vol. 2, pp. 205–223
Ed. K.L. Mittal
© VSP 2003

Synthesis of aromatic benzoxazole polymers for high T_g, low dielectric properties

THUY D. DANG,[*,1] PATRICK T. MATHER,[2] MAX D. ALEXANDER,[1]
MARLENE D. HOUTZ,[3] ROBERT J. SPRY[1] and FRED E. ARNOLD[1]

[1]*Polymer Branch, AFRL/MLBP, Wright Patterson Air Force Base, 2941 P Street, Building 654, WPAFB, OH 45433-7750*
[2]*Institute of Materials Science, University of Connecticut, 97, N. Eagleville Rd., Storrs, CT 06269*
[3]*University of Dayton Research Institute, Dayton, OH 45469*

Abstract—Next generation microelectronic packaging requirements are driving the need to produce increasingly lower dielectric constant materials while maintaining high thermal stability and ease of processing. Important material requirements for these applications include low dielectric constant, high thermal, thermooxidative and dimensional stabilities, T_g above 350°C, low thermal expansion coefficient and low moisture uptake. Required dielectric constants will be between 2.0-2.4, with a progressive need for a figure of merit below 2.0. Our research efforts have focused on the synthesis and characterization of new polymer compositions with a view to meeting the objectives of imparting high thermal stability (degradation temperatures substantially higher than 350°C), low moisture uptake (less than 1%), solubility for processing in common organic solvents, as well as obtaining dielectric constants less than 2.5 and low thermal expansion coefficients. The logical extension of such an approach has recently resulted in some unique post-polymerization chemistry involving intramolecular rearrangement and crosslinking of the candidate polymeric materials to provide resistance to chemicals encountered during microelectronic packaging.

Keywords: Low dielectric constant; low-k; benzoxazole polymers; fluorinated aromatic polymers; hydrophobic polymers; microelectronic packaging.

1. INTRODUCTION

An attractive approach to lowering the dielectric constants of polymers is based on the concept of nanofoams, whereby air (dielectric constant of 1.0) replaces a portion of the polymer. This is accomplished by the use of microphase-separated ABA tri-block copolymers, in which the B blocks are thermally stable materials and the A blocks are thermally labile segments [1]. Upon thermal treatment, the A blocks undergo thermolysis leaving behind nano-sized pores. The current ther-

---

*To whom all correspondence should be addressed. Phone: (937) 255-0042, Fax: (937) 255-9157, E-mail: thuy.dang@wpafb.af.mil

mally stable B block candidate is a fluorine-containing aromatic polyimide fabricated via its poly(amic acid) precursor.

Our initial synthesis research effort was to provide an alternative structural paradigm to the thermally stable B block. The materials that were selected for this study were fluorinated aromatic benzoxazole polymer systems [2]. The benzoxazole polymer systems have the advantage of lower initial dielectric constants (3.0 versus 3.4 for polyimides), improved thermal stability, and significantly lower moisture uptake. This effort dealt with a complex balance of factors involving the T_g, solubility for processing, and low dielectric properties. Hexafluoroisopropylidene groups (6F) in the backbone are effective in lowering the dielectric constant by mechanisms that include hydrophobicity, increased free volume and decreased electronic polarizability [3-6]. In addition, fluorination is also known to improve thermal stability because of the inherently higher strength of the C-F bond over that of the C-H bond [3]. A series of new fluorinated benzoxazole homo- and copolymer systems were prepared utilizing 2,2-bis(3-amino-4-hydroxyphenyl)hexafluoropropane (BAHH), monohydroxyterephthalic acid (MHT), 2,5-dihydroxyterephthalic acid (DHT) and 2,2-bis(4-carboxyphenyl) hexafluoropropane (BCPH) monomers (Scheme 1). The purpose of the pendant hydroxyl groups is to form intramolecular hydrogen bonds with the C=N of the benzoxazole heterocyclic ring, thus forming either four or seven fused rings, (Scheme 2), thereby providing the mechanism for T_g enhancement. The fluorine-containing backbone segment can provide increased solubility as well as confer lower dielectric constant properties. Indeed, these materials exhibited very high T_g values (362-426°C), low moisture uptake (< 0.2%), low dielectric constant values (2.0-2.4), and practical solubilities in organic solvents such as tetrahydrofuran, chloroform, and o-dichlorobenzene. Thus, the successful implementation of this molecular design has obviated the use of the nanofoam approach, which is more elaborate in its materials and processing needs.

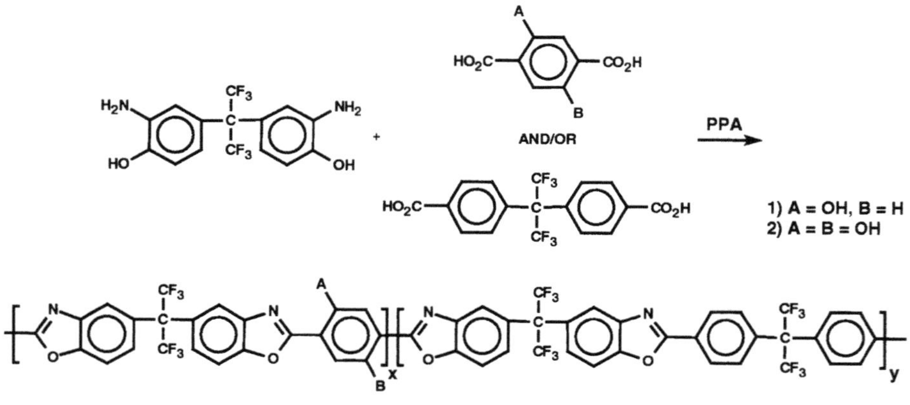

Scheme 1. Hydroxy-6F-benzoxazole homo- and copolymer synthesis.

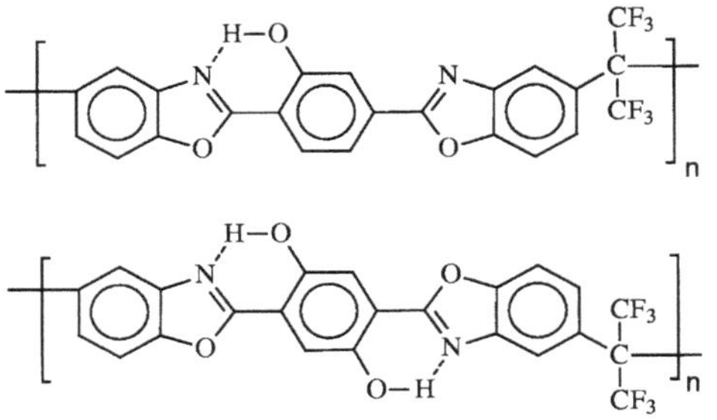

Scheme 2. Fused ring hydroxy-6F-benzoxazole structures for T_g enhancement.

Although they have met the microelectronic packaging material requirements with regard to glass-transition temperatures, dielectric properties and solution-processability, these materials need to be solvent-resistant after the fabrication process in order to be device qualified. The following effort was directed toward designing chemical routes to impart insolubility to the hydroxyl-pendant homo- and copolymer systems after processing by incorporating a crosslinking site, viz., an O-allyl moiety, directly onto the polymer backbone [7]. A unique concept that entails the sequential utilization of thermally-driven Claisen rearrangement and allyl-mediated crosslinking reactions has been investigated. Briefly described, the polymers undergo an intramolecular Claisen rearrangement (typically occurring at ~ 200°C), thus allowing further T_g enhancement, by increasing the number of fused rings via intramolecular hydrogen bonding between the *in-situ* formed OH group and the nitrogen atom of the nearby benzoxazole group (Scheme 3). At elevated temperatures (250-300°C), crosslinking of the allyl groups should occur, thus providing a mechanism for promoting insolubility as well as the dimensional stability to the polymer system. To demonstrate this concept, three model compounds containing pendant allylether and bis(allylether) groups were synthesized. They were analyzed via differential scanning calorimetry (DSC) to determine whether the expected Claisen rearrangement had taken place and to observe what effects this had upon the thermal properties of the compounds. The initial DSC trace of the diallyloxybenzoxazole model compound did indeed demonstrate that a Claisen rearrangement occurred first, followed by the thermal cure of the allyl group in the 300-400°C range.

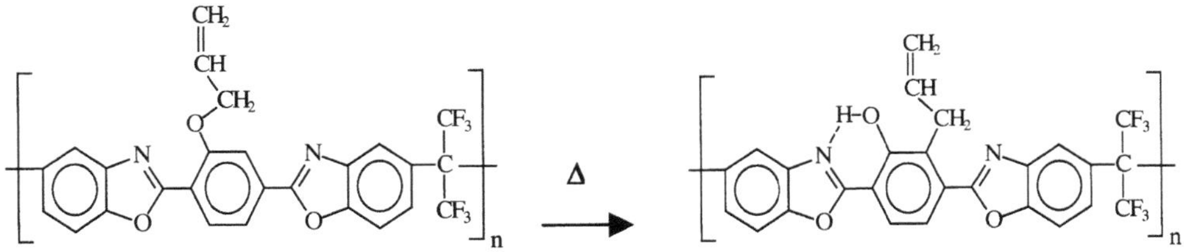

Scheme 3. Intramolecular Claisen rearrangement for the O-allyl-6F-benzoxazole unit.

The overall objective of this phase of our efforts was geared toward creating a polymer system that retained the low dielectric constant and the high glass transition temperature of the hydroxy-6F-PBO polymer while becoming solvent resistant and mechanically robust via post-polymer chemistry. The reaction paradigm to achieve this goal was examined with select copolymer compositions incorporating hydroxy-6F/12F benzoxazole structural units, as will be described later. A further step in our investigation also sought to control crosslink density by the partial allylation of the hydroxyl groups attached to the aromatic benzoxazoles. To accomplish this objective, two random tri-block aromatic 6F/12F benzoxazole copolymers were successfully synthesized using 5 and 10 mole percents of allyl bromide to effect partial O-allylation of the previously synthesized hydroxyl pendant-6F/12-F-PBO through a post-polymerization reaction. The properties of the resulting partially-allylated tri-block copolymer systems were also studied and are reported here.

2. EXPERIMENTAL

2.1. Materials

Monohydroxyterephthalic acid (MHT) was prepared by the procedure described in the literature [8]. 2,5-dihydroxyterephthalic acid (DHT) was purchased from Aldrich and recrystallized from ethanol (mp = 319-320°C). Polyphosphoric acid (assay > 83%; Fluka) was used as received, as a polymerization solvent. Ethylsalicylate, 2,2-bis(3-amino-4-hydroxyphenyl)hexafluoropropane (BAHH) and 2,2-bis(4-carboxyphenyl)hexafluoropropane (BCPH) were obtained from commercial sources or through custom synthesis and used as received.

2.2. Monomer synthesis

The required monomer, monohydroxyterephthalic acid [8], was prepared from the displacement reaction of 2-bromoterephthalic acid and sodium hydroxide in the presence of sodium acetate and copper powder. Recrystallization from ethanol provided white crystals, m.p. 325-326°C.

2.3. Polymer synthesis

In an effort to establish what effect intramolecular hydrogen bonding might have on the dielectric properties, 2-hydroxybenzobisoxazole rigid-chain polymer was synthesized (Scheme 4). The polymer from monohydroxyterephthalic acid (MHT) and 1,3-dihydroxy-4,6-diaminobenzene dihydrochloride (DHDAB) was prepared in polyphosphoric acid (PPA) following the P_2O_5 polymerization method [9]. The polymerization was carried out at a polymer concentration of 14 wt% to promote the formation of the anisotropic rigid-rod polymer composition. It was shown that the intramolecular hydrogen bonding had a minimal effect, if any, on the dielectric constant of polybenzobisoxazole (PBO) backbone (2.93 for pristine PBO) since a dielectric constant of 2.73 (at 10^6 Hz) was measured for 2-hydroxy-polybenzobisoxazole polymer.

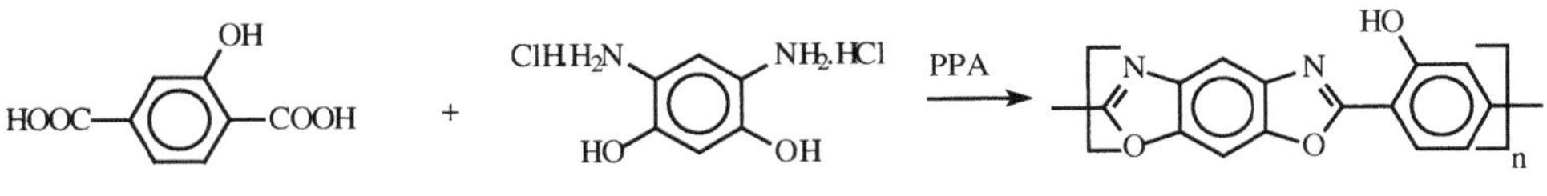

Scheme 4. Synthesis of 2-hydroxypolybenzobisoxazole.

The fluorinated homopolymers were prepared by the polycondensation of MHT or DHT with BAHH in polyphosphoric acid. Polymerizations were carried out at polymer concentrations of 12 wt% and reaction temperatures up to 180°C. Copolymers were prepared in an analogous fashion using BCPH as the co-acid monomer (Scheme 1). All the polymers were isolated by precipitation in water, treated with ammonium hydroxide and soxhlet extracted with water until they were acid free. All the polymers prepared were soluble in organic solvents and were fabricated into films for evaluation with the exception of 2-hydroxypolybenzobisoxazole, which was fabricated from methane sulfonic acid (MSA).

2.4. Model compound synthesis

A post-reaction allyl substitution was employed as the method of choice to synthesize all the model compounds for this study. In this approach, O-allylation was carried out as the final step, after the hydroxy-6F-benzoxazoles had formed (Scheme 5).

2.5. Post-polymer synthesis

The methodology, eminently successful in the case of the model compounds, was extended to the synthesis of the homo- and copolymers in this study, as described in Scheme 6. A typical procedure is described for the post-derivatization of monohydroxy-6F/12F (50/50) benzoxazole copolymer.

0.26 g of potassium carbonate was suspended in 100 ml anhydrous dimethylformamide (DMF) under a nitrogen atmosphere and heated to 60°C. 1.0 g of the copolymer was added to the solution and stirred until the polymer dissolved to form a red, homogeneous solution. Allyl bromide (0.3 g) was added and the mixture was stirred under nitrogen for 16 hours; the solution turned yellow, indicating completion of the reaction. The polymer was precipitated in water and agitated into a fine suspension. The light yellow solid was soxhlet-extracted with hot hexane to remove any unreacted allyl bromide. The polymer was finally vacuum-dried at 56°C for 24 hours to afford 0.98 g polymer (95% yield).

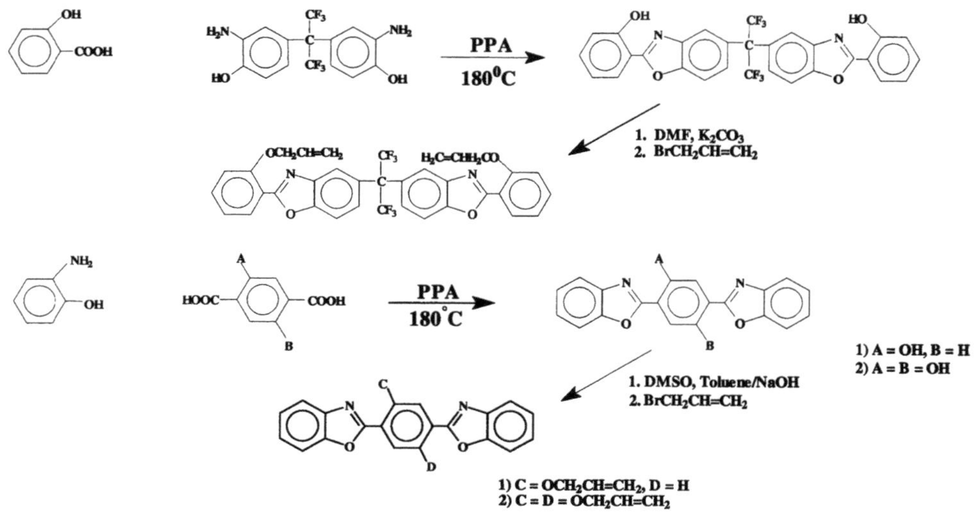

Scheme 5. Synthesis of allyloxy benzoxazole and allyloxy 6F-benzoxazole model compounds.

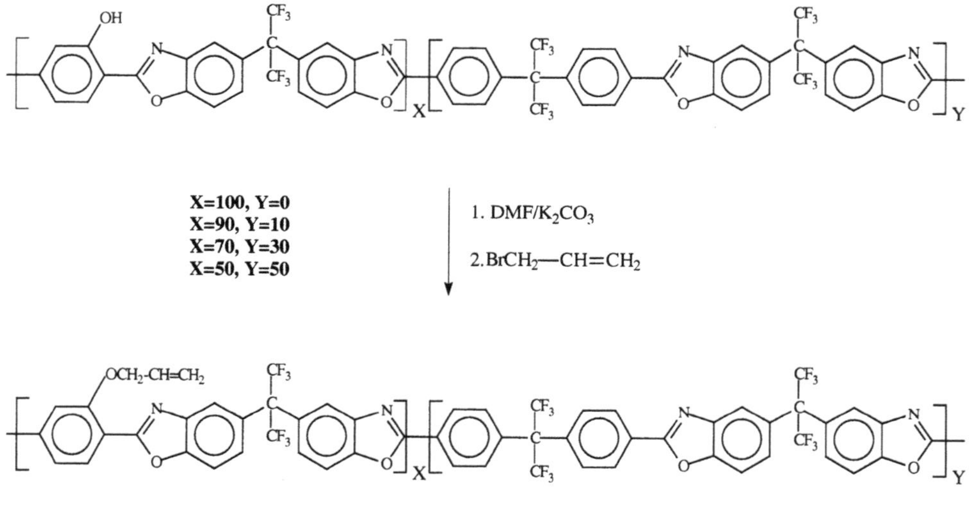

Scheme 6. Post-polymer O-allylation reactions for the monohydroxy 6F-benzoxazoles.

2.6. Elemental analysis

Elemental analysis and mass spectrometry were performed by Chemsys, Inc. at Wright-Patterson Air Force Base. A Finnegan 4500 mass spectrometer was used to obtain mass spectra (solid state). Melting points were obtained from a MelTemp capillary melting point apparatus and used without correction. The water uptake of the polymers was measured by determining the weight difference of several film samples before and after submersion in boiling water for three days.

2.7. Thermal analysis

Differential scanning calorimetry (DSC) was employed for the measurement of the glass transition temperatures and detection of any other thermal transitions. For this purpose, a TA Instruments, DSC 2910, was used with a heating rate of 10°C/min for samples weighing 5-15 mg. Decomposition onset temperatures were determined by thermogravimetric analysis (TGA) employing a TA Instruments, TGA 2950. The in-plane coefficients of thermal expansion (CTE) of the copolymers were measured as a function of temperature via thermomechanical analysis (TMA). The technique utilized a TA Instruments, TMA 2940, in tensile film mode with a heating rate of 4°C/min and typical sample dimensions of 100 μm thick x 15 mm long x 5 mm wide. The CTE's were determined from the slopes of the resultant expansion-temperature plots.

2.8. Dynamic mechanical analysis

Dynamic mechanical analysis (DMA) was performed in order to determine the influence of the copolymer constitution on tensile modulus and on mechanical relaxation behavior. For this purpose a Perkin Elmer DMA-7 was run in tensile mode, at an oscillation frequency of 1 Hz with a static stress level of 5×10^5 Pa and a superposed oscillatory stress of 4×10^5 Pa. With this stress controlled instrument, the strain and phase difference between stress and strain are the measured outputs. Typically, the resulting strain levels ranged from 0.05% to 0.2% while sample dimensions were 8 mm long, 2 mm wide, and 0.1 mm thick. A gaseous helium purge and a heating rate of 3°C/min were employed. The temperature scale was calibrated with indium, and the force and compliance calibrations were performed following conventional methods.

2.9. Dielectric spectroscopy

Dielectric measurements were made mainly with an EG&G Model 398 Impedance System over a frequency range from 0.1 Hz to 5 MHz, using an amplitude of 5 mV rms. The sample holder consisted of two stainless steel circular plates of 12.7 mm diameter pressed against each side of the polymer film. Sample thicknesses as measured by a digital micrometer, an optical microscope and scanning electron microscopy ranged from 38 μm to 134 μm, insuring that the dielectric constants were bulk values.

Certain samples were also selected for the in-plane dielectric constant determination by the interdigitated comb method [10, 11]. These measurements were performed on a TA Instruments, DEA 2970 Dielectric Analyzer, over a frequency range from 3×10^{-2} Hz to 10^5 Hz.

Waveguide coupling was employed to measure the refractive index of selected polymer films of 4 μm thickness at optical frequencies and to determine the anisotropy of the polymer films [12-14]. For these measurements, a Metricon Model 2010 Prism Coupler was employed at 632.8 nm and 1152 nm wavelengths. The data also provided an additional verification of film thickness. The dielectric constant, ε, was obtained from the refractive index, n, through the Maxwell relation:

$$\varepsilon = n^2$$

3. RESULTS AND DISCUSSION

3.1. Solution properties and thermal analysis

High molecular weight materials were obtained as evidenced by their solution viscosities, η, in MSA, listed in Table 1 and by the tenacity of the films fabricated. Moisture uptake for the fluorine containing polymers was less than 0.2%. Solvents for the polymers ranged from chloroform and tetrahydrofuran (THF) for the highly fluorinated polymers to o-dichlorobenzene and MSA for the polymers with a high content of the aromatic hydroxyl pendants. The T_gs of the polymers, determined by differential scanning calorimetry (DSC), are listed in Table 1, which also lists T_g values for selected samples from dynamic mechanical analysis (DMA). The values from the two different methods are in substantial agreement. A typical example of DSC data obtained on sample **6** is shown in Figure 1. The T_g value (361°C) is obtained from the inflection point (I).

The introduction of intramolecular hydrogen bonding via the interaction of pendant hydroxyl groups with the C=N of the benzoxazole unit greatly enhances

Table 1.
Solution and thermal properties of various fluorinated benzoxazole homo- and copolymers with hydroxyl pendants

Polymer number	A	B	X	Y	$[\eta]^a$	$T_g^{\ b}$ (°C)	$T_g^{\ c}$ (°C)
1	–H	–H	0	100	2.30	325	315
2	–H	–H	50	50	1.37	346	336
3	–OH	–H	100	0	2.06	426	390
4	–OH	–H	90	10	1.30	415	410
5	–OH	–H	70	30	2.34	373	371
6	–OH	–H	50	50	1.35	362	353
7	–OH	–OH	100	0	3.10	>450	454
8	–OH	–OH	75	25	1.34	>450	–
9	–OH	–OH	50	50	0.88	>450	430
10	–OH	–OH	25	75	0.47	>450	–

^{a}MSA, 30°C, 0.12 g/dL
^{b}DSC data
^{c}DMA, loss tangent peak

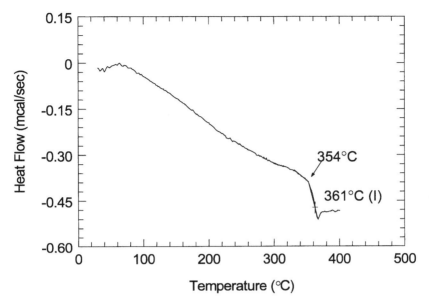

Figure 1. DSC trace of 50/50 6FPBO-OH-co-12FPBO (polymer **6**) showing T_g onset at 354°C and an inflection point at 361°C.

the polymer T_g. This can be ascertained from the data in Table 1 by comparing sample groups **7-10**, **3-6**, **1-2**, and by comparison among samples **3-6**. The T_g enhancement is clearly borne out to be a function of the number of consecutive fused rings in the hydroxy 6F-benzoxazole systems as envisaged in Scheme 2.

Thermogravimetric analysis (TGA) showed that the onset temperature of degradation of sample **6** occurred at approximately 533°C in helium (Figure 2). This high onset degradation temperature is representative of all the polymers listed in Table 1.

3.2. Dynamic mechanical analysis

A comparison of the tensile storage moduli among different compositions are shown in Figure 3, for polymers **2**, **4**, **5**, and **6**, revealing the impact of the hydroxyl functionality on the mechanical behavior of the polymers. It is observed that while the tensile moduli are all in the range of 1.5 to 2 GPa (for temperatures near room temperature), the fall-off in tensile modulus with temperatures near T_g shows sensitivity to the copolymer composition. In particular, it should be noted that the hydroxyl content of polymers in Figure 3 ranges from polymer **2** with no hydroxyl content, to polymer **4** with 90% content of the hydroxyl-bearing repeat unit. The glass transition temperature *increases* substantially while the negative slope of log (modulus) versus temperature *decreases*. Intermediate levels of hydroxyl content in the polymers, i.e. polymers **5** and **6**, lead to mechanical behavior intermediate between the two extremes in hydroxyl content.

T.D. Dang et al.

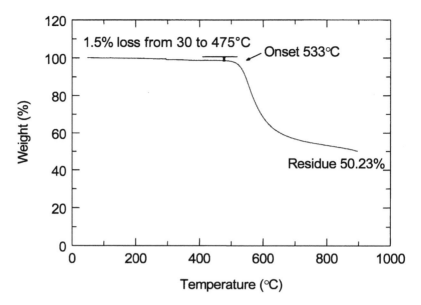

Figure 2. TGA trace, in helium, of 50/50 6FPBO-OH-co-12FPBO (polymer **6**).

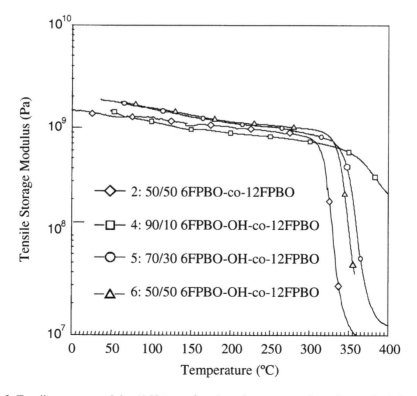

Figure 3. Tensile storage modulus (1 Hz) as a function of temperature for polymers **2**, **4**, **5** and **6**.

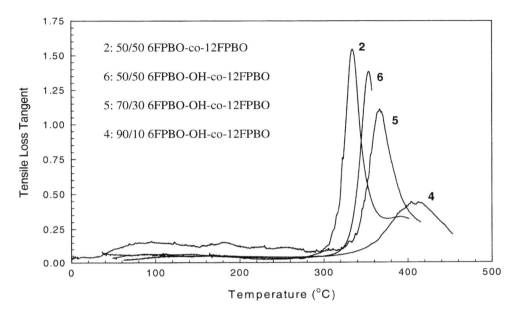

Figure 4. Tensile loss tangent (1 Hz) as a function of temperature for polymers **2**, **4**, **5** and **6**.

The effect of copolymer composition on the glass transition behavior is perhaps best seen in Figure 4, where the tensile loss tangent (loss modulus/storage modulus) is plotted versus temperature for the same copolymers appearing in Figure 3. The T_g results from the peaks of the DMA loss tangents in Figure 4 are compared to the T_g values from DSC measurements (Table 1). The agreement is very good, except for the DMA value for polymer sample **3** (Table 1), the 6F-PBO-OH homopolymer. This slight discrepancy is not understood but is observed reproducibly. Increasing the hydroxyl content of the polymers dramatically increases the glass transition temperature, as indicated by the observed rightward shift of the loss tangent peak (Figure 4) with hydroxyl content: from a value of 336°C for polymer **2** to 410°C for polymer **4**. This shift in T_g of ~ 75°C by simply substituting a hydroxy-functionalized monomer is a clear example of the concept of altering thermal properties of polymers using fused rings.

In addition to the observed increase in T_g in Figure 4, both the magnitude and shape of the loss tangent peak through the glass transition change significantly enough to be commented upon. More specifically, the width of the peak increases from 23°C full-width-at-half-maximum (FWHM) for polymer **2** to 80°C FWHM for polymer **4**. This large transition in the peak width is accompanied by a substantial decrease in the peak value of the loss tangent with increasing hydroxyl content. Polymer **2** features a peak loss tangent of 1.54, while polymer **4** shows a smaller peak loss tangent of 0.44. This observed decrease in the peak value of the loss tangent together with the observation in Figure 3 that the storage moduli for temperatures below the glass transition are roughly the same is an indication that the modulus of the rubbery material (i.e., for $T > T_g$) increases with increasing hydroxyl content. The origin of this behavior in unclear, but suggests that some *interchain* hydrogen bonding begins to play a distinct role above T_g.

3.3. Dielectric analysis

The dielectric constants for all the polymers at 100 Hz, 10 kHz and 1 MHz are listed in Table 2, with values at 1 MHz ranging from 2.09 to 2.95. The frequency dependencies are typical for all of the polymers, i.e., a small monotonic decrease in ε as a function of increasing frequency. The dielectric constants for polymers **5** and **6** are exceptionally low for non-foamed cast polymers, making them desirable candidates to be evaluated for the high temperature processing of high-speed integrated circuits.

Other constituents being equal, the polymers or copolymers with the largest fraction of 6F units should have the lowest dielectric constant. This was observed to hold true by comparing the dielectric constants of polymers **1** and **2**. For polymers **3-6**, the order should be $\varepsilon(6) < \varepsilon(5) < \varepsilon(4) < \varepsilon(3)$. The correct trend is present overall, but there is some deviation from the perfect order, probably due to small measurement errors or variations among sample properties (e.g., density). The dielectric constant of polymer **9** would be expected to be smaller than polymer **7**, but it is not. Again this may be caused by small measurement errors or sample variability. However, another factor which changes with composition is the interchain conformation whose effect upon dielectric constant is complex, a subject to be discussed in a future publication.

The effect of the OH groups upon dielectric constant is most clearly ascertained by comparing polymers **2**, **6**, and **9** and polymers **3** and **7**. Polymers with two OH units have a higher dielectric constant than those with one OH unit. On the other hand, the polymers with one OH unit have a lower dielectric constant than those having no OH units. The possession of two OH units produces a planar segment with seven consecutive fused rings, while the single OH-containing segment would have four fused rings. The greater planarity of the seven-ring system would allow closer packing of the polymer chains, reducing the free volume and increasing the dielectric constant. The reason for the lower dielectric constants for polymers containing one OH group compared to those with no OH groups is not understood at this point.

The optical frequency dependence and anisotropy of the dielectric constant of polymer **6** were investigated by prism coupling, and the results are presented in Table 3. The out-of-plane dielectric constant at the 632.8 nm wavelength is larger than the dielectric constant at 1152 nm, and both are larger than the out-of-plane impedance results. The wavelength dependence is to be expected because of the dispersion associated with the HOMO-LUMO absorption in the UV. The dielectric anisotropy is common to very thin cast polymer films and is caused by preferential alignment to the first deposited molecules parallel to the casting dish surface [3].

Table 2.
Dielectric constants at selected frequencies for the fluorinated benzoxazole polymers with hydroxyl pendants

Structure	Dielectric constant		
	100 Hz	10 kHz	1 MHz
1	2.41	2.36	2.32
2	2.44	2.42	2.40
3	2.38	2.37	2.35
4	2.49	2.46	2.42
5	2.14	2.12	2.09
6	2.23	2.19	2.15
7	3.02	2.95	2.88
9	3.14	3.04	2.95

Table 3.
Dielectric constant anisotropy data for 50/50 6FPBO-OH-co-12FPBO polymer

Waveguide results						
λ (nm)	$n_{x,\,y}$	n_z	Δn	$\varepsilon_{x,\,y}$	ε_z	$\Delta\varepsilon$
632.8	1.5965	1.5523	0.044	2.549	2.410	0.139
1152	1.5613	1.5230	0.038	2.438	2.320	0.118

Electrical measurements		
Transverse dielectric constant	$(10^5$ Hz$)\ \varepsilon_z = 2.19$	
Longitudinal dielectric constant	$(10^5$ Hz$)\ \varepsilon_{x,\,y} = 2.33$	$\Delta\varepsilon = 0.14$

The anisotropy of the dielectric constant of polymer **6** was also investigated by comparing the dielectric constant from transverse impedance measurements with that from interdigitated electrode measurements. The results at 10^5 Hz are also listed in Table 3, this frequency being the upper frequency limit of the interdigitated electrode method. It is seen that the anisotropy from the electrical method is 0.14, which is consistent with the optical anisotropy values. These values are within acceptable standards for integrated circuit packaging [3, 15, 16].

A number of polymers and copolymers with very low, practical dielectric constants have been synthesized and characterized. A likely candidate for the high temperature processing of high-speed integrated circuits would be polymer **6**, 50/50 6FPBO-OH-co-12FPBO. The T_g is 362°C, T_d is 532°C, moisture uptake is < 0.2% and the coefficient of thermal expansion is 53 ppm from −127°C to 323°C. The dielectric constant is 2.15 at 1 MHz, low enough to be considered suitable for microelectronic devices. The dielectric anisotropy of 0.14 is also within acceptable limits for such an application. If a higher T_g is required, the candidate would be the homopolymer **3** with a T_g of 426°C, but with a somewhat higher dielectric constant at 1 MHz of 2.35.

3.4. Post-polymer reactions via O-allylation of monohydroxy-6F-benzoxazole polymers

Aromatic benzoxazole homo- and copolymers containing allyl ether pendent groups were synthesized in quantitative yields and with high molecular weights as evidenced by their intrinsic viscosities (Table 4) and the tenacity of the films fabricated from chloroform or THF. The allylation of the hydroxyl pendants in the polymers was generally very extensive as indicated by IR spectra and elemental analysis. Corroboratively, it was found that model O-allylated compounds in this study were pure materials obtained in moderate to good yields, characterized by sharp melting points, mass spectra, and elemental analysis.

The proof of the concept for achieving enhancement of the T_g of O-allylated 6F-benzoxazole polymers via Claisen rearrangement and imparting solvent resis-

Table 4.
Thermal properties of O-allylated 6F/12F benzoxazole polymers

Polymer number	X	Y	$[\eta]^a$	Exotherm 1 (Max, °C)	Exotherm 2 (Max, °C)	T_g (°C)	$T_d^{\ b}$ (onset, °C)
1	100	0	1.20	220	395	371	527
2	90	10	1.41	240	386	307**	514
3	70	30	1.38	228	398	434	516
4	50	50	1.42	257	363	350**	517

	X	Y	Z				
5	10	50	40	1.40	244	385	324	514
6	5	50	45	1.43	248	390	390	519

^{a}dL/g, MSA, 30°C ^{b}re-scan after heating to 400°C

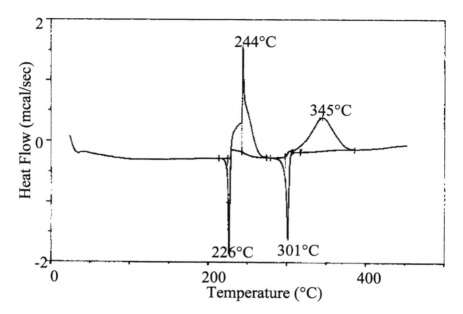

Figure 5. DSC scan of the model compound, 2,5-bis(2-benzoxazolyl)-1,4-di(allyloxy)benzene.

tance to the polymers through thermal cure of the allyl functionality was provided by model compound chemistry. DSC characterization of the model compound (Figure 5), 2,5-bis(benzoxazolyl)-1,4-di(allyloxy)benzene exhibited a clear trend. It showed an endotherm from 224-226°C, which corresponded well with the capillary melting point (222-223°C) of the model compound. This is followed by an exotherm in the 225-275°C range, centered at 244°C. This is presumably due to the formation of the product of a Claisen rearrangement reaction of the starting material. A second endotherm occurs at 298-301°C, which should presumably correspond to the melting point of the new product. A final exotherm in the 303-400°C temperature range and centered at 345°C is attributable to the crosslinking reaction of the allyl groups.

By analogy, the DSC scans of a candidate O-allyl-6F-12F (50/50) benzoxazole copolymer (Figure 6) exhibit two distinct exotherms centered at 252°C and 363°C, attributable to thermal rearrangement of the phenylallyl ether and crosslinking reaction, respectively. The T_g of the cured polymer is 350°C, as can be seen by DSC.

Table 4 also summarizes the maximum temperatures for the exotherms that correspond to the Claisen rearrangement, crosslinking reactions, the T_g of the thermally cured materials as well as the onset temperatures of degradation of the aromatic benzoxazole homo- and copolymers containing monoallylether groups. Also included in the Table are the thermal characteristics of the random triblock copolymers which are partially allylated. High glass transition temperatures after

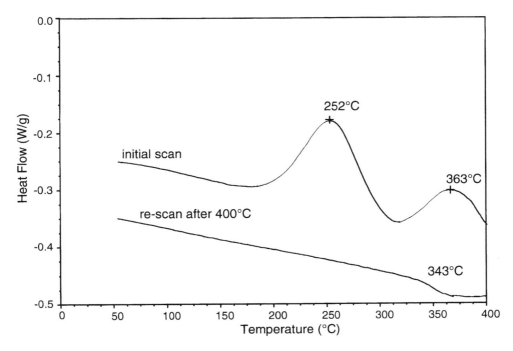

Figure 6. DSC scan of 50/50 6FPBO-O-allyl-co-12FPBO.

cure were obtained for all the polymers and the solvent resistance, as well as the toughness and creasability of the polymer films, were evident upon curing around 350°C for two hours.

Scheme 7 is illustrative of the changes in the glass transition temperature and dielectric constant values as a function of the successive chemical transformations that occur in the candidate O-allylated 6F/12F (50/50) benzoxazole copolymer when heat-treated under an inert atmosphere. The T_g enhancement (from 278°C to 312°C) as a consequence of the thermal rearrangement at 260°C and the final cure at 360°C is quite evident from the scheme. The values of the dielectric constant measured on the polymer films remain overall low during different stages of processing, with an ε of 2.5 obtained for the cured polymer. This demonstrates that the complex materials requirements for microelectronic packaging can be met by utilizing the post-polymer reaction strategy without compromising thermal and dielectric properties.

Current research efforts are directed towards initiating the Claisen rearrangement reaction of the model compounds in high boiling solvents to facilitate their isolation and characterization by IR and NMR. *In situ* monitoring of the changes in the chemical structures of fabricated films of the homo- and copolymers will be performed using variable temperature infrared spectroscopy to examine the rearrangement and crosslinking reactions at elevated temperatures.

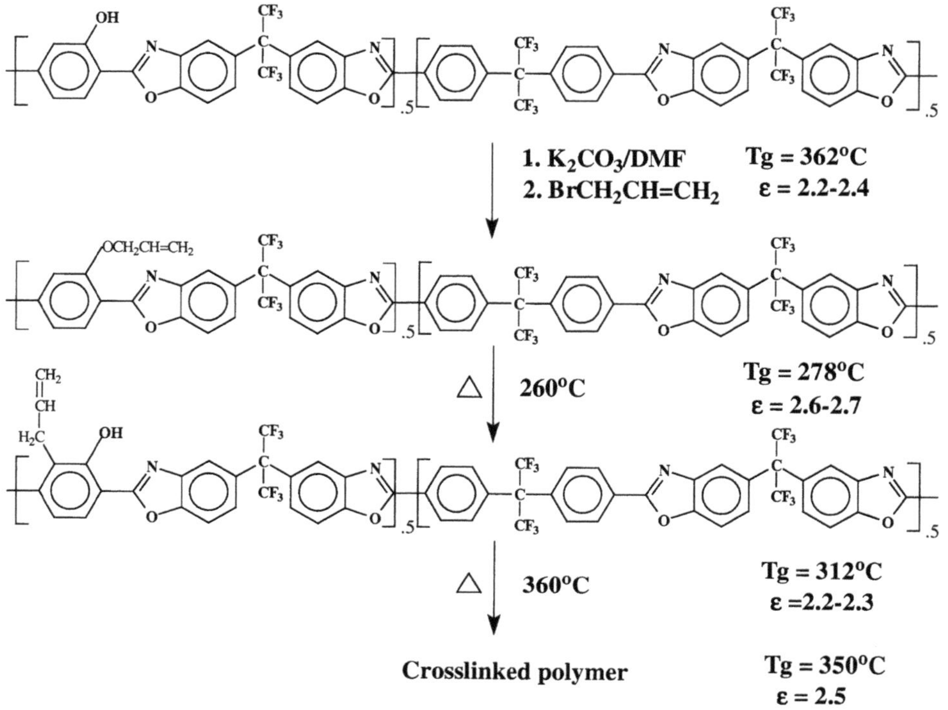

Scheme 7. Thermal and dielectric properties of O-allylated 6F/12F (50/50) benzoxazole copolymer as a function of intramolecular rearrangement and crosslinking.

4. CONCLUSIONS

Aromatic benzoxazole polymers carrying perfluoroisopropyl groups in their backbone and with hydroxyl pendants were found to exhibit high T_g and low dielectric characteristics. The dielectric constants were lowered as a function of the fluorine content in these polymers and the high glass transition temperatures could be attributed to the presence of fused rings due to intramolecular hydrogen bonding. Their high thermo-oxidative stabilities and low moisture uptake ensured that these materials were highly compatible with integrated circuit processing. The coefficient of thermal expansion and dielectric anisotropy were also within acceptable limits for such applications. A post-polymer reaction paradigm was also examined in which Claisen rearrangement and polymer crosslinking were sequentially employed on O-allylated 6F-benzoxazole polymers, for T_g enhancement and for imparting solubility, respectively. The desired solvent-resistance of the polymer films after the fabrication process along with overall retention of mechanical integrity and low dielectric properties render these materials eminently suitable for future evaluation in microelectronic packaging technology.

Acknowledgement

The authors acknowledge Dr. Narayanan Venkat (University of Dayton Research Institute) for useful technical discussions during the preparation of the manuscript.

REFERENCES

1. J.L. Hedrick, L.D. Russell, D. Hofer and V. Wakharker, *Polymer* **34**, 4717 (1993).
2. B.A. Reinhardt, *Polym. Commun.*, **31**, 453 (1990).
3. E.T. Ryan, A.J. McKerrow, J. Leu and P.S. Ho, *MRS Bull.* **22(10)**, 49 (1997).
4. G. Hougham, G. Tesoro and J. Shaw, *Macromolecules* **27**, 3642 (1994).
5. G. Hougham, G. Tesoro, A. Viehbeck and J.D. Chapple-Sokol, *Macromolecules* **27**, 5964 (1994).
6. G. Hougham, G. Tesoro and A. Viehbeck, *Macromolecules* **29**, 3453 (1996).
7. T.D. Dang, L.S. Hudson, W.A. Feld and F.E. Arnold, *Polym. Preprints* (American Chemical Society) **41**(1), 103 (2000).
8. Y. Miura, E. Torres and C.A. Panetta, *J. Org. Chem.* **53**, 439 (1998).
9. J.F. Wolfe, in: *Encyclopaedia of Polymer Science and Engineering*, H.F. Mark and J.I. Kroschwitz (Eds.), vol. 11, p. 601 (1988).
10. J.P. Runt and J.J. Fitzgerald (Eds.), *Dielectric Spectroscopy of Polymeric Materials: Fundamentals and Applications*, American Chemical Society, Washington, D.C. (1997).
11. A. Deutch, M. Swaminathan, M.H. Ree, C.W. Surovic, G. Arjavalingan, K. Prasad, D.C. McHerron, M. McAllister, G.V. Kopcsay, A.P. Giri, E. Perfecto and G.E. White, *IEEE Trans. Components, Packaging, and Manufacturing Technology*, **B17**(4), 486 (1994).
12. R. Ulrich and R. Torge, *Appl. Opt.* **12**, 2901 (1973).
13. S.J. Bai, R.J. Spry, D.E. Zelmon, U. Ramabadran and J. Jackson, *J. Polym. Sci. B: Polym. Phys. Ed.*, **30**, 1507 (1992).
14. S.J. Bai, R.J. Spry, M.D. Alexander, Jr. and J.R. Barkley, *J. Appl. Phys.*, **79**, 9326 (1996).
15. N.P. Hacker, *MRS Bull.*, **22(10)**, 33 (1997).
16. H.S. Nalwa, M. Suzuki, A. Takabashi and A. Kageyama, *Appl. Phys. Lett.*, **72**, 1311 (1998).

Polyimides and Other High Temperature Polymers, Vol. 2, pp. 225–239
Ed. K.L. Mittal
© VSP 2003

Polybenzobisthiazoles – Critical issues in their performance and properties

ROLI SAXENA* and L.D. KANDPAL

Polymer Science Division, Defence Materials & Stores Research & Development Establishment, G. T. Road, Kanpur-208 013, India

Abstract—In the last few years researchers on high performance rigid-rod polymers have developed a variety of polymers, which have been studied for their tensile, compressive, and thermal properties. We describe here some recent research developments on PBZTs with regard to the synthesis of the polymers and their properties. In doing this some properties of polymers as well as the problems related to the compressive strength of PBZT fibres are discussed.

Keywords: Rigid-rod polymers; polybenzobisthiazoles (PBZT); compressive properties; tensile properties; fibres.

1. INTRODUCTION

Poly (p-phenylenebenzobisthiazole) (PBZT) is a member of a new class of highly rigid, linear, thermally-stable aromatic heterocyclic polymers. These aromatic heterocyclic polymers with high molecular weight have para-catenated backbone, and this renders them to be rod-like polymers. The only conformational flexibility is provided by the rotation of bonds between alternating phenylene and heterocyclic groups [1, 2]. PBZTs are environmentally stable, high strength and high modulus materials.

PBZTs have main chain, repeat-unit structures containing one or more aromatic heterocyclic moieties. The backbone bonds that link the heterocyclic units in each repeat unit are positioned regularly thereby defining a specific catenation angle, i.e., the angle between the two exocyclic bonds of each rigid unit. In wholly aromatic systems, these exocyclic bonds are formally single bonds but may have appreciable double bond character if adjacent rings are coplanar. Resonance structures of this type contribute to increased backbone rigidity, which is a characteristic of PBZTs.

---

*To whom all correspondence should be addressed. E-mail: rolls342000@yahoo.com
Present address: WI-134, Samtel Centre for Display Technology, I.I.T, Kanpur-208016, India.

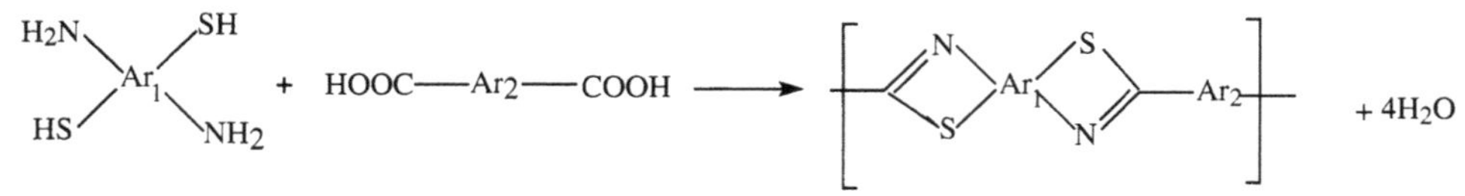

Diaminodithiol   Diacid   Polybenzobisthiazole   Water

Scheme 1. General method for synthesis of PBZT.

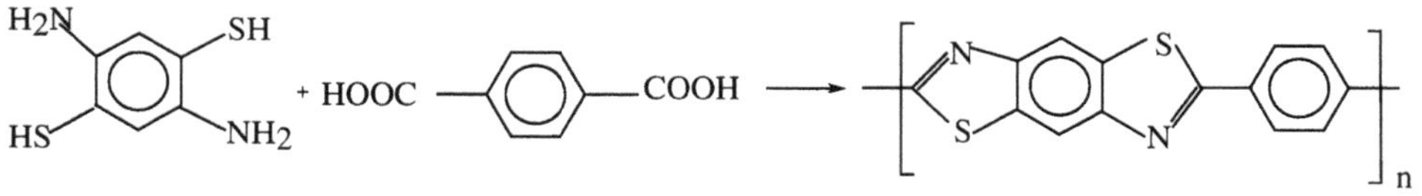

2,5-diamino-1,4 - benzene   Terephthalic acid   Poly{(benzo[1,2-d;4,5-d']-bisthiazole
dithiol dihydrochloride   -2,6-1,4-phenylene} (PBZT)

Scheme 2. Synthesis of parent PBZT.

Since the late 1970s the Air Force Materials Laboratory, Wright Patterson Air Force Base had program in lyotropic liquid crystalline rigid-rod polymers. The program has encompassed theoretical aspects, synthesis, solution processing, mechanics and morphology in an effort to exploit the potential of using the ordered polymers in structural materials for aerospace application.

The synthesis of an aromatic heterocyclic PBZT polymer is illustrated by the reaction of an aromatic diaminodithiol and an aromatic dicarboxylic acid to form a polybenzobisthiazole (Scheme 1). While the synthesis scheme looks simple but it is quite complex in practice. A number of parameters have to be considered when carrying out the polymerisation reaction. These include,

(1) Monomer reactivity

(2) Reaction rate

(3) Yield

(4) Polymer molecular weight

(5) Molecular weight distribution

Good monomer reactivity, reaction rate and yield are desired for economic reasons and the high molecular weight and narrow molecular weight distribution are needed for fibre preparation. However, the reaction time for these polymers is considerably long because of low monomer reactivity. Wolfe and colleagues [3-5] synthesized a polybenzobisthiazole (PBZT) from the reaction of 2,5-diamino-1, 4-benzenedithiol dihydrochloride [6, 7] and terephthalic acid as shown in Scheme 2. Wolfe and Sybert [8] demonstrated that rod-like heterocyclic polymers of high molecular weight could be obtained by solution polymerisation in polyphosphoric acid (PPA). These polymers are soluble in methanesulphonic acid (MSA) and chlorosulphonic acid (CSA).

2. SYNTHESIS OF PBZTs FROM DIFFERENT ACIDS

A number of polymers having different modifications in the structures have been synthesized or are currently being synthesized to obtain better properties. Two fluorine containing aromatic polybenzobisthiazoles (**I, II**) [9] were synthesized by the direct polycondensation of 4,4'-(hexafluoroisopropylidene) dibenzoic acid and tetrafluoroterephthalic acid with 2, 5-diamino-l, 4-benzenedithiol dihydrochloride. These PBZTs were amorphous and showed good resistance to organic solvents, excellent mechanical properties, and high thermal stability. PBZTs and copolymers having 2,5-dihydroxy bicyclo [2.2.2] octane (**III**) [10] and triarylamino [11] moieties have also been synthesised via diacid monomers.

(**I**) PBZT with hexafluoroisopropylidene unit

(**II**) PBZT with bicyclo [2.2.2] octane moiety

(**III**) PBZT with tetrafluoroterephthalic acid

A PBZT was also synthesized using 1,4-bis (trichloromethyl) benzene [12]. A PBZT with tetramethylbiphenyl was synthesized and was found to be more stable at elevated temperatures than the parent PBZT. Although the weight loss starts at 505°C, i.e., 135°C less than the parent PBZT, but 80% weight is retained up to 900°C. All these efforts on the rigid-rod polybenzobisthiazole polymers have been devoted to structural modifications of the rod-like polymers in order to get better properties for these materials. It is known that compressive properties of these materials are not at par with other mechanical properties. Structural tailoring with pendent groups has provided various approaches to investigate and improve compressive properties of high modulus PBZT films and fibres. Polybenzobisthiazole containing substituted p-terephenylene dicarboxylic acid (**IV**) [14] was utilized to disrupt the packing order and potentially increase the compressive properties.

(**IV**) Terphenylene unit

(**V**) Random copolymer containing terephthalic / terphenylene unit

Random copolymers were prepared [15] in PPA by replacing a small portion (1-2 mol%) of the terephthalic acid with terphenylene diacid (**V**). Polymerisations were carried out at 10% by weight concentrations to give copolymers. PBZT containing pendent labile methyl groups were prepared [16] from 2-methyl and 2,5-dimethylterephthalic acids by condensation with the appropriate amino monomers. Benzobisthiazole random copolymers [17] were also prepared using 2-methylterephthalic acid and terephthalic acid with 2,5-diamino-l, 4-benzenedithiol dihydrochloride. Sweeny [18] developed a method for cross-linking of stiff chain polymers such as Kevlar and PBZT. The method was based on the thermolysis of active aryl halides contained in the polymer unit and coupling of the free radicals formed. Intermolecular cross-linking has been considered as an effective approach to improve the compressive strength of polymeric fibres. On this basis Mehta *et al.* [19] studied the cross linking in methyl pendent PBZT fibre (**VI**).

(VI) Cross-linking mechanism in methyl pendent PBZT

(VII) PBZT containing thienylene units

The fibres were treated at 400-550°C in air and in nitrogen for varying times to achieve intermolecular cross-linking. Other structural variations of the rigid rod PBZTs have encompassed a variety of changes that affected the backbone geometry resulting in deviation from 180° (i.e. para-catenation). For example, substitution of the 1,4-phenylene unit by 2,5-thiophene units **(VII)** has been investigated as a backbone geometry deviation (i.e. 180° to 148°) in PBZT [20].

We have made efforts towards the synthesis and characterization of the various polybenzobisthiazoles. Several PBZTs have been synthesized in our laboratory using different dicarboxylic acids with 2,5-diamino-1, 4-benzenedithiol dihydrochloride. We have synthesized 2,5-diamino-1, 4-benzenedithiol dihydrochloride by the known method of Wolfe *et al.* [5] with some modification.

A PBZT was synthesized by the polycondensation reaction of 2,5-diamino-1, 4-benzenedithiol dihydrochloride with cyclohexane 1,4 dicarboxylic acid [21]. The reaction was carried out for 75 hr. The reaction temperature was gradually raised from room temperature to 200°C. Finally the reaction mixture was poured into water and neutralized by ammonium hydroxide and washed thoroughly by hot water. The polymer obtained was dried in vacuum at 110°C for 24 hr. The co-polymers of PBZT using cyclohexane 1,4 dicarboxylic acid and terephthalic acid were also obtained under similar conditions. PBZTs containing halogens (fluoro and bromo groups) were synthesized by polycondensation of 2,5-diamino-1, 4-benzenedithiol dihydrochloride with tetrafluorophthalic acid and 4-bromo-isophthalic acid [22]. As reported in the literature, halogen containing aromatic condensation polymers such as polycarbonates [23], polyformals [24], polyketones [25], polyazomethines [26] and their copolymers have some unique properties such as good solubility in organic solvents, toughness and flexibility at low temperature, water and oil repellency, low refractive index, and high thermal stability in addition to the prominent characterstics inherent in the corresponding polymers containing no fluorine atom. Thus the introduction of fluoro and bromo groups into the existing aromatic condensation polymers makes them soluble, processable and gives them several additional characteristics. Solubility in some solvents and thermal properties of these polymers have been studied.

Benzothiazoles have been prepared from isophthalic acid and 5-hydroxy isophthalic acid (VIII) with 2,5-diamino-1, 4-benzenedithiol dihydrochloride [27]. It is reported that a high compressive strength could be achieved with lateral chain supports such as an inter-digitation network formed by suitable side chains at regular intervals along the rigid polymer backbone. An obvious approach to rendering such lateral support at the molecular level is to cross-link the rigid-rod polymer chains. Research along this line, using potentially cross-linkable units such as methyl [28-31], fluorine [32] and benzocyclobutene [33] groups has been reported. Cross-linking of PBZT fibers via high-energy radiation has also been explored [34]. PBZT with a hydroxy functional group was selected because of its ability to form hydrogen bonds, coordination bonds and to undergo some nucleophilic substitution reactions. It was later found to be relatively inert under the polymerisation conditions required to form high molecular weight PBZT. As reported by Tan et al. [35] the PBZT synthesized using 3,6- dihydroxy1, 4-phenylene dicarboxylic acid (IX) had propensity to form a ladder structure, driven by strong intramolecular hydrogen bonds between the hydroxyl moiety and the nitrogen atoms of the benzothiazole units. The ladder structure was similar to the classical ladder polymer poly [(7-oxo-7, 1 OH benz [de] imidazo [4',5':5, 6] benzimidazo [2, 1-a] isoquinoline-3, 4:10, 11]-tetrayl)-10-carbonyl] (BBL) (X).

3. SPECTROSCOPIC STUDIES

The FT-IR spectrum of monohydroxy PBZT did not show any strong and broad band around 3300-3200 cm^{-1} where normally intermolecularly bonded –OH group of phenol appears [36]. Instead, it showed weak and broad absorption bands in the region 2700-3000 cm^{-1} that is not attributed to intermolecular hydrogen bonding

(VIII) PBZT with hydroxy isophthalic acid

(IX) PBZT with dihydroxyterephthalic acid

(X) BBL, Poly [(7-oxo-7, 1 OH-benz[de]imidazo[4,5;5,6]
benzimidazo[2,1-a]isoquinoline-3,4:10,11]-tetrayl)-10-carbonyl]

in monohydroxy PBZT. The model compounds were also synthesized using 2-aminothiophenol and dicarboxylic acids, e.g., cyclohexane 1,4-dicarboxylic acid, tetrafluorophthalic acid, 4-bromoisophthalic acid, isophthalic acid, and 5-hydroxy isophthalic acid in order to compare the chemical structures of synthesized polymers. Polybenzobisthiazole copolymer was also synthesized using thiophene -2,5-dicarboxylic acid with cyclohexane 1,4 dicarboxylic acid and 2,5-diamino-1, 4-benzenedithiol dihydrochloride.

The synthesized PBZTs were soluble in strong protic acids such as polyphosphoric acid (PPA), methanesulphonic acid (MSA) and chlorosulphonic acid (CSA) [37-39]. Further, the structures of these synthesized polymers were established by elemental analysis and FT-IR. As these polymers were insoluble in common solvents so the solid-state NMR was carried out. The four intense bands at 1602, 1454, 1308 and 1093 cm^{-1} in the FT-IR spectra were assigned to benzodithiazole hetero-ring for PBZT containing cyclohexyl moiety in the main chain. These stretching vibrations are comparable to those reported by Shen and Hsu [40] for benzodithiazole ring. Similarly the ^{13}C solid-state NMR spectra of PBZT show the integrals of bands, which basically reflect corresponding carbon ratios in the repeat unit of the polymer chain. The band integral ratio of PBZT with cyclohexyl moiety in the main chain is approximately (1:1: 1:1:1: 2) which is quite close to the theoretical value.

4. PROPERTIES OF PBZT

4.1. Comparison of properties with other fibres

Many high strength and high modulus-fibres, including metallic and organic polymer fibres, have been developed. In the wide variety of novel high-performance fibres reported, carbon, glass, Kevlar®, and ultra-high molecular-weight extended-chain polyethylene (i.e. Spectra®) are the commercially most important ones. The great potential of rigid rod polymer materials is only partially exploited. Among all these fibres the fibres of benzazole family (i.e. polybenzothiazoles (PBZT), polybenzimidazoles (PBZI) and polybenzoxazoles (PBZO)) have received special attention. PBZT and PBZO are the two strongest and most thermally stable synthetic organic fibres to date (as shown in Table 1). Their tensile moduli are above 300 GPa and tensile strength reaches 6 GPa and their onset of thermal degradation is in the range 600-700°C. Superior stiffness and tenacity combined with low-density makes these fibres among the best materials when compared on the basis of specific strength and modulus.

Table 1.
Properties of fibres

Fibres	Tensile modulus (GPa)	Tensile strength (GPa)	Compressive strength (GPa)	Density (g/cm^3)
Steel	200	2.8	–	7.8
Alumina	350-380	1.7	6.9	3.7
Silicon carbide	200	2.8	3.1	2.8
S-glass	90	4.5	>1.1	2.46
E-glass	76	3.4	4.2	2.58
Pitch based carbon P-100	725	2.2	0.48	2.15
Pitch based carbon P-120	827	2.2	0.45	2.18
Kevlar 49	125	3.5	0.39-0.48	1.45
Kevlar 149	185	3.4	0.32-0.46	1.47
PBZT	325	4.1	0.26-0.41	1.58
PBZO	360	6.0	0.2-0.4	1.56-1.58

4.2. Thermal properties

Among all the high temperature, high performance polymers, the polybenzobis-thiazole rigid-rod polymers are one of the most thermally stable systems known. A variety of thermal techniques have been used to investigate their thermal stability such as thermogravimetric analysis (TGA), thermogravimetric mass spectrometry (TG-MS) and isothermal aging studies. Rigid rod polymers have shown good thermal stability because of their aromatic backbone, rigid molecular chains and high order. PBZT fibres, in particular, have excellent thermal behaviour. Figure 1 shows the thermogravimetric analysis (TGA) of PBZT, PBZI, and PBZO polymers [41]. The early weight loss exhibited by these materials from ambient temperature to approx 200°C is the result of water loss. The degradation temperature of PBZT in air is around 620°C. PBZT fibres do not melt below the decomposi-tion temperature and do not support combustion but will char at high temperature. They generate only low quantities of smoke on burning [42]. The thermogravimetric mass spectra in Figure 2 show the evolution of different gases as PBZT was heated to high temperature [43]. The main gases evolved are H_2S, HCN and CS_2, which are the characteristic degradation products of polymer backbone. It was reported that the total weight loss of 28% at 1000°C was consistent with the loss of 1 mol of H_2S, 1 mol of HCN and 0.25 mol of CS_2 per mol of PBZT repeat unit [44].

The thermal stability of various PBZT derivatives synthesized in our laboratory is comparable to that of parent PBZT containing para phenylene unit. The TGA analysis was performed in a nitrogen atmosphere at 10°C/min. The thermal stability of various PBZT derivatives synthesised in our laboratory is summarized in Table 2.

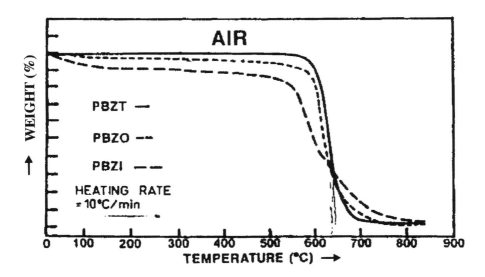

Figure 1. TGA of PBZT, PBZO and PBZI.

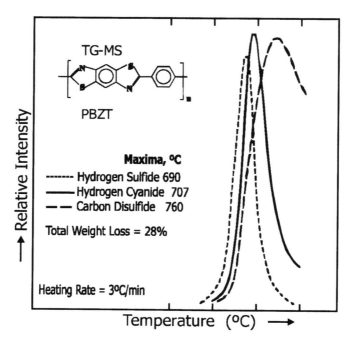

Figure 2. Thermogravimetric mass spectra (TG-MS) of PBZT.

Table 2.
Thermal stability of various PBZT derivatives

	PBZT-tereph-thalic	PBZT-cyclohexyl	PBZT-tetrafluoroph-thalic	PBZT-isophthalic	PBZT-4-bromoisoph-thalic	PBZT-monohydroxy-isophthalic
T_0 (°C)	547.15	495.32	506.81	350.75	386.23	483.39
T_m (°C)	595.67	523.93	678.25	409.79	425.86	503.17

Where T_0 is onset temperature, T_m is maximum temperature.

As reported [41] in the literature our polymer systems also do not exhibit any glass transition before their decomposition temperature in differential scanning calorimetery (DSC). This is due to the overwhelming effect of the fused ring structure retaining the softening of the benzothiazole-based polymers.

4.3. Tensile properties

Compared to all other synthetic polymer fibres, the rigid-rod polymer fibres display the highest tensile strength and tensile modulus Table 1 shows the tensile properties of various fibres. Tensile modulus of PBZT fibre is 50-60 times higher than that of high tenacity nylon, 25-30 times higher than polyester, 2.5 times higher than Kevlar-49, 4 times higher than glass fibres and 1.5 to 2 times higher than steel fibres. PBZT fibres are superior due to their tensile properties and low density.

From a comparison of Kevlar fibres and typical rigid-rod fibres, it is clear that in PBZT the rigid rings are connected almost co-axially along the chain axis. During tensile deformation, the strain energy is consumed directly by the deformation of the stiff para bonds and rings; therefore PBZT can have a very high theoretical modulus. Whereas in the Kevlar, the chain exists in a zigzag conformation and the chain bonds are inclined to the chain axis so that the strain energy is distributed partly to the reorientation of the benzene ring and, thus the rigidity is not so high as in PBZT [45]. The theoretical tensile strength of PBZT fibres is more than 20 GPa [46]. The tensile stress is determined by covalent bonds, and lateral packing of the main chain. Calculation of the tensile strength of the fibres is complicated as it depends on several factors, such as chain length distribution, orientation of the chain with respect of the fibre axis, interrelation between atoms (covalent bonds and secondary forces), non-uniformity and anisotropy of the fibre structure and morphology, and the presence of impurities and voids.

The tensile properties, especially the tensile strength of the high performance polymer fibres, generally depend on their molecular weight. It has been observed that with increasing molecular weight both the tensile strength and elongation-at-break of the fibres increase. But after reaching a certain level the molecular weight does not affect the fibre tensile modulus.

4.4. Compressive properties

Compared with the superior tensile properties, the axial compressive strength of a rigid-rod polymer fibre is relatively low. One important application of the high performance polymeric fibres is in reinforced composites. Many aerospace applications require good composite compressive strength. As shown in Table 1 the compressive strength of rigid-rod polymer fibres is comparatively low compared to boron, alumina and silicon carbide fibres. The relatively low compressive performance is an obstacle to the use of these fibres in structural composites.

Since the compressive strength of high modulus organic fibres appears to be the limiting factor in their structural application, considerable research effort is being directed [47, 48] towards enhancing the compressive strength of rigid-rod polymeric fibres.

4.4.1. Factors governing compressive strength of PBZTs
Various factors responsible for limiting the use of polymers in composites are:

(1) Fibre buckling

(2) Microfibril/fibril buckling

(3) Low glass transition temperature

(4) Poor fibre morphology/microstructure

(5) Poor intermolecular interaction

Compressive failure of rigid-rod polymer fibres is always associated with the formation of kink bands. Kink bands are nucleated in a localized region probably initiated somewhere near the fibre surface at a critical stress and then propagate away from the initiation point and continue to grow in toward the centre and the other side of the fibre. In a certain range of compressive strain, kink band density is proportional to the applied compressive strain, but at larger compressive strain, the kink band density tends to saturate [49, 50]. It has been reported that the kink band angle is closely related to fibre structure and properties [51, 52]. There are several arguments about the mechanism of kink formation and the compressive failure. However, many unanswered questions still remain in this field, for many different deformation processes, such as micro-buckling, shearing and bending, etc.

4.4.2. Approaches to enhance the compressive properties
The following approaches have been taken to enhance the compressive properties of rigid-rod polymer fibres.

(1) Changing process conditions

(2) Changing morphological structure

(3) Modifying synthesis routes

(4) Increasing the shear strength by bonding molecular chains more strongly. Modifications include change of rigid-rod backbones, incorporation of active groups and diffusion of external agents into the fibre, to form intermolecular cross-links.

(5) Radiation cross-linking and coating of the fibres as well as infusing reactive monomers or metals into the fibres [53-59].

One suggested route is to crosslink PBZT fibres based on thermal elimination or radiation degradation of activated aryl halogen or methyl pendent groups followed by combination of aryl/benzyl free radicals leading to ring coupling [19, 28]. In other case a thermally reactive cross-linking fluorine moiety was incorporated into PBZT [31]. Cross-linking benzobisthiazole rigid-rod copolymers via labile methyl groups, making rigid-rod random copolymers of phenyl benzobisthiazole with phenyl benzobisthiazole pendant -p-terphenyl units changes the polymer packing order. As shown in Table 3 there is no significant change in the compressive properties of cross-linked fibre compared to parent PBZT. The potential advantage of studying the effect of Lewis acid complexation on intermolecular interactions in order to understand the role of the latter in influencing compressive properties is clear. It was also observed that the presence of glass within PBZT sol-gel microcomposite films increased the resistance of the films to compression.

Unfortunately, no significant success has yet been reported. There may be a good potential for improving the compressive strength of rigid-rod polymer fibres by coating them with suitable metal or metal compounds [60, 61]. The PBZT fibres immersed in molten aluminium-silicon alloy or vapour deposited with aluminium were investigated. It was observed that the molten aluminium formed an

Table 3.
Compressive properties of modified PBZT fibres

Modified PBZT fibres	E_t (GPa)	σ_t (GPa)	σ_c (GPa)
Fluorine moiety in PBZT	240	2.1	0.32-0.55
Terphenyl PBZT	220	2.4	0.21-0.49
Methyl pendent PBZT	200	2.3	0.3-0.4
Hydroxy PBZT	180	2.1	0.14-0.41
Unmodified PBZT	325	4.1	0.21-0.46

Where E_t = tensile modulus, σ_t = tensile strength, σ_c = compressive strength.

oxidized layer at the fibre surface that protected PBZT from degradation. Therefore, it is possible to produce PBZT/metal composites using liquid aluminium alloy to improve compressive properties.

The compressive strength could also be enhanced using other techniques, e.g.,

(1) Introducing some H-bonding groups in the polymer chain.

(2) Introducing nano-metal or nano-ceramic materials in the fibre at the spinning stage.

(3) Blending of PBZT fibre with high compressive strength fibres. Hydrogen bonding groups are expected to enhance the intermolecular interaction and contribute to increased compressive properties. Similarly nano-metal/nano-ceramic particles in appropriate proportion should enhance compressive properties because these particles would be attached chemically to co-ordinating groups. Blending of fibres is an established art in textile technology and fibres such as graphite or alumina could be considered for this purpose.

5. SUMMARY

PBZT is the most thermally stable organic fibre synthesised to date and compares well with PBZO in this respect. The tensile strength of these fibres is also of high order and is higher than the tensile strength of carbon fibre. As for density it is only slightly higher than Kevlar and is lower than all other reinforcing fibres.

PBZT has one disadvantage that its compressive properties are lowest among all reinforcing fibres. Many efforts have been made in this regard such as variation in the backbone, introduction of cross-linking via radiation method, introduction of hydrogen bonding groups etc but all these methods have yielded insignificant improvements. It is, therefore, concluded that this area is open for active research.

REFERENCES

1. W.J. Welsh, D. Bhaumik and J.E. Mark, Macromolecules **14**, 947 (1981).
2. M.W. Wellam, W.W. Adams, R.A. Wolff, D.R. Wiff and A.V. Fratini, Macromolecules **14**, 935 (1981).
3. J.F. Wolfe and B.H. Loo, U.S.Pat. 4, 225,700 (1980).
4. J.F. Wolfe, B.H. Loo and F.E. Arnold, Polymer Prepr. **19**, 1 (1978).
5. J.F. Wolfe, B.H. Loo and F.E. Arnold, Macromolecules **14**, 915 (1981).
6. S.L. Solar, R.J. Cox, N.J. Cleack and R. Ettinger, J. Org. Chem. **33**, 2132 (1968).
7. A.W. Chow, J.F. Sandell and J.F. Wolfe, Polymer **29**, 1307 (1988).
8. J.F. Wolfe and P.D. Sybert, U.S.Pat. 4,533693 (1985); 4,703,103 (1987).
9. Y. Saegusa, M. Horikiri, D. Sakai and S. Nakamura, J. Polym. Sci Pt A: Polym Chem. **36**, 429 (1998).
10. M.H. Delynn and I.I. Harruna, J.Polym Sci Pt A: Polym Chem. **36**, 277 (1998).
11. L.S. Tan, K.R. Srinivasan and S.J Bai, J. Polym. Sci Pt A: Polym Chem. **35**, 1909 (1997).
12. Y.-H. So, J. Polym. Sci Pt A: Polym Chem. **35**, 2143 (1997).
13. X. Hu, S. Kumar, M.B. Polk and D.L. VanderHart, J. Polym. Sci Pt A: Polym Chem. **36**, 1407 (1998).
14. J. Burkett and F.E. Arnold, Polym. Prepr. ACS. Polym Div. **28**, 1 (1987).
15. C.S. Wang, J. Burkett, S. Bhattacharya, H.H. Chuah and F.E. Arnold, Proc. ACS. Div Polym. Mater. Sci. **60**, 767 (1989).
16. T.T. Tsai and F.E. Arnold, Polym. Prepr. ACS. Polym. Div. **29**, 324 (1988).
17. H.H. Chauh, T.T. Tsai, K.H. Wei, C.S. Wang and F.E. Arnold, Proc. ACS. Div. Polym. Mater. Sci. Eng. **60**, 175 (1989).
18. W. Sweeny, J Polym Sci Pt A: Polym Chem. **30**, 1111 (1992).
19. V.R. Mehta, Satish Kumar, M.B. Polk, D.L. VanderHart, F.E. Arnold and T.D. Dang, J. Polym. Sci Pt B: Polym. Phys. **34**, 1881 (1996).
20. M. Dotrong, R.C. Tomlison, M. Sinsky and R.C. Evers, Polym. Prepr. ACS. Polym Div. **32**, 85 (1991).
21. R. Saxena, L.D. Kandpal and G.N. Mathur, in *Recent Advances in Polymers and Composites*, G.N. Mathur, L.D. Kandpal and A.K. Sen (Eds), p. 54, Allied Publishers, Delhi (2000).
22. R. Saxena, L.D. Kandpal and G.N. Mathur, J. Polym. Sci. Pt A: Polym. Chem. **40**, 3959 (2002).
23. Y. Saegusa, M. Kuriki, A. Kawai and S. Nakamura, J. Polym. Sci. Pt A: Polym. Chem. **28**, 3327 (1990).
24. Y. Saegusa, M. Kuriki, A. Kawai and S. Nakamura, J.Polym.Sci. Pt A: Polym. Chem. **32**, 57 (1994).
25. Y. Saegusa, M. Kuriki, A. Kawai and S. Nakamura, Makromol. Chem. **194**, 777 (1993).
26. Y. Saegusa, M. Kuriki, A. Kawai and S. Nakamura, Macromol. Chem. Phys. **195**, 1877 (1994).
27. R. Saxena, Ph.D. Thesis, CSJM University Kanpur, India (2001).
28. H.H. Chuah, T.T. Tsai, K.H Wei, C.S. Wang and F.E. Arnold, Proc. ACS. Polym. Mater. Sci. Eng. **60** 517 (1989).
29. M. Dotrong, M.H. Dotrong and R.C. Evers, Proc. ACS. Polym. Mater. Sci. Eng. **65**, 38 (1991).
30. M. Dotrong and R.C. Evers, J.Polym. Sci. PtA: Polym. Chem. **28**, 3241 (1990).
31. S. Bhattacharya, H.H. Chuah, M. Dotrong, K.H. Wei, C.S. Wang, D. Vezie A. Day and W.W. Adam, Proc. ACS. Polym. Mater. Sci. Eng. **60**, 512 (1989).
32. M. Dotrong and R.C. Evers, Proc. ACS. Polym. Mater. Sci. Eng. **60**, 507 (1989).
33. L.J. Markosi, K.A. Walker, G.A. Deeter, G.E. Spilman, D.C. Martin and J.S. Moore, Chem. Mater. **5**, 248 (1993).
34. R. Kovar, "Radiation-Induced Modification of Ordered Polymers for Compressive Strength" Air Force Materials Laboratory Technical Report, WL-TR-91-4110 (1992).
35. L.S. Tan, F.E. Arnold, T.D. Dang, H.H. Chuah and K.H. Wei, Polymer. **35**, 3091 (1994).
36. N.B. Colthup, L.H. Daly and S.E. Wiberly, *Introduction to Infrared and Raman Spectroscopy* 3rd edn, p. 332, Academic Press, New York (1990).

37. G.C. Berry, W.W. Adams, R.K. Eby and D.E. McLemore, Mater. Res. Soc. Symp. Proc. **134**, 181 (1989).
38. C.P. Wong, H. Ohnuma and G.C. Berry, J. Polym. Sci. Polym. Symp. **65**, 173 (1978).
39. D.B. Cotts and G.C. Berry, Macromolecules **14**, 930 (1981).
40. D.Y. Shen and S.I. Hsu, Polymer. **23**, 969 (1982).
41. L.R. Denny, I.J. Goldfarb and E.J. Soloski, Mater. Res. Soc. Symp. Proc. **134**, 345 (1989).
42. I.J. Goldfarb, H. Reale, S. Wierschke and J. Medrano, Mater. Res. Soc. Symp. Proc **134**, 609 (1989).
43. E.G. Jones and D.L. Pedrick, Mater. Res. Soc. Symp. Proc. **134**, 407 (1989).
44. J.F. Wolfe, B.H. Loo and F.E. Arnold, Macromolecules **14**, 915 (1981).
45. K. Tashiro and M. Kobayashi, Polymer. **32**, 454 (1991).
46. T. Ohata, T. Kunugi and K. Yobuki, *High Tenacity and High Modulus Fibers*, Kyo Ritsu, Tokyo (1998).
47. H.D. Ledbetter, S. Rosenberg and C.W. Hurtig, Mater. Res. Soc. Symp. Proc. **134**, 253 (1989).
48. M.G. Northolt, J. Mater. Sci. **16**, 2025 (1981).
49. S.J. Detersa, Ph.D Thesis, University of Massachusetts (1985).
50. W. Huh, S. Kumar, T.E. Helminiak and W.W. Adams, SPE Annual Technical Conference Proceeding, 1245 (1983).
51. T. Takahashi, M. Miwa and S. Sukari, J. Appl. Polym Sci. **28**, 580 (1983).
52. S.V. Zwag and G. Kampschoer, in *Integration of Fundamental Polymer Science and Technology*, 2nd ed., vol. 2, p. 545, Elsevier Applied Sciences, London (1988).
53. S.V. Zwagg, S.J. Picken and C.P. Vansluijs, in *Integration of Fundamental Polymer Science and Technology*, 2nd ed., vol. 3, p. 199, Elsevier Applied Sciences, London (1989).
54. R.F. Kovar, R.R. Haghighat and R.W. Lusingnea, Mater. Res. Soc. Symp. Proc. **134**, 389 (1989).
55. U. Santosh, M.H. Dontrong, H.H. Song and C.Y.C. Lee, Proc. ACS. Polym. Mater. Sci. Eng. **64**, 40 (1991).
56. C.S. Wang, J. Barkett, S.B. Bhattacharya, H.H. Chauh and F.E. Arnold, Proc. ACS Polymer. Mater. Sci. Eng. **60**, 767 (1989).
57. T. Dang, L.S. Tan, K.H. Wei, H.H. Chuah and F.E. Arnold, Proc. ACS Polym. Mater. Sci. Eng. **60**, 424 (1989).
58. M.C.G. Jones, T. Jiang and D.C. Martin, Bull. Amer. Phys. Soc. **39**, 697 (1997).
59. I.K. Gillie, M. Newsham, S.J. Nolau, V.S. Jear and R.A. Bubeck, Bull. Amer. Phys. Soc. **38**, 292 (1993).
60. K.E. Newman, P. Zhang, L.J Cerddy and D.L. Auara, J. Mater. Res. **6**, 1580 (1991).
61. U. Santhosh, K.E. Newman and C.Y.C. Lee, J. Mater. Sci. **30**, 1894 (1995).

Polyimides and Other High Temperature Polymers, Vol. 2, pp. 241–254
Ed. K.L. Mittal
© VSP 2003

Electrical breakdown and electrostatic phenomena in ultra-thin polyimide Langmuir-Blodgett films

MASAHIRO FUKUZAWA and MITSUMASA IWAMOTO[*,1]

Department of Electrical Engineering, Kyushu-Sangyo University, 2-3-1 Matsuka-dai, Higashi-ku, Fukuoka 813-8503, Japan
[1] *Department of Physical Electronics, Tokyo Institute of Technology, 2-12-1 O-okayama, Meguro-ku, Tokyo 152-8552, Japan*

Abstract—Electrostatic phenomena occurring at insulator-metal interfaces are interesting to the fields of electronics and electrical engineering. In this study, we examined the surface potential of polyimide (PI) Langmuir-Blodgett (LB) films on metal (Al, Ag and Au) electrodes, which were charged at various voltages under a needle-plane electrode system. It was found that PI LB films were negatively charged as the biasing voltage applied to the needle-electrode increased. The surface potential saturated when the number of deposited layers was 20-30, and it was found to be dependent on the nature of metal electrode. We concluded that the interfacial electrostatic space charge at the as-deposited film/metal interface made a significant contribution to the creation of the additional electrostatic potential, and the dielectric breakdown of ultra-thin films was controlled by the interfacial space charge.

Keywords: Space charge; polyimide; LB; electrical breakdown.

1. INTRODUCTION

The study of electrostatic interfacial phenomena occurring at organic film/metal interfaces is obviously important to the fields of electronics and electrical insulation [1], and it has been a continuous research subject since the discovery of contact electrification [2, 3]. It is known that the interfacial electrostatic charges make a significant contribution to electrical breakdown, electrical conduction, electrical treeing and others [4]. From contact-charge exchange experiments using polymers, it has been revealed that excess charges are injected from metals into films, and micrometer thickness space charge layer is formed at the interface due to the existence of electronic states contiguous to the Fermi energies of contacting metals [5]. Duke, Fabish and other researchers carried out many experiments for elucidating the interfacial electrostatic charge phenomena in polymer films, and many interesting and important models of surface states, e.g., acceptor states model, have been proposed [6-11].

---

[*]To whom all correspondence should be addressed. Phone & Fax: +81-3-5734-2191,
E-mail: iwamoto@pe.titech.ac.jp

All of these studies made a significant contribution to the development of electrical insulation engineering. However, the knowledge from these studies is still not sufficient to support the discussion on nano-interfacial electrostatic phenomena. The surface of polymers used for the experiments was mechanically destroyed by the repeated application of contact and separation between polymers and metals. As a result, it is difficult to obtain information on the nanometer thickness space charge layer at an organic film-metal interface. Furthermore, recent progress in the field of molecular electronics requires information on nanometer thickness scale electrostatic phenomena. For a better understanding of the interfacial electrostatic phenomena at an organic film-metal interface, it is helpful to study ultra-thin films whose thickness is less than the electrostatic double layer. The information on the distribution of the electronic density of states as well as the space charge distribution of excess charges in the films will be very instructive [12-14]. However, this study is not easy owing to the difficulty in the preparation of sophisticated organic ultrathin films. Furthermore, the nanometer thickness film surface must not be destroyed during the experiment. Thus the nature of the organic film-metal contacts and the organic film band structure are much less known than those for inorganic semiconductors. Furthermore, the non-crystalline structure of insulating organic films makes the theoretical analysis of the electronic structure of organic films much more complicated [15]. However, the situation has improved step by step in recent years. By means of novel film preparation methods such as Langmuir-Blodgett (LB) technique, self-assembly monolayer (SAM) method, organic molecular beam epitaxy (OMBE) technique and others [16, 17], it has become possible to prepare very thin films. Furthermore, along with the development of these techniques, much attention has been paid to build up tunnel junctions and molecular rectifying junctions using organic materials with the hope of observing novel and useful electrical and optical properties [18]. Thus it is an urgent task to clarify the interfacial electronic phenomena occurring at film-metal interfaces for the intended operation of these junctions, because the interfacial phenomena play a dominant part in organic ultrathin film devices. Needless to say, the information on the nanometric dielectric phenomena including interfacial electrostatic phenomena in organic ultrathin films will help in elucidating the electrical breakdown process in organic materials [2, 3, 15].

Some polymers such as polyimide (PI) are thermally and chemically stable, and they have been used in microelectronics technology because of their excellent dielectric and other properties [19, 20]. During the past ten years, electrically insulating PI LB films with a monolayer thickness of 0.4 nm have been successfully prepared onto solid substrates [21, 22]. The PI LB films were found to have the excellent electrical insulating property with a resistance higher than $10^{15}\,\Omega\cdot\text{cm}$ and an electrical breakdown strength higher than $10^7\,\text{V/cm}$. As such they function as electrical insulating barriers in metal-insulator-metal (MIM) structures and as good tunneling spacers in tunnel junctions. Thus PI LB films are suitable for use for a better understanding of the interfacial phenomena at interfaces.

In the present study, we employed the surface potential measurement based on the Kelvin-probe method, and then measured the surface potential of PI LB Films on Al, Ag and Au electrodes, which were charged under a needle-plane electrode system by applying a dc voltage to the needle electrode. Here the needle-plane electrode system was very similar to that used in the experiments on the electrical breakdown of insulating polymers [23]. Electrical breakdown voltage was also examined. We then studied the interfacial electrostatic phenomena in PI LB films, and electrical breakdown process in ultrathin PI LB films.

2. EXPERIMENTAL

Polyimide LB films (denoted as PI), whose chemical structure is shown in Fig. 1, were used in this study. PI has a large electron affinity and thus it has a strong tendency to accept electrons from metal electrodes such as Al and Au. PI LB films were prepared onto Al and Au evaporated glass slides (10 mm x 50 mm) by means of the LB technique coupled with a precursor method developed by Kakimoto *et al.* [22], in a manner as described in our previous study [24, 25]. Briefly, amphiphile monolayers of poly (amic acid) that have a chain of long alkyl-amine salt, a precursor of PI, were transferred onto the glass slides by the vertical dipping method at a surface pressure of 35 mN/m. After the chemical imidization with acetic anhydride and pyridine, PI LB films with a chemical structure shown in Fig. 1 were obtained [22]. The number of deposited layers was between 10 and 70, and the monolayer thickness was 0.4 nm.

The electrode configuration for the sample is shown in Fig. 2. All samples were placed in a vacuum vessel with a pressure of 1×10^{-3} Torr, and they were then annealed at a temperature of 100°C for 1 hour to remove extrinsic charges and adsorbed water molecules because these charges and molecules affect the electrostatic properties of PI LB films. The surface potentials of deposited PI LB films biased under a needle tungsten-electrode were measured by means of the Kelvin method (TRek Model 320B), where the detector D of the surface potential meter and the needle electrode moved parallel to the sample (see Fig. 2). The surface potential of PI LB films (position Q) was measured with reference to the potential of the clean metal electrodes (position P).

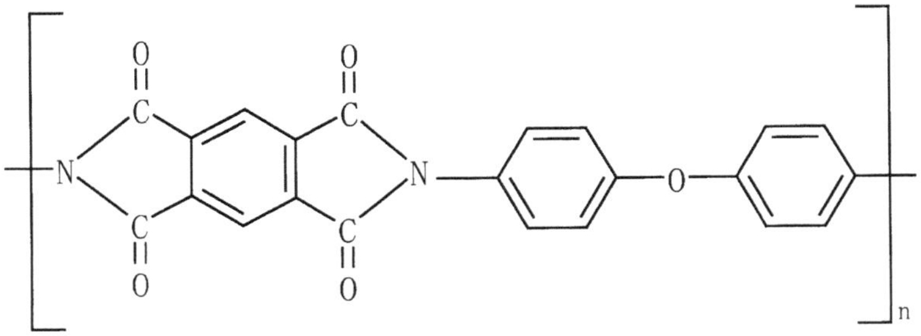

Figure 1. Chemical structure of polyimide used.

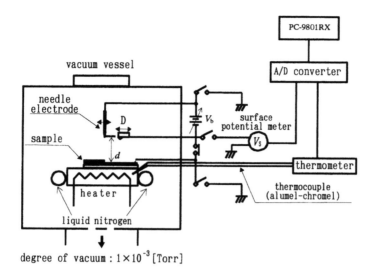

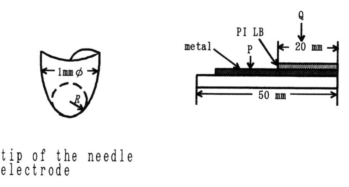

tip of the needle
electrode

Figure 2. Experimental system and configuration of the samples. The tip of the needle electrode is also shown.

The needle electrode was placed above the sample in a vacuum vessel as shown in Fig. 2, where the distance d between the tip of the electrode and the sample was 1 mm. The needle electrode used in the experiments had a point angle of 30° and a point radius of curvature R of 10 μm. The samples were charged by applying a *d.c.* voltage between −2.5 kV and +2.5 kV to the needle electrode for 20 min, and then the surface potential of the charged samples was examined. Whilst the samples were charged, some of them were electrically broken. We defined the *d.c.* biasing voltage as the breakdown voltage when 50% of the samples were electrically broken by the application of this biasing voltage to the samples. To clarify the electrical breakdown process, experiments were carried out on PI LB films deposited on Al, Cu and Au.

3. RESULTS

Figures 3 (a) and (b) show the surface potential of PI LB films on Al and Au electrodes, respectively. The numbers in the figures represent the number of deposited layers. As we can see in the figure, in all cases the surface potential decreases as

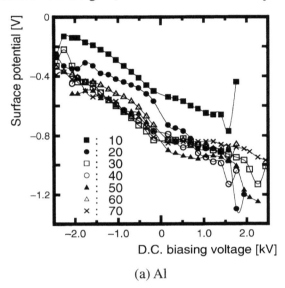

(a) Al

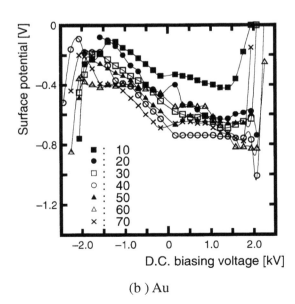

(b) Au

Figure 3. Surface potential of PI LB films on Al electrodes (a), and Au electrodes (b). The number represents the number of deposited layers.

the *d.c.* biasing voltage increases, and the surface potentials are created even at a biasing voltage of 0 V. As we described in our previous papers [24, 25] the surface potential at a zero biasing voltage is due to excess electrons displaced from the metal electrodes. Usually electrically insulating polymer films are charged with electronic homo-charges when the polymers are non-polar and do not possess extrinsic ionic impurities; that is, the surfaces of films are negatively charged by the application of negative voltages to the needle electrode with respect to the plane electrode, whereas these are positively charged by the application of positive voltages. However, the experimental results plotted in Fig. 3 are opposite to this prediction. This is a main characteristic found for ultra-thin polyimide LB films. PI LB films do not contain impurities because the films are prepared from monolayers on pure water surface. The experimental results are also dependent on the biasing voltage. Thus the electronic charge is a main contributor to the potentials observed in Fig. 3 [26]. The abrupt change at the biasing voltage around 2.0 kV is either due to the electrical breakdown of the PI LB films or the unexpected discharge in our experimental system by the application of such high voltages of the order of kV.

Figures 4 (a) and (b) show the relationships between the surface potentials and the number of deposited layers for PI LB films on Al and Au electrodes, respectively, where the results were obtained by re-plotting the experimental results shown in Fig. 3. The surface potential saturates when the number of deposited layers is 30-40. In other words, the surface potential saturated when the film thickness was 12-16 nm. It is easy to expect that the surface potential created by the application of a *d.c.* biasing voltage should increase in proportion to the number of deposited layers if the main contribution to the additional surface potential is due to orientational motion of polar molecules. Thus we may again expect that the main contribution is the excess electronic charges.

It is interesting here to plot the relationship between the surface potential and workfunction of electrodes. Figure 5 shows the results for 50-layer PI LB films, where PI LB films were more negatively charged as the workfunction of the electrode decreased; that is, the PI LB films were negatively charged in the order of workfunction of metals, Al, Ag and Au. This result again suggests that the main contribution to the surface potential is due to excess electronic charges and not polar molecules.

Figure 6 shows the relationship between the breakdown voltage and the number of layers for PI LB films deposited on Al, Au and Cu electrodes. As we can see in the figure, the breakdown voltage increases as the workfunction of metals increases (workfunction, Au > Cu > Al). Further, the electrical breakdown voltage depends on the polarity of biasing voltages. The breakdown voltage obtained by positively biasing is higher than that by negatively biasing. The results obtained here again suggest that the interfacial electronic charges also make a significant contribution to the electrical breakdown.

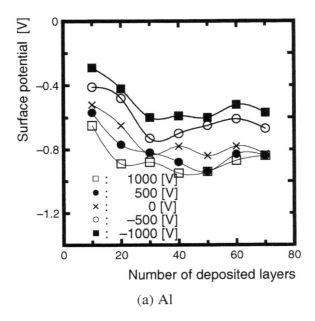

(a) Al

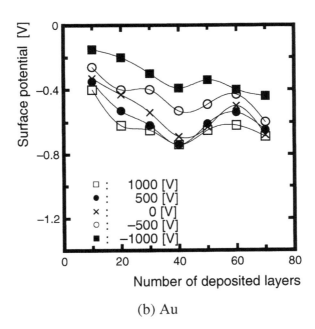

(b) Au

Figure 4. Relationship between the surface potential and the number of deposited layers for PI LB films on Al electrodes (a), and Au electrodes (b).

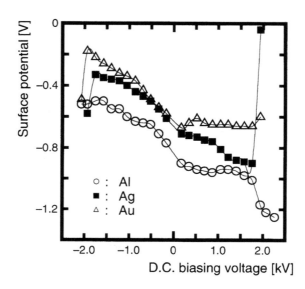

Figure 5. Relationship between the surface potential and *d.c.* biasing voltage for different metal electrode. 50-layer PI LB films were used.

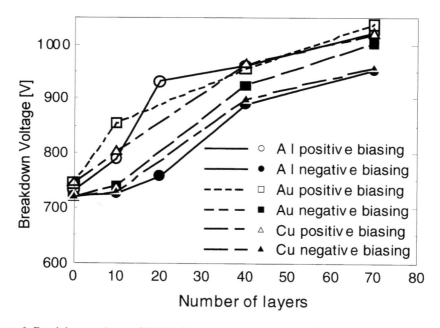

Figure 6. Breakdown voltage of PI LB films on metal electrodes. "Positive" and "negative" represent the polarities of biasing applied to the needle electrode with respect to the metal electrodes.

4. DISCUSSION

As has been mentioned in Section 3, the surface potential of PI LB films shows an interesting behavior. That is, the surface potential created by the application of the voltage depends on the polarity of the applied voltage, the number of deposited layers, and workfunction of metals. These results suggest that the main contribution to the creation of the surface potential is the excess electronic charges. Similarly, the electrical breakdown voltage of PI LB films depends on the workfunction of metals used as substrates for the LB film preparation. In this section, firstly we explain our experimental results phenomenologically, and then discuss the possibility of charge transfer created by the application of biasing voltage.

As described in our previous papers [13, 25], very high density of electron acceptor states, which can acquire excess electrons displaced from the metal electrode, exists in PI LB films. These acceptor states exist contiguous to the Fermi energies of contacting metal electrode at the PI LB film/metal interface. Excess electronic charges are transferred from metals to these electronic states in PI LB films until a thermodynamic equilibrium is established at the interface. Figures 7 (a) and (b) illustrate the space charge distribution and the distribution of density of states at the interface. As shown in Fig. 7 (a), the excess charges are injected into PI LB films from the metal electrodes. Thus electric flux diverging from the positive charges on the surface of metal electrodes falls on the excess charges in PI LB films. Therefore, the surface potential V_s built across the PI LB films is given by [24, 26]

$$V_S = \int_0^D \frac{x\rho(x)}{\varepsilon_0 \varepsilon_r} dx = \frac{\overline{x}Q}{\varepsilon_0 \varepsilon_r}, \tag{1}$$

$$\text{with } \overline{x} = \frac{\int_0^D x\rho(x)dx}{\int_0^D \rho(x)dx} \text{ and } Q = \int_0^D \rho(x)dx.$$

Here, ε_0 is dielectric permittivity of vacuum, $\varepsilon_r (= 3)$ is the relative dielectric constant of PI, D is the film thickness, x is the distance from the metal electrode, $\rho(x)$ is the space charge density, $\overline{x}$ is the mean location of the excess charges injected from the electrodes, and Q is the total charge injected from the metal electrodes. As described in our previous study [24, 25], most of the excess charges exist in PI LB films within a distance of 4 nm from the electrodes, and about 1 to 10% of monomer units of PI accept electrons from the metal electrodes in this region. At equilibrium, it is expected that the surface Fermi level of PI LB films and the Fermi level of metals coincide at the interface as shown in Fig. 7 (b). Therefore, the electronic states of PI whose electronic energy is higher than the Fermi

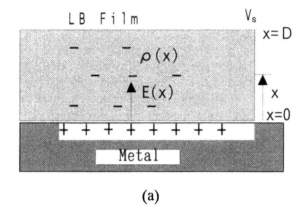

(a)

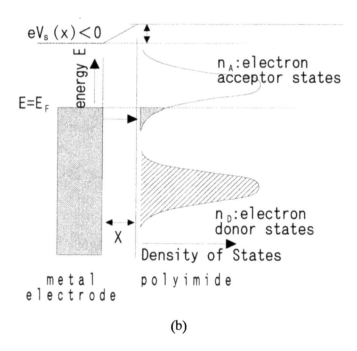

(b)

Figure 7. (a) Space charge distribution in PI LB films; (b) Energy diagram of the PI LB films at the film/metal interface. V_S: surface potential.

energy of the metal can donate electrons to the metal if the electronic states of PI are filled with electrons before the PI is in contact with the metal (electron donor states n_D). By contrast, the electronic states of PI whose electronic energy is lower than the Fermi energy of the metal can accept electrons from the metal if the states are empty (electron acceptor states n_A). As shown in Fig. 3, PI LB films are charged negatively even when the biasing voltage is zero. We may conclude that electron-acceptor states, n_A, play a dominant part in the charge exchange phenomena at the PI LB film/metal interface. On the basis of the above discussion, we may explain our experimental results as follows: As shown in Fig. 3, PI LB films are negatively charged even when biasing voltage is zero. The PI LB films are more negatively charged by applying the positive biasing voltage to the needle electrode, whereas it is gradually positively charged as the biasing voltage becomes more negative. One possible explanation is that both the mean location of electronic charges $\bar{x}$ and the total charge Q given by eq. (1) increase by the application of positive biasing voltage, whereas they decrease by the application of negative biasing voltage. Thus we may expect that electrons are injected from base electrodes to PI LB films via interfacial states at the film/metal interface when the needle electrode is positively biased, whereas holes are injected into PI LB films when the needle is negatively biased. In other words, the main contribution of excess charges are the electrons injected from metal electrodes, and not the electronic charges deposited onto the film surface from the side of needle electrode. Based on this model, we could explain the experimental results shown in Figs. 4 and 5 in the same way, where the surface potential depends on the number of deposited layers (see Fig. 4) and on the nature of metal electrode (see Fig. 5). As has been discussed above, all of our experimental results could be explained phenomenologically by assuming the presence of electronic charges at the film/metal interface, and that the additional surface potential created by the application of voltage is due to electronic charges displaced from the metal electrodes via interfacial surface states. In the following, we calculate the electric field formed at the film/metal interface, assuming film is placed on the surface of plane metal electrode under a needle-plane electrode system [26], and then discuss the possibility of the electron (hole) charge injection from the metal electrodes into PI LB films, which leads to the electrical breakdown.

For simplicity, the experimental electrode arrangement of the needle-plane electrode system shown in Fig. 2 is represented using prolate spheroidal coordinates (η, θ, ψ) which satisfy the following relationship with Cartesian coordinates (x, y, z) [27-29] (see Fig. 8):

$$\begin{cases} x = A\sinh\eta\sin\theta\cos\psi \\ y = A\sinh\eta\sin\theta\sin\psi \\ z = A\cosh\eta\cos\theta \end{cases} \text{ with } 0 < \eta < \infty \text{ , } 0 < \theta < \pi \text{ , and } 0 < \psi < 2\pi. \qquad (2)$$

Here, $A = d / \cos\theta_0$. The angle θ_0 is the coordinate representing the surface of the needle electrode, and d the distance between the needle tip and the plane electrode. The location of the surface of needle electrode is expressed by $\theta = \theta_0$. The location of plane electrode is expressed as $\theta = \pi/2$, using the coordinates

(η,θ,ψ). The electric field formed under a needle-plane electrode system is given by

$$E = \frac{V}{d}F_1, \quad \text{with } F_1 = \frac{\cos\theta_0}{\sqrt{\sinh^2\eta + \sin^2\theta}\,\sin\theta\ln\cot\theta_0/2}, \tag{3}$$

assuming $d>R$. Here R is the radius of curvature at the tip of needle electrode, and d is the distance between the needle tip and the plane electrode. θ_0 is approximately expressed as [29]

$$\theta_0 = \arcsin\sqrt{\frac{R}{R+d}}. \tag{4}$$

Since the V/d represents the electric field formed between the parallel two-electrode systems, the term F_1 represents the effect that the parallelly arranged two-electrode system is replaced by the needle-plane electrode system. Thus we find that the electric field formed on the plane electrode just under the tip of the needle electrode ($\theta = \pi/2$ and $\eta = 0$) is estimated as

$$E = \frac{V}{d}\cdot\frac{\cos\theta_0}{\ln\cot(\theta_0/2)} < \frac{V}{d}. \tag{5}$$

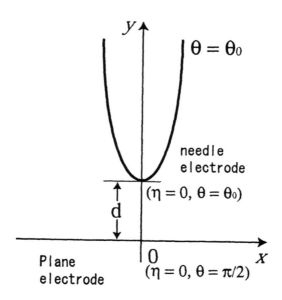

Figure 8. Electrode arrangement and relationship between prolate coordinates (η,θ,ψ) and Cartesian coordinates (x, y, z).

Using equations (4) and (5), we find that the electric field is of the order of 10^3-10^4 V/cm for $R = 10$ μm and $d = 1$ mm by the application of V = 0.5 kV-5 kV. The electric field formed at the film/metal interface is assumed to be of the order of 10^3 V/cm to 10^4 V/cm. In our previous studies, we showed that the electronic charge injection was expected to occur via interfacial surface states at the film/metal interface when the electric field was of this order [30]. Further, it is more informative to note that electroluminescence (EL) from the polymer-metal interface was observed under very low electric field due to the presence of the interfacial states [31]. Based on the above information, we expect that electronic charges existing at the polymer film/metal interface may be pulled into or pushed out of the films by the application of biasing voltage. The amount of interfacial charge is higher for PI LB films on Al electrodes as compared to the films on Au. Thus the electrical breakdown is also influenced by the amount of interfacial charge. As we can see in Fig. 6, the breakdown voltage of PI LB films on Al electrodes is lower than for films on Au electrodes. Further the electric field given by eq. (5) assists the injection of holes and electrons from metal electrode into PI LB films. As we can see in Figure 6, the breakdown voltage of PI LB films that are negatively biased by a needle electrode with reference to metal electrodes is lower than that of the films positively biased. Thus it is expected that the main contribution to the electrical breakdown is the injection of holes from the metals into films. Further investigation is required to clarify the details.

5. CONCLUSION

The surface potential of PI LB films under a needle-plane electrode system was examined. It was concluded that the main contribution to the creation of the surface potential was excess electronic charges injected from the metal electrodes via interfacial states into PI LB films at the film/metal interface. Further, the interfacial space charge also makes a significant contribution to the electrical breakdown of ultrathin films.

REFERENCES

1. H. Meier, *Organic Semiconductors, Dark-and Photo-conductivity of Organic Solids*, Verlag Chemie, Weinheim (1974).
2. J. Lowell and A.C. Rose-Innes, Adv. Phys. **29**, 947 (1980), and references therein.
3. L.H. Lee, J. Electrostatics **32**, 1 (1994), and references therein.
4. T.J. Lewis, IEEE. Trans. Dielectr. & Elect. Insul., **1**, 812-822 (1994).
5. T.J. Fabish and C.B. Duke, J. Appl. Phys. **48**, 4256-4266 (1976).
6. C.B. Duke and T.J. Fabish, Phys. Rev. Lett. **37**, 1075-1078 (1976).
7. W.R. Harper, *Contact and Frictional Electrification*, Oxford Univ. Press (1967).
8. J.M. Pochan, H.W. Harry, F.C. Bailey and D.F. Hinman, J. Electrostatics **8**, 183-194 (1980).
9. T.J. Fabish, C.B. Duke, M.L. Hair and H.M. Saltsburg, J. Appl. Phys. **51**, 1247-1249 (1980).
10. T.J. Fabish, H.M. Saltsburg and M.L. Hair, J. Appl. Phys. **47**, 930-939 (1976).

11. T.J. Fabish, H.M. Saltsburg and M.L. Hair, J. Appl. Phys. **47**, 940-948 (1976).
12. H. Ishii, K. Sugiyama, E. Itoh and K. Seki, Adv. Mater. **11**, 605-625 (1999).
13. M. Iwamoto and E. Itoh, Thin Solid Films **331**, 15-24 (1998).
14. M. Iwamoto, J. Mater. Chem. **10**, 99-106 (2000).
15. K.C. Kao, Proceedings of the 6th inte. Conf. on Properties and Applications of Dielectric Materials, held in Xi'an, China pp.1-17 (2000).
16. A. Ulman, *An Introduction to Ultrathin Organic Thin Films from Langmuir-Blodgettt Films to Self-assembly*, Academic Press, San Diego (1991).
17. Special Issue on Functional Organic Materials for Devices: J. Mater. Chem. **9**, 1853-2276 (1999).
18. Special Issue on Organic Thin Films: Materials Chemistry Discussion No. 2: J. Mater. Chem. **10**, 1-206 (2000).
19. M.K. Ghosh and K.L. Mittal (Eds.), *Polyimides: Fundamentals and Applications*, Marcel Dekker, New York (1996).
20. K.L. Mittal (Ed.), *Polyimides and Other High Temperature Polymers: Synthesis, Characterization and Applications*, Vol. 1, VSP, Utrecht (2001).
21. M. Iwamoto, M. Kakimoto, in: M.K. Ghosh and K.L. Mittal (Eds.), *Polyimides: Fundamentals and Applications*. Marcel Dekker, New York (1996) and references therein.
22. M. Kakimoto, M. Suzuki, T. Konishi, Y. Imai, M. Iwamoto and T. Hino, Chem. Lett. 823-826 (1986).
23. M. Fukuzawa and M. Iwamoto, IEEE Trans. Dielectri. and Electric. Insul. **6**, 858-863 (1999).
24. M. Iwamoto, A. Fukuda and E. Itoh, J. Appl. Phys. **75**, 1607-1610 (1994).
25. E. Itoh and M. Iwamoto, J. Appl. Phys. **81**, 1790-1797 (1997).
26. G.M. Sessler, *Electrets*, Springer Verlag, New York (1987).
27. M. Iwamoto, M. Fukuzawa and T. Hino, IEEE Trans. Elect. Insul. **22**, 419-424 (1987).
28. H. Prinz, *Hochspannungsfelder*, R. Oldenbourg Verlag, Munchen (1969).
29. M. Fukuzawa and M. Iwamoto, IEEE Trans. Dielectr. Electrical Insul. **8**, 832-837 (2001).
30. E. Itoh and M. Iwamoto, J. Appl. Phys. **85**, 7239-7243 (1999).
31. T. Mizuno, Y.S. Liu, W. Shinoyama, K. Yasuoka, S. Ishi, H. Miyata and A. Yokoyama, IEEE Trans. Dielectrics Electrical Insul. **4**, 433-438 (1997).

Polyimides and Other High Temperature Polymers, Vol. 2, pp. 255–266
Ed. K.L. Mittal

Humid ageing of polyetherimide: Chemical and physical interactions with water

F. THOMINETTE,* I. MERDAS and J. VERDU

ENSAM, 151 Bd de l'Hôpital, 75013 Paris, France

Abstract—The study of humid ageing of polyetherimide (PEI) in an aqueous environment (neutral, or acidic solution, pH 2) at 100, 170 and 180°C shows a very good hydrolytic stability of this material, despite the presence of imide groups which are sensitive to hydrolysis.

The behavior of the polymer in the presence of water was studied at long term: The sorption kinetics of water into polyetherimide (ULTEM 1000) was investigated at various temperatures ranging from 20 to 100°C. The water equilibrium concentration increased slightly with temperature from 1.39% (by weight) at 20°C to 1.51% at 100°C. The solubility coefficient, S, calculated from these data and the water vapor pressure decresed with temperature. The calculated heat of dissolution H_s was close to -43 kJ.mol^{-1}, which explains the slight effect of temperature on the equilibrium concentration.

The diffusion coefficient D varied from about 1.10^{-12} m^2.s^{-1} at 20°C to about 50.10^{-12} m^2.s^{-1} at 120°C. The apparent activation energy of diffusion, E_D and the heat of dissolution H_s of water in the polymer have opposite values (respectively +43 and -42 kJ.mol^{-1}). These results and the data on water diffusion characteristics in other glassy polar polymers indicate that the transport rate of water is kinetically controlled by the dissociation of water-polymer complexes.

Keywords: Polyetherimide; water; sorption; solubility; diffusion; hydrolysis.

1. INTRODUCTION

Aromatic polyimides display interesting properties due to their high glass transition temperature (T_g). They can, however, undergo "humid ageing" because they are capable of absorbing typically 1 to 6% by weight of water at equilibrium in saturated atmosphere or in immersion. Water can induce plasticization [1] and sometimes hydrolysis [2, 3]. Swelling can be a problem because it induces stresses or dimensional changes [4, 5].

Despite their high industrial interest, thermoplastic polyetherimides [6] have not been widely studied from the point of view of water absorption, probably because the first experimental studies showed that they displayed a relatively low

---

*To whom all correspondence should be addressed. Phone: (33) 01 44 24 62 51,
Fax: (33) 01 44 24 63 82, E-mail: francette.thominette@paris.ensam.fr

hydrophilicity compared to other members of the polyimide family. Values close to 1.5% were, for instance, reported for the equilibrium mass gain in saturated conditions [7]. As it has been shown in a previous paper [8], such water concentrations are, however, sufficient to induce significant changes in the mechanical behaviour. This is the reason why it was interesting for us to study in detail the water absorption in saturated conditions in PEI samples in the 20-100°C temperature interval. On the other hand, the PEI chemical structure, in particular the presence of imide functions sensitive to hydrolysis, allows us to study the eventual PEI-water chemical interaction.

2. EXPERIMENTAL

2.1. Material

The material examined in this study was an extruded plaque (3 mm thick) of thermoplastic polyetherimide (Ultem 1000) supplied by General Electric (number average molar mass 20 kg.mol^{-1}, glass transition temperature in dry state 217°C and density 1.27 g.cm^{-3}).

2.2. Ageing tests

Ageing tests were performed in a thermostated (±0.1°C) bath with distilled water at 100°C, in acidic solution (pH 2) at 100°C, and into hermetically sealed reaction vessels at 170°C and 180°C.

In order to avoid any other polyimide degradation process than hydrolysis, especially oxidation, oxygen was carefully removed from the reaction vessel using 30 min nitrogen bubbling before heating.

2.3. Size exclusion chromatography (SEC)

SEC measurements were carried out in chloroform solution, at a 0.5 ml.min^{-1} flow rate. The calibration was made with polystyrene standards.

2.4. Infrared spectroscopy

IR spectroscopic measurements were made on a Brücker IFS 28 spectrophotometer at a resolution of 4 cm^{-1}.

2.5. Sorption tests and gravimetric study

After drying at 50°C under vacuum until constant weight, the samples were immersed in a thermostated (±0.1°C) bath with distilled water at temperatures ranging between 20 and 100°C. They were periodically taken out and weighed with an electronic microbalance (Mettler Mark 3) with relative accuracy of 10^{-4}.

The average duration of weighing operation was less than 3 min, during which the sample was exposed to ambient atmosphere (temperature: 20°C, relative humidity: 50% RH). Indeed such stopping period is expected to delay sorption and thus to lead to underestimation of the diffusion coefficient. It can be, however, remarked that the whole time for weighing ranges from about 4% at 100°C to about 0.2% at 20°C of the time to reach sorption equilibrium. Thus, the relative error in diffusion coefficient is expected to be of the same order. At 20°C, the drying effect of the time of weighing, in the extreme case of dry air, would be, according to Fick's law, a decrease of the water mass fraction of $1.8.10^{-4}$ for 200 seconds. This effect is pratically undetectable in the chosen experimental conditions.

3. RESULTS

3.1. Hydrolysis

SEC chromatograms of samples immersed for one and two days in water at 180°C are compared to that of the virgin sample in Figure 1. The results show no significant variation of the average molecular weight after 2 days in immersion at 180°C because no other peaks relative to oligomers resulting from the hydrolysis reaction were observed. (Similar results were obtained after 21 days immersion at 170°C).

The FTIR results presented in Figure 2 show no modification of the PEI chemical structure after immersion in water.

The gravimetric results indicate a low mass loss, close to 1% after 6 and 8 days immersion time at 180°C and 170°C, respectively.

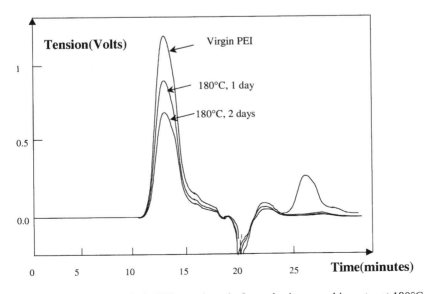

Figure 1. Chromatograms of a virgin PEI sample and of samples immersed in water at 180°C.

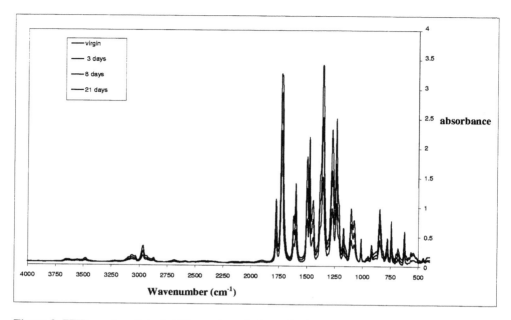

Figure 2. FTIR spectra of virgin PEI sample and of samples immersed in water at 180°C.

3.1.1. Discussion

The hydrolysis process can be described as:

This process is accompanied by:

— Mass gain (18 g/mol corresponding to the incorporation of water during the chain scissions),

— Formation of acid and amide groups.

At 100°C, pH 2, the viscosimetric results show chain scissions resulting from the following two hydrolysis reactions:

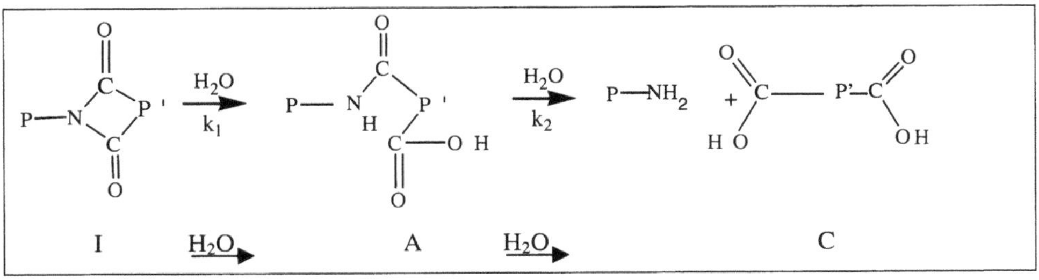

A kinetic scheme can be written as:

$$I + W \quad \rightarrow \quad A \qquad (k_1)$$

$$A + W \quad \rightarrow \quad C \qquad (k_2)$$

where I is the imide group, A is the (amic acid) group, C is the acid chain resulting from the scission process and W is the water. One can write:

$$\frac{dA}{dt} = k_1 I.W - k_2 A.W \qquad (1)$$

and

$$\frac{dC}{dt} = k_2 A.W \qquad (2)$$

At low conversions (at the beginning of the hydrolysis reaction) one can consider in a first approximation that I≫A and thus I is almost constant. So one obtains:

$$A = \frac{k_2}{k_1} I(1 - \exp - k_2 Wt) \qquad (3)$$

Equations (2) and (3) lead to

$$\frac{dC}{dt} = \frac{k_2^2}{k_1} I.W(1 - \exp - k_2 Wt) \qquad (4)$$

Thus

$$\frac{\dfrac{dC}{dt}}{\dfrac{dA}{dt}} = \frac{1 - \exp - k_2 Wt}{\exp - k_2 Wt}$$

At low conversions, $\dfrac{dC}{dt} \ll \dfrac{dA}{dt}$, the hydrolysis rate $\left(\dfrac{dC}{dt} = 2v_H \right)$ is low ($v_H \leq$ 5.3.10^{-8} mol.l^{-1}.s^{-1} as determined from IR results).

So after 350 days at 100°C, pH 2: $\frac{[\eta](350days)}{[\eta](0day)} \leq \frac{38.3}{42.8} = 0.89$

According to the Mark-Houwink law: $[\eta] = KM^{\alpha}$ where K and α are constants depending on the polymer, solvent and temperature of viscosimetric measurements; and M is the number average molecular weight. Thus

$$\frac{M(350days)}{M(0day)} = \left[\frac{[\eta](350days)}{[\eta](0day)}\right]^{1/\alpha}$$

As $\alpha \sim 0.7$ one obtains: $\frac{M(350days)}{M(0day)} \leq 0.85$. The number of chain scissions, n, after 350 days immersion time at 100°C, pH 2 is given by:

$$n = \frac{1}{M} - \frac{1}{M_o} \leq \frac{1}{M_o}\left(\frac{1}{0.85} - 1\right) = 9.5.10^{-3}\ mol.kg^{-1}.$$

So the average rate of chain scissions v_c is: $v_c \leq 3.2.10^{-10}\ mol.kg^{-1}.s^{-1}$. As expected, v_c is lower than v_H.

The conclusion is that all the experimental data are consistent and indicate that hydrolysis is extremely low at 100°C, pH 2. One possible explanation is that the amide and acid resulting from the first hydrolysis reaction remain "trapped" in a cage favoring their immediate recombination (as shown in the following scheme).

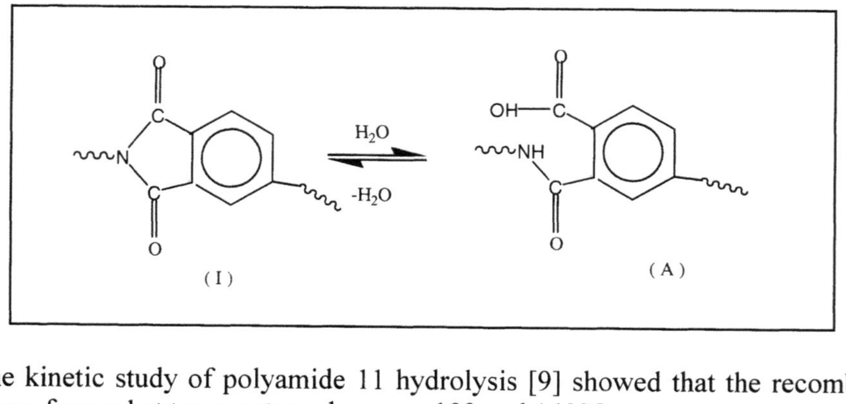

The kinetic study of polyamide 11 hydrolysis [9] showed that the recombination was favored at temperatures between 100 and 140°C.

Between 170-180°C, the two main experimental facts are:

(i) Crosslinking: gelation occurs after 21 days at 170°C and after 6 days at 180°C.

(ii) A significant mass gain for PEI (close to 20%) in wet state.

Crosslinking could result from intermolecular condensation between an amine and an acid as shown in the following scheme:

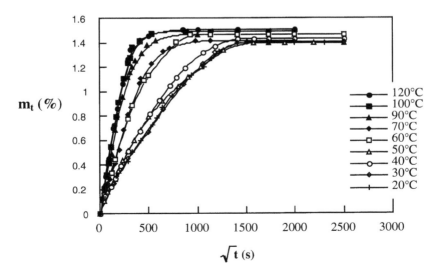

These results show the high hydrolytic stability of this polymer even in very extreme immersion conditions (neutral or acidic medium) at 100, 170 and 180°C. Hydrolysis effects can, therfore, be neglected in the study of PEI.

3.2. Water sorption characteristics of PEI

The essentials of the experimental results have already been reported [10, 11].

An example of sorption curves (mass gain m_t against time t) [12] is shown in Figure 3. The sorption curves are linear in $t^{1/2}$ in their initial part ($0 \leq m_t \leq 0.6$ m_∞,

Figure 3. Sorption curves of water in PEI from 20°C to 120°C (100% R.H). m_t (%) is the water mass gain.

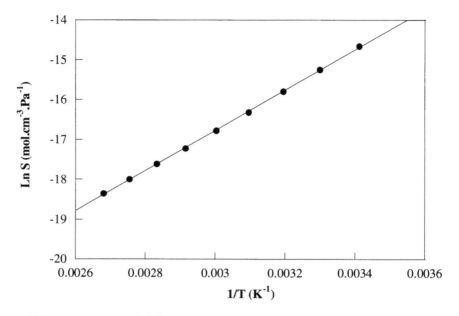

Figure 4. Variation of water solubility in PEI with temperature.

m_∞ being the equilibrium mass gain), showing that the diffusion process is apparently Fickian.

3.2.1. Solubility: effect of temperature

Since, in the studied case, the concentration remains low (water mass fraction, w ≤ 0.07) and the polymer remains in the same state, one can, in a first approximation, consider that the equilibrium water concentration C obeys the Henry's law:

$$C = Sp \tag{5}$$

where S is the solubility coefficient and p the water vapor pressure. S varies with the temperature according to Arrhenius law:

$$S = S_0 \exp\left(\frac{-H_s}{RT}\right) \tag{6}$$

with H_s: heat of dissolution.

S$_o$: pre-exponential factor of solubility.

Figure 4 shows the Arrhenius plot for solubility. The heat of dissolution: $H_s = -43$ kJ.mol^{-1}.

However, the sorption tests are made in saturated vapor pressure at various temperatures. Between 20 and 100°C the vapor pressure also obeys Arrhenius law:

$$p = p_o \exp\left(\frac{-H_p}{RT}\right) \tag{7}$$

with $H_p = 43$ kJ.mol^{-1}.

The combination of equations (5), (6) and (7) gives:

$$C = C_o \exp\left(\frac{-H_c}{RT}\right) \tag{8}$$

with $C_o = S_o \, p_o$ and $H_c = (H_s + H_p)$.

All the experimental results can be explained if H_s is negative and close to H_P in absolute value. Thus, C will increase or decrease with temperature depending on whether H_s is lower or higher than H_P in absolute value, as previously established in recently published works [10, 11, 13].

3.2.2. Diffusion

Diffusivity is analyzed assuming a Fickian mechanism. The coefficient of diffusion D was determined from the classical relationship:

$$D = \frac{\pi}{16}\left(\frac{m}{m_\infty}\right)^2 \frac{e^2}{t} \quad \text{for } m_t \leq 0.6 \, m_\infty \tag{9}$$

where e is the sample thickness.

The Arrhenius plot for diffusion coefficient (Fig. 5) displays a non-linear shape suggesting two possible ways of interpretation. The first explanation considers

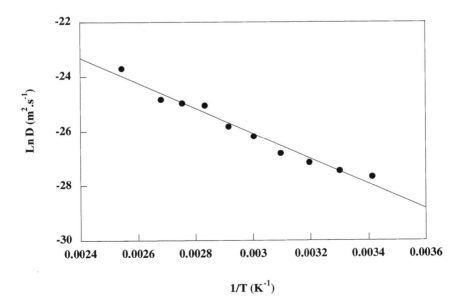

Figure 5. Variation of water diffusivity in PEI with temperature.

that the temperature dependence of D can be described by two plateaus, at $T \geq$ 80°C and $T \leq 40$°C separated by a transition zone. There is, however, no physical argument for this interpretation. The second one consider that non-linearities result from experimental uncertainties and that D obeys, in fact, the Arrhenius law.

A complementary test in an autoclave at 120°C, 0.2 MPa gave results in favor of the latter hypothesis (Fig. 5). The regression coefficient is R = 0.9861. The apparent activation energy is: $E_D = 43$ kJ.mol^{-1}.

3.2.3. Discussion

The heat of dissolution H_s is made up of two contributions: heat of mixing H_m and heat of condensation H_p. Two extreme hypotheses can, therefore, be envisaged:

(a) Water occupies only the free volume. In this case, the "dissolution" is essentially due to the re-formation of water-water bonds and, then, $H_m = 0$ and $H_s = H_p$. There is, in fact, no dissolution of water in the matrix.

(b) Water is entirely dissolved into the matrix. In this case, the dissolution is essentially due to the formation of water-polymer bonds, and there are no water-water bonds so that $H_p = 0$, $H_m = H_s$.

The fact that $H_s \sim H_p$ results from the similarity of water-water and water-polymer bonds.

Certain authors, for instance Adamson [14] have suggested that water absorption by polar polymers such as epoxies would correspond to case (a); however, the following arguments are rather in favor of case (b):

(i) The fact that dissolution is very exothermic (H_m closer to H_p than to 0) can be predicted from the structure-property relationships available for polymer-solvent interactions [12]. The hypothesis that $H_m = 0$ would disagree with these relationships.

(ii) H_s increases, as expected in case (b), with the polymer polarity [10, 11, 13]. In the case where water occupies only the free volume, one would expect $H_s = H_p$, independently of the polymer structure.

(iii) In case (a), water and polymer would form two phases; in this case, DSC is expected to show the exotherm (or endotherm) of water crystallization (or melting) at 0°C. This exotherm was never observed in our experiments.

(iv) In case (b) water plasticizes PEI and the extent of plasticization, as determined from the T_g variations, agrees with the hypothesis that all the water and polymer form one phase into the matrix [8]. In case (a) there would be no plasticization.

(v) More generally, hypothesis (a) fails to explain why certain materials in which the free volume fraction is high, for instance apolar liquids, are almost hydrophobic.

It can thus be concluded that water is essentially dissolved into the matrix and that the dissolution is due to the strong polarity of the water molecules. In other words, there is a critical polymer polarity, for which $H_s = H_p$. For polymers of lower polarity, $|H_s| < |H_p|$ the equilibrium concentration increases with the temperature, and the opposite effect is observed when $|H_s| > |H_p|$.

It is noteworthy that the apparent activation energy of diffusion E_D is practically equal to the absolute value of the heat of dissolution H_s.

This experimental observation and our own results obtained in the case of linear and crosslinked polar polymers [10, 11, 13], in order to establish the structure/hydrophilicity relationships, indicate clearly that the water-polymer interactions play a role in diffusion. The diffusion of a water molecule (W) from site P_1 to site P_2 could be tentatively shown by the following sequence involving three distinct steps:

(I) $[P_1...W]$ $\rightarrow$ $P_1 + W$ Dissociation of the polymer-water complex

(II)   W $\rightarrow$ migration from P_1 to P_2

(III) $W + P_2$ $\rightarrow$ $[P_2...W]$ Formation of the polymer-water complex

The water dissolution is strongly exothermic when the polymer is strongly hydrophilic; in other words, the dissociation of the polymer-water complex is more difficult when the polymer is hydrophilic. In fact, we have to compare two rate constants k_1 and k_2 linked, respectively, to process (I) and (II): k_1 is a decreasing function of polymer-water bond energy; and $k_2 \sim D/d^2$ where D is the diffusivity of water in the polymer and d is the distance between two neighboring sites. For high D or low d, the diffusion is controlled by the dissociation of the polymer-water complex.

4. CONCLUSION

The study of PEI humid ageing for long times in an aqueous environment (neutral or acidic solution, pH 2) at 100, 170 and 180°C shows a very good hydrolytic stability of this material, despite the presence of imide groups which are sensitive to hydrolysis.

The physical sorption study indicates that the activation energy of diffusion E_D is opposite to the heat of dissolution H_s. From the experimental results and our previous works on other polar linear and crosslinked polymers, one can suggest that the water diffusivity in these polymers is essentially governed by the strength of polymer-water hydrogen bonds. In other words, the dissociation of the polymer-water complex would be the slowest step in a diffusion event.

REFERENCES

1. E. Sacher and J.R. Susko, J. Appl. Polym. Sci., **23**, 2355-2364 (1979).
2. R. Deiasi and J. Russell, J. Appl. Polym. Sci., **15**, 2965-2974 (1971).
3. D.K. Yang, W.J. Koros and V.T Stannett, J. Appl. Polym. Sci., **31**, 1619-1629 (1986).
4. A. Tcharkhtchi, P.Y. Bronnec and J. Verdu, Polymer, **41**, 5777-5785 (2000).
5. W.P. Paplham, R.A. Brown and I.M. Salin, J. Appl. Polym. Sci., **57**, 133-137 (1995).

6. T. Takekoshi, J.E. Kochanowski and M.J. Webber, J. Polym. Sci: Polym. Chem. Ed., **23**, 1759-1769 (1985).

7. O. Robert and S.H. Burlhis, J. Appl. Polym. Sci: Polym. Symp., **70**, 129-143 (1983).

8. I. Merdas, F. Thominette and J. Verdu, J. Appl. Polym. Sci., **77**, 1445-1451 (2000).

9. B. Jacques, M. Werth, I. Merdas, F. Thominette and J. Verdu, Polymer, In press.

10. I. Merdas, F. Thominette and J. Verdu, J. Appl. Polym. Sci., **77**, 1439-1444 (2000).

11. I. Merdas, F. Thominette, A. Tcharkhtchi and J. Verdu, Composites Sci. Technol., **62**, 487-492 (2002).

12. G.J. Van Amerongen, J. Appl. Phys, **17**, 972 (1946); J. Polym. Sci., **5**, 307 (1950).

13. I. Merdas, A. Tcharkhtchi, F. Thominette, J. Verdu, K. Dean and W. Cook, Polymer, **43**, 4619-4625 (2002).

14. M.J. Adamson, J. Mater. Sci., **15**, 1736-1745 (1980).

Polyimides and Other High Temperature Polymers, Vol. 2, pp. 267–285
Ed. K.L. Mittal
© VSP 2003

Transport of water in high T_g polymers: A comparison between interacting and non-interacting systems

G. MENSITIERI,[*,1] S. COTUGNO,[1] P. MUSTO,[2] G. RAGOSTA[2] and L. NICOLAIS[1]

[1] *Dept. of Materials and Production Engineering, University of Naples Federico II, P.le Tecchio 80, 80125 Naples, Italy*
[2] *Institute of Research and Technology of Plastic Materials, National Research Council of Italy, Via Toiano 6, 80072 Arco Felice (Naples), Italy*

Abstract—Molecular interactions of absorbed water molecules with two different high performance polymer matrices have been investigated using classical gravimetric analysis as well as *in situ, time-resolved* FTIR spectroscopy. Sorption and desorption kinetics at several activities and equilibrium isotherms have been evaluated by both experimental techniques, and useful information has been gathered by comparing the results of the two approaches. These results have been discussed with respect to the molecular interactions and the state of aggregation of the penetrant molecules in the two systems. In fact, various species of absorbed water molecules have been identified spectroscopically, depending on the molecular structure of the two matrices. In the case of epoxy resin, free and dimeric water molecules coexist with water molecules strongly bound to the polymer network through H-bonding interactions. Conversely, in the case of polyimide, because of the absence of strongly interacting sites along the polymer backbone, mostly free and self-associated water has been detected. The coupled analysis of gravimetric and spectroscopic sorption isotherms has shed light on the different sorption mechanisms operating in the two matrices. The difference in the molecular interactions between the penetrant and the polymer matrices causes a much faster water diffusivity in the case of polyimide as compared to the epoxy.

Keywords: Sorption; water; epoxy; polyimide; infrared spectroscopy.

1. INTRODUCTION

Mass transport of low molecular weight compounds in polymeric materials has caused a great scientific and technological interest in the past 40 years. There are several engineering applications where diffusion behaviour has a major impact, e.g. gas mixtures separation with membranes, drug delivery, barrier structures for food packaging, environmental resistance of polymer based composites, and devolatilization. In this respect, the durability of high performance matrices for composites is of considerable technological relevance. In fact these matrices,

---

*To whom all correspondence should be addressed. Phone: +39/081/7682512, Fax: +39/081/7682404, E-mail: mensitie@unina.it

when exposed to humid environment, absorb significant amounts of water which adversely affects their physical-mechanical properties.

The main documented effects of water on polymer matrices are, among others, plasticization, which occurs by different mechanisms depending on the level of interaction of sorbed water molecules with the matrix; changes of physical properties, i.e. decrease of elastic moduli, decrease of yield strength, change of yield/deformation mechanisms; hygrothermal degradation, i.e., microcracks, ageing, chain scission through hydrolysis, degradation of fibre/matrix interface in composites and swelling stresses.

To explain the anomalous sorption features of some glassy polymers several authors have proposed that the water is made up of two species, i.e., those forming a molecular solution and those confined into areas of abnormally large free volume, often referred to as microvoids.

Recently, direct evidence of the presence of nanopores by Positron Annihilation Life-time Spectroscopy (PALS) has been reported by Soles and coworkers for a series of epoxy systems [1-3]. It has been found that these nanopores have an average diameter of around 5-6 Å and can form a continuous network of channels across the polymer matrix. Water traverses the epoxy through the network of nanopores until it reaches polar hydroxyl and amine groups where it forms strong hydrogen bonding interactions. Thus, the nanopores provide access routes to the water molecules to reach the specific interaction sites. In this description the topology, polarity and molecular motions combine to control transport. Aging effects on water sorption equilibrium and kinetics observed in the investigated samples have been interpreted by assuming that sorbed water is capable of reorganizing the molecular topology of the matrix in a way that reveals more potentially interacting polar sites.

Also Zhou and Lucas [4, 5] have invoked the interplay between network topological modification and molecular interactions to interpret the water sorption results. They assumed the existence of two types of molecular interactions on the basis of solid state NMR, gravimetric measurements, and calorimetry.

With respect to the issue of molecular interactions, an earlier NMR investigation by Jelinski and coworkers [6, 7] concluded that i) water was impeded in its movements, jumping from one site to another with a characteristic residence time of $7 \cdot 10^{-10}$ s; ii) no free water existed, iii) there was no evidence of tightly bound water, and iv) the disruption of the secondary hydrogen-bonded network in the epoxy matrix by sorbed water was unlikely.

Furthermore, several investigations on difunctional epoxies based on dielectric relaxation spectroscopy (DRS) support the idea of water molecules partly bound to polymer chains and partly clustered into microvoids [8-10]. Grave et al. [10] were able to calculate the relative amounts of bound and free water from the magnitude of the relaxation at 10^5 Hz and 10^9 Hz, respectively. It is worth noting that the relaxations at 10^9 Hz are in good agreement with the characteristic jumping time for the water molecules to move from one site to the other as determined by NMR analysis [6, 7].

Combined DRS and FTIR measurements by Mijovic and coworkers [11, 12] on difunctional epoxy resin and bismaleimide (BMI) confirmed the existence of non-hydrogen bonded and hydrogen-bonded water. However, molecular dynamics simulation performed by Mijovic and coworkers [11], indicated that in the epoxy resin free water was absorbed in the form of isolated single molecules and that a water molecule participated in hydrogen bonding to only one site on the network.

In connection with the mass transport and equilibrium sorption in interacting media, *in-situ, time-resolved* FTIR spectroscopy provides not only a means for monitoring accurately the penetrant sorption kinetics and equilibrium sorption isotherm, but also offers the possibility of investigating in detail the concurrent development of molecular interactions [13, 14]. This experimental approach has been adopted in the present contribution, along with gravimetric analysis, to investigate water sorption and transport properties in two high performance polymer matrices: tetraglycidyl-4,4' diamino diphenylmethane (TGDDM) cross-linked by 4,4'-diamino-diphenylsulphone (DDS) and a commercial pyromellitic dianhydride-oxydianiline (PMDA-ODA) polyimide (PI). The tetrafunctional epoxy formulation, TGDDM-DDS, is the matrix of choice for advanced aerospace applications and may absorb up to 7% by weight of water [15-17], due to the high concentration of polar groups into the network. The PMDA-ODA PI is used as a matrix for carbon fibre microcomposites and silica based nanocomposites [18] and is characterised by a lower level of water sorption.

2. EXPERIMENTAL

2.1. Materials

The epoxy resin was a commercial grade tetraglycidyl-4,4' diamino diphenyl-methane (TGDDM) supplied by Ciba Geigy (Basel, Switzerland), and the curing agent was 4,4'-diamino-diphenylsulphone (DDS) from Aldrich (Milwaukee, WI). A total of 30 g of DDS were dissolved in 100 g of TGDDM [corresponding to a concentration of 30 parts per hundred resin by weight (phr)] at 120°C, degassed under vacuum, and poured in a stainless steel mould. The temperature of the mould was then raised to 140°C and kept at this value for 16 hours; afterwards the temperature was increased to 200°C and kept for 4 hours. Thin films were obtained with a thickness of about 9 μm. The glass transition temperature (T_g) of the material, as evaluated from the position of the tan δ peak in the dynamical-mechanical spectrum, was 278°C.

The polyimide was a condensation product of pyromellitic dianhydride (PMDA) and oxydianiline (ODA). Thin films (about 20 μm thick) were obtained from a poly(amic acid) solution (14 wt%) in N-methyl pyrrolidone and xylene (trade name Pyre ML RK692 from DuPont, Wilmington, DE). The films cast on glass slides were kept at 80°C for 1 hour to eliminate most of the solvent and then cured in successive isothermal steps of 1 hour each, at 100, 150, 200, 250 and 300°C. Full imidization was confirmed spectroscopically.

2.2. Techniques

2.2.1. Water sorption by in situ time-resolved FTIR spectroscopy

A vacuum tight FTIR cell was designed and used to record the FTIR transmission spectra of the polymer films exposed to a controlled humidity environment. The cell was connected by service lines to a water reservoir and to vacuum and pressure transducers (see Figure 1). The details of the experimental apparatus are reported elsewhere [13].

The instrument used for the collection of IR spectra was a Perkin-Elmer System 2000 interferometer equipped with a germanium/KBr beam splitter and a wide-band DTGS detector. Instrumental parameters were as follows: resolution = 4 cm^{-1}, optical path difference (OPD) velocity = 0.2 cm/s and spectral range 4000-400 cm^{-1}. A single data collection was performed for each spectrum (3551 data points) which, under the selected instrumental conditions, took 6 seconds to complete. The signals were acquired as single beam spectra at specific time intervals. A typical sorption run lasted about 30 minutes and acquisition time intervals were 6 seconds during the first 10 minutes of the experiment and 60 seconds afterwards.

Tests were performed at 24°C and at 5 activity levels, a, (0.08, 0.2, 0.4, 0.6 and 0.8) of water vapour on epoxy samples of thickness 9 ± 0.5 μm, and at 30°C and at

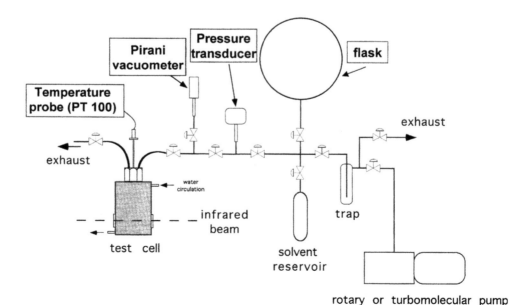

Figure 1. Schematic representation of the apparatus for sorption measurements by *in situ* FTIR spectroscopy.

7 activity levels (0.1, 0.2, 0.3, 0.4, 0.5, 0.6 and 0.75) on polyimide samples of thickness 20 ± 0.5 µm. Water vapour activity is defined as p/p_O, where p is the actual pressure of water vapour and p_O is the water vapour pressure at the test temperature.

2.2.2. Gravimetric analysis

The equipment used to determine weight gain of samples exposed to a controlled humidity environment was essentially the same as the one used for the on-line FTIR spectroscopy measurements, but in place of the FTIR measuring cell an electronic microbalance, model D200 (by CAHN Instruments, Madison, WI), with a sensitivity equal to 0.1 µg was used.

2.3. Data analysis

The single beam spectra were processed, to obtain the absorbance spectra, using as background the empty cell at the test conditions. An example of absorption spectra of 'wet' and 'dry' samples is reported in Figure 2a for the epoxy resin and in Figure 2b for the polyimide. In both cases the absorbed water brings about an increase in absorbance in the v_{OH} range (about 3400 cm^{-1}) and in the in-plane bending interval (about 1600 cm^{-1}). The spectra of sorbed water at different times were obtained by eliminating the interference due to the polymer matrix using subtraction spectroscopy [19, 20], i.e.:

$$A_d = A_s - K \cdot A_r \tag{1}$$

where A is the absorbance, the subscripts d, s and r denote, respectively, the difference, sample ('wet' specimen) and reference ('dry' specimen) spectra and K is an adjustable parameter used to compensate for possible thickness differences between 'wet' and 'dry' samples [19, 20]. In our case, since no detectable thickness changes occur upon water sorption, K is consistently equal to unity. The frequency region analysed in the present study is the 3800-2800 cm^{-1} range where the fundamental stretching vibrations of water are located. As an example of the results obtained by subtraction analysis, in Figure 3 are reported the spectra of sorbed water in the 3800-2800 cm^{-1} range, collected at different sorption times at a = 0.4 in the polyimide.

In order to separate the individual peaks in the case of unresolved, multicomponent bands, a curve resolving algorithm was employed, based on the Levenberg-Marquardt method [21]. In order to reduce the number of adjustable parameters the baseline, the band shape and the number of components were fixed. The number of components was evaluated by visual inspection of the abrupt changes in the slope of the experimental line-shape. The program was then used to calculate, by a non-linear curve fitting of the data, the height, the full width at half height (FWHH) and the position of the individual components.

The peak function used was a mixed Gaussian-Lorentzian line shape of the form [22]:

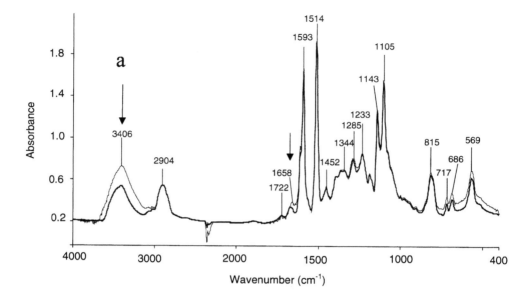

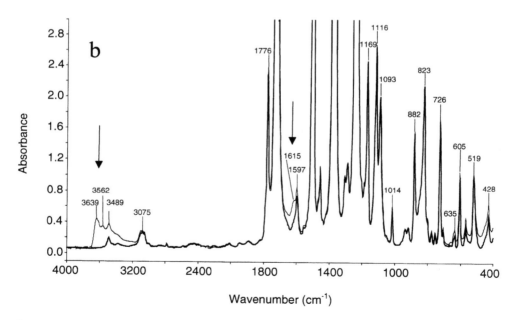

Figure 2. Transmission FTIR spectra in the 4000-400 cm^{-1} interval for (a) epoxy and (b) polyimide in dry state (thick line) and after equilibrium water sorption at water vapour activity, a, equal 0.4 (thin line).

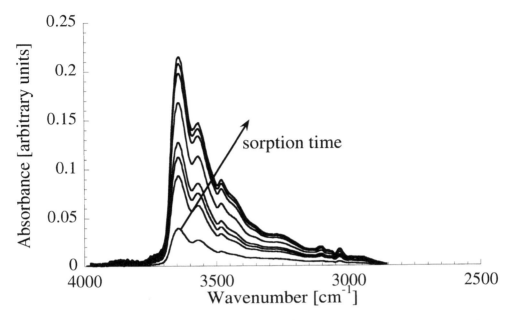

Figure 3. Subtraction FTIR spectra in the 4000-2500 cm⁻¹ range collected at different sorption times for water vapour sorption at a = 0.4 in the PMDA-ODA matrix, at 30°C.

$$f(x) = (1-L)H\exp-\left[\left(\frac{x-x_0}{w}\right)^2(4\ln 2)\right] + L\frac{H}{4\left(\frac{x-x_0}{w}\right)^2 + 1} \qquad (2)$$

where: x_0 = peak position; H = peak height; w = FWHH; L = fraction of Lorentz character.

3. RESULTS AND DISCUSSION

The subtraction spectrum of the epoxy exposed to liquid water in the 3800-2800 cm⁻¹ region, representative of the absorbed penetrant, has been already reported in a previous contribution [14]. This spectrum was interpreted on the basis of a simplified association model, whereby three different water species (S_0, S_1 and S_2) could be spectroscopically distinguished, according to Figure 4. Spectral interpretation was aided by analysis of the NIR region [14]. The results obtained in the present study for water vapour sorption in the epoxy resin are reported in Figure 5, which displays the experimental subtraction spectrum along with its curve-fitting results for the case of a water activity equal to 0.8 at 24°C; similar profiles were obtained at the other investigated activities. According to the vibrational analysis reported in ref. [14], the peak centred at 3623 cm⁻¹ is related to the asymmetric O-H stretching vibration (v_3) of unassociated water (i.e. water which does not establish any H-bond, S_0); the broad band at lower frequencies is due to hydrogen-bonded water molecules (S_1, S_2). In particular, referring to the components identified by the curve fitting results, the peak at 3555 cm^{-1} can be associated with S_1 molecules: on the basis of previous FT-NIR investigation [14, 23], these species can be most likely identified as water molecules self-associated in the form of dimers or as water molecules interacting with the network through a hydrogen bond involving only one hydrogen atom of the water molecule. Conversely, the two components at 3419 and 3264 cm^{-1} are associated with S_2 species displaying stronger H-bond interactions, i.e., most likely, water molecules interacting with proton-acceptor sites on polymer backbone, although a limited contribution from self-associated water cannot be entirely ruled out.

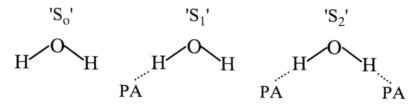

Figure 4. Simplified association model for water. 'PA' stands for 'proton acceptor'.

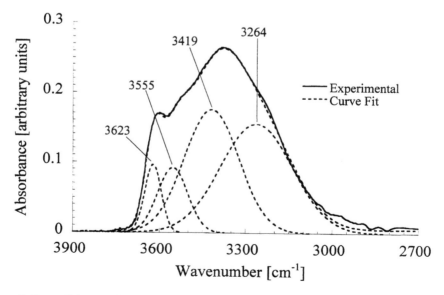

Figure 5. Curve fitting results of the spectrum representative of water absorbed at equilibrium (a = 0.8) in the epoxy resin. The figure displays the resolved components, the curve fitted and the experimental profiles.

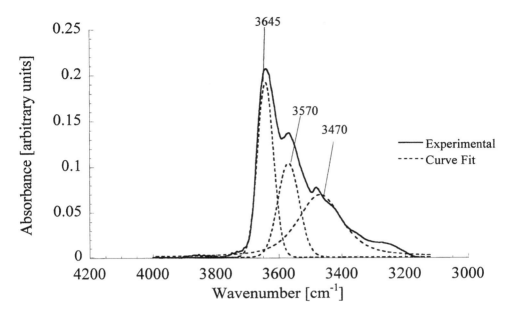

Figure 6. Curve fitting results of the spectrum representative of water absorbed at equilibrium (a = 0.4) in the polyimide resin. The figure displays the resolved components, the curve fitted and the experimental profile.

Figure 6 displays the experimental subtraction spectrum, along with its curve-fitting results, for the case of the PMDA-ODA film exposed to a water vapour activity equal to 0.4 at 30°C. The profile is significantly different from the one obtained for the epoxy: the high frequency peak, associated with the S_0 species, becomes the dominant component. Other two peaks, related, respectively, to S_1 and S_2 species, are present, even though their relative contribution is much lower than in the case of the epoxy. For the PMDA-ODA matrix, FT-NIR results [14, 23] suggest that most of the water is free (S_0), while the remaining small fraction (S_1 and S_2) is likely to be self-associated water in the form of dimers (S_1) and multimers (i.e. clusters of more than two molecules, S_2). In fact, it has been shown [14, 23] that a polyimide network with no other functional groups than the imide carbonyls (a thermosetting bismaleimide, BMI), do not interact with absorbed water. This is consistent with the very low tendency of imide carbonyls to act as proton acceptors [24]. However, in the present case, there are ether linkages along the backbone and, possibly, trace amounts of unreacted carboxylic and/or amide groups. Therefore, a limited fraction of water interacting with the polymer backbone cannot be entirely ruled out.

Further support for the spectral interpretation proposed so far is provided by comparing the positions of the components in the spectra of Figures 5 and 6. The two high frequency peaks, in the case of PMDA-ODA, are both displaced at higher frequencies by about 20 cm^{-1} with respect to their positions in the epoxy resin; this effect, which is likely due to differences in the dielectric constant of the matrices, confirms the common origin of the peaks in the two systems. On the other hand, the lower frequency component for PMDA-ODA is displaced by at least 50 cm^{-1} compared to its counterpart in the epoxy: this indicates that the water absorbed in the polyimide forms hydrogen bonding interactions of a lower strength as compared to those formed in the epoxy matrix.

In the case of the epoxy resin, the water species not interacting with the network (S_0) are expected to be characterised by high molecular mobility and limited plasticizing efficiency. These species correspond to free water molecules confined into nanopores, according to the physical picture put forward by Soles and coworkers [1-3], or to molecularly dispersed water with no H-bonding interactions, located in pockets surrounded by hydrophobic moieties, where the likelihood of hydrogen bonding is low, according to Mijovic and coworkers [12]. The latter species are those responsible for γ-relaxations observed through DRS in epoxy systems and in BMI [11, 12].

Conversely, S_2 molecules are firmly bound to specific sites along the polymer network, thus exhibiting a much lower mobility and a higher plasticizing efficiency.

With regard to the nature of the S_1 species, as already pointed out they can either represent dimers or water molecules interacting with the polymer network through a hydrogen bond which involves only one hydrogen atom of the water molecule. These two adducts cannot be distinguished on the basis of the experimental evidence provided by FTIR spectroscopy. Since these different kinds of interactions are expected to promote different plasticizing effects on the matrix, no conclusion can be made about the mobility and plasticizing efficiency of the S_1 water species.

This scenario is somehow different from the one proposed by Zhou and Lucas [4] which assumes that two types of absorbed water molecules exist, both of them residing on specific interaction sites: type I molecules, interacting with only one site of the epoxy network and type II molecules, forming interconnective hydrogen bonds with two different sites on the epoxy network. The latter lead to a secondary physical crosslink network, contributing an antiplasticization effect [5]. In this view, water molecules may act as proton donors as well as acceptors. In our opinion, at least for TGDDM-DDS system, this physical picture is unlikely. In fact, the above model does not consider the presence of free water, which is clearly identified in both MIR and NIR spectra. Moreover, FTIR spectroscopy indicates that most of the interacting water molecules are S_2 species: molecular geometry considerations suggest that intramolecular S_2 adducts are formed at single sites along the network, which are characterized by the presence of two closely located proton acceptors (e.g. amino-alcohol moieties or sulphone groups). Possible structures of these complexes formed by hydrogen-bond associations are reported in ref. [14]. This hypothesis is also supported by the results of molecular dynamics simulations reported by Mijovic *et al.* [11].

Furthermore, Zhou and Lucas [5] argue that a proof for the existence of water molecules forming multi-site interconnective bonds is the increase of T_g of the water/epoxy system for long exposure time in sorption experiments, which follows the initial abrupt T_g depression related to the so-called type I water molecules. Actually, this T_g increase can be alternatively accounted for by invoking standard free volume arguments [25]. The addition of a penetrant increases the free volume, lowering the T_g but, at the same time, reduces the excess free volume associated with the glassy state. If the addition of free volume from the penetrant does not outweighs the loss of excess free volume associated with lowering of T_g (as is the case for molecules that significantly relax the glassy matrix structure) a progressive densification sets in, promoting an antiplasticization effect.

In the case of PMDA-ODA, most of the water molecules are either confined in microvoids and/or molecularly dispersed with no H-bonding interaction with the polymer backbone. Also in this case, a lower mobility is expected for S_2 species; however, the origin of this effect is now related to the increased volume of the clusters or, alternatively, to the increase of activation energy due to a diffusive jump (detachment of a single water molecule from the cluster) [26]. On the basis of the above considerations, a lower plasticizing efficiency of absorbed water molecules is expected in the case of the polyimide as compared to the epoxy.

The relative contributions, at sorption equilibrium, of the different water species as derived by the curve fitting results of the subtraction profiles, are reported as a function of water vapour activity in Figure 7 for the epoxy and Figure 8 for the polyimide.

In the case of the epoxy resin, the relative amounts of S_0 and S_2 species decrease with water vapour activity, while that of S_1 increases. Although these absorbance data cannot be directly transformed into relative concentrations, due to the dependence of molar absorptivity on frequency [14], nevertheless they provide useful qualitative information. The above findings are consistent with the physical picture of a system where the relative population of interacting water species at equilibrium changes as a function of the external water activity and hence of the total content of absorbed water. In fact, at low activities, strong interactions are more likely to develop for enthalpic reasons, while, as the activity increases, strongly interacting sites get closer to saturation thereby favouring the formation of weakly interacting or dimeric species.

In the case of the PMDA-ODA matrix, an opposite behaviour was observed: the relative contributions of both S_0 and S_1 species decreased with water activity while the contribution of S_2 species increased. This trend is likely related to the tendency of forming water clusters, which increases with the concentration of absorbed molecules [27].

The absorbance-concentration curves for the two matrices relative to the v_{OH} band are compared in Figure 9. In agreement with the Beer-Lambert relationship,

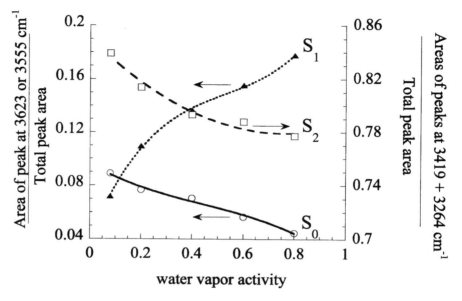

Figure 7. Relative absorbance areas of the different components of the subtraction spectra collected at water vapour sorption equilibrium, reported as a function of water vapour activity, for epoxy matrix at 24°C. Lines are drawn to guide the eye.

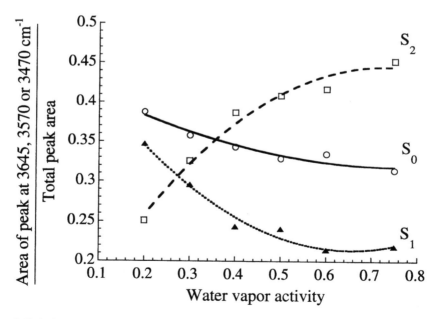

Figure 8. Relative absorbance areas of the different components of the subtraction spectra collected at water vapour sorption equilibrium, reported as a function of water vapour activity, for polyimide matrix at 30°C. Lines are drawn to guide the eye.

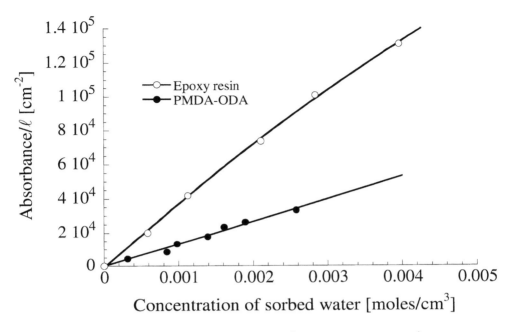

Figure 9. Absorbance normalized by sample thickness (ℓ) of the band at 3408 cm^{-1} as a function of concentration of sorbed water at equilibrium for the epoxy (24°C) and PMDA-ODA polyimide matrices (30°C).

a linear behaviour passing through the origin is found for polyimide. A slight downward concavity is detectable in the case of the epoxy matrix, whose origin will be discussed later. The higher initial slope of the absorbance-concentration curve for the epoxy matrix as compared to the polyimide is likely due to the more pronounced contribution of S_2 species: in fact, strongly interacting water molecules show larger absorptivity values and, consequently, the contribution of each S_2 species to the total absorbance area is higher than that of non-interacting species.

The spectroscopic findings are reflected in the shape of gravimetric sorption isotherms for the two matrices (see Figure 10): both display a slight sigmoidal shape. The initial downward concavity is related, in the case of the epoxy matrix, to saturation of both microvoids and interacting sites; while, for the polyimide, this effect is only due to saturation of excess free volume. The subsequent upward concavity sets in at lower activities for polyimide than for epoxy: this is due to the occurrence of clustering in the former system, which enhances the slight upward concavity generally observed for bulk dissolution in non-interacting polymeric systems [28].

The high sampling rate and quantitative accuracy of time-resolved FTIR spectroscopy allows a reliable monitoring of sorption and desorption kinetics. Exam-

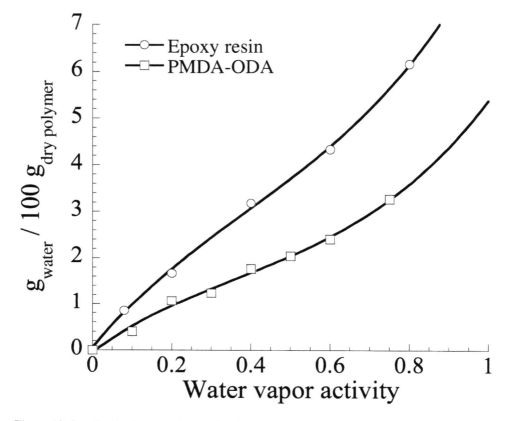

Figure 10. Sorption isotherms in terms of water sorbed amount vs. water vapour activity for epoxy matrix (at 24°C) and polyimide matrix (30°C).

ples of these curves are shown in Figure 11a for the epoxy and in Figure 11b for the polyimide. For both matrices, at all the investigated activities, desorption kinetics is slower than sorption kinetics, which is a typical feature of Fickian systems where mutual diffusivity is an increasing function of penetrant concentration [29]. In the case of epoxy this effect is more pronounced and is related to the progressive saturation of absorption capacity of microvoids and of specific interaction sites of the network. In fact, both of these adsorption phenomena slow down the diffusion process [26]. On the other hand, for the polyimide, only the absorption in microvoids plays a role, while clustering, which occurs at higher activities, has an opposite effect (i.e. depression of mutual diffusivity). Therefore, it is reasonable to expect that the occurrence of the two opposing effects produces a negligible dependence of diffusivity upon concentration.

When mutual diffusivity increases with concentration, the shape of sorption curves does not depend significantly on the functional dependence of the diffusion coefficient [29] and is not substantially different from the case of constant

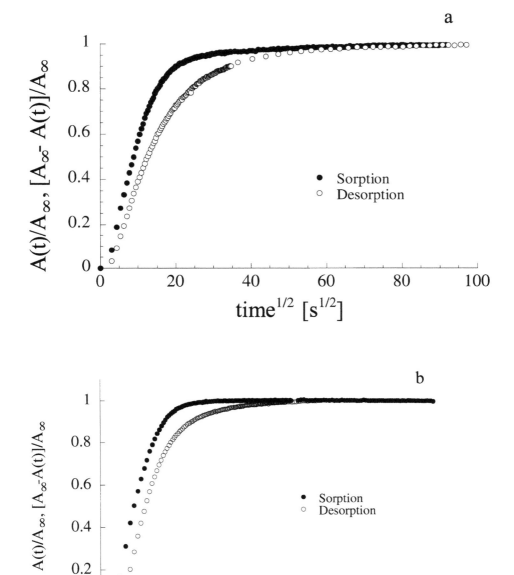

Figure 11. (a) Sorption and desorption kinetics plots (a = 0.4) for epoxy matrix at 24°C in terms of normalized absorbance; (b) Sorption and desorption kinetics plots (a = 0.4) for polyimide matrix at 30°C in terms of normalized absorbance. (A(t) = absorbance at time t, A_∞ = absorbance at sorption equilibrium).

diffusivity. Good curve fittings of sorption kinetics data are obtained, for both PMDA-ODA and TGDDM-DDS, by assuming a Fickian behaviour with constant mutual diffusivity (D). In fact, the time evolution of absorbed mass, normalized by the equilibrium mass uptake, is described for sorption in a plane sheet by the following relationship [29]:

$$\frac{M_t}{M_\infty} = 1 - \frac{8}{\pi^2} \sum_{m=0}^{\infty} \frac{1}{(2m+1)^2} \cdot \exp\left[\frac{-D(2m+1)^2 \pi^2 t}{\ell^2}\right] \tag{3}$$

where M_t is the amount of penetrant absorbed at time t, M_∞ is the amount of penetrant absorbed at equilibrium and ℓ is the thickness of the sheet. This equation has been used to fit sorption data to determine the values of diffusivity. It is worth noting that the values of D determined in this way are actually averages over the ranges of water concentration in each sorption test. The diffusion coefficients evaluated in this way are reported, as a function of water vapour activity, in Figure 12. Consistent with the previous discussion, the dependence is more pronounced in the case of epoxy. The lower values of D measured for the TGDDM-DDS resin are partly due to the lower test temperature and partly to the strong molecular interactions between the penetrant and the matrix.

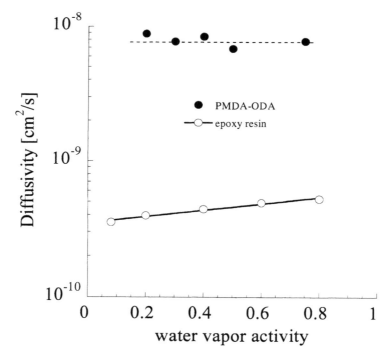

Figure 12. Water diffusivity evaluated from sorption tests reported as a function of water vapour activity for the case of epoxy matrix at 24°C and of polyimide matrix at 30°C.

Sorption and desorption kinetics evaluated spectroscopically compare well with those evaluated by means of the gravimetric method. Such a comparison is reported in Figures 13a and 13b for epoxy and polyimide matrices, respectively, for sorption tests at a = 0.4. It is worth noting that in the case of the epoxy, the gravimetric and spectroscopic curves are close but not exactly superimposable: this

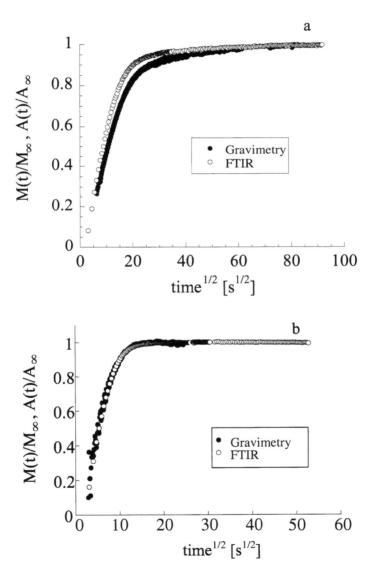

Figure 13. (a) Comparison of gravimetric and FTIR sorption kinetics at a = 0.4 for epoxy matrix at 24°C; (b) Comparison of gravimetric and FTIR sorption kinetics at a = 0.4 for polyimide matrix at 30°C. Sorption data are reported in terms of water mass and absorbance normalized with their respective values at equilibrium [M_∞, A_∞].

effect is related to the deviation from linearity of the absorbance-concentration curve for the epoxy (see Figure 9). The observed nonlinearity originates from the fact that the molar absorptivity of water increases significantly with the degree of interaction (i.e. $\varepsilon_{S_0} < \varepsilon_{S_1} << \varepsilon_{S_2}$). Thus, when S_2 species are predominant, as for the epoxy, a reduction of their relative population with overall water concentration (see Figure 7) induces a downward concavity in the absorbance-concentration curve. Conversely, in the case of the polyimide, the gravimetric and spectroscopic curves are superimposable since the contribution of S_2 species is considerably lower in terms of absorbance and even more so in terms of concentration; thus the nonlinearity effect in the absorbance-concentration curve is minimized (see Figure 9).

4. CONCLUSIONS

Interactional aspects related to sorption and transport of water molecules in high T_g matrices have been investigated by coupling classical gravimetric analysis with *in situ, time-resolved* FTIR spectroscopy. As model systems for interacting and non-interacting matrices, a tetrafunctional epoxy resin (TGDDM-DDS) and a polyimide (PMDA-ODA), respectively, have been used. The presence of different species of absorbed water has been evidenced in both materials, characterized by different levels of interaction with the matrices. In the case of epoxy, free and dimeric water molecules coexist with water molecules strongly bound to the polymer network through H-bonding interactions. Conversely, in the case of polyimide mostly free and self-associated water has been detected.

Based on the different degrees of molecular interactions, the two systems behave differently with respect to their equilibrium sorption and transport properties.

REFERENCES

1. C.L. Soles, F.T. Chang, B.A. Bolan, H.A. Hristov, D.W. Gidley and A.F. Yee, *J. Polym. Sci., Part B Polym. Phys.*, **36**, 30-35 (1998).
2. C.L. Soles, F.T. Chang, D.W. Gidley and A.F. Yee, *J. Polym. Sci., Part B Polym. Phys.*, **38**, 776-791 (2000).
3. C.L. Soles and A.F. Yee, *J. Polym. Sci., Part B Polym. Phys.*, **38**, 792-802 (2000).
4. J. Zhou and J.P. Lucas, *Polymer*, **40**, 5505-5512 (1999).
5. J. Zhou and J.P. Lucas, *Polymer*, **40**, 5513-5522 (1999).
6. L.W. Jelinski, J.J. Dumais, R.E. Stark, T.S. Ellis and F.E. Karasz, *Macromolecules*, **16**, 1019 (1983).
7. L.W. Jelinski, J.J. Dumais, A.L. Chiolli, T.S. Ellis and F.E. Karasz, *Macromolecules*, **18**, 1091 (1985).
8. R.A. Pethrick, E.A. Hollins, P. Johncock, I. McEwan, E.A. Pollock and D. Hayward, *Polymer*, **38**, 1151 (1996).

9. R.A. Pethrick, E.A. Hollins, I. McEwan, E.A. Pollock and D. Hayward, *Polym. Int.*, **39**, 275 (1996).

10. C. Grave, I. McEwan and R.A. Pethrick, *J. Appl. Polym. Sci.*, **69**, 2369 (1998).

11. J. Mijovic, N. Miura and S. Soni, *J. Adhesion*, **76**, 123 (2001).

12. J. Mijovic, N. Miura, H. Zhang and Y. Duan, *J. Adhesion*, **77**, 323 (2001).

13. S. Cotugno, D. Larobina, G. Mensitieri, P. Musto and G. Ragosta, *Polymer*, **42**, 6431 (2001).

14. P. Musto, G. Ragosta and L. Mascia, *Chem. Mater.* **12**, 1331 (2000).

15. E.L. McKague, J.D. Reynolds and J.E. Halkies, *J. Appl. Polym. Sci.*, **22**, 1643 (1978).

16. J. Mijovic and K. Lin, *J. Appl. Polym. Sci.*, **30**, 2527 (1985).

17. P. Musto, L. Mascia, G. Ragosta, G. Scarinzi and P. Villano, *Polymer*, **41**, 565 (2000).

18. L. Mascia, *Trends in Polymer Science*, **3**, 61 (1995).

19. J.L. Koenig, *Spectroscopy of Polymers*. American Chemical Society, Washington D.C. (1992).

20. K. Krishnan and J.R. Ferraro, in: *Fourier Transform Infrared Spectroscopy*, J.R. Ferraro and L.J. Basile (Eds.), Vol. 3, p. 198. Academic Press, New York (1982).

21. D.W. Marquardt, *J. Soc. Ind. Appl. Math.*, **11**, 441 (1963).

22. W.F. Maddams, *Appl. Spectroscopy*, **34**, 245 (1980).

23. P. Musto, G. Ragosta, G. Scarinzi and L. Mascia, *J. Polym. Sci.: Part B: Polym. Phys. Ed.*, **40**, 922 (2002).

24. P. Musto, F.E. Karasz and W.J. MacKnight, *Polymer*, **30**, 1012 (1989).

25. J.L. Duda and J.M. Zielinski, in: *Diffusion in Polymers*, P. Neogi (Ed.), chap. 3, p. 143, Marcel Dekker, New York (1996).

26. W.J. Koros and M.W. Hellums, in: *Ecyclopedia of Polymer Science and Engineering, Supplement Volume*, H.F. Mark, N.M. Bikales, C.G. Overberger, G. Menges and J.I. Kroschwitz (Eds.), Wiley, New York (1990).

27. J.A. Barrie and D. Machin, *Trans. Faraday Soc.*, **67**, 224 (1971).

28. P.J. Flory, *Principles of Polymer Chemistry*, Cornell University Press, Ithaca, NY (1953).

29. J. Crank, *The Mathematics of Diffusion*. 2nd edition, Clarendon Press, Oxford (1975).

Part 2

Surface Modification, Interfacial or Adhesion Aspects and Applications

Polyimides and Other High Temperature Polymers, Vol. 2, pp. 289–314
Ed. K.L. Mittal
© VSP 2003

New developments in the adhesion promotion of electroless Ni or Cu films to polyimide substrates

M. CHARBONNIER, Y. GOEPFERT and M. ROMAND[*]

Laboratoire de Sciences et Ingénierie des Surfaces, 43 Boulevard du 11 Novembre 1918, Université Claude Bernard - LYON 1, 69622 Villeurbanne Cedex, France

Abstract—Electroless metallization (or electroless plating) of polymers and composite materials is a multi-step chemical process requiring: (i) the pre-conditioning of the sample surface to render it chemically reactive, (ii) the activation of the so-modified surface through the adsorption of a palladium-based catalyst, and (iii) the metallization operation itself. In the present work, electroless plating was carried out on polyimide substrates and step (i) was performed via plasma or VUV irradiation-assisted treatments operating especially in nitrogenated atmospheres (NH_3 or N_2). In addition, step (ii) was accomplished using two different routes. The one is based on the chemisorption of Pd^{+2} species (from a $PdCl_2$ solution) on surfaces which are previously grafted with nitrogenated functionalities. The other involves deposition (e.g. by spin-coating) of a palladium acetate (Pdac) thin film on the same substrates functionalized in the same way. In both cases, the so-grafted Pd^{+2} species have to be subjected to a reduction to Pd^0 before becoming catalytically active towards the metallization initiation. In laboratory-made Ni plating baths in which there are no stabilizers this occurs through the reducing agent (sodium hypophosphite NaH_2PO_2). On the other hand, in laboratory-made Cu plating baths, such a reduction by formaldehyde (HCHO), although thermodynamically possible, does not take place because of a too slow kinetics. Note that a similar behavior is observed with both commercial Ni or Cu plating baths due to the presence of stabilizers which probably mask the potential catalytic sites when the Pd^{+2}-grafted surfaces are dipped in these baths, and therefore prevent Pd^{+2} species from being reduced. Under these conditions, the reduction of the Pd^{+2} species has to be performed prior to the surface immersion in the Ni or Cu plating baths. For this, one simple method consists in dipping the Pd^{+2}-grafted surfaces in a hypophosphite solution. Another route consists in subjecting the same samples to a VUV irradiation under vacuum. However, the experiments show that the kinetics of the Pd^{+2} reduction to Pd^0 is clearly more favorable in the first case and under the experimental conditions used. In short, the work described in this paper has been successful in leading to a better understanding of the chemisorption mechanisms of the palladium-based catalysts on functionalized surfaces and of their role in initiating the electroless deposition, as well as in allowing the development of new approaches for electrolessly metallizing polymer-based materials.

Finally, this work has shown that electrical measurements carried out in conjunction with a fragmentation test appear to be able to distinguish the effects of different surface treatments of polymer substrates on the practical adhesion of the metal film on the polymer surface.

---

[*]To whom all correspondence should be addressed. Phone: +33 4 72 44 81 68, Fax: +33 4 72 43 12 06, E-mail: Maurice.Romand@univ-lyon1.fr

Keywords: Polyimide; plasmas; excimer lamps; VUV irradiation; surface functionalization; surface energy; surface catalysis; electroless metallization; electroless plating; chemical metallization; XPS analysis; thin metal film adhesion.

1. INTRODUCTION

Polyimides (PIs) are a class of engineering plastics which possess many outstanding characteristics including, among others, high thermal and chemical resistance, good tensile strength and modulus, low dielectric constant and ease of processing into films or coatings [1-4]. As a result, PIs have been extensively used in recent years in particular in microelectronics, both in packaging and integrated circuit fabrication [1-18]. In some applications where samples must be coated with a metal (e.g. by deposition or lamination) it is highly desirable to develop reliable metal / substrate systems with enhanced long-term chemical and mechanical stability. This means that metal / polymer adhesion plays a critical role in the performances of many manufactured devices used in the field. Regardless of the nature of the metallization process used (dry physical or wet chemical methods), the surface of polymer substrates (PI substrates in this work) has to be chemically, and even in same cases morphologically, modified. For this purpose, numerous procedures have been studied and put into practice. A review of theses approaches can be found in the recent literature [9, 10, 19, 20] and references therein.

In the present paper, special attention has been focused on the development of simplified surface preparation procedures allowing, finally, the metallization of PIs substrates through electroless technique [21]. Electroless metallization or electroless plating (EP) consists in a redox process in which ions of the metal to be deposited (generally Ni or Cu) are chemically reduced to the metallic state at the substrate surface due to the presence of a strong reducer in the plating bath. A key aspect of this process is that metallization of non-conducting samples, and thus of polymers such as PIs, requires making the substrate surface catalytically active. Palladium-based catalysts are commonly used to initiate the redox reaction and for this purpose these must be adsorbed on the surface to be metallized. Once the metallization is initiated, it progresses through the catalytic action of the deposit itself, hence the term "autocatalytic" is also used to describe the EP process. As will be seen further, the traditional approaches leading to metallization require multi-step procedures involving hazardous and environmentally harmful chemicals. The alternative approaches considered in this work are studied with the objective to put in practice procedures which are technologically, economically and ecologically attractive. These procedures are based on the use of tin-free processes leading to the seeding of palladium species on PI surfaces which are previously plasma or VUV treated in nitrogenated (NH_3 or N_2) environments [22-33]. For the sake of comparison, some surface modifications are also carried out using plasma or VUV treatments in oxygenated (O_2 or air) environments. Regardless of the applications developed in the present work, it should be mentioned that numerous researches have previously been carried out dealing with the gas-phase activation of PI surfaces using plasma or UV/VUV treatments in various atmospheres (for example, see refs [34-39]). In addition, adhesion characteristics of Ni or Cu films electrolessly deposited on the PI substrates from both laboratory-made and commercially available plating baths are also investigated in the present work.

2. EXPERIMENTAL

2.1. Material and treatment procedures

In the present work, surface pretreatments and subsequent electroless metallizations were carried out on 125 μm thick polyimide (PI) substrates (Kapton® 500 HN foils) produced by DuPont, USA. Before processing, these foils were ultrasonically cleaned in ethanol and blown dry with N_2.

Plasma treatments were performed in a parallel plate RF reactor operating in the Reactive Ion Etching mode (RIE 80 reactor from Plasma Technology, Bristol, England). The samples to be surface-modified were placed on the powered electrode. Unless otherwise stated, experiments were conducted under the following conditions: working pressure: 100 mTorr, gas-flow: 100 sccm, power density: 0.5 W cm^{-2}.

VUV treatments were carried out using an Xe_2* excimer lamp (Excivac Laboratory System from Heraeus Noblelight, Kleinostheim, Germany) emitting an incoherent radiation centered at 172 nm. This lamp is mounted in an experimental setup already described elsewhere [33]. Briefly, the latter consists of two contiguous chambers which can be evacuated individually under moderate vacuum. The upper chamber in which the excimer lamp is located is separated from the reactor by a 5 mm thick CaF_2 window. In these studies and unless otherwise stated, the experimental conditions were the following: working pressure in the reactor: 3.5 Torr, window-sample distance: 1 cm, photon flux estimated between 30 and 40 mW cm^{-2} [31].

The seeding of PI surfaces with palladium species first utilized two different methods. The one consisted in immersing the plasma or VUV-treated substrates for 2 min in a solution containing 0.1 g/liter $SnCl_2$ and 0.1 ml/liter concentrated HCl, rinsing them in DI water (sensitization step) for 1 min, then immersing them for 2 min in a solution containing 0.1 g/liter $PdCl_2$ and 3.5 ml/liter concentrated HCl and rinsing (activation step). The other consisted in immersing the pretreated substrates in a simple solution containing 0.1 g/liter $PdCl_2$ and 3.5 ml/liter concentrated HCl for 2 min and rinsing them (direct activation step).

In a second set of experiments, the seeding of PI surfaces with palladium species was carried out using palladium acetate (Pdac) thin films which were deposited by spin-coating (10^3 rpm) from a Pdac solution (0.015 M) in chloroform

(CHCl$_3$), then rinsed with this same solvent for 1 min. During the spin-on process, the chloroform evaporates spontaneously and leaves a palladium-based thin film whose thickness depends on the spin-speed used and number of drops of solution deposited on the substrate.

Both laboratory-made and commercial Ni and Cu plating baths were used in the present study. The laboratory-made Ni plating bath contained NiSO$_4$·6H$_2$O (0.14 M), NaH$_2$PO$_2$·H$_2$O (sodium hypophosphite) (0.09 M), lactic acid (0.27 M) and operated at pH 5 and 85°C. The laboratory-made Cu plating bath contained five parts of the following solution CuSO$_4$·5H$_2$O (0.28 M), KNaC$_4$H$_4$O$_6$·4H$_2$O (0.6 M), NaOH (0.14 M) and EDTA (0.03 M) and one part of formaldehyde HCHO 37% and operated at pH 12 and room temperature. In addition, commercial Europlate Ni 520 electroless nickel (Frappaz, Neyron, France) and Enplate Cu 872 electroless copper (Enthone-OMI, France) baths were employed according to manufacturer's instructions. In all these cases, sodium hypophosphite and formaldehyde were the reducing agents present in the Ni and Cu plating baths, respectively. These agents are required to reduce Ni^{2+} and Cu^{2+} ions to Ni^0 and Cu^0 state, respectively.

2.2. Characterization techniques

XPS analyses were performed with a Riber SIA 200 spectrometer using a non-monochromatic Al K$_\alpha$ photon source and a take-off angle of 65° with respect to the sample surface. Spectra were referenced to the C 1s signal at a binding energy of 285.0 eV characteristic of the C–C and C–H bonds.

Adhesion of thin metal films to their substrates was first characterized using a simple adhesive tape peel test (cross-cut tape test according to ASTM D 3359). In addition, a stretch deformation test (also known as fragmentation test) was employed in conjunction with electrical measurements in order to distinguish the influence of the polymer surface conditioning. For this purpose, dog-bone shape metal-PI systems were strained under uniaxial elongation at a constant strain rate of 0.05 mm min^{-1} using a DY 25 Adamel Lhomargy machine. Some details about the approach using electrical measurements can be found in recent papers [33, 40, 41].

3. RESULTS AND DISCUSSION

3.1. Surface "activation" through plasma or VUV-assisted techniques and chemisorption of Pd-based species

Figure 1 represents schematic diagrams of the different preparation processes: (A) which are conventionally used, and (B) which were explored in our laboratory to make the surfaces of polymers and polymer-based materials catalytically active towards the electroless metallization [22-33]. Note that the first experiments carried out in the afore-mentioned works were mainly performed with laboratory-made Ni plating baths.

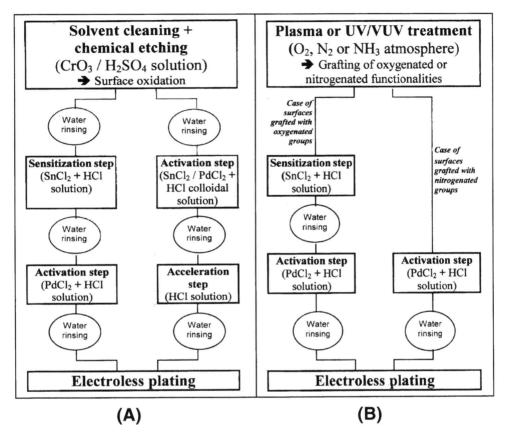

Figure 1. Schematic diagram of the electroless plating process using: (A), a conventional oxidizing chemical treatment and the two-step (on the left) or one-step (on the right) route leading to surface "activation", and (B), the newly-proposed gas-phase pretreatments and the two-step (on the left) or one-step (on the right) route also leading to surface "activation".

As can be seen in Figure 1 (A), the conventional process first requires substrate cleaning with appropriate solvents to remove surface contaminants, and then chemical treatment (generally in chromic acid / sulfuric acid mixtures) to produce a micro-roughened oxidized surface. Chronologically, the subsequent activation process was first accomplished by using a two-step procedure involving the substrate immersion successively in acidic $SnCl_2$ (sensitization step) and $PdCl_2$ (activation step) solutions. Further, a one-step procedure using a colloidal suspension containing both Sn and Pd species (an acidic $SnCl_2/PdCl_2$ solution) has been developed and is in common industrial use today. Note that, in this case, the Pd/Sn colloidal particles anchored onto polymer surfaces must be exposed to an accelerator (acceleration step). This is generally a hydrochloric solution which is used to remove the hydrolyzed stannous hydroxide which surrounds the core of the so-formed colloidal particles and inhibits the action of the electroless bath. As a result, this operation bares the palladium sites which then become available for the initiation of the metal deposition in the electroless bath. On the other hand, the alternative approaches described in Figure 1 (B), only involve, before activation, gas phase treatments through plasma or VUV-assisted processes, the activation being carried out successively in acidic $SnCl_2$ and $PdCl_2$ solutions or in a simple acidic $PdCl_2$ solution when the polymer surfaces are functionalized in oxygenated or nitrogenated environments, respectively. In these experiments, various polymer surfaces were subjected either to plasma treatments in O_2, N_2 or NH_3 atmosphere [22-24, 27-33] or to UV/VUV irradiation in O_2, air or NH_3 atmosphere using Nd:YAG (5th harmonic, $\lambda = 213$ nm) and ArF* excimer ($\lambda = 193$ nm) lasers [25, 26] or an Xe_2* excimer lamp [32, 33]. Note that, in the case of the UV/VUV irradiation, the molecular nitrogen N_2 cannot be used as a gas source allowing the grafting of nitrogenated species. Indeed, the coherent and incoherent photon sources emitting at 213, 193 and 172 nm are not energetic enough (5.8, 6.4 and 7.2 eV, respectively) to break in the gas-phase the $N{\equiv}N$ triple bond of the nitrogen molecule (9.8 eV). On the contrary, they are quite able to break the N-H single bond (4.0 eV) and O=O double bond (5.1 eV) of ammonia and oxygen molecules, respectively.

For illustrating the new approaches to the EP process described in Figure 1 (B) some XPS survey spectra relative to Kapton® HN substrates plasma-treated in NH_3 and O_2 environments are shown in Figures 2 and 3, respectively. Figure 2 represents XPS survey spectra of Kapton® samples, after cleaning in ethanol (a) then after NH_3 plasma for 1 min (b) and immersion either in a simple acidic $PdCl_2$ solution (c) or successively in acidic $SnCl_2$ and $PdCl_2$ solutions (d). In a similar way, Figure 3 represents XPS survey spectra of substrates subjected to the same treatments as those described for Figure 2 except that the plasma treatment was carried out for 1 min in O_2 instead of NH_3. As can be seen from (b) spectra, O_2 plasma causes grafting of oxygenated functionalities (Fig. 3) while NH_3 plasma is responsible for the uptake of nitrogenated functionalities (Fig. 2) which add to those naturally present at the surface of PI substrates. In other respects, spectra (c) relative to plasma-treated samples dipped in a simple acidic $PdCl_2$ solution show that only traces of palladium are attached on the O_2 plasma-treated surface (Fig. 3) while a significant Pd amount (1.4 at.% in this experiment) is anchored onto the NH_3 plasma-treated surface (Fig. 2). Finally, when the plasma-treated samples are dipped successively in the acidic $SnCl_2$ and $PdCl_2$ solutions (spectra (d)), a significant amount of tin is attached onto the O_2-plasma treated surface (Fig. 3) and not at all on the NH_3-plasma treated surface (Fig. 2). Note also that Sn is not found on NH_3-plasma treated surfaces when the latter are treated either in a simple acidic $SnCl_2$ solution or simultaneously in an equimolecular $SnCl_2$ and $PdCl_2$ acidic solution. On the other hand, palladium is attached at a relatively low surface concentration (0.8 at.% in this experiment) on samples which are O_2 plasma-treated, and then sensitized by immersion in the acidic $SnCl_2$ solution (Fig. 3, spectrum (d)). This palladium attachment on tin species is in agreement with the mechanism of the sensitization / activation two-step procedure. More interesting for the topic un-der investigation is the fact that palladium is attached in a large amount (~ 1.5 at.% in this experiment) on the NH_3-plasma treated surface (Fig. 2, spectrum (d)). Note again the complete absence of adsorbed tin species in this case.

Before drawing some conclusions from the afore-described experiments, we must add that:

(i) the presence of palladium traces on PI substrates which are O_2 plasma-treated and then immersed in a simple acidic $PdCl_2$ solution results from the pres-ence of nitrogen atoms naturally available at the sample surface. In addition, this statement is supported by the fact that polymer materials containing no nitrogen atoms in their chemical structure do not adsorb, under similar conditions, any pal-ladium species [22-24].

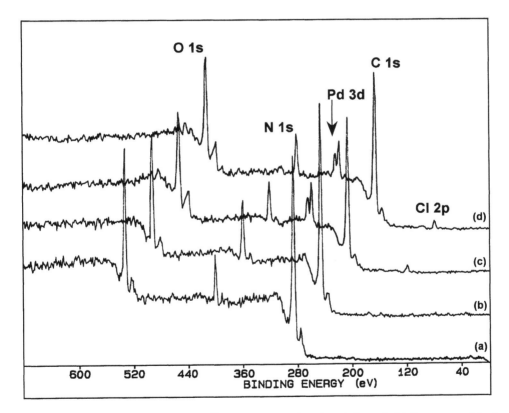

Figure 2. XPS survey spectra of Kapton® HN substrates: (a) ethanol cleaned, (b) the same as (a) af-ter plasma treatment in NH_3 (working pressure: 100 mTorr; flow-rate: 100 sccm; treatment time: 1 min), (c) the same as (b) after a simple "activation" process (immersion in an acidic $PdCl_2$ solution), and (d) the same as (b) after the two-step sensitization / activation process (immersion successively in acidic $SnCl_2$ and $PdCl_2$ solutions). Note that the abscissa scale (energy scale) is associated with the bottom spectrum (a). Other spectra are shifted for sake of convenience.

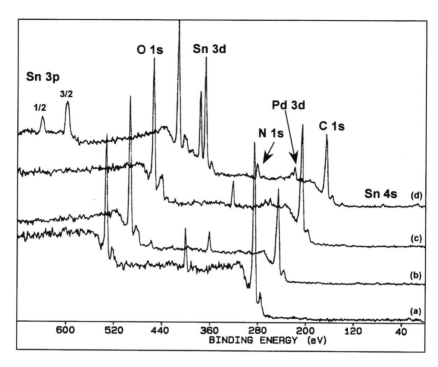

Figure 3. XPS survey spectra of Kapton® HN substrates: (a) ethanol cleaned, (b) the same as (a) after plasma treatment in O_2 (working pressure: 100 mTorr; flow-rate: 100 sccm; treatment time: 1 min), (c) the same as (b) after a simple "activation" process (immersion in an acidic $PdCl_2$ solution), and (d) the same as (b) after the two-step sensitization / activation process (immersion successively in acidic $SnCl_2$ and $PdCl_2$ solutions). Note that the abscissa scale (energy scale) is associated with the bottom spectrum (a). Other spectra are shifted for sake of convenience.

(ii) concerning the Pd adsorption, results quite similar to those obtained by NH_3 plasma treatments are obtained when PI substrates are surface-functionalized using N_2 plasma or VUV irradiation in NH_3 at reduced pressure, and subsequently activated by immersion in a simple acidic $PdCl_2$ solution. The only noticeable difference is that VUV irradiation in NH_3 requires a longer treatment time (~ 3 min) for grafting a palladium amount similar to that adsorbed after plasma treatment in N_2 or NH_3 for 1 min [32, 33]. In addition, similar results concerning the Sn and Pd uptakes are also obtained when the PI substrates are surface-functionalized using VUV-irradiation in O_2 at reduced pressure, and subsequently sensitized / activated through the two-step procedure (immersion successively in acidic $SnCl_2$ and $PdCl_2$ solutions).

(iii) regardless of the nature of the gas-phase method used to functionalize the PI surface with nitrogenated species (NH_3 or N_2), the change in the Pd surface concentration resulting from the activation process follows the same trend as that of the grafted nitrogenated species when plasma or VUV treatment time varies

[33]. As an example, Figure 4 is relative to PI substrates NH_3 plasma-treated, and then activated in a simple acidic $PdCl_2$ solution. As can be seen, the surface concentration of Pd varies as a function of the plasma treatment time similarly to that of the nitrogenated species.

Given the previous results and remarks, the following conclusions can be drawn:

1 - Pd^{+2} species have a strong chemical affinity towards nitrogenated functionalities and no chemical affinity towards oxygenated functionalities as they are both grafted through plasma or VUV-assisted treatments. Under these conditions, Pd-based catalysis of PI surfaces (and more generally of polymer surfaces) can be carried out using a simplified process involving the surface functionalization with nitrogenated species and the subsequent surface "activation" by immersion in a simple acidic $PdCl_2$ solution. In other words, the creation of Pd-based catalytic sites occurs selectively onto grafted nitrogenated groups via the chemisorption of Pd^{+2} ions and the formation of covalent C-N-Pd bonds.

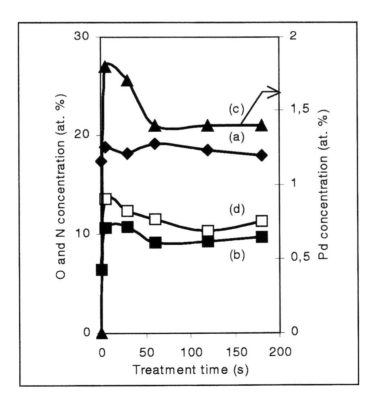

Figure 4. Plasma treatments of Kapton® HN samples in NH_3. Influence of the plasma treatment time on the oxygen (a), nitrogen (b) and palladium (c) surface concentrations (at.%) after the "activation" step in the $PdCl_2$ acidic solution. For comparison, curve (d) is relative to the nitrogen surface concentration before this step, i.e. just after the plasma treatment. The other treatment conditions were: working pressure = 100 mTorr, flow-rate = 100 sccm, power density = 0.52 W cm^{-2}.

2 - Sn^{+2} species have a strong chemical affinity towards oxygenated functionalities and no chemical affinity towards nitrogenated functionalities as they are both grafted through plasma or VUV-assisted treatments. Under these conditions, the creation of Pd-based catalytic sites on PI surfaces (and more generally on polymer surfaces) requires the use of the conventional sensitization / activation processes, i.e., the use of both tin and palladium species. In this case, the Pd-based catalyst chemisorbs on Sn^{+2} ions which are themselves "attached" onto oxygenated species leading to the formation of covalent C-O-Sn(Pd) bonds.

Obviously, these results raise the question why the nature of the specific chemical interactions between tin or palladium species and polymer surfaces has not been established earlier. Two reasons can be emphasized. The one is associated with the fact that, in the present-day technology, the EP process makes use only of colloidal solutions containing both $SnCl_2$ and $PdCl_2$. The other stems from the fact that only a systematic XPS study on many polymer substrates with or without oxygenated groups could lead to understanding of the mechanisms of the chemisorption of tin and palladium species when the two-step sensitization / activation process is used. However, in interpreting the results described above, one should keep in mind that they confirm the fact that Pd^{+2} ions can be easily covalently bonded to ligands containing electron-donating nitrogen atoms [42]. As far as electroless metallization is concerned, note that Calvert and coworkers [43-45] have coated a variety of non-conducting substrates including polymers with self-assembled monolayers (SAMs) of organosilanes bearing alkylamine or pyridine-terminated groups for chemisorbing selectively Pd^{+2} species from an aqueous Pd(II) catalyst solution. In addition, these authors have shown that such a chemisorption does not exist when the organosilanes possess no nitrogenated functionality. More recently, Kang and coworkers have used the same concept as the one developed in our studies for activating directly in $PdCl_2$ or $Pd(NO_3)_2$ solutions polyimide substrates which are surface-modified by UV-induced graft-copolymerization of 1-vinylimidazole [46]. Similarly, they have graft-copolymerized poly(tetrafluoroethylene) (PTFE) surfaces using different N-containing vinyl monomers or acrylamide [47] before a subsequent surface activation in a $PdCl_2$ solution. About PTFE, it should also be mentioned that our own electroless metallization approach involving surface functionalization (N-group grafting) and activation ($PdCl_2$ route) operates quite well [29].

From a practical point of view, consider now only the simplified process involving surface amination of PI substrates (e.g. through plasma treatment in ammonia) and their subsequent activation by immersion in a simple acidic $PdCl_2$ solution, and observe what occurs when the said "activated" surfaces are dipped either in the laboratory-made or in the industrial Ni and Cu plating baths. As shown in the schematic diagram represented in Figure 5, deposition of nickel (in fact a Ni-P alloy in these experiments) starts in the case of the laboratory-made Ni bath after only a short initiation time (10-15 s). During this period, some Pd^{+2} ions which were grafted onto the PI surfaces are reduced to metallic palladium Pd^0 by the reducing agent (hypophosphite ions $H_2PO_2^-$) present in the Ni plating bath. On

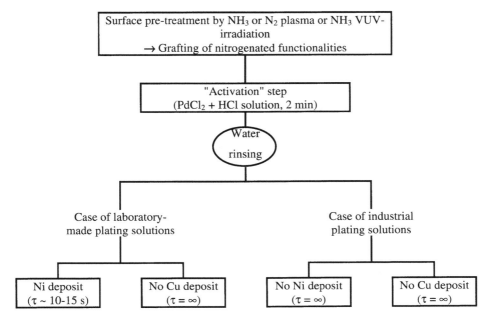

Figure 5. Schematic diagram representing the sequences of the newly proposed electroless plating (EP) process and showing the differences in the initiation time (τ) of the metal deposition when laboratory-made and commercial Ni and Cu plating baths were used.

the other hand, no metal deposition was observed when the "activated" surfaces were immersed in the laboratory-made Cu plating bath as well as in the industrial Ni or Cu plating baths. This means that, even though the reducing agent (formaldehyde HCHO) present in all the Cu baths is thermodynamically capable of reducing Pd^{+2} ions, it is inoperative for kinetic reasons (infinitely long initiation time). This is supported by the XPS spectra given in Figure 6. Spectrum (a) represents the Pd 3d spin-doublet of a PI substrate plasma-treated in NH_3, and then immersed in the acidic $PdCl_2$ solution. The binding energy of the Pd $3d_{5/2}$ peak at 338.0 eV confirms that palladium is surface-grafted as Pd^{+2}. Spectra (b) and (c) are relative to the same sample as (a) after dipping in the laboratory-made Cu plating bath for 4 min and in the laboratory-made Ni plating bath for 12 s, respectively. In addition, spectrum (d) (Pd $3d_{5/2}$ peak at about 335.5 eV) is a reference spectrum characteristic of the Pd^0 state. Compared to spectrum (a), the Pd $3d_{5/2}$ peak is slightly more shifted and broadened towards the low binding energy side for the Pd^{+2}-grafted substrate immersed in the Ni plating bath than for that immersed in the Cu plating bath, even though in the latter case the immersion time is significantly longer. These results clearly indicate that some Pd^0 species are rapidly formed during the initiation time of the metal deposition in the Ni plating bath due to the action of the $H_2PO_2^-$ reducing agent, and that the kinetics of the Pd^{2+} reduction is notably slower in the Cu plating bath. Also note that quite simi-

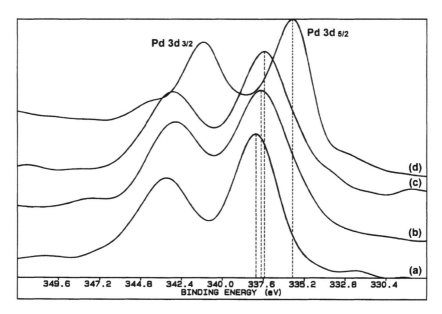

Figure 6. Pd 3d XPS spectra of Kapton® HN samples (a) after surface functionalization by NH₃-plasma treatment and "activation" by an acidic PdCl₂ solution, (b) the same as (a) after immersion in the laboratory-made electroless Cu bath for 4 min, and (c) the same as (a) after immersion in the laboratory-made electroless Ni bath for 12 s. Spectrum (d) characteristic of the Pd⁰ state is shown for comparison purpose.

lar results concerning the reduction of Pd^{2+} ions are obtained when the Pd^{+2}-grafted surfaces are immersed in aqueous solutions containing only the reducing agent (hypophosphite or formaldehyde) and operating under the same conditions (temperature, pH) as those used in their respective plating bath [33]. Finally, all these experiments show that only the Pd^0 species operate as the actual electroless catalyst. Given the results mentioned above, it appears interesting to look for what happens when the Pd^{2+}-grafted surfaces are treated in a hypophosphite ($H_2PO_2^-$) solution prior to their immersion in the Ni or Cu plating bath. As shown in the schematic diagram given in Figure 7, both Ni and Cu platings take place without any initiation time when laboratory-made plating solutions are used. In the case of the Cu plating, note that as soon as the Cu deposition is initiated due to the formation of a sufficient surface density of Pd^0 species, then the Cu deposition progresses by autocatalysis due to the action of the formaldehyde HCHO, the only reducing agent capable of reducing Cu^{2+} ions to Cu^0. Figure 7 also shows what happens when the commercial electroless baths are used. Experiments show that Ni or Cu deposition requires that the grafted Pd^{+2} species be reduced to the Pd^0 state. For this purpose, the reduction was performed in a hypophosphite solution before immersion of the "activated" surface in the relevant electroless bath. These results and the comparison with those obtained by using laboratory-made plating

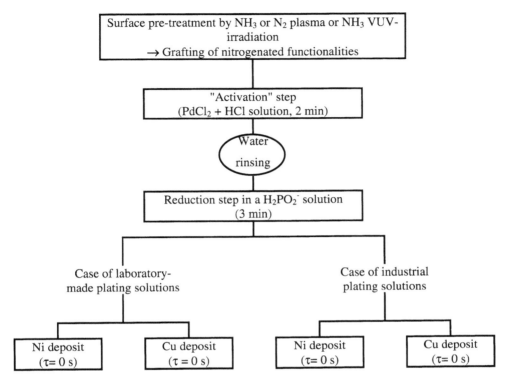

Figure 7. Schematic diagram representing the sequences of the newly proposed electroless plating (EP) process and showing that no initiation time is observed when the Pd^{+2}-grafted species are chemically reduced prior to the sample immersion in both laboratory-made and commercial Ni or Cu plating baths.

solutions raise the question as to what are the effects of stabilizers contained in industrial plating baths on the grafted Pd^{+2} species. Progress in the understanding of the corresponding phenomena would need further investigations.

Note that when the Pd-catalyst is adsorbed on tin ions at the polymer surface through the conventional process, the Pd^{+2} reduction is enhanced due to the presence of the Sn^{+2} ions [30]. Indeed, a redox reaction happens between Sn^{+2} and Pd^{+2} ions, resulting in the reduction of Pd^{+2} to Pd^{0} and the formation of Sn^{+4} ions. A proof of the Sn^{+2} activity is brought by the fact that the deposit initiation is always taking place, irrespective of the nature of the electroless plating bath (with or without stabilizers).

3.2. Surface "activation" through plasma or VUV-assisted techniques and photolytic decomposition of a Pd-based metal-organic compound

The second part of the present work is devoted to PI substrates which were first plasma or VUV-pretreated in ammonia (or in an oxygenated atmosphere for a comparison purpose), then spin-coated with a palladium acetate (Pdac) thin film obtained from a Pdac solution in chloroform, and finally rinsed in this same solvent. As in the previous section, the first step of this process was carried out to substitute an attractive gas phase pretreatment for a traditional treatment in liquid phase which uses hazardous chemicals. In addition, the second step of this process aims at producing on the polymer surface a palladium-containing thin film which, under specific conditions, can be used for initiating electroless deposition. Indeed, the metal-organic thin films obtained after solvent evaporation can be photo-chemically and / or photo-thermally decomposed (this depends on the nature of the radiation source and energy flux used), and consequently converted into their metal constituent. As demonstrated in the recent years, a variety of polymer substrates (as well as other insulating substrates) on which palladium catalytic sites have been created in this way can be processed in electroless plating solutions for obtaining Cu or Ni deposits [31, 32, 48-59].

In a first step, consider the effects of surface pretreatments of PI substrates on the attachment of palladium when use is made of the Pdac route. Data in Table 1 compare the C, O, N and Pd surface concentrations (as measured by XPS) of PI substrates which are similarly plasma or VUV-pretreated in different atmospheres, subsequently palladium-activated through either the Pdac or the $PdCl_2$ routes, and finally rinsed (in chloroform or water, respectively) for removing the weakly bonded species. These data show that palladium, irrespective of its origin, is not strongly adsorbed onto oxygenated surfaces but is adsorbed in significant amounts onto surfaces grafted with nitrogenated species. In what follows, only the simplified routes using surface "amination" of PI substrates through plasma or VUV irradiation in NH_3, deposition of Pdac thin films by spin-coating, and their rinsing in chloroform are investigated. Consider now what happens when such thin films are immersed directly in the plating baths (laboratory-made or commercial Ni plating baths in this set of experiments). In the case of labora-

Table 1.
Surface atomic concentrations (at.%) of Kapton® HN substrates after functionalization (plasma or VUV irradiation processes), then "activation" through the $PdCl_2$ or Pdac route and final water (WR) or chloroform (CR) rinsing

Surface functionalization	Subsequent surface treatment before metallization	Surface composition (at.%)			
		C	O	N	Pd
NH_3 plasma, 1 min	$PdCl_2$ + WR	68.7	20.0	9.8	1.5
NH_3/VUV, 3 min	$PdCl_2$ + WR	62.7	18.1	18.0	1.2
NH_3 plasma, 1 min	PdAc + CR	68.8	19.1	11.0	1.2
NH_3/VUV, 3 min	PdAc + CR	56.8	23.7	16.8	2.8
O_2 plasma, 1 min	$PdCl_2$ + WR	69.4	23.8	6.6	0.2
air/VUV, 3 min	$PdCl_2$ + WR	70.8	22.7	6.1	0.3
O_2 plasma, 1 min	PdAc + CR	65.4	23.3	5.2	0.2
air/VUV, 3 min	PdAc + CR	67.8	22.2	6.1	0.2

tory-made Ni plating baths, the schematic diagram given in Figure 8 shows that the Ni deposits grow without any initiation time. However, these deposits are not at all adherent to the substrate. On the other hand, with this same plating bath, bright and well-adherent Ni deposits are obtained when the rinsed thin films are previously either chemically reduced in an aqueous solution of hypophosphite or photo-chemically reduced via VUV-irradiation under vacuum [30, 32]. Note that, in these two examples, Ni deposition starts also without any initiation time ($\tau = 0$ s) when the polymer surface with adsorbed Pd^0 species is immersed in the plating baths. Figure 8 represents also what happens when the industrial Ni plating bath is used. No Ni deposit is observed when the rinsed Pdac thin film is immersed directly in the plating bath. On the other hand, bright and well adherent Ni films are obtained when the rinsed Pdac thin films are previously chemically or photo-chemically reduced. However, in contrast to the results obtained when the laboratory-made Ni plating bath is used, there is here clear evidence that the starting of metal deposition requires longer incubation times. This experimental fact again brings out the influence (not understood) of the stabilizers contained in the industrial plating baths. Regardless of these considerations, it also appears that the reduction of the Pdac molecules (i.e. of Pd^{+2} species) in the (0.09 M) hypophosphite solution used in the relevant experiments is significantly more efficient than that resulting from the VUV-irradiation carried out under vacuum using the Xe_2* excimer lamp and experimental conditions described previously. Note that in the latter case the significant initiation time ($\tau = 100$ s) has allowed to observe the formation of Ni clusters and the subsequent lateral growth of the Ni deposit before becoming a homogeneous metal film.

Despite the fact that one of the processes described above and involving deposition of a Pdac precursor by spin-coating (or possibly by dipping or spraying) and then photo-reduction of the so-formed Pdac film appears to be rather simple, there are many parameters which can influence the quality of the interphase created between the electrolessly deposited metal and polymer substrate. Indeed, besides the nature of the substrate pretreatment, the experimental data reported in the literature mention the use of solutions of Pdac or other metal-organic precursors having various concentrations [52-58], resulting in precursor films of different thicknesses (e.g. from 0.01 to 1 μm). In addition, the rinsing step, when it is used (it is not always the case), is performed after the photo-reduction step under conditions which are generally non-specified and undoubtedly very different. Lastly, the photo-reduction of the precursor films is undoubtedly almost complete depending on the photon source and energy flux used. Obviously, all the parameters involved in the implementation of the Pdac deposit and reduction influence the various properties of the metal / polymer interfaces such as adhesion. Unfortunately, such a study is very difficult to perform and has not been considered in the main works cited above. One exception is the investigation carried out by Bauer and coworkers [51]. These authors have studied the applicability of the laser-assisted deposition process for electrolessly coating a Cu film on polymers

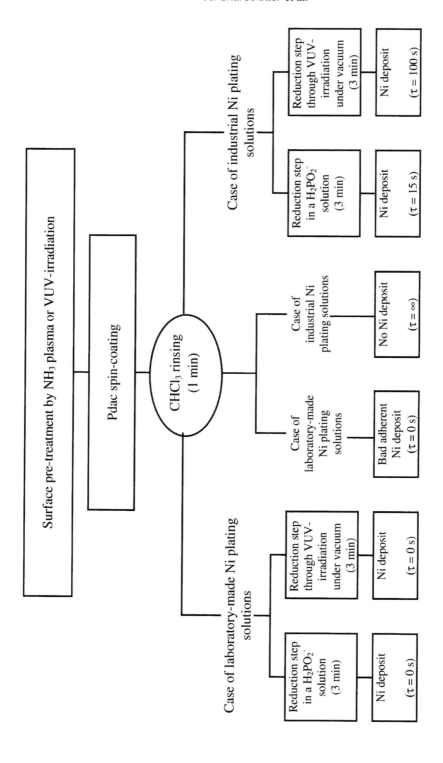

Figure 8. Schematic diagram of the sequences of the electroless plating (EP) process involving the spin-coating of a palladium acetate (Pdac) thin film on a surface pretreated by plasma or VUV-irradiation in NH$_3$. Chemical and photo-chemical reduction of the adsorbed Pd^{+2} species is required to deposit adherent Ni films from both laboratory-made and commercial plating baths. Note the difference in the initiation times in the different baths.

such as polyimide (PI), poly(etherimide) (PEI) and poly(tetrafluoroethylene) (PTFE) and have shown that the increase of the Pdac concentration leads to an increase of the Pd nuclei density, and finally to an increase of the practical adhesion of the Cu deposit (peel test).

3.3. Adhesion characteristics of electroless films

The last part of the present work aimed at investigating the adhesion characteristics of Ni thin films electrolessly deposited on Kapton® 500 HN substrates. In all the cases which were previously qualified as providing bright and adherent Ni deposits, the use of the cross-cut peel test showed that no metal part was removed by the Scotch® tape, and therefore that a good practical adhesion had been obtained between the metal and the polymer. This also means that the effects of the different efficient treatments used to prepare the PI surface before metallization cannot be distinguished through this test.

For a similar purpose, a fragmentation test has also been employed. It should be pointed out that such a test has already been used for studying both fracture and delamination of metallic or ceramic thin, brittle films adhering to ductile substrates [60-65]. Usually, the coated specimens are strained under uniaxial tension and the development of parallel cracks which are transverse to the straining direction is observed (by electron or optical microscopy) as the applied strain progresses. According to the theoretical approach by Wheeler and Osaki [60], the maximum interfacial shear stress as it is inferred from experimental data (crack density at saturation) can be taken as a parameter related to the interfacial adhesion. In the present experiments, an original approach has been employed [33, 40, 41]. It consists in subjecting the coated specimens (in dog-bone shape) to the uniaxial tension and measuring the resulting changes in the electrical resistance of the metal films between two metallic contacts symmetrically located at the ends of the narrow part of the investigated specimens. As the main objective of these experiments is to show the possible effects of surface treatments on the practical adhesion of the metal films to their substrate, it is necessary, for a comparison purpose, to deposit Ni films of same composition, structure and thickness. Consequently, such films must be deposited from the same plating bath (here the commercial Europlate Ni 520 solution) and their thickness determined. This means that calibration curves must be established. As an example, Figure 9 illustrates the dependence of X-ray fluorescence data (Ni K_α intensity measurements) on the initial conductance ($1/R_o$) of the Ni films. In the thickness range investigated, the Ni K_α intensity and film initial conductance vary linearly with the film thickness. As a result, a mere measurement of the initial electrical resistance R_o of the Ni films provide access to their thickness.

Coming back to the fragmentation test itself, Figure 10 is plotted for 200 nm thick Ni films deposited on PI substrates which were previously subjected to different surface treatments. The (a) and (b) curves show how the relative electrical resistance $\delta R/R$ varies as a function of the applied strain, i.e. in fact against the

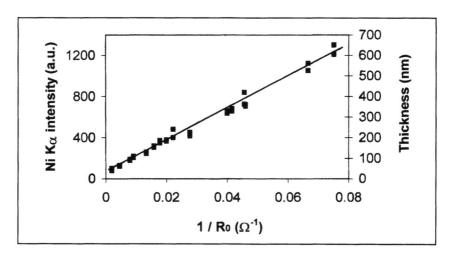

Figure 9. Calibration diagram showing the linear dependence of the Ni K_α intensity (XRFS analysis) and thickness of the electrolessly deposited Ni films on the initial conductance $1/R_0$ of these films. For films obtained from the same plating bath, a simple electrical measurement provides an indication of the film thickness.

relative elongation ε (%) = $\Delta L/L_0$ where L_0 is the initial distance between the two metallic contacts. These curves are characteristic of Ni / PI systems in which the polymer surface has been plasma-treated in NH_3 and subsequently "activated" either (a) by immersion in a simple acidic $PdCl_2$ solution or (b) through the Pdac route. In both cases, it is possible to distinguish different zones associated with the progress of the fragmentation process of the Ni thin film. Indeed, there is a first fragmentation step during which many random cracks appear (upward-sloping parts of curves (a) and (b)) and form parallel metal strips. As the applied strain increases, the initial step of the fragmentation process progressively gives rise to a second step which consists of the division of the widest segments at their midpoint (downward parts of curves (a) and (b)). Note that the boundary between the two corresponding zones cannot be precisely determined since the random cracks formed probably coexist with the cracks resulting from the division of the widest segments. When curves (a) and (b) tend to a minimum ($\varepsilon \sim$ 12-14%) crack density saturation is reached. This means that, at this stage, the primary process of segment division can be considered as nearly completed. Eventually, a secondary fragmentation process resulting from the contraction of the metal coating perpendicular to the straining direction and controlled by Poisson's ratio of the polymer substrate begins to appear. This results in the formation of cracks parallel to the straining direction, and, therefore, in the formation of secondary fragments. In the experiments relative to curves (a) and (b) of Figure 10, this secondary cracking mode leads only to a very slight increase of the $\delta R/R$ values. This means that the conduction mode is practically not changed and that most of the secondary frag-

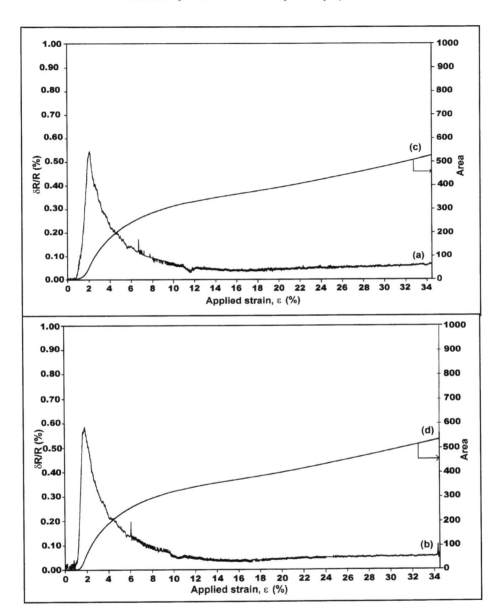

Figure 10. Fragmentation test. Dependence of the relative electrical resistance of 200 nm thick Ni films on the applied strain. Curve (a) is relative to a Kapton® HN substrate which was plasma-treated in NH_3, then "activated" in an acidic $PdCl_2$ solution and immersed in a $H_2PO_2^-$ reducing bath. Curve (b) is relative to the same NH_3 plasma-treated substrate after its activation via the Pdac route and its reduction in a $H_2PO_2^-$ bath. Curves (c) and (d) were obtained by integration of curves (a) and (b), respectively. Area [in $(\%)^2$] on the right ordinate scales is the result of these integrations.

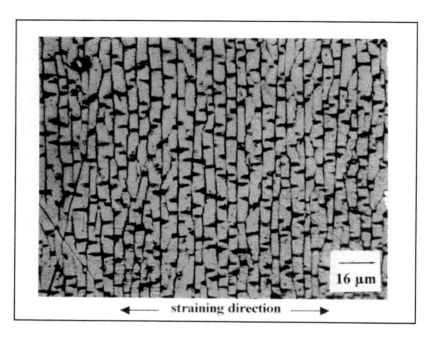

Figure 11. Optical photomicrograph (× 625) of a fragmented zone of a 200 nm thick Ni film deposited on a NH₃ plasma-treated and PdCl₂-activated Kapton® HN substrate (125 μm thick), after application of a 30% strain. The straining direction is indicated by the arrows.

ments remain electrically connected. Figure 11 shows a typical fragmentation pattern (optical photomicrograph) from which we can distinguish the coating fragmentation into parallel strips, and breakage of these strips into small rectangular fragments. In this example, the fragmentation pattern is obtained from a coated specimen (plasma-treated in NH₃ / curve (a)) subjected to 30% strain. Note that the observed ruptures are only "apparent" because a careful comparison of the optical photomicrographs obtained by lighting the coated specimen from both underneath and above shows that they do not all reach the metal / polymer interface and this explains the electrical continuity. This also means that curves (a) and (b) are indicative of a good adhesion of the metal coating to the so-treated PI substrates. It is clear from these experiments that no segment delamination occurs before the breakage of the PI substrates. Note also that curves (a) and (b) are obtained with an excellent reproducibility. In Figure 10, curves (c) and (d) result from integration of curves (a) and (b), respectively. Curves (c) and (d) can also be used to determine, on the one hand, the onset of the primary fragmentation process and, on the other hand, the strain at which crack density saturation is reached.

Figure 12 represents PI substrates which were chemically treated in a H₂SO₄/CrO₃ mixture and subsequently sensitized / activated by immersion successively in SnCl₂ and PdCl₂ solutions. Contrary to the cases described in Figure 10, the conventional method used here for conditioning the polymer surface be-

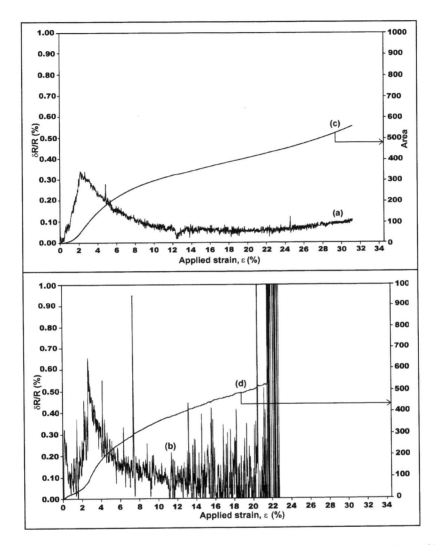

Figure 12. Fragmentation test. Dependence of the relative electrical resistance of 200 nm thick Ni films on the applied strain. Curves (a) and (b) are for Kapton® HN substrates which were chemically pretreated (H_2SO_4 / CrO_3 solution), and then sensitized / activated by immersion successively in acidic $SnCl_2$ and $PdCl_2$ solutions. A reasonable adhesion at the metal / polymer interface is obtained in the case (a) while a poor adhesion is achieved in the case (b). Curves (c) and (d) were obtained by integration of curves (a) and (b), respectively. Area [in (%)²] on the right ordinate scales is the result of these integrations.

fore the electroless metallization does not provide (a) or (b)-type curves with a good reproducibility. In the present example, curve (a) is characteristic of a reasonable adhesion of the Ni film to its substrate, despite a nearly immediate onset of the primary fragmentation process. On the other hand, curve (b) bears witness

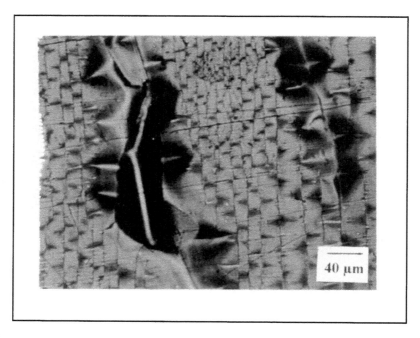

Figure 13. Optical photomicrograph (× 250) of a fragmented zone of a 200 nm thick Ni film electrolessly deposited on a Kapton® HN substrate (125 μm thick) through a conventional process involving chemical treatment in a H_2SO_4 / CrO_3 solution and sensitization / activation by immersion successively in acidic $SnCl_2$ and $PdCl_2$ solutions. This photomicrograph for a coated specimen after application of a 22% strain shows the presence of many fragments which no more adhere to the substrate. The straining direction is as shown in Figure 11.

of a very poor adhesion with a complete loss of the electrical continuity for an applied strain of about 22%. Note that in this last case a negative value of δR/R results from the overlapping at time t + δt of secondary fragments which were delaminated at time t. In addition, Figure 13 represents an optical photomicrograph of a fragmented zone in which several delaminated metal segments can be observed.

Obviously, interpretation of curves shown in Figures 10 and 12 is still partial and progress in this area requires further investigations.

4. CONCLUSIONS

In this work, new results were obtained concerning the surface functionalization, activation and electroless metallization of polyimide substrates (Kapton® 500 HN foils). The main conclusions of this paper be summarized as follows:
1. Grafting of nitrogenated functionalities (through the use of plasma or VUV treatments operating in NH_3 and/or N_2 atmospheres at reduced pressure) con-

fers to PI surfaces a specific chemical reactivity allowing a direct attachment of palladium-based catalysts. Two routes were used for this purpose. The former involves the chemisorption of the Pd^{+2} ions present in a dilute acidic $PdCl_2$ solution onto the grafted nitrogenated functionalities, and consequently the formation of strong covalent C-N-Pd bonds. The latter uses the deposit of a metal-organic thin film (here a palladium acetate thin layer) from a solution containing the precursor.

2. Grafting of oxygenated functionalities (through the use of plasma or VUV treatments in an O_2 atmosphere at reduced pressure) also confers to PI surfaces a specific chemical reactivity allowing, in this case, a direct attachment of tin species. Under these conditions, the creation of Pd^{+2} sites requires the use of the conventional sensitization / activation method (involving both tin and palladium species). This route suggests the formation of strong C-O-Sn (Pd) covalent bonds.

3. Regardless of the process used to attach the palladium-based catalyst ($PdCl_2$ or Pdac route) on surfaces grafted with nitrogenated groups, the catalytic species are chemisorbed in the Pd^{+2} state while the triggering of the electroless metallization occurs only when these bonded Pd^{+2} species are, at least, partly reduced to the Pd^0 state. In this study, both laboratory-made and industrial Ni and Cu plating baths were studied. In the case of the laboratory-made Ni plating solution, the Pd^{+2} reduction occurs spontaneously in the solution due to the presence of the hypophosphite reducing agent. In all the other cases (industrial Ni bath and laboratory-made or industrial Cu bath) the Pd^{+2} reduction does not take place spontaneously in the solutions. This is due to the presence, in the industrial baths, of stabilizers which probably block the Pd^{+2} catalytic sites. Furthermore, in the case of Cu baths, the Pd^{+2} reduction by formaldehyde is very slow and does not take place in a reasonable period of time. Under these conditions, it is necessary to carry out the Pd^{+2} reduction either in a hypophosphite solution or under VUV-irradiation under vacuum before substrate immersion in the previously cited plating baths.

4. Electroless Ni or Cu films deposited onto PI substrates through the simplified process involving surface functionalization in a nitrogenated atmosphere, and a subsequent "activation" through the $PdCl_2$ or Pdac route, and qualified in this work as bright and adherent, exhibit good adhesion characteristics when the relevant metal / polymer systems are subjected to a cross-cut peel test. However, the effects of different surface treatments of the polymer surface cannot be distinguished using this test. On the other hand, an original approach based on measurements of the change in the electrical resistance of the metallic film during a fragmentation test on the coated specimens appears to provide interesting prospects. Experiments carried out in this work dealt only with electroless Ni deposits. Nevertheless, they show that the tin-free processes described in the present study can be advantageously compared in terms of practical adhesion of Ni deposits to those involving the conventional sensi-

tization / activation processes requiring the use of both tin and palladium species. In addition, note that these approaches to the EP process do not need microroughening of the polymer substrates as required for the implementation of the conventional method using chemicals in the surface preparation step.

REFERENCES

1. K.L. Mittal (Ed.), *Polyimides: Synthesis, Characterization and Applications*, Vols. 1 and 2, Plenum Press, New York (1984).
2. D. Wilson, H.D. Stenzenberger and P.M. Hergenrother (Eds.), *Polyimides*, Chapman and Hall, New York (1990).
3. M.K. Ghosh and K.L. Mittal (Eds.), *Polyimides: Fundamentals and Applications*, Marcel Dekker, New York (1996).
4. M.J.M. Abadie and B. Sillion (Eds.), *Polyimides and Other High Temperature Polymers*, Elsevier, Amsterdam (1991).
5. W.D. Weber and M.R. Gupta (Eds.), *Recent Advances in Polyimide Science and Technology*, Mid-Hudson Section, The Society of Plastics Engineers, Wappingers Falls, NY (1987).
6. C. Feger and H. Franke, in: *Polyimides: Fundamentals and Applications*, M.K. Ghosh and K.L. Mittal (Eds.), pp. 759-814, Marcel Dekker, New York (1996).
7. E.M. Wilson, in: *Polyimides: Synthesis, Characterization and Applications*, K.L. Mittal (Ed.), Vol. 2, pp. 715-733, Plenum Press, New York (1984).
8. G. Samuelson and S. Lytle, in: *Polyimides: Synthesis, Characterization and Applications*, K.L. Mittal (Ed.), Vol. 2, pp. 751-766, Plenum Press, New York (1984).
9. L.J. Matienzo and W.N. Unertl, in: *Polyimides: Fundamentals and Applications*, M.K. Ghosh and K.L. Mittal (Eds.), pp. 629-696, Marcel Dekker, New York (1996).
10. K.-W. Lee and A. Viehbeck, in: *Polyimides: Fundamentals and Applications*, M.K. Ghosh and K.L. Mittal (Eds.), pp. 505-532, Marcel Dekker, New York (1996).
11. L.P. Buchwalter, in: *Polyimides: Fundamentals and Applications*, M.K. Ghosh and K.L. Mittal (Eds.), pp. 587-628, Marcel Dekker, New York (1996).
12. E. Sacher, J.J. Pireaux and P. Kowalczyk (Eds.), *Metallization of Polymers*, ACS Symp. Series No. 440, American Chemical Society, Washington, DC (1990).
13. K.L. Mittal and J.R. Susko (Eds.), *Metallized Plastics 1: Fundamental and Applied Aspects*, Plenum Press, New York (1989).
14. K.L. Mittal (Ed.), *Metallized Plastics 2: Fundamental and Applied Aspects*, Plenum Press, New York (1991).
15. K.L. Mittal (Ed.), *Metallized Plastics 3: Fundamental and Applied Aspects*, Plenum Press, New York (1992).
16. K.L. Mittal (Ed.), *Metallized Plastics: Fundamentals and Applications*, Marcel Dekker, New York (1998).
17. K.L. Mittal (Ed.), *Metallized Plastics 5&6: Fundamental and Applied Aspects*, VSP, Utrecht, The Netherlands (1998).
18. K.L. Mittal (Ed.), *Metallized Plastics 7: Fundamental and Applied Aspects*, VSP, Utrecht, The Netherlands (2001).
19. E. Sacher, in: *Metallization of Polymers*, E. Sacher, J.J. Pireaux and P. Kowalczyk (Eds.), p. 1, ACS Symp. Series No. 440, American Chemical Society, Washington, DC (1990).
20. M. Romand and M. Charbonnier, in: *Polyimides: Fundamentals and Applications*, M.K. Ghosh and K.L. Mittal (Eds.), Vol. 2, Marcel Dekker, New York, to be published.
21. G.O. Mallory and J.B. Hajdu (Eds.), *Electroless Plating: Fundamentals and Applications*, American Electroplaters and Surface Finishers Society, Orlando, FL (1990).
22. M. Charbonnier, M. Alami and M. Romand, *J. Electrochem. Soc.* **143**, 472 (1996).
23. M. Alami, M. Charbonnier and M. Romand, *J. Adhesion* **57**, 77 (1996).

24. M. Alami, M. Charbonnier and M. Romand, *Plasmas and Polymers* **1**, 113 (1996).
25. M. Charbonnier, M. Alami, M. Romand, J.P. Girardeau-Montaut and M. Afif, *Appl. Surface Sci.* **109/110**, 206 (1997).
26. M. Romand, M. Charbonnier, M. Alami and J. Baborowski, in: *Metallized Plastics 5 & 6: Fundamental and Applied Aspects*, K.L. Mittal (Ed.), p. 3, VSP, Utrecht, The Netherlands (1998).
27. M. Charbonnier, M. Alami and M. Romand, *J. Appl. Electrochem.* **28**, 449 (1998).
28. M. Charbonnier, M. Romand, G. Stremsdoerfer and A. Fares-Karam, *Recent Res. Devel. Macromol. Res.* **4**, 27 (1999).
29. M. Charbonnier, M. Romand and M. Alami, in: *Polymer Surface Modification: Relevance to Adhesion*, K.L. Mittal (Ed.), Vol. 2, p. 29, VSP, Utrecht, The Netherlands (2000).
30. M. Charbonnier, M. Romand, E. Harry and M. Alami, *J. Appl. Electrochem.* **31**, 57 (2001).
31. M. Charbonnier, M. Romand, U. Kogelschatz, H. Esrom and R. Seeböck, in: *Metallized Plastics 7: Fundamental and Applied Aspects*, K.L. Mittal (Ed.), p. 3, VSP, Utrecht, The Netherlands (2001).
32. M. Charbonnier, M. Romand, H. Esrom and R. Seeböck, *J. Adhesion* **75**, 381 (2001).
33. M. Romand, M. Charbonnier and Y. Goepfert, in: *Metallization of Polymers 2*, E. Sacher (Ed.), Kluwer Academic / Plenum Publishers, New York (2002).
34. F.D. Egitto, V. Vukanovic, F. Emmi and R.S. Horwath, *J. Vac. Sci. Technol.* **B 3**, 898 (1985).
35. L.J. Matienzo and F.D. Egitto, *Polym. Degrad. Stabil.* **35**, 181 (1992).
36. F.D. Egitto, L.J. Matienzo, K.J. Blackwell and A.R. Knoll, *J. Adhesion Sci. Technol.*, **8**, 411 (1994).
37. J.E. Klemberg, L. Martinu, O.M. Küttel and M.R. Wertheimer, in: *Metallized Plastics 2: Fundamental and Applied Aspects*, K.L. Mittal (Ed.), p. 315, Plenum Press, New York (1991).
38. J.E. Klemberg-Sapieha, O.M. Küttel, L. Martinu and M.R. Wertheimer, *J. Vac. Sci. Technol.* **A 9**, 2975 (1991).
39. N. Inagaki, S. Tasaka and K. Hibi, *J. Polym. Sci.: Polym. Chem.* **30**, 1425 (1992).
40. Y. Goepfert, M. Charbonnier and M. Romand, in: *Proc. 24th Annual Meeting of the Adhesion Society*, J.A. Emerson (Ed.), p. 183, The Adhesion Society, Blacksburg, VA (2001).
41. Y. Goepfert, M. Charbonnier, and M. Romand, in: *Proc. 25th Annual Meeting of the Adhesion Society and 2nd World Congress on Adhesion and Related Phenomena (WCARP-II)*, A.V. Pocius and J.G. Dillard (Eds.), pp. 51-53, The Adhesion Society, Blacksburg, VA (2002).
42. F.R. Nartley, *Chemistry of the Platinum Group Metals. Recent Developments; Studies in Inorganic Chemistry II*, Elsevier, Amsterdam (1991).
43. J.M. Calvert, W.J. Dressick, C.S. Dulcey, M.-S. Chen, J.H. Georger, D.A. Stenger, T.S. Koloski and G.S. Calabrese, in: *Polymers for Microelectronics, Resists and Dielectrics*, C.F. Thomson, C.G. Willson and S. Tagawa (Eds.), pp. 210-219, ACS Symp. Series No. 537, American Chemical Society, Washington, DC (1994).
44. W.J. Dressick, C.S. Dulcey, J.H. Georger, G.S. Calabrese and J.M. Calvert, *J. Electrochem. Soc.* **141**, 210 (1994).
45. M.-S. Chen, S.L. Brandow, C.S. Dulcey, W.J. Dressick, G.N. Taylor, J.F. Bohland, J.H. Georger, E.K. Pavelchek and J.M. Calvert, *J. Electrochem. Soc.* **146**, 1421 (1999).
46. Y. Zhang, K.L. Tan, G.H. Yang, E.T. Kang and K.G. Neoh, *J. Electrochem. Soc.* **148**, C574 (2001).
47. G.H. Yang, E.T. Kang and K.G. Neoh, *Langmuir* **17**, 211 (2001).
48. H.S. Cole, Y.S. Liu, J.W. Rose and R. Guida, *Appl. Phys. Lett.* **53**, 2111 (1988).
49. Y.S. Liu and H.S. Cole, *Chemtronics* **4**, 209 (1989).
50. H. Esrom and G. Walh, *Chemtronics* **4**, 216 (1989).
51. A. Bauer, J. Ganz, K. Hesse and E. Köhler, *Appl. Surface Sci.* **46**, 113 (1990).
52. H. Esrom, J. Demny and U. Kogelschatz, *Chemtronics* **4**, 202 (1989).
53. H. Esrom and U. Kogelschatz, *Appl. Surface Sci.* **46**, 158 (1990).
54. H. Esrom and U. Kogelschatz, *Thin Solid Films* **218**, 231 (1992).
55. G. Shafeev, W. Marine, H. Dellaporta, L. Bellard and A. Cross, *Thin Solid Films* **241**, 52 (1994).

56. J.-Y. Zhang, H. Esrom and I.W. Boyd, *Appl. Surface Sci.* **96-98**, 399 (1996).
57. J.-Y. Zhang and I.W. Boyd, *Appl. Phys.* **A 65**, 379 (1997).
58. J.-Y. Zhang and H. Esrom, *Appl. Surface Sci.* **54**, 465 (1992).
59. H. Esrom, R. Seeböck, M. Charbonnier and M. Romand, *Surface Coat. Technol.* **125**, 19 (2000).
60. D.R. Wheeler and H. Osaki, in: *Metallization of Polymers*, E. Sacher, J.J. Pireaux and S.P. Kowalczyk (Eds.), pp. 500-512, ACS Symp. Series No. 440, American Chemical Society, Washington, DC (1990).
61. E.K. Chong and G. Stevens, in: *Metallized Plastics 5&6: Fundamental and Applied Aspects*, K.L. Mittal (Ed.), pp. 409-422, VSP, Utrecht, The Netherlands (1998).
62. P.H. Wojciechowski and M.S. Mendolia, *J. Vac. Sci. Technol.* **A 7**, 1282-1288 (1989).
63. Y. Pitton, S.D. Hamm, F.-R. Lang, H.J. Mathieu, Y. Leterrier and J.-A. Månson, *J. Adhesion Sci. Technol.* **10**, 1047-1065 (1996).
64. Y. Leterrier, Y. Wyser, J.-A. Månson and J. Hilborn, *J. Adhesion* **44**, 213 (1994).
65. B. Brogier, G. Nansé, M. Nardin, G. Baud, M. Jacquet and J. Schultz, *J. Adhesion Sci. Technol.* **14**, 339 (2000).

Polyimides and Other High Temperature Polymers, Vol. 2, pp. 315–329
Ed. K.L. Mittal
© VSP 2003

Surface modification of polyimide to improve its adhesion to deposited copper layer

MANFRED DANZIGER* and WINFRIED VOITUS

FRACTAL AG, Vor dem Gröperntor 20, 06484 Quedlinburg, Germany

Abstract—Flexible copper-polyimide laminates with improved stability under rapidly changing thermal and mechanical stresses have been obtained by a pre-treatment of the polyimide surface comprising irradiation of the polyimide foil with heavy ions and a subsequent chemical etching of the latent ion tracks. During the etching step a special "surface-depth relief" is created on the surface of the polyimide. The metallisation of this polyimide (so-called "ion track foil") by chemical electroless deposition and electrodeposition of copper leads to a copper-polyimide laminate with improved peel strength and high resistance to both thermal and mechanical stresses.

Keywords: Polyimide; copper; peel strength; adhesion; surface modification; ion irradiation.

1. INTRODUCTION

Commercially available copper-polyimide laminates have been used for several decades as base materials for flexible printed circuit boards. Laminates with properties customised to different applications have been developed. However, it is well known that the copper-polyimide interface has some severe weaknesses, especially if the laminate is used under harsh operating conditions. Such conditions include high temperature, high humidity, thermal stress, vibrational stress or combinations thereof. Especially, the automobile industry needs a flexible base material which is able to withstand the high temperatures and vibration loads which are present close to the engine.

Most of the commercially available laminates are adhesive laminates, i.e. an adhesive is applied between the polyimide and copper foil to ensure a sufficiently high peel strength. The thermal resistance of these adhesives is generally lower than that of polyimides. Polyimides can withstand continuous operating temperatures of about 200°C and short-term soldering processes of more than 300°C. The coefficient of thermal expansion of the adhesive differs from that of polyimide

*To whom all correspondence should be addressed. Phone: +49 3946 68990,
Fax: +49 3946 689921, E-mail: danziger@fractal-ag.de

and copper. Therefore, the range of the operating temperature is limited because of mechanical stress created.

Successful attempts have been made to develop adhesiveless laminates. Most of these techniques include coating the polyimide surface with a partially cured polyimide and laminating a copper foil to the coated polyimide under elevated pressure and temperature. During this lamination process the partially cured polyimide film is fully cured and an adhesiveless copper-polyimide laminate is formed. Additionally, attempts have been made by KANEKA Corp. to match the coefficient of thermal expansion of polyimide to that of copper, and thus mechanical stress is reduced. This has led to the development of the new polyimide compositions, e.g., the Kapton® E, H films by DuPont [1].

Adhesiveless copper-polyimide laminates show improved thermal stability, but suffer from relatively low peel strength and high cost. An increase of the peel strength has been achieved through the application of thin tie-coats of reactive metals, e.g. chromium or monel (a nickel-copper alloy), between copper and polyimide [2]. Typical thicknesses of such tie-coats are between 75 nm and 200 nm. A further drawback of tie-coats is that they complicate the subsequent etching process of the conductor pattern. In many cases an additional etching step is necessary to remove the tie-coat completely between the conductor lines to ensure the required insulation resistance.

Therefore, a copper-polyimide laminate which is based on an adhesiveless bonding technology without applying tie-coats is needed to enable easy processability. A promising approach to improve adhesion is the surface modification of the polyimide foil. The surfaces of polymer foils may be modified physically (i.e. mechanical roughening) or chemically (e.g. by introduction of reactive functional groups onto polymer surfaces) [3, 4]. Both approaches are widely used by laminate manufacturers.

This paper presents a new approach to manufacture adhesiveless laminates with improved peel strength at lower cost. It is based on an innovative *Ion-Track Technology* developed by FRACTAL AG to create the so-called "surface-depth relief" on polyimide films. This technique can be applied not only to the manufacture of flexible circuits, but also to any purposes where polyimide surfaces with well-defined profiles are needed. It can also be applied to other polymers.

2. BASICS OF ION-TRACK TECHNOLOGY

The surface modification by ion-track technology comprises two steps. First, the irradiation of the polyimide foil with high-energy ions to produce latent ion tracks and second, the etching of the irradiated surface to obtain the required surface morphology (the so-called "surface-depth relief").

2.1. Irradiation

Fig. 1 shows the schematic of the irradiation unit. Multiple ionised heavy ions, for instance $^{40}Ar^{n+}$, impinge on the polyimide foil which is continuously moved through the ion beam using a reel-to-reel equipment. The ion beam is defocussed to a size of several square centimetres and is rapidly moved over the whole width of the polymer foil to obtain a uniform irradiation of the whole foil surface. The angle of incidence can be varied from normal incidence to about 60°. Especially, symmetric incidences are possible, as shown in the figure. Ions of noble gases like argon, krypton and xenon are preferably used. Typical doses are between 10^7 and 10^8 ions per square centimetre.

This "ion bombardment technique" has been used for several decades for the manufacture of so-called "ion-track membranes" [5]. Such membranes are polymer foils with micro- and nano-sized channels which penetrate the whole thickness of the foil. They have found a wide range of applications in the field of filtration.

The ion-track technology, which is presented here, relies on a careful control of the penetration depth of the ions into the polymer foil. The ions transfer their kinetic energy to the polymer material in the close environment of their paths. Thus latent ion tracks with a well-defined length and a diameter of 10 to 30 nm are created. The amount of energy deposited along the track and the track length are determined by the species and the kinetic energy of the incoming ions.

It is possible to create latent tracks which penetrate the whole thickness of the foil (as in the case of membranes) as well as latent tracks with very short lengths. A foil with such "dead-end" tracks is called an "ion-track foil". The term "latent track" is used, as these ion tracks are difficult to be observed using scanning electron microscopy.

2.2. Etching

After irradiation the polymer film is submitted to a chemical etching process, which is shown schematically in Fig. 2. Because of the structural changes resulting from irradiation, the etching rate in the ion tracks is much higher than in the unaffected bulk. This effect is used to produce differently shaped microholes that superpose onto different "surface-depth reliefs". It is possible to produce cylindrical through-holes, blind holes or cones of different, but well-defined, size and shape. Examples are given in Fig. 3. The hole sizes and shapes are precisely controlled by a careful choice of parameters related to the irradiation, i.e. ion species, ion initial energy and incidence angle and parameters related to the etching step, i.e. etching medium, pH-value, temperature and time. The etching process is completed by neutralization and rinsing in pure water as shown in Fig. 2.

Examples of modified polyimide surfaces are shown in Fig. 4. On the left, microholes are visible which were obtained by irradiation under normal incidence and subsequent etching. On the right, microholes are shown which were obtained by inclined incidence. From the shadows in the enlarged view d) it is clear that two symmetric angles of incidence have been used.

M. Danziger and W. Voitus

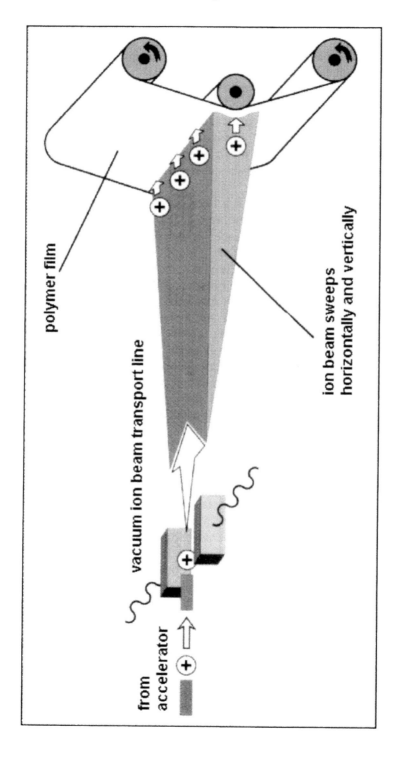

Figure 1. Schematic of the irradiation unit.

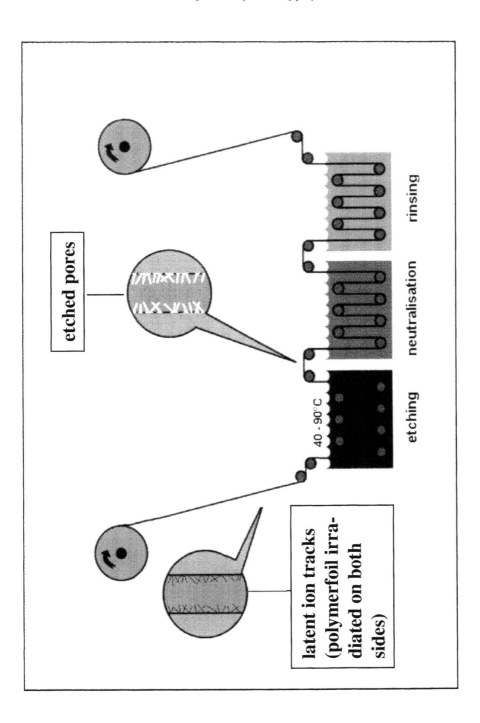

Figure 2. Scheme of the etching process.

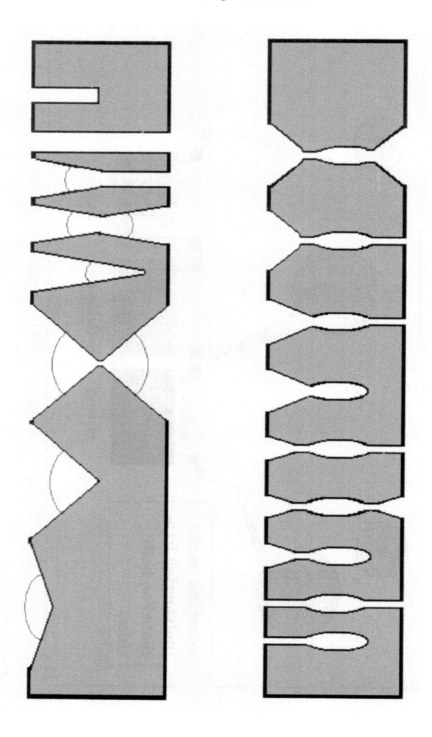

Figure 3. Scheme (top and bottom) of structures that can be produced with ion track technology (height and width of the structures are not to scale).

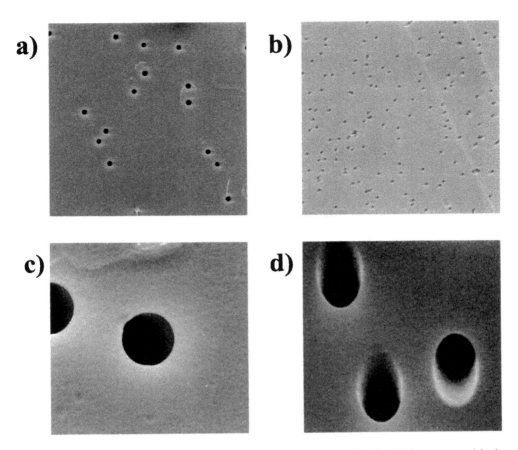

Figure 4. SEM micrographs of modified polyimide surfaces. (a, c) after irradiation at normal incidence (b, d) after inclined irradiation under symmetric angles of incidence (all holes shown have a diameter of 2 μm).

The SEM-micrographs in Fig. 5 illustrate the superposition of microholes onto well-defined "surface-depth reliefs". On the left, a surface which was irradiated with a low ion dose and etched for a short time is shown. Here the ion tracks are well separated from each other and their superposition is negligible.

In the middle, a medium ion dose which resulted in a considerable superposition of the microholes after etching was used. On the right, a surface which was irradiated with a high ion dose and more intensively etched is shown. This leads to a strong superposition of the microholes and a significant overetching of the surface (formation of an Alpine-like profile).

The criterion low, medium or high ion dose is determined by the relation of the size and number of the pores. Single pores stand for a low ion dose, pores which

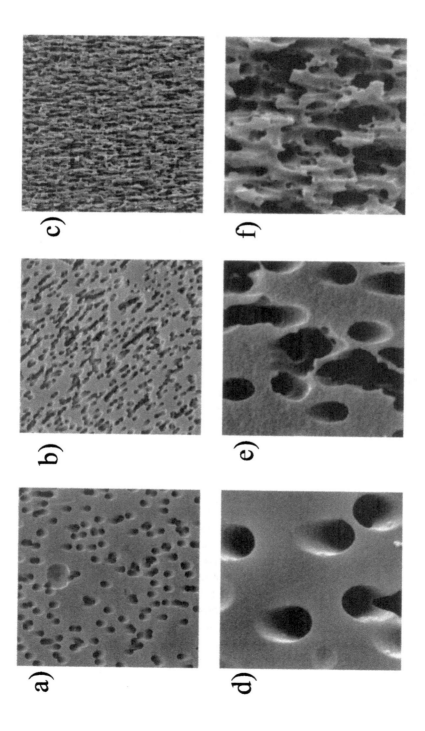

Figure 5. SEM micrographs showing microholes onto different surface-depth reliefs. (a, d) after irradiation with low ion dose – negligible overetch (b, e) after irradiation with medium ion dose – weak overetch (c, f) after irradiation with high ion dose – strong overetch (all holes shown have a diameter of 2 μm).

are partly superposed stand for a medium ion dose, and pores which are totally superposed stand for a high ion dose.

Between the extreme structures given in Fig. 5 an infinite number of different surface profiles can be obtained reproducibly. In this way, the polymer surface profile can be matched to the requirements of any functional coating which is subsequently deposited. These functional coatings may include conductive or insulating layers, wear-resistant layers, simple decorative layers and others. Any deposited layer is reliably anchored in the optimized surface profile of the polymer which ensures high peel strength.

The ion-track technology presented here is useful to modify the surface morphology of most commercially available polymers. Especially, it is useful for polyimide foils like Kapton, Upilex, Apical and others.

3. APPLICATION EXAMPLE: ADHESIVELESS COPPER-POLYIMIDE LAMINATE

As an example, the application of ion-track technology in the manufacture of adhesiveless all-polyimide laminates will be shown here. Such laminates are used as base materials for flexible printed circuit boards. In Fig. 6, cross sections of a polyimide foil which has been subjected to an ion irradiation and etching treatment are shown. It is clearly visible that latent ion tracks have been expanded through etching into two bundles of microholes which are inclined symmetrically to the surface.

The microholes affect only a thin surface layer of about 6 μm thickness due to the limited penetration depth of the ions. This modified surface layer represents the "surface-depth relief" mentioned above.

Applying a special vacuum-coating process developed by FRACTAL AG the "surface-depth relief" of the polyimide film is covered completely with a thin copper seed-coat. The seed-coat may be deposited on one or both sides of the polyimide film. After depositing the seed-coat an electroplating process is performed. This process allows to fill the pores uniformly with copper and to deposit a uniform copper layer up to the desired thickness.

The two different groups of symmetrically oriented microholes filled with copper ensure a high peel strength of the copper coating resulting in an outstanding stability of the copper-polyimide laminate to thermal and mechanical stresses. This is depicted in Fig. 7a showing the cross section of a copper-polyimide laminate which has been forced to separate partially by applying a very high peel force of about 5 N/cm. This gives an impression of the highly effective mechanical interlocking between the copper layer on the top and the polyimide foil on the bottom. Fig. 7b presents the copper surface profile which is obtained after complete removal of the polyimide substrate by etching.

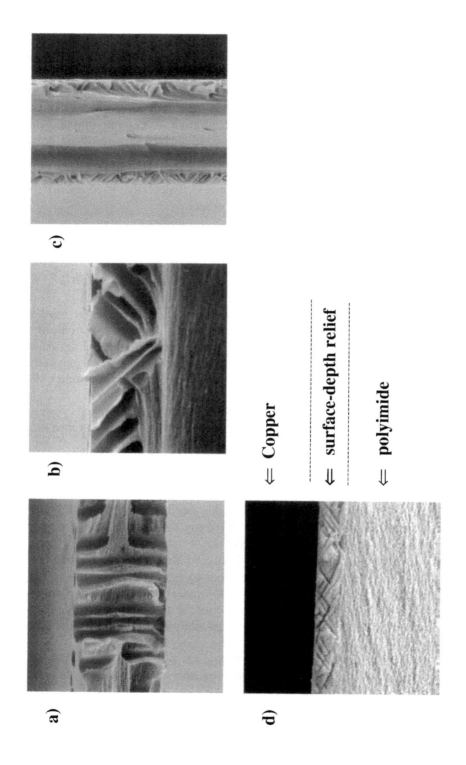

Figure 6. SEM micrographs of a polyimide foil after ion irradiation and etching treatment. (a, b, c, d) etched polyimide foil irradiated with heavy ions of different angles of incidence a) 90°, b), c) and d) 45° (all holes shown have a diameter of 2 μm and the thickness of the copper layer is 6 μm).

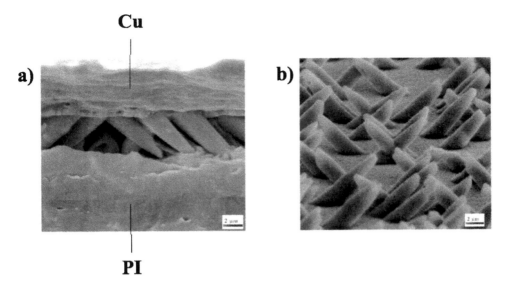

Cu

a)

b)

PI

Figure 7. SEM micrograph of copper-polyimide laminate and copper surface profile. (a) Copper-polyimide laminate, partially separated after applying a very high peel force (b) Copper surface profile after complete removal of the polyimide substrate.

4. PEEL STRENGTH MEASUREMENTS ON COPPER-POLYIMIDE LAMINATES

The peel strength of four copper-polyimide laminates prepared by ion-track technology has been measured as a function of storage temperature and time. The samples were prepared using identical ion irradiation conditions, but different subsequent processing steps, i.e. the etching, the seed-coat deposition and the electroplating process, as shown in Fig. 8. The different conditions and parameters are given on the right-hand side.

In Fig. 9 the evolution of the peel strength with storage time at ambient temperature is shown. The value at $t = 0$ is the peel strength measured immediately after finishing the electroplating process. Different preparation parameters result in different behaviours of the peel strength. Generally, with extended storage time the peel strength is improved. However, after very long storage times of about 150 hours or more a peel strength of at least 1.8 N/mm was obtained in all four experiments.

It is clearly seen that different processing parameters result in very different initial peel strengths. However, if the processing parameters are chosen properly, an initial peel strength of more than 0.6 N/mm can be obtained as shown in Figure 9.

In Fig. 10, the evolution of peel strength with storage time of the samples at 75°C is shown. It is clear that the peel strength improves more rapidly here compared to the storage at ambient temperature shown before. The aim is to achieve a value for t = 0 which lies in the target range as illustrated in Fig. 10.

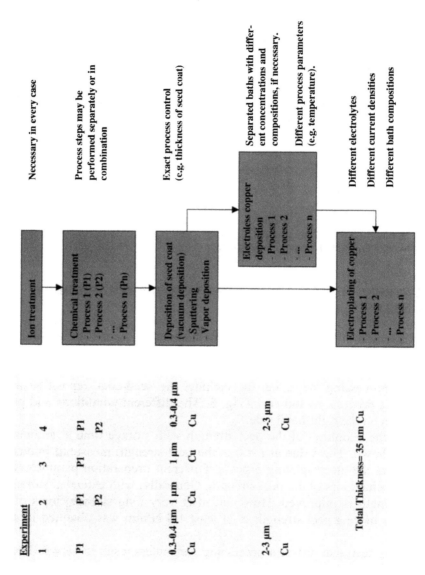

Figure 8. Scheme of the manufacturing process for copper-polyimide laminates.

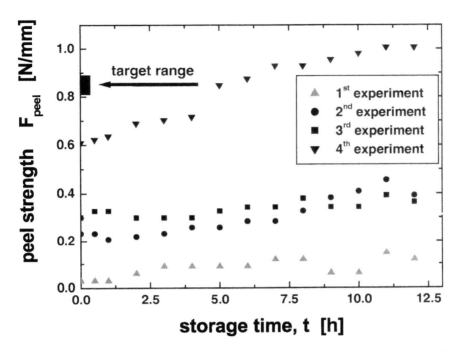

Figure 9. Peel strength of copper-polyimide laminates (samples stored at ambient temperature).

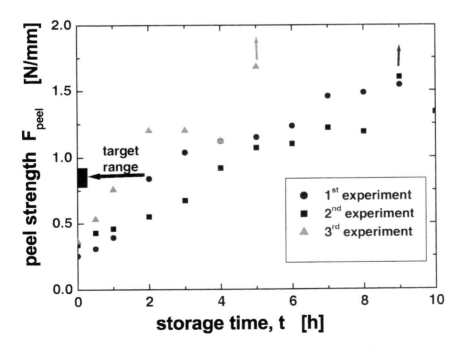

Figure 10. Peel strength of copper-polyimide laminates (samples stored at 75°C).

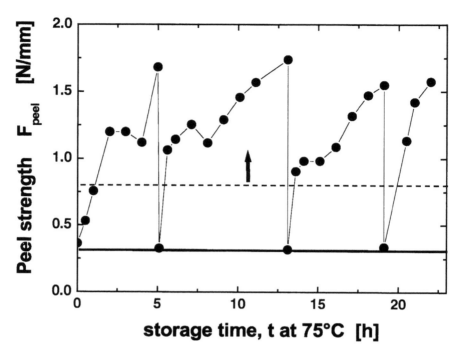

Figure 11. Peel strength of copper-polyimide laminates after dipping into boiling water. Horizontal solid line = minimum value of the peel strength required for printed circuit boards. Dotted line = target range, aim to achieve a higher initial value of the peel strength. Arrow = increasing peel strength.

In Fig. 11, the results of an additional storage experiment at 75°C are shown. As expected, the peel strength increases with extended storage time. After reaching a value higher than 1.5 N/mm the sample was dipped into boiling water for 5 minutes. During boiling the peel strength dropped rapidly to the initial value. This process was repeated three times. In all cases the same initial peel strength value was observed. The dashed line represents the minimum value of the peel strength desired for printed circuit boards.

These test results give direction to formulate the next Research & Development goals: Firstly, a manufacturing process which enables a steeper increase of the peel strength should be developed. Secondly, and perhaps more important, a process which gives a higher initial value of the adhesion has to be developed.

5. SUMMARY

The innovative ion-track technology presented in this paper allows the formation of "surface-depth reliefs" on polymer foils which are optimised for the subsequent deposition of functional or decorative coatings. An important application is the manufacturing of copper-polyimide laminates for flexible printed circuit boards. These laminates do not contain any tie-coats, adhesives or other intermediate layers. They show high peel strength and improved stability under rapidly changing thermal and mechanical stresses.

REFERENCES

1. J.A. Kreuz, S.N. Milligan and R.F. Sutton, "Advanced Flexible Dielectric Substrates for FPC/TAB Applications", (http://www.dupont.com/kapton/applications/fpc-tab.pdf).
2. T. Bergstresser and J. Sallo, *Printed Circuit Fabrication*, (April 2001) (http://www.pcfab.com /db_area/archive/2001/0104/gould.html).
3. K.L. Mittal (Ed.), *Polymer Surface Modification: Relevance to Adhesion*, VSP, Utrecht (1996).
4. K.L. Mittal (Ed.), *Polymer Surface Modification: Relevance to Adhesion*, Vol. 2, Utrecht (2000).
5. R.L. Fleischer, T.B. Price and R.M. Walker, *Nuclear Tracks in Solids*, University of California Press, Berkeley, CA (1975).

Polyimides and Other High Temperature Polymers, Vol. 2, pp. 331–343
Ed. K.L. Mittal
© VSP 2003

Study on the structure and adhesion of copper thin films on chemically modified polyimide surfaces

LEE LEE LAU, YIN XIONG and LIANG HONG[*]

Department of Chemical and Environmental Engineering, National University of Singapore, 10 Kent Ridge Crescent, Singapore 119260

Abstract—Upilex-s® [BPDA-PDA, poly(biphenyldianhydride-p-phenylenediamine)] and 6FDA-ODA [4,4'-(hexafluoroisopropylidene)diphthalic anhydride-4,4'-oxydianiline] polyimide sheets were subjected to wet treatment using ethylenediamine (EDA) and hydrazine as the base. The results of wet treatment were evaluated in terms of elemental analysis (N/C atomic ratio) obtained using XPS as well as surface morphologies determined using AFM and SEM. Furthermore, a copper layer was deposited onto the modified surface via electroless plating. The adhesion strength between the copper layer and the modified surfaces was assessed by the tape-peel test. A large adhesion enhancement was achieved when copper was plated onto the modified surfaces of PI sheets as compared to the untreated samples.

Keywords: Polyimide; surface modification; wet treatment; adhesion; copper plating.

1. INTRODUCTION

Polyimides (PIs) have been widely used in microelectronic packaging due to their low dielectric constant, good thermal and mechanical stability and excellent chemical resistance, for example, its use as an insulator in the multi-chip module (MCM) is being investigated [1-2]. However, the adhesion of the polyimide to metals is normally rather poor, this is probably due to its rigid molecular structure and the fact that low surface energy component (aromatic rings) overwhelms the high surface energy component (polar groups) in the surface phase. The smooth surface also contributes to the low adhesion property [3].

In the past decades, many surface treatment methods have been developed to enhance the interfacial adhesion between metals or polymers and a polyimide film. The aim of modification is to introduce polar functional groups as well as micro-roughening to the polyimide surface. This can be done by employing either the dry process, in which the surface is subjected to reactions with gaseous species, such as plasma, particle beams, corona discharge, X-rays, ion beams or γ-

---

*To whom all correspondence should be addressed. Phone: (65) 6874-5029, Fax: (65) 6779-1936, E-mail: chehongl@nus.edu.sg

irradiation [4-7], or the wet process in which the surface is modified using chemical solutions [8]. The adhesion enhancement could be attributed to the occurrence
of certain types of chemical bonds or/and mechanical interlocking at the interface
between the plated metal and the polyimide substrate [9-10].

Recently, the wet treatment method has received special attention since it relies
on the opening of the imide ring in an alkaline solution, which is apparently cost
effective in contrast to the dry process. The alkaline aqueous solution has been the
often-used medium for performing the surface modification [11]. With the aim of
altering the surface functionalities, we used aqueous solutions of organic bases
(EDA and hydrazine) instead to open the surface imide-rings of two types of PIs
(Fig. 1). The grafted amino-groups at the PI surface have been known to have
affinity to metal atoms [10, 12]. A copper thin film was deposited electrolessly
onto the modified PI surface. We have studied the adhesion strength of the copper
film to the PI sheets. The conclusions drawn from adhesion testing have been
correlated with the surface composition and morphology of the PI sheets. Both
surface composition and morphological changes at different stages of surface
treatments were studied by X-ray photoelecron spectroscopy (XPS), atomic force
microscopy (AFM) and scanning electron microscopy (SEM).

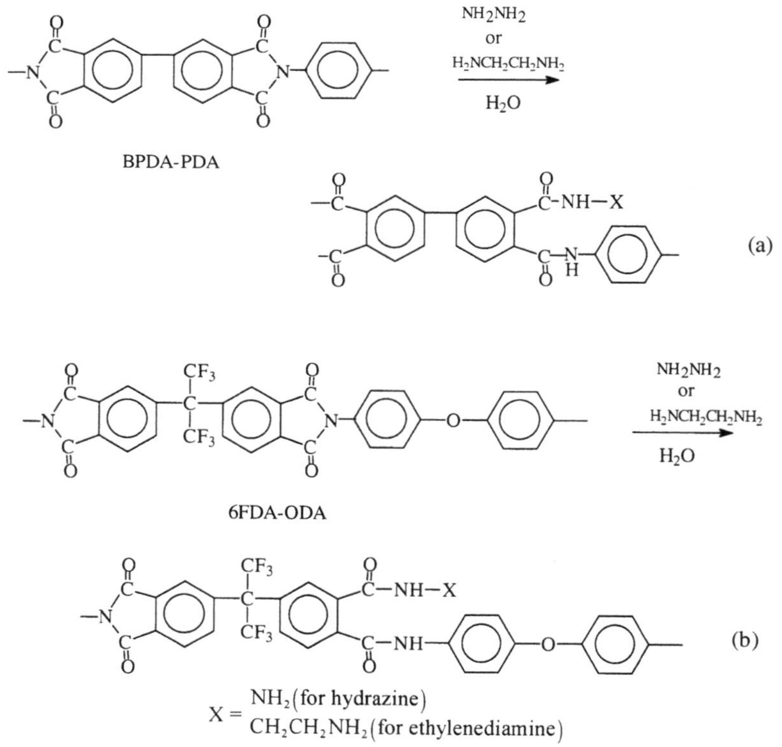

Figure 1. The ring-opening reaction of (a) BPDA-PDA polyimide; (b) 6FDA-ODA polyimide with
EDA or hydrazine.

2. EXPERIMENTAL

2.1. Synthesis of 6FDA-ODA film

The 6FDA-ODA PI was synthesized via a two-step method: First, dried 4-4'-oxydianiline powder (>99%, Aldrich) was dissolved in freshly distilled NMP (N-methyl-2-pyrrolidinone) in a round-bottom flask under a nitrogen atmosphere at room temperature. After the diamine was totally dissolved, an equimolar amount of 4,4'-(hexafluoroisopropylidene)-diphthalic anhydride was added to the solution with stirring to obtain a viscous poly(amic acid) (PAA) after eight hours of reaction. The PAA solution was then spread on a Petri dish and the PAA was imidized by heating in a forced-air oven at 100°C, 200°C and 300°C for one hour each with a ramping rate of 2.0°C/min. The imidization extent of the PAA of 6FDA-ODA at the three temperature points was characterized by FT-IR spectroscopy. The IR peak at 1776 cm^{-1} (the stretching vibration frequency of carbonyl group of imide) was used as the main identification for the formation of the imide ring.

2.2. Samples preparation

Upilex-S® sheet (BPDA-PDA, thickness = 50 μm, obtained from UBE Industries, Tokyo, Japan), EDA (99.5%, Aldrich), hydrazine hydrate (55%, Aldrich) and 2-propanol (99%, Fisher Scientific) were used as received. The 6FDA-ODA PI sheet was synthesized as mentioned before. The PI sheets were cut into small pieces (about 2 cm×2 cm) and washed in 2-propanol under ultrasonication for 15 min at room temperature, followed by rinsing with DI water and drying in air. The purified PI pieces were soaked in an aqueous solution of either EDA or hydrazine at a given temperature for a certain period of time. The treated PI pieces were washed successively with DI water and 2-propanol. Finally, the samples were dried under vacuum for 12 hours at room temperature.

2.3. Electroless copper plating and tape-peel test

The modified PI sheets were immersed in SnCl$_2$-HCl solution for 10 minutes and then in DI water for 2 minutes. The PI sheet was then immersed in PdCl$_2$ solution for another 10 minutes and washed finally in DI water for 2 minutes. After this two-step pretreatment, metallic Pd colloidal particles were embedded into the surface of PI sheet as the initiation sites for electroless copper deposition [13]. A home-made electroless plating bath was prepared using the composition shown in Table 1. Electroless copper plating was carried out at room temperature for 4 minutes and a copper layer with thickness of about 0.3 μm was obtained according to the SEM image. Extended plating will cause delamination of copper film; this phenomenon has been studied recently by Choi *et al.* [14]. The plated PI sheets were washed with DI water and dried in vacuum before characterization and testing. The tape-peel test was conducted on a rectangular area (1×3 cm) of Cu layer on a PI sheet. The examined surface was carefully cleaned and then

Table 1.
Electroless copper plating bath composition

Constituent	Concentration
Copper sulphate ($CuSO_4 \cdot 5H_2O$)	10 g/L
Potassium sodium tartarate ($KO_2CCH(OH)CH(OH)CO_2Na \cdot 4H_2O$)	50 g/L
Formaldehyde (HCHO, 37%)	50 ml/L
Sodium carbonate (Na_2CO_3)	10 g/L
Sodium hydroxide (NaOH)	10 g/L

fixed onto a flat platform before testing. An aluminum tape (Aldrich, High temperature aluminum tape) was pasted evenly on the copper layer and then peeled, this process was repeated on the same sample until pieces of copper peeled off from the sample surface and the number of times of repetition were recorded as the number of times of T-peel test. The surface was then examined under optical and electron microscopes.

2.4. Instrumental analysis

FTIR spectra of the PI sheets were obtained on a Bio-Rad FTS-135 spectrophotometer (Cambridge, MA, USA). XPS spectra were obtained on a Kratos Axis HSI (Manchester, U.K.) instrument, using a monochromatic excitation radiation from Al-K_α at 1486.6 eV. A pass-energy of 80 eV was used for the wide scans and 40 eV for the high resolution scans. The photoelectron take-off angle with respect to the polymer surface was set at 90° for the analysis of surface composition. All spectra were calibrated against the C 1s peak at 284.6 eV. Elemental peaks were deconvoluted by curve fitting using the Gaussian distribution. The elemental concentrations were calculated by dividing the areas under the peaks by the atomic sensitivity factors of the respective elements. The sensitivity factors used throughout this work were: O=0.736, C=0.318, N=0.505 and F=1.00; all calculated values have a maximum error of ±10%. The sectional analysis and 3D images were obtained using an atomic force microscope (AFM, Digital Instruments, U.S.A) in the taping mode. The average roughness (R_a) was calculated by averaging the roughnesses obtained from five arbitrarily selected locations each with a scanning area of 10 μm×10 μm. Surface morphologies were investigated using a scanning electron microscope (JEOL JSM-5600LV).

3. RESULTS AND DISCUSSION

3.1. Reactions of N-phenylphthalimide with EDA and hydrazine

In order to confirm the feasibility of the ring-opening reaction at the surface of a PI sheet and the species produced by the reactions, a low molecular-weight imide,

N-phenylphthalimide, was employed as a model to react with EDA or hydrazine. This is because the reflective IR spectroscopy does not have enough sensitivity to detect the imide ring-opening products if their concentrations are far lower than the polyimide background. The reaction took place very fast under the reaction conditions (by mixing the imide with 55 vol% of hydrazine aqueous solution at 50°C or 55 vol% of EDA aqueous solution at 80°C for 1 minute). The FTIR spectra of the products and N-phehylphthalimide are shown in Figure 2. The pristine N-phenylphthalimide structure shows the following characteristic peaks: 1776 cm^{-1} (w) being the asymmetric stretching vibration of the two carbonyl groups in the imide ring and 1418 cm^{-1} and 1175 cm^{-1} peaks belonging to the asymmetric vibration of the C-N-C bond in the imide ring. The peak at 1710 cm^{-1}(s) appearing in all the three spectra is characteristic of the stretching vibration of carbonyl group. In addition, new peaks appearing at 3260-3280 cm^{-1}, 1640 cm^{-1} and 1540 cm^{-1} in both (b) and (c) spectra belong to the amide groups generated. On the other hand, the fact that the intensities of the peaks at 3280, 1640 and 1540 cm^{-1} in spectrum (b) are stronger than the corresponding ones in spectrum (c) suggests that hydrazine reacts more rapidly than EDA with N-phenylphthalimide even at lower reaction temperature. The FTIR results support the reactions proposed in Figure 1.

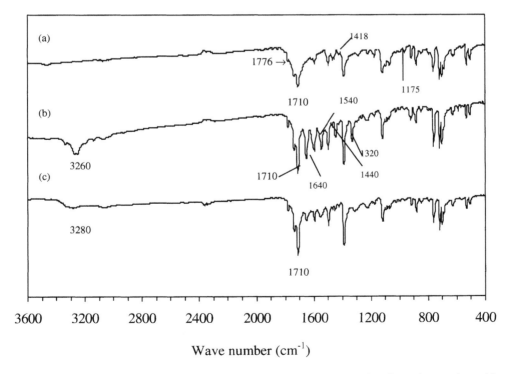

Figure 2. FTIR spectra of (a) pristine N-phenylphthalimide (b) the product from the reaction with hydrazine and (c) the product from the reaction with EDA.

3.2. XPS analysis of the modified PI surfaces

The modified surfaces were analyzed at 90° take-off angle to investigate the surface composition. Tables 2 and 3 show the surface elemental concentrations of different surface states including the pristine PIs, the degreased (as there is a wax layer on Upilex-s® sheet) and the modified PIs. There was also a small amount of silicone contamination on the pristine PI that originated from the silicone rollers used in the production. The contamination may affect the treatment process and was removed using 2-propanol before performing the modification. From the elemental concentration data in Table 2 (which is followed by an explanation for the symbols used), the atomic percentage of N at the surface increases with the increase in soaking time in the EDA solution at 50°C. The same trend has also been observed in the samples treated in hydrazine solution. The XPS results also show that hydrazine is more reactive than EDA, for instance, the treatment condition 30haz50-10 brought about a larger N/C ratio than the relatively harsh condition 55eda50-10. This is consistent with the previous FT-IR result on the hydrolysis of N-phenylphthalimide.

Figures 3 and 4 show that the N/C ratio of pristine Upilex-s® PI is 0.09 and this ratio increases with the treatment time. However, the ratio increased slightly after soaking in solutions for about 30 minutes at 50°C. Prolonged treatment time also causes a decrease in the N/C ratio in the cases of 99.5eda50 and 55haz50. A simi-

Table 2.
XPS-surface elemental concentrations (%) of the modified Upilex-s® (BPDA-PDA)

Treatment conditions[a]	Elemental concentration[b]			
	C	N	O	Si
Stoichiometric value	78.6	7.1	14.3	0
Undegreased	77.3	5.8	14.7	2.2
Degreased	79.1	6.3	14.6	0
30eda80-10	78.3	9.3	12.4	0
30eda80-30	81.3	6.6	12.1	0
55eda50-10	80.6	8.2	11.2	0
55eda50-30	78.7	9.3	12.0	0
99.5eda50-10	78.8	10.8	10.4	0
99.5eda50-30	75.22	12.1	12.7	0
30haz50-10	72.3	9.6	18.1	0
30haz50-30	71.8	10.9	17.3	0
55haz50-10	75.1	13.0	11.9	0
55haz50-30	74.1	13.6	12.3	0

Note: [a] The symbol (e.g. 30eda80-10) stands for the 30 vol% of EDA aqueous solution, the treatment temperature = 80°C and duration of treatment = 10 minutes. The abbreviation "haz" stands for hydrazine. [b] Hydrogen atom is not included in the calculation as it does not appear in the XPS spectrum.

Table 3.
XPS-surface elemental concentrations of the modified 6FDA-ODA PI

	Elemental concentration (%)				
	C	N	O	F	Si
Stoichiometric value	70.0	4.5	11.4	13.6	0
Undegreased 6FDA-ODA	66.1	4.5	20.3	8.2	0.9
Degreased 6FDA-ODA	69.2	4.3	12.4	14.1	0
80eda50-30	67.2	8.5	12.4	11.9	0
80eda50-90	71.0	7.2	10.5	11.3	0
99.5eda50-30	61.8	8.1	14.1	16.0	0
30haz50-10	60.1	7.9	16.0	16.0	0
30haz50-30	62.9	6.6	16.8	13.7	0
55haz50-10	63.1	10.0	13.4	13.5	0
55haz50-30	61.9	9.4	15.0	13.7	0

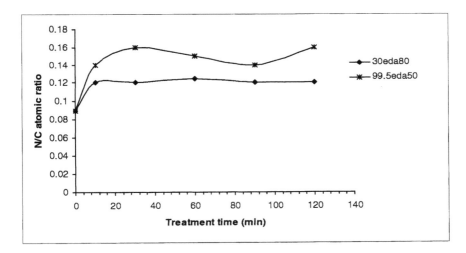

Figure 3. N/C ratios of the modified BPDA-PDA surface resulted from the treatment with the EDA.

lar outcome is that 30eda80-30 reveals a lower N concentration than 30eda80-10 (Table 2). This phenomenon can be explained in terms of the degradation of PI chains in the outermost surface layer which yields lower-molecular-weight fragments. The dissolution of these fragments in the rinsing step results in a fresh PI surface [15].

As illustrated in Figure 1, the major difference in using EDA or hydrazine than using hydroxides [16] as the base to open the imide ring lies in that the former gives rise to amides while the latter to carboxylate. Based on the monomer unit of Upilex-s®, the N/C ratio increases from 0.09 to 0.23 (after being hydrolyzed by

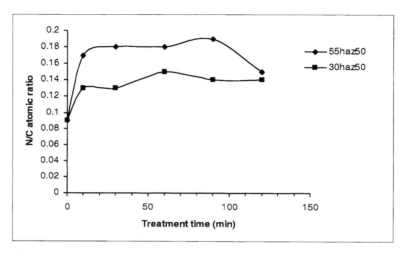

Figure 4. N/C ratios of the modified BPDA-PDA surface resulted from the treatment in hydrazine solution.

EDA) and to 0.27 (after being hydrolyzed by hydrazine). According to the experimental results shown in Figures 3 and 4, the stoichiometric N/C ratios were not obtained, which suggests that only part of the surface imide-rings could be converted to the corresponding amides. The fact that breaking down of the aromatic amide bond (as indicated in Fig. 1) took place simultaneously with opening of the imide ring is likely responsible for this outcome, which leads to shorter segments. In Figure 3 the optimum conditions for modification in EDA should be 30 min of soaking time in a 30 vol% solution at 80°C. Despite the fact that the 99.5 vol% of EDA could give rise to a higher N/C ratio, it is not economical due to the use of much higher concentration of EDA. In Figure 4 where hydrazine was used as the base, it also shows that an extension of reaction time beyond 30 min did not significantly affect the N/C ratio.

As far as the modification of 6FDA-ODA surface is concerned, the results are shown in Table 3; the nitrogen atomic concentration seemed to reach its highest limit after being immersed in EDA solution for 30 minutes at 50°C regardless of the base concentration. In the case of using hydrazine as the reagent, the surface nitrogen atomic concentration reached the limit after 10 minutes of treatment at 50°C. A higher concentration of hydrazine could result in a higher N concentration within the same period of time. However, the N concentration dropped with treatment time; this is again due to the removal of the outermost layer consisting of short polyimide segments, resulting in exposure of the layer underneath.

3.3. AFM analysis

The topological changes on PI surfaces before and after the modifications were recorded by AFM. The images were obtained using the tapping mode. Both the image and the mean roughness (R_a) were determined by scanning a surface area of 10 μm×10 μm. Figure 5 shows the mean roughness values of all the modified Upilex-s® samples with different treatment time. The curves show that hydrazine is much stronger than EDA in etching the PI surface.

The surface of pristine 6FDA-ODA was investigated by both AFM and SEM. It was found that a large number of big holes (around ~ 3 μm) were present on the surface scanned by SEM (Figure 6); this may be caused by the release of water vapor during imidization. The AFM image of pristine 6FDA-ODA showed a smooth surface with a roughness (R_a) of 0.888 nm as shown in Figure 7b. The analysis of the mean roughness (Figure 7a) of the modified 6FDA-ODA showed that when using EDA temperature had a stronger effect than concentration on the generation of surface roughness. Taking for example the two sets of treatment conditions: 99.5eda50 and 80eda80 (the latter one was performed at a higher temperature but at a lower concentration of EDA than the former), it turned out that the latter one had a greater extent of roughness. This scenario differs from that occurring on BPDA as shown in Fig. 5, in which the concentration effect overtook the temperature effect as shown by the curves of 99.5eda50 and 55eda80 in the initial 30 min of treatment. It seems that the imide ring-opening reaction at the more hydrophobic 6FDA-ODA surface (due to the presence of 6 fluorine atoms in each repeat unit) requires a higher activation-energy level than that at the more hydrophilic BPDA-PDA surface. Besides this, we also observed that the reaction at 6FDA-ODA film surface preferred to take place at the pores resulting in bigger pores, which was found by AFM cross-sectional analysis. Another noticeable effect is the nature of base used. Hydrazine is more effective than EDA to create a rough surface on 6FDA-ODA, for instance, the mean roughness of 55haz50 sample is higher than that of 80eda80 as shown in Fig. 7a.

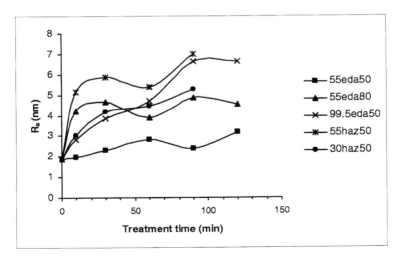

Figure 5. Change in mean surface roughness of BPDA-PDA with the treatment time for different treatment conditions.

Figure 6. SEM image of the pristine 6FDA-ODA.

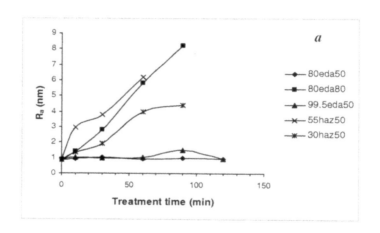

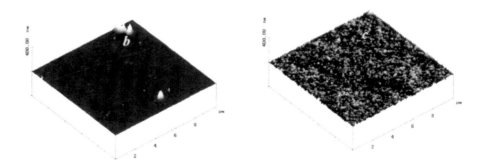

Figure 7. (a) Change in mean surface roughness of 6FDA-ODA under various treatment conditions; (b) AFM-3D image of the pristine 6FDA-ODA; (c) AFM-3D image of the 6FDA-ODA after being treated using condition 55haz50-30.

3.4. Adhesion between the copper layer and PI substrate

The samples for tape-peel test were prepared by electroless plating the modified PI surfaces with Cu. Table 4 shows the tape-peel test results on some selected samples with similar thickness of Cu layer (~ 0.3 μm), in which the data in the column number of times of Peel-Test represent the number of times of taping-peeling operation before pieces of copper were peeled off. Figure 8 is an optical photograph showing the surface after the taping-peeling test. Albeit the conventional approach of measuring adhesion strength must be carried out on an Instron instrument with a defined peel angle and rate [17-18], the tape-peel test was applied in this work as a simple and approximate measure. The results of peel-test show that both PIs in their virgin state have poor adhesion. As far as the relationship between the mean roughness of the PI surface and the adhesion strength is

Table 4.
Results of tape-peel test

Samples	Surface roughness of the substrate (R_a) (nm)	Number of times of peel-test for delamination of copper overlayer
BPDA-PDA (unmodified)	1.9	1
30eda80-30	2.5	120
55eda50-30	2.3	110
55eda80-30	4.7	60
99.5eda50-30	3.8	100
55haz50-30	5.9	120
6FDA-ODA (unmodified)	0.9	3
99.5eda50-120	1.0	70
55haz50-30	3.8	170
80eda50-30	2.3	130

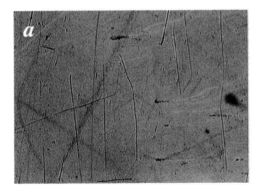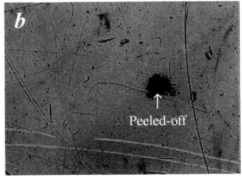

Figure 8. Optical images (80×) of copper layer deposited on modified BPDA-PDA surface (55haz50-90): (*a*) before T-peel test; and (*b*) after 120 times of T-peel test.

concerned, the roughness, in general, favors the adhesion though no direct quantitative relationship could be established. As an example, Figure 9 presents the
changes in the surface morphology of Cu layers deposited on the modified surfaces (80eda50-30 and 55haz50-30) of 6FDA-ODA before and after tape-peel
test. It is known that these two PI substrates have different roughness as shown in
Fig. 7a, 55haz50-30 having a rougher surface than 80eda50-30. In connection
with this, copper layer on the rougher surface achieves much better adhesion
strength. Although the surface roughness is an essential element for the enhancement of adhesion through mechanical anchoring, chemical bonding between the
copper and the organic functional groups at the surface of PI substrate also plays a
key   role.   Different   chemical   functionalities   at   the   two   surfaces
($-CONHCH_2CH_2NH_2$ versus $-CONHNH_2$ in Fig. 1) and the length of the polymer-
chains after the reaction are very likely to affect the adhesion. In fact, the data in
Table 4 show no direct dependence of adhesion strength on the surface roughness
of the substrates. This can be interpreted as the involvement of the organic functional groups. In light of this, increasing the surface concentration of the organic
functional groups that have affinity for copper atom is a strategy in pursuit of im-

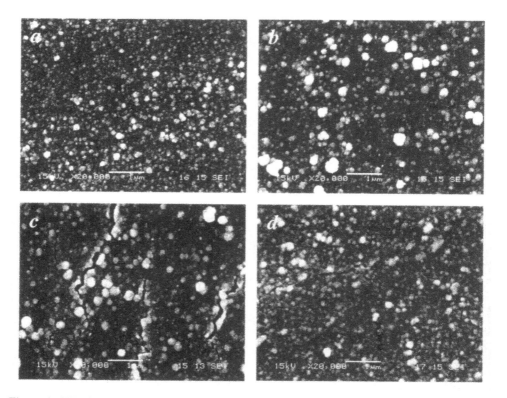

Figure 9. SEM images of the copper layer (*a*) on the modified 6FDA-ODA surface (55haz50-30);
(*b*) on the modified 6FDA-ODA surface (80eda50-30); (*c*) after 130 times of T-peel test; and (*d*) after 170 times of T-peel test.

proving interfacial adhesion strength. In this regard, the graft polymerization approach [18] allows to achieve higher interfacial adhesion strength than other wet treatment methods as it can introduce a high density of functional groups onto the PI surface. However, the surface graft polymerization often utilizes physical methods (e.g. plasma sputtering) to generate reactive species at the surface, which obviously increases the operation cost.

4. CONCLUSIONS

Both hydrazine and ethylenediamine are able to open the five-member imide-ring of polyimides. Two types of polyimides were used to perform the surface modification and it was found that 6FDA-ODA required more vigorous reaction conditions to carry out the surface modification than BPDA-PDA. Based on the AFM investigation, the average roughness (R_a) of the modified polyimide increased apparently after the modification. From the results of tape-peel test, it was found that modified polyimides surfaces exhibited much higher adhesion strength than pristine polyimides to copper overlayers. The present method has its advantage in simplicity of manipulation; however, it causes degradation of the surface PI chains when more severe treatment conditions are applied.

REFERENCES

1. D. M. Mechtel, H. K. Charles, Jr. and A. S. Francomacaro, *Int. J. Microcircuits Electron. Packag.*, **22(3)**, 242-247 (1999).
2. D. J. Kovach, N. S. Amirgulyan, C-P. Chien and M. H. Tanielian, *Int. J. Microcircuits Electron. Packag.*, **23(1)**, 70-77 (1999).
3. C. Feger and H. Franke, in: *Polyimides: Fundamentals and Applications*, M. K. Ghosh and K. L. Mittal (Eds.), pp. 759-814, Marcel Dekker, New York (1996).
4. G. Rozorski, J. Vinkevicius and J. Jaciauskiene, *J. Adhesion Sci. Technol.*, **10**, 399-406 (1996).
5. G. S. Chang, K. H. Chae, C. N. Whang, E. Z. Kurmaev, D. A. Zatsepin, R. P. Winarski, D. L. Ederer, A. Moewes and Y. P. Lee, *Appl. Phy. Lett.*, **74**, 522-524 (1999).
6. R. Waichenhain, D. A. Wesner, W. Pfleging, H. Horn and E. W. Kreutz, *Appl. Surf. Sci.*, **109-110**, 264-269 (1997).
7. L. E. Stephans, A. Myles and R. R. Thomas, *Langmuir*, **16**, 4706-4710 (2000).
8. C. E. Baumgartner and L. R. Scott, *J. Adhesion Sci. Technol.*, **9**, 789-799 (1995).
9. K. W. Lee and A. Viehbeck, *IBM. J. Res. Develop.*, **38**, 457-474 (1994).
10. G. H. Yang, E. T. Kang, K. G. Neoh, Y. Zhang and K. L. Tan, *Colloid Polym. Sci.*, **279**, 745-753 (2001).
11. H. K. Yun, K. Cho, J. K. Kim, C. E. Park, S. M. Kim, S. Y. Oh and J. M. Park, *Polymer*, **38**, 827-834 (1997).
12. I. Ghosh, J. Konar and A. K. Bhowmick, *J. Adhesion Sci. Technol.*, **11**, 877-893 (1997).
13. H. Esrom, R. Seebock, M. Charbonnier and M. Romand, *Surf. Coat. Technol.*, **125**, 19-24 (2000).
14. J. W. Choi, K-E. Lee and T-S. Oh, *Mater. Res. Soc. Symp. Proc.*, **629**, 1-6 (2001).
15. L. J. Matienzo and W. N. Unertl, in: *Polyimides: Fundamentals and Applications*, M. K. Ghosh and K. L. Mittal (Eds), pp. 629-698, Marcel Dekker, New York (1996).
16. W. Yu and T-M. Ko, *Eur. Polym. J.*, **37**, 1791-1799 (2001).
17. N. Inagaki, S. Tasaka and T. Baba, *J. Adhesion. Sci. Technol.*, **15**, 749-762 (2001).
18. M. C. Zhang, E. T. Kang, K. G. Neoh, C. Q. Cui and T. B. Lim, *Polymer*, **42**, 453-462 (2000).

Polyimides and Other High Temperature Polymers, Vol. 2, pp. 345–358
Ed. K.L. Mittal
© VSP 2003

Chemical interaction of Fe, Ni and Au with poly(vinyl chloride) and poly(tetrafluoroethylene) during thermal evaporation and the effect of post-metallization X-ray irradiation studied by *in situ* X-ray photoelectron spectroscopy

C.C. PERRY, S.R. CARLO,* J. TORRES, A.J. WAGNER
and D. HOWARD FAIRBROTHER†

*Department of Chemistry and Department of Materials Science and Engineering,
The Johns Hopkins University, 3400 N. Charles Street, Baltimore, MD 21218*

Abstract—The interaction of Au, Ni and Fe with poly(vinyl chloride) (PVC) and poly(tetrafluoro-ethylene) (PTFE) during thermal evaporation was investigated using *in-situ* X-ray photoelectron spectroscopy (XPS). On PVC, Fe, Ni and Au were all found to be reactive during thermal evaporation, with the extent of reaction varying in the order: Fe > Ni > Au. In contrast, only Fe and Ni formed metal fluoride bonds on PTFE. The differing metal halide yields on PVC and PTFE have been rationalized in terms of the relative strengths of the M-X (X = F, Cl) and C-X bonds. In the Fe/PTFE, Fe/PVC and Ni/PTFE systems, post-metallization X-ray surface modification also increased the yield of metal halides which was postulated to be due to the generation of reactive halogen species.

Keywords: Metallization; physical vapor deposition; X-ray modification; poly(tetrafluoroethylene); poly(vinyl chloride).

1. INTRODUCTION

The production of metallic thin films on polymeric substrates by vacuum deposition (metallization) is of both technological and scientific interest. In particular, developing an understanding of metal/polymer interface is relevant to both microelectronics and packaging industries where the performance of the metallized structure depends upon the microscopic nature of the metal/substrate bonding [1, 2]. Thus, for electronic devices such as transistors and capacitors, metal interconnects should exhibit low resistivity and interconnection times. Lower interconnection times (defined by the product of resistance and capacitance) may be achieved

---

*Current address: Naval Research Laboratory, Code 5613 Washington, DC 20375.
†To whom all correspondence should be addressed. Phone: +1-410-5164-328,
Fax: +1-410-5168-420, E-mail: howardf@jhu.edu

by using a combination of a low-k dielectric polymeric substrate and a low resistivity metal such as Cu. This has led to investigations of the interfacial phenomena at the metal/polymer interfaces in the case of polyimides (PI) [1, 3-6] and fluoropolymers [7-9]. Fluoropolymers, such as poly(tetrafluoroethylene) (PTFE) are potentially attractive substrates for microelectronic devices because of their chemical inertness, high thermal stability and low dielectric constants [7, 10]. Previous studies have shown that metals such as Cr, Ti and Al undergo reactive metallization upon deposition onto PI and PTFE, resulting in better adhesion characteristics than noble (non-reactive) metals such as Cu, Ag, and Au. This illustrates that, in general, the formation of new chemical species between the metallic overlayer and the polymer during metallization is correlated with improved adhesion characteristics. Consequently, surface modification of halogenated polymers for weakly reactive metallic overlayers is often employed in situations where good adhesion is required. For example, in the case of the Cu/PTFE system, copper adhesion characteristics were improved when the PTFE surface was pre-treated using an Ar^+ ion plasma [11-13].

Although there are many studies investigating the effects of pre-treatments on native polymers prior to metallization, much less is known about the effect of post-metallization modification of metal/polymer systems. Wang *et al.* [14] in a study of high energy (1.1 GeV) Bi^{2+} ion post-metallization irradiation of the Cu/PTFE system found that the peel strength increased after ion beam irradiation, an effect that was attributed to increased Cu diffusion into PTFE combined with the formation of CuF_2. In a related study by Shi *et al.* [15], post-metallization X-ray irradiation of Teflon PFA (tetrafluoroethylene–perfluoroalkyl vinyl ether copolymer) and FEP (fluoroethylenepropylene) did not increase the chemical interaction of reactive (Cr, Ti, Al) or noble metals (Cu, Ag, Au) at the metal/fluoropolymer interface. In contrast, experiments in our laboratory on poly(vinyl chloride) (PVC) and PTFE have shown evidence of increased iron [16] and copper [17] halogen bonding after post-metallization X-ray irradiation and Ar^+ bombardment.

In this investigation, we report results on the interaction of Fe, Ni and Au with the halogenated polymers PVC and PTFE. These metals were chosen to encompass a range of reactivity. Although all of the metals studied were reactive towards PVC during thermal evaporation the extent of reaction varied considerably. In contrast, only Ni and Fe were found to be reactive towards PTFE with Au depositing as a metallic overlayer without any compound formation. Results are also presented on the effects of post-metallization modification of these systems by X-ray irradiation. In certain metal/polymer systems, this treatment strategy was found to be an effective means for promoting the amount of metal halide formed at the metal/polymer interface.

2. EXPERIMENTAL

Metal deposition and X-ray photoelectron spectroscopy (XPS) measurements were carried out in the same ultrahigh vacuum (UHV) chamber. A typical base pressure of 1 x 10^{-8} Torr was maintained during data collection and of 5 x 10^{-7} Torr during metal deposition. All XP spectra were acquired at 15 kV and 300 W with a 45° take-off angle from the sample normal using a Mg K_α source. Elemental scans used a pass energy of 44.75 eV and 0.125 eV/step. Binding energy scales were referenced to literature values of known peaks in each elemental region (i.e. CHCl in native PVC and CF_2 in native PTFE). XPS data fitting was performed using 100% Gaussian line shapes using Shirley background subtraction [18]. *In situ* post-metallization X-ray irradiation was performed using the same irradiation source as used for data collection. PVC and PTFE substrates were prepared as described in previous publications [16, 19].

The metal (Au, Ni, Fe) evaporation sources were prepared as described previously using a shrouded tungsten filament wrapped with the deposition metal and heated using a DC power supply [16, 19]. The source was degassed until pure metal (free of oxygen and carbon contamination) was deposited on a Au sample, as determined by XPS. $NiCl_2$ (Aldrich, 98%), $FeCl_2$ and $FeCl_3$ (Aldrich, 98%) reference samples were mounted on carbon tape prior to XPS analysis. In these experiments, the position of the halide peak was used to calibrate the XPS binding energy scale.

3. RESULTS

3.1. PVC

Figure 1 shows comparative spectra of the Cl(2p) region as well as the major relevant metal transitions for the Au/PVC, Ni/PVC and Fe/PVC systems following metallization of native PVC. In the native polymer, the Cl(2p) region consists of a Cl($2p_{3/2}/2p_{1/2}$) doublet at 200.0 eV and 201.6 eV respectively, associated with the CHCl groups present (not shown). During Au, Ni and Fe deposition, a reduction in the CHCl component in the C(1s) region was observed (not shown), indicative of C-Cl bond cleavage during metallization. Metal deposition on PVC also resulted in a broadening of the Cl(2p) envelope to lower binding energy consistent with the formation of a new set of Cl($2p_{3/2}/2p_{1/2}$) peaks between 198–200 eV, indicative of Cl^- formation (Figure 1) [20].

Although additional bands associated with Au^+ in the Au(4f) region are obscured by the metallic envelope during Au deposition on PVC, the presence of Cl^- ions in the Cl(2p) region indicates production of AuCl [20, 21]. During Ni deposition on PVC, the Ni($2p_{3/2}$) region exhibits two peaks (Figure 1(b)). On the basis of literature values, the 852.0 eV peak is assigned to metallic nickel and the peak at 855.0 eV to Ni^{2+} [20]. In conjunction with the changes observed in the Cl(2p) region, this indicates the formation of $NiCl_2$ during metallization. Figure 1(c) shows

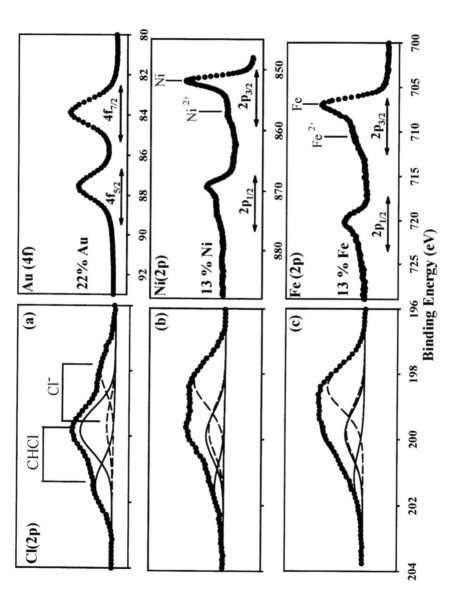

Figure 1. Illustrative plots of (a) Au (b) Ni and (c) Fe on PVC following thermal evaporation. Metal concentrations are given as percentages based on the total C, Cl, O and metal signal determined from regional XPS scans. Raw XPS data are shown as filled circles (●). In the Cl(2p) region, the $Cl(2p_{3/2}/2p_{1/2})$ doublets corresponding to unreacted PVC and chloride ions are shown as solid (–) and dashed lines (– –) respectively.

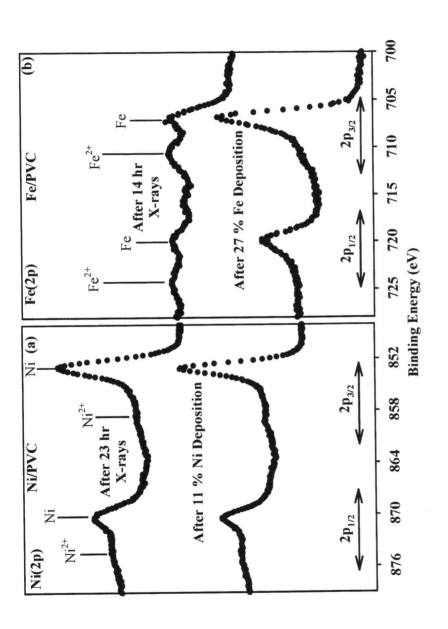

Figure 2. Comparative plots of the Ni(2p) and Fe(2p) regions before and after 300 W X-ray irradiation. (a) 11% nickel coverage on PVC before (lower spectrum) and after 23 hours X-ray irradiation (upper spectrum). (b) 27% iron coverage on PVC before (lower spectrum) and after 14 hours (upper spectrum) X-ray irradiation.

that when Fe is deposited on PVC, two peaks appear in the $Fe(2p_{3/2})$ region during metallization, one centered at 707.0 eV and another at 711.0 eV. On the basis of literature values, the lower $Fe(2p_{3/2})$ binding energy peak at 707.0 eV was assigned to metallic iron and the higher $Fe(2p_{3/2})$ binding energy peak at 711.0 eV to Fe^{2+}, indicative of $FeCl_2$ formation [20, 22]. At higher iron exposures, the metallic $Fe(2p_{3/2})$ peak at 707.0 eV became the dominant spectral feature. The identity of the metal chlorides formed during Ni and Fe deposition was confirmed using the peak positions of $NiCl_2$, $FeCl_2$ and $FeCl_3$ standards.

The extent of reaction was determined using the metal halide: metal ratio by deconvolution of $Ni(2p)$, $Fe(2p)$ and $Au(4f)$ XPS regions [23]. Based on this analysis, Ni and Fe have comparable reactivities during thermal evaporation on PVC, while Au is significantly less reactive. This is also clearly evidenced by differences in the $Cl(2p)$ regions during thermal evaporation of Au and Fe on PVC (Figure 1).

Figure 2(a) shows the effect of 23 hours of X-ray irradiation on the Ni/PVC system with an initial Ni coverage of 11%. Changes in the $Ni(2p)$ region as a result of X-ray irradiation comprised a slight decrease of overall spectral intensity, mainly due to a reduction in the signal associated with Ni metal. Figure 2(b) shows the effect of post-metallization X-ray modification on the Fe/PVC system. In contrast to the Ni/PVC system, pronounced changes are observed in both $Cl(2p)$ and $Fe(2p)$ regions. The $Fe(2p)$ region in Figure 2(b) shows that X-ray irradiation induced a loss of metallic iron and a concomitant increase in Fe^{2+} at ≈ 711.0 eV, indicative of $FeCl_2$ formation.

3.2. PTFE

In contrast to PVC, no evidence of AuF formation was observed during thermal evaporation of Au on PTFE. Au-F bond activation was also not observed during subsequent Ar^+ or X-ray irradiation experiments (not shown). In the case of Ni, although C-F bond cleavage was observed, the extent of reaction with PTFE is less pronounced than on PVC. Figure 3 (lower trace) shows evidence of Ni-F bonding on PTFE as indicated by the appearance of a small feature in the $F(1s)$ region at ≈ 686 eV due to F^- and a concomitant peak in the $Ni(2p)$ region at ≈ 857 eV due to Ni^{2+}. This production of Ni-F bonds upon evaporation is consistent with previous Ni/PTFE investigations [24, 25]. Upon X-ray irradiation, the metallic Ni peak decreased relative to the Ni-F band (Figure 3) while in the $F(1s)$ region, the F^- band increased relative to CF_x ($x = 0$–3) indicating the production of NiF_2.

The reactivity of Fe with PTFE has been shown to lead to FeF_2 formation [16]. The effect of post-metallization X-ray irradiation on the Fe/PTFE system has been examined by irradiating the sample at 300 W concurrently with data collection (Figure 4). The lower two traces show the effect of post-metallization X-ray irradiation following 22% Fe deposition (where Fe^{2+} dominates) before and after 16 hours irradiation (Figure 4(a)). The $Fe(2p)$ region is dominated by Fe^{2+} and in the $F(1s)$ region by F^- (not shown) at ≈ 686 eV. Irradiation resulted in the decrease in

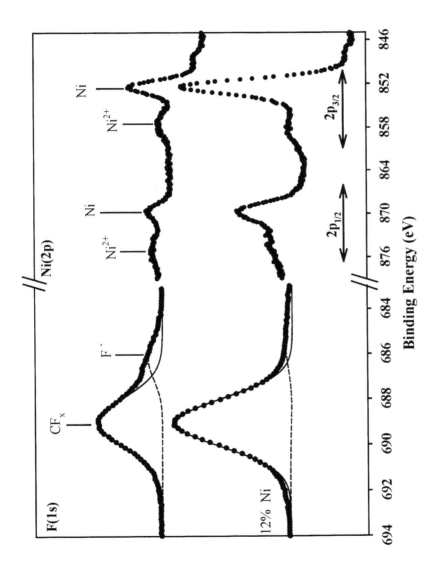

Figure 3. Variation in the F(1s) and Ni(2p) regions before (lower spectra) and after (upper spectra) 23 hours of 300 W X-ray irradiation on the Ni/PTFE system. Raw XPS data are shown as filled circles (●). In the F(1s) region, the CF$_x$ and F$^-$ species are shown as solid (−) and dashed lines (---) respectively.

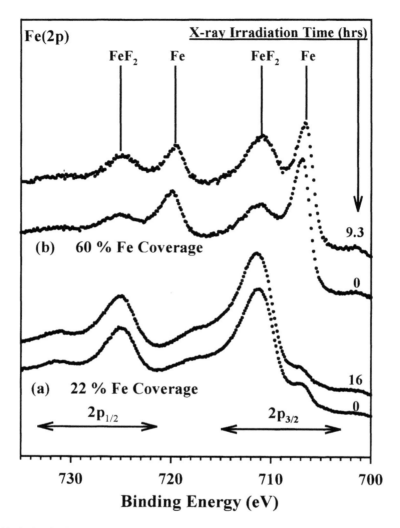

Figure 4. Variation in the Fe(2p) region as a function of X-ray irradiation at 300 W in the regimes where (a) FeF_2 (lower panel) and (b) metallic iron (upper panel) are the dominant species at the Fe/PTFE interface. All spectra have been normalized to the total Fe(2p) photoelectron intensity.

the intensity of the CF_x (x = 0–3) band at 689 eV (not shown) and a small decrease in the metallic peak in the Fe(2p) region. Figure 4(b) shows the effect of X-ray irradiation of a much thicker overlayer (60% coverage of PTFE by Fe) where metallic iron dominates. In this case, there was a significant loss of metallic iron and a concomitant production of Fe^{2+}. In the C(1s) region (not shown) CF-C and C-CF_n intermediate species were observed at all Fe coverages between 288–286 eV along with CF_3 groups (293.0 eV).

4. DISCUSSION

4.1. Metallization

The results obtained in this investigation indicate that thermal evaporation of Au, Ni and Fe on PVC results in chemical reaction (reactive metallization) leading to the formation of the lowest stable valence metal chloride (Table 1). In contrast, on PTFE, only Ni and Fe form metal fluoride salts upon thermal evaporation. At higher exposures, metallic overlayers are observed for all the metal/polymer systems studied in the present investigation.

To rationalize the different reactivities of various metals on polymers during thermal evaporation, thermodynamic arguments have been previously invoked based on different metal condensation energies and relative carbon-halogen bond strengths [7]. These arguments, however, isolate just one contribution to the overall energetics. Thus, the use of condensation energy alone as a measure of relative reactivity does not provide information about the nature and strength of chemical bonds within the polymeric substrate. Au for example, whose condensation energy is greater than either Cu or Ag, has a lower reactivity than either Cu or Ag on both PVC and PTFE (Table 1). This highlights the fact that any empirical relationship used to determine relative reactivity trends needs to consider both bond breaking and forming reactions that must accompany the formation of new chemical compounds during metal evaporation.

Specifically, during metallization, a vapor-phase metal atom (M) can either condense onto the polymer or undergo chemical reaction (reactive metallization). In its simplest form, reactive metallization with either PVC or PTFE can be repre-

Table 1.
Thermodynamic terms associated with metallization of PVC (BDE (C-Cl) = 327 kJ mol^{-1}) and PTFE (BDE(C-F) = 485 kJ mol^{-1}) compiled from reference [34]. The metals are placed in decreasing order of reactivity. The observed metal halides are in bold type [16, 17, 19, 23]

Metal	Condensation energy ΔH_c (kJ mol^{-1})	Metal chloride	BDE M-Cl (kJ mol^{-1})	Metal fluoride	BDE M-F (kJ mol^{-1})
Ti	473	**TiCl$_2$** (TiCl$_3$) (TiCl$_4$)	504.6 (460.2) (429.3)	**TiF$_3$** (TiF$_4$)	~ 700 584.5
Fe	415	**FeCl$_2$** (FeCl$_3$)	400 (341.4)	**FeF$_2$** (FeF$_3$)	481 (456)
Ni	425	**NiCl$_2$**	370.7	**NiF$_2$**	462.3
Cu	335	**CuCl** (CuCl$_2$)	360.7 (293.7)	CuF$_2$	365
Ag	290	**AgCl**	314	AgF (AgF$_2$)	348.9 (268.9)
Au	355	**AuCl**	289	AuF	305

sented by the following reaction [17]:

$$M_{(gas)} + X\text{-}C_{(solid)} \rightarrow M^+X^-_{(solid)} + C_{(solid)} \; (X = F \text{ or } Cl)$$

Scheme 1. Proposed reaction scheme for reactive metallization.

Consequently, as long as the fate of the carbon containing fragment ($C_{(solid)}$) is the same for a common metal thermally evaporated onto different halogen-containing polymers (e.g. PVC and PTFE), it should be possible to rationalize differences in metal-polymer reactivity from the thermodynamics associated with the bond breaking (C-F or C-Cl) and forming (M-F or M-Cl) reactions that accompany reactive metallization. Scheme 1 illustrates that if we assume that the bond breaking and forming reactions are constant within a particular polymer, the heat of reaction associated with reactive metallization will depend on the difference between the bond dissociation energy (BDE) of M-X and C-X (ΔH_{BDE} = BDE(M-X)-BDE(C-X)). Table 1 shows that ΔH_{BDE} does indeed provide a useful diagnostic for a metal's overall reactivity with PTFE and PVC, with a high degree of reactivity favored when ΔH_{BDE} is large and positive. Conversely, in less reactive metal/polymer systems, ΔH_{BDE} is small or negative. Table 1 also illustrates that for a common polymer, the reactivity trend within a series of metals scales with the magnitude of BDE(M-X). For example, Figure 1 illustrates that although both Fe and Au are reactive with PVC, the extent of reaction is much more pronounced for Fe (BDE(Fe-Cl) = 400 kJ mol^{-1}) compared to Au (BDE (Au-Cl) = 289 kJ mol^{-1}) (Figure 1).

The thermodynamic argument derived from Scheme 1 can also rationalize the fact that in situations where different halides (e.g. FeCl$_2$ vs FeCl$_3$) can be formed, the halide with the highest BDE(M-Cl) value is produced. Similarly, Table 1 shows that upon moving between PVC and PTFE the change in BDE(C-X) (X = Cl, F) is always greater than the corresponding change in BDE(M-X). This is responsible for the generally lower reactivity of metals during thermal evaporation on PTFE compared to PVC; for example, the formation Au and Cu halides on PVC only (Table 1).

In a more general sense, the success of ΔH_{BDE} in predicting reactivity trends implies that the extent of reaction is sensitive to the heat released during reaction. This effect could be a consequence of the large increase in surface temperature at the polymer surface during evaporation anticipated in situations when ΔH_{BDE} is large. An increased surface temperature is expected to correlate with greater morphological changes at the polymer surface that are, in turn, expected to enhance metal diffusion, leading to a greater extent of reaction. This is supported by the fact that in related studies, morphological changes on PVC and PTFE as evidenced by AFM, were generally observed to be more pronounced during metallization with more reactive metals (Ti and Fe) compared to more noble metals (Ni, Cu, and Au) [16, 19, 23].

Another potential explanation for the correlation between ΔH_{BDE} and the extent of reaction relates to the possible activation of the polymer surface at high surface temperatures. Evidence of polymer surface activation at high temperatures during metallization has been observed in our laboratory by the formation of CuF_2 during Cu deposition on a PTFE surface that was indirectly heated by the evaporation source in the vacuum chamber. In contrast, Cu thermally evaporated on PTFE at room temperature was chemically inert [12, 26]. Similarly, studies of the reactions between perfluoropolyalkylether (PFPAE) and iron surfaces have shown that FeF_2 production increases with temperature [27].

4.2. Post-metallization surface modification by X-ray irradiation

Figures 2–4 show that post-metallization treatment using X-ray irradiation can be used to enhance metal halide formation at the metal/polymer interface in some metal/polymer systems. For example, post-metallization X-ray irradiation of Fe/PVC or Fe/PTFE systems resulted in an increased concentration of Fe^{2+} and a loss of metallic Fe (Figure 4). In contrast, the Au/PVC and Au/PTFE systems were inert to the effect of post-metallization X-ray irradiation (not shown), while an increase in the amount of nickel halide was observed only in the Ni/PTFE system (Figure 3).

4.2.1. Mechanistic implications

In principle, the conversion of metal to the metal halide during post-metallization could be a result of X-ray mediated metal diffusion into the polymer or as a result of reaction between the metallic overlayer and reactive chlorine/fluorine species generated by the X-ray irradiation process. Alternatively, a combination of these two processes could also be occurring. The attenuation of metallic signal upon X-ray irradiation was observed by Shi *et al.* for Cr, Ti, Al, Ag, Au and Cu metals on Teflon-PFA and FEP during post-metallization X-ray irradiation, although no additional chemical interaction in the metal/polymer interfacial region was observed [15]. The reduction in metallic signal in this case was attributed to metallic diffusion into the bulk.

The anticipated lack of mobility associated with metals at room temperature, however, leads us to conclude that the increase in metal halide content observed during post-metallization X-ray irradiation is principally due to the migration of halide species through the metal/polymer interfacial region and subsequent reaction with the metallic overlayer. The clearest evidence for the importance of X-ray irradiation in generating reactive halogen species from reactions within the polymeric substrate is evidenced in the Cu/PTFE system [17]. In this case, no reaction was observed during initial thermal evaporation although CuF_2 was formed as a result of subsequent post-metallization X-ray irradiation. Given the lack of reactivity associated with copper evaporation on PTFE, increased Cu diffusion resulting from post-metallization alone would not be expected to initiate Cu-F bond formation.

The production of metal halides by X-rays is postulated to be a consequence of the secondary electrons produced from X-ray-substrate interaction [28, 29]. These secondary electrons initiate C-X (X = Cl, F) bond breaking in the polymer, producing "active" halogen species, including F⁻ and Cl⁻ which can migrate into the metal/polymer interface. Subsequent reactions between halogen-containing free radicals e.g. $-\dot{C}F-$, $-CH_2-\dot{C}H-$ as well as ionic species with the metal lead to the production of MX_n (X = Cl, F; n = 1 or 2) (Scheme 2).

$$R\text{-}X + e^- \rightarrow R^{\bullet} + X/X^- \ (R = -\dot{C}F-, -CH_2-\dot{C}H-, X = F, Cl)$$

$$n\ X/X^- + M \rightarrow MX_n \ (n = 1 \text{ or } 2)$$

Scheme 2. X-ray induced metal halide production.

The results shown in Figures 2–4 also illustrate that the conversion of nickel and iron to MX_n (X = Cl or F: n = 1 or 2) during post-metallization X-ray treatment of the PVC and PTFE system does not proceed to completion. One possible explanation for this behavior is that the metal halide salts formed (Scheme 2) inhibit the diffusion of active halogen species to the metallic overlayer. As the thickness of the MX_n interlayer increases, the rate of metal halide formation will decrease and the maximum concentration of metal halide that can be produced will be limited. The importance of a metallic overlayer in post-metallization X-ray irradiation is also illustrated in Figure 4. This clearly shows that in the absence of a metallic overlayer little change is observed in the chemical composition of the metal/polymer interface. In contrast, significant changes are evidenced in the presence of a metallic overlayer.

Based on a comparison of the results obtained in the present investigation and those of previous studies [16, 17, 19], it appears that the effects of both X-ray and Ar⁺ ion post-metallization treatment strategies are sensitive to the chemical nature of both the metal and the polymer. For example, the Ni/PTFE system is sensitive to post-metallization X-ray irradiation, while the Ni/PVC system is relatively unaffected. Results of other related studies in our laboratory on the Ti/PTFE system indicate that metallic titanium overlayers can be almost completely converted from Ti to TiF_3 as a result of post-metallization X-ray treatment [19]. This illustrates the fact that, in general, reactive metals such as Fe and Ti are most susceptible to modification under the influence of post-metallization X-ray treatment.

Kinetic control may also be important in certain systems where the production of metal fluoride is an activated process requiring post-metallization modification by X-rays or energetic ions. In this context, it should be noted that CuF_2 formation has been observed during sputter deposition onto another fluoropolymer, Teflon AF1600 [30]. The kinetic energy of the copper atoms, however, (several thousand kJ mol⁻¹) is much larger than when Cu is thermally evaporated, and is of sufficient magnitude to activate C-F bond cleavage within the fluoropolymer.

The results from this investigation show that X-ray modification of the Ni/PFTE and Fe/PTFE interfaces can increase the extent of compound formation in the metal/polymer interfacial region. Based on the empirical observation [12, 31-33] that compound formation in the metal/polymer interfacial region is correlated with improved adhesion characteristics, the effect of post-metallization X-ray irradiation in these systems may have technological significance. It should be pointed out, however, that any strategy for increasing the interlayer thickness derived from X-ray induced modification is only viable if the metallic overlayer is less than the X-ray penetration depth.

5. CONCLUSIONS

The reactivity of Fe, Ni and Au was investigated on PVC and PTFE as well as the effect of post-metallization X-ray irradiation. On PVC, Fe, Ni and Au were all found to be reactive, although the extent of reaction varied, while on PTFE only Fe and Ni were reactive. These results, as well as those from other metal/polymer systems are rationalized in terms of the difference between M-X (X = F, Cl) and C-X bond strengths. Post-metallization modification by X-rays was found to result in the formation of metal chloride at the Fe/PVC interface and metal fluoride at the Fe/PTFE and Ni/PTFE interfaces. The enhanced metal halide production during post-metallization X-ray irradiation is postulated to arise from the creation of active chlorine/fluorine species in the polymer and their subsequent reaction with the metallic overlayer rather than an effect associated with metal migration into the bulk.

Acknowledgements

Support for this research was provided by a National Science Foundation CAREER award (# 9985372). This work was carried out in the surface analysis laboratory at Johns Hopkins as part of the Materials Research Science and Engineering Center, funded through the National Science Foundation.

REFERENCES

1. R.L. Opila, K. Konstadinidis and S.O. Conner, in *Polymer Surfaces and Interfaces: Characterization, Modification and Application*, K.L. Mittal and K.-W. Lee (Eds.), p. 179, VSP, Utrecht (1997).
2. V. Zaporojtchenko, T. Strunskus, K. Behnke, C.V. Bechtolsheim, M. Kiene and F. Faupel, *J. Adhesion Sci. Technol.*, **14**, 467 (2000).
3. E. Sacher, J.J. Pireaux and S.P. Kowalczyk (Eds.), *Metallization of Polymers, ACS Symp. Ser. 440*, Amer. Chem. Soc., Washington, DC (1990).
4. K.L. Mittal (Ed.), *Metallized Plastics 2: Fundamental and Applied Aspects*, Plenum Press, New York (1991).
5. K.L. Mittal and J.R. Susko (Eds.), *Metallized Plastics 1: Fundamental and Applied Aspects*, Plenum Press, New York (1989).

6. T. Strunskus, M. Grunze, G. Kochendoerfer and C. Woll, *Langmuir*, **12**, 2712 (1996).
7. E. Sacher, *Prog. Surf. Sci.*, **47**, 273 (1994).
8. M. Du, R.L. Opila, V.M. Donnelly, J. Sapjeta and T. Boone, *J. Appl. Phys.*, **85**, 1496 (1999).
9. M. Du, R.L. Opila and C. Case, *J. Vac. Sci. Technol. A*, **16**, 155 (1998).
10. G. Maier, *Prog. Polym. Sci.*, **26**, 3 (2001).
11. C.-A. Chang, Y.-K. Kim and A.G. Schrott, in *Metallization of Polymers, ACS Symp. Series No. 440*, E. Sacher, J.-J. Pireaux and S.P. Kowalczyk (Eds.), p. 416, Amer. Chem. Soc., Washington, DC (1990).
12. C.-A. Chang, Y.-K. Kim and A.G. Schrott, *J. Vac. Sci. Technol. A.*, **8**, 3304 (1990).
13. C.-A. Chang, Y.-K. Kim and A.G. Schrott, *J. Appl. Phys.* **67**, 251 (1990).
14. L. Wang, N. Angert, C. Trautman and J. Vetter, *J. Adhesion Sci. Technol.*, **9**, 1523 (1995).
15. M.-K. Shi, B. Lamontagne, L. Martinu and A. Selman, *J. Appl. Phys.*, **74**, 1744 (1993).
16. S.R. Carlo, A.J. Wagner and D.H. Fairbrother, *J. Phys. Chem. B.*, **104**, 6633 (2000).
17. C.C. Perry, J. Torres, S.R. Carlo and D.H. Fairbrother, *J. Vac. Sci. Technol. A*, **20**, 1690 (2002).
18. N. Fairley, CASA-XPS, 1.0, Casa Software Ltd (2000).
19. S.R. Carlo, C.C. Perry, J. Torres, A.J. Wagner, C. Vecitis and D.H. Fairbrother, *Appl. Surf. Sci.*, **195**, 93 (2002).
20. G.E. Muilenberg, *The Handbook of X-ray Photoelectron Spectroscopy*, Perkin Elmer Corporation, Eden Prairie, MN (1979).
21. K. Kishi and S. Ikeda, *J. Phys. Chem.*, **78**, 107 (1974).
22. J.C. Carver and G.K. Schweitzer, *J. Chem. Phys.*, **57**, 973 (1972).
23. S.R. Carlo, C.C. Perry, J. Torres and D.H. Fairbrother, *J. Vac. Sci. Technol. A*, **20**, 350 (2002).
24. D.R. Wheeler and S.V. Pepper, *J. Vac. Sci. Technol.*, **20**, 442 (1982).
25. T.C.S. Chen and S.M. Mukhopadhyay, *J. Appl. Phys.*, **78**, 5422 (1995).
26. A.J. Pertsin and Y.M. Pashunin, *Appl. Surf. Sci.*, **47**, 115 (1991).
27. G. John, J.S. Zabinski and V.K. Gupta, *Appl. Surf. Sci.*, **93**, 329 (1996).
28. D.R. Wheeler and S.V. Pepper, *J. Vac. Sci. Technol.*, **20**, 226 (1982).
29. D.T. Clark and W.J. Brenan, *J. Electron. Spectr. Rel. Phenom.*, **41**, 399 (1986).
30. D. Popovici, J.E. Klemberg-Sapieha, G. Czeremuzkin, E. Sacher, M. Meunier and L. Martinu, *Microelectron. Eng.*, **33**, 217 (1997).
31. M.K. Shi, A. Selmani, L. Martinu, E. Sacher, M.R. Wertheimer and A. Yelon, *J. Adhesion Sci. Technol.*, **8**, 1129 (1994).
32. J.M. Burkstrand, *J. Appl. Phys.* **52**, 4795 (1981).
33. Y. Fujinamia, H. Hayashi, A. Ebe, O. Imai and K. Ogata, *Materials Chemistry and Physics*, **54**, 102 (1998).
34. J.E. Huheey, *Inorganic Chemistry*. Harper & Row, New York (1983).

Polyimides and Other High Temperature Polymers, Vol. 2, pp. 359–387
Ed. K.L. Mittal
© VSP 2003

Plasma polymer adhesion promoters for metal-polymer systems

J. FRIEDRICH,* G. KÜHN, R. MIX, I. RETZKO, V. GERSTUNG,
ST. WEIDNER, R.-D. SCHULZE and W. UNGER

Bundesanstalt für Materialforschung und -prüfung (BAM) / Federal Institute for Materials Research and Testing, Unter den Eichen 87, D-12205 Berlin, Germany

Abstract—The retention of chemical structure and functional groups during plasma polymerisation was investigated. Usually plasma polymer layers, prepared by continuous wave radio-frequency plasma, are often chemically irregular in their structures and chemical compositions. To minimise these irregularities, low wattages and the pulsed plasma technique were applied to avoid fragmentations. The polymerisation of vinyl and acryl-type monomers was strongly enhanced in the dark phase (plasma-off) of a pulsed r.f. plasma caused by the reactivity of the vinyl or acryl-type double bonds. Bifunctional monomers with acryl or allyl double bonds and also polar groups such as OH, NH_2, and COOH were used to produce plasma polymers with defined (regular) structures and a high density of a single type of functional groups. The maximum yields were 30 OH, 18 NH_2, 24 COOH groups per 100 C atoms. To vary the density of functional groups a chemical copolymerisation with "chain-extending" comonomers such as butadiene and ethylene was initiated in the pulsed plasma. The composition of these copolymers was investigated by XPS and IR spectroscopy.

Homopolymers and copolymer layers were deposited on polypropylene (PP) foils and then aluminium was thermally evaporated. The peel force increased considerably and showed a dependence on the density of functional groups. The plasma polymer deposition was also monitored *in situ* by the Self-Exciting Electron Resonance Spectroscopy (SEERS) to show correlations between plasma parameters and properties of the deposited plasma polymer layers measured "*quasi-in situ*" by coupling the plasma chamber with an XPS spectrometer.

Keywords: r.f. pulsed plasma; plasma polymers with functional groups; copolymers; adhesion promoting interlayers; metal-polymer systems.

1. INTRODUCTION

Acetylene, ethylene, butadiene and polystyrene have been deposited as thin polymer films by pulsed plasmas of low wattages as described elsewhere [1]. It is known that acetylene can react by opening the triple bond. A substitution of H at the triple bond by organic residues was observed [1]. Ethylene does not form a

---

*To whom all correspondence should be addressed. Phone: 0049 30 8104 1630,
Fax: 0049 30 8104 1637, E-mail: joerg.friedrich@bam.de

plasma polymer in the continuous wave plasma that is comparable to chemically polymerised polyethylene [2-14]. The different structure of plasma polyethylene can be illustrated by the absence of the $\rho(CH_2) \geq 4$ vibration, which is characteristic for $(CH_2) \geq 4$ sequences in commercial polyethylene [15]. The butadiene plasma polymerisation has not been so intensively studied, however, a highly branched and crosslinked polymer was found to be formed. Styrene polymerisation needs a low activation energy of < 1 eV to form a regularly structured polymer by an ordinary radical polymerisation. However, in the continuous wave plasma, especially on exposure to high-energy doses (high wattage), an undesirable fragmentation of monomers in the plasma is observed. Moreover, the deposited plasma polymer layer is exposed to high vacuum-uv doses which additionally crosslink the polymer and form additional radical sites in the polymer.

A pulsed low-energy plasma should help in avoiding the monomer fragmentation by enhancing the pure chemical radical polymerisation in the gas phase. Pulsed plasma polymerisation was first introduced by Tiller in 1972 [16], later continued by Yasuda [17], Shen and Bell [18, 19] and then further developed by Timmons [20, 21]. Theoretically, one short plasma pulse should be enough to start the radical chain polymerisation. However, the number of monomers sticking at radical sites of the growing chains is limited because of the low pressure (1–20 Pa). Therefore, the radical chain reaction is disrupted and disproportionation (or a radical-radical recombination or a reaction with oxygen molecules from residual gas) occurs. Therefore, pulsed plasma with long plasma-off periods and short plasma pulses offers a good compromise to produce polymer structures with a minimum of irregularities which are produced during the plasma-on periods. In Fig. 1 the theoretically expected structures of pulsed plasma (homo) polymerised bifunctional monomers with OH, NH$_2$ or COOH groups are shown.

The reactivity of monomers is very different. The most active monomers, characterised by small activation energies for starting the radical polymerisation, are vinyl and acryl compounds, and less active are allyl monomers. Dienes are also of high activity, and olefinic double bonds are of lower activity. It should be noted

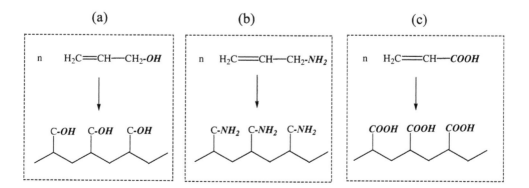

Figure 1. Structures of pulsed plasma polymerised allylalcohol (a), allylamine (b), and acrylic acid (c).

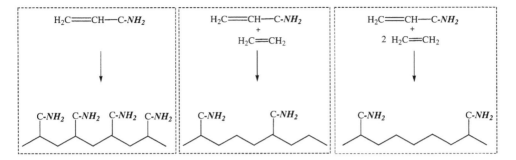

Figure 2. Scheme of pulsed plasma copolymerisation of allylamine with ethylene.

that non-activated olefinic, acetylenic and aromatic multiple bonds are similar in their (low) reactivity and do not graft spontaneously in the plasma-off period. The ability of radical polymerisation also depends on the reactivity or stabilisation of the radical intermediate; especially, resonance stabilisation can hinder the chain propagation.

Therefore, this process can be considered as a pulsed-plasma initiated radical copolymerisation of two organic monomers in the gas phase. Using monomers with functional groups and linear olefins or dienes the density of functional groups should be adjustable in a broad range (Fig. 2). Both the kinetics and the structures of copolymers were investigated. This predominantly chemical co-polymerisation should be distinguished from a non-chemical "plasma copolymer-isation". This plasma copolymerisation does not consider the chemical reactivity of comonomers. Here, the copolymerisation of chemically reactive or unreactive comonomers is enforced. Using the continuous wave mode and high wattage the comonomers are completely fragmented. Then, a random recombination to irregu-larly formed copolymers follows as described by German/Austrian scientists in the 1950's or by Yasuda in the 1970's [22-30].

In Table 1 the reactivities of functional groups bearing (bifunctional) and chain-extending comonomers are listed. It was predicted that a similar reactivity of co-monomers was necessary to form copolymers over the full range of possible co-monomer mixtures. Otherwise, a homopolymerisation of the more reactive compo-nent would occur. Problems are introduced by the side reactions of both the bifunctional monomers and the chain-extending monomers. Bifunctional monomers may split off their functional groups and the acetylenic and olefinic monomers may become branched, crosslinked and oxidised as described before. Such copolymers were deposited on polypropylene (PP) foils without and with oxygen plasma treat-ment. This plasma pre-treatment of PP should improve the adhesion to the plasma polymer layer. Using OH, NH_2 and COOH functionalised layers as adhesion promoting layers and varying the density of functional groups, a specific dependence of peel strength of Al-PP-systems was expected. Here, OH functionalised surfaces were investigated. Allylalcohol and ethylene or butadiene were used in different molar ratios to produce copolymers with different densities of OH groups.

Table 1.
Proposed reactivity of comonomers in radical gas phase polymerisation during the plasma-off period in a pulsed plasma polymerisation

	Chemical reactivity	Side reactions
Functional group bearing monomers		
Allylalcohol	moderate	low (formation of >C=O)
Allylamine	moderate	oxidation
Acrylic acid	high	COOH loss
Acrylonitrile	high	Reactions at CN
"Chain-extending" monomers		
Acetylene	very low	crosslinking
Ethylene	low	low
1,3-butadiene	moderate	crosslinking
Styrene	high	low

The plasma activation of monomers was monitored during the plasma pulses by time-resolved SEERS (Self-Exciting Electron Resonance Spectroscopy).

2. EXPERIMENTAL

2.1. Plasma polymerisation

Plasma polymerisations were performed in a vacuum system with a base pressure of 10^{-3} Pa or lower. The principal design of the plasma reactor was described earlier [1]. The pumping system comprised an oil-free membrane pump connected to a turbomolecular pump. A pulsable r.f. generator with an automatic matching unit was used to ignite the plasma. The duty cycles (plasma-on [ms] / plasma-off [ms] + plasma-on [ms]) were varied from 0.1 to 1 with frequencies from 10 to 10^5 Hz. The power input was varied from a few Watts to 300 W. The effective power applied was calculated by taking into account the plasma-on time. Therefore, it was possible to realize, e.g., an effective input power lower than 1 W.

The plasma deposited layers were usually analysed "*ex situ*", i.e. after exposure of the plasma polymer films to ambient air. Usually the stability of deposited polymer layers produced by the pulsed plasma was high enough to allow a fast transfer of samples (i.e. within 20 minutes) to XPS or IR spectrometers without changes in their chemical composition and structure. However, with XPS "*in situ*" experiments were preferred to avoid such exposure to air.

Monomers used in homo and copolymerisation were allylalcohol, allylamine and acrylic acid as carriers of different functionalities and 1,3-butadiene, hexadiene, ethylene and styrene as "chain-extenders". All gases and monomers used in this study had the highest commercially available purity. Liquid monomers were introduced into the reactor by means of a piston pump and a lamellar vaporizer. All liquids were carefully degassed. Monomer fluxes were measured by mass flow controllers.

2.2. Surface analysis

The XPS data acquisition was performed with a SAGE 150 Spectrometer (Specs, Berlin, Germany) using non-monochromatized MgK_α or AlK_α radiation with 12.5 kV and 250 W settings at a pressure 10^7 Pa in the analysis chamber. This instrument is equipped with a plasma reactor separated by a gate valve from the UHV system, where surface treatments can be carried out at 10^1–10^{-7} Pa. XPS spectra were acquired in the constant analyser energy (CAE) mode at 90° take-off angle. Peak analysis was performed using the peak fit routine from Specs.

Near Edge X-ray Absorption Fine Structure (NEXAFS) and XPS measurements on *in situ* plasma treated surfaces have been described in detail elsewhere [31-33]. The plasma polymerisation was performed in a UHV preparation chamber connected to the spectrometer. The FTIR spectra were recorded with a Magna 550 instrument (Nicolet, USA) using the ATR technique with Ge crystals or a diamond cell (Specac, Kent, UK). IR-GIR (Grazing Incidence Reflectance) measurements were performed on a GIR unit (Specac, Kent, UK) where the angle of incidence was variable. Most often these IR measurements were made at an angle of incidence of 70°. The GIR method was convenient and helped us to avoid influences of moisture, as shown by parallel measurements with plasma polymer layers on KBr or NaCl.

2.3. Analysis of plasma polymer layers

2.3.1. ThFFF analysis

Molecular masses higher than ca. 500,000 Da and their distribution were determined by Thermal Field Flow Fractionation (ThFFF, for details see [34]). Additionally, slightly crosslinked polymer particles (gels) with the radius of gyration up to ca. 100 µm were analysed. This range of radii corresponds to molar masses of 10^5 to $\geq 10^{12}$ Da. The separation principle is based on the different thermal diffusion behaviours of molecules and particles between a high temperature gradient. One side of the channel is heated up to 150°C and the other side is cooled. It was possible to analyse all pulsed plasma polymerised vinyl, acrylic and allyl monomers without filtering. A ThFFF apparatus supplied by Wyatt Technology (Hamburg, Germany) including Refractive Index (RI), UV, and Multi-Angle Laser Light Scattering (MALLS) detectors as well as an Evaporative Light Scattering Detector (ELSD) was used in these experiments.

2.3.2. AFFFF analysis

In contrast to the separation mechanism used in ThFFF, the Asymmetric Flow Field Flow Fractionation (AF^4, also called Crossflow FFF) uses a (second) cross flow of the solvent. Particles, suspensions, colloids/gels and molecules up to 100 µm in diameter are separated. Due to the cross flow through a membrane, low-molar mass impurities are separated from the polymer solution prior to analysis. An instrument from Postnova Analytics (Munich, Germany) was used for the AF^4 experiments. Refractive index (RI), UV, Multi-Angle Laser Light Scattering (MALLS) and an Evaporative Light Scattering Detector (ELSD) were used as detectors.

2.3.3. TG/DTG and IR analyses

Thermogravimetric measurements were performed with a DTA 220 balance from Seiko Instruments (USA). IR spectra were recorded on a Vector 22 instrument, made by Bruker (Karlsruhe, Germany), or on a Nexus 550 instrument made by Nicolet (Offenbach, Germany).

2.4. Labelling of surface functional groups

2.4.1. Derivatization of OH groups

Samples with surface OH groups were exposed to trifluoroacetic anhydride (TFAA) vapour for 10–15 min in order to react as follows:

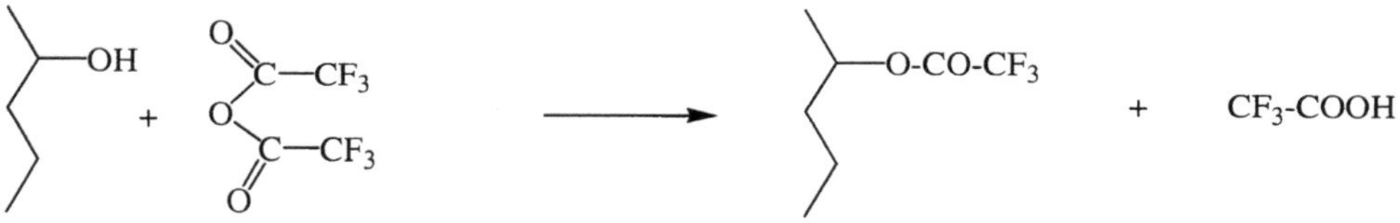

Afterwards these samples were degassed at a pressure of 10^{-5} Pa to remove the produced trifluoroacetic acid [35, 36]. The derivatization of OH groups proved to be highly selective (ca. 90% [36]). F1s peak intensities in the XPS spectra as well as $>C=O$ (1785 cm^{-1}) and CF_3 (1223 cm^{-1}) stretching vibrations were used to evaluate the number of surface OH groups [35].

2.4.2. Derivatization of NH_2 groups

Primary amino groups react with pentafluorobenzaldehyde (PFBA) or trifluoro-methyl benzaldehyde (TFBA) by formation of a Schiff's base:

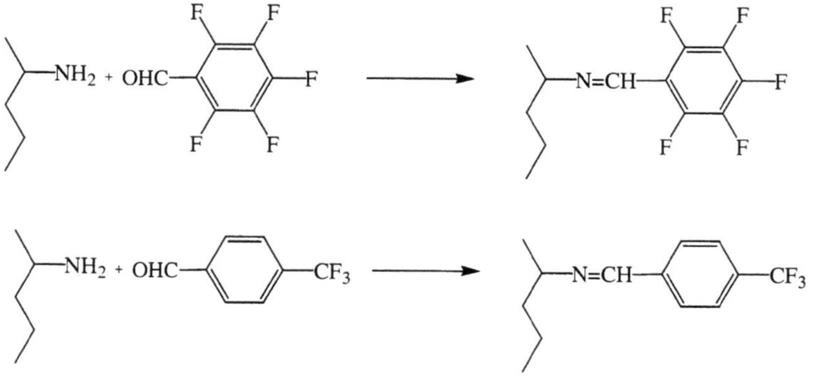

Similar to the OH derivatization, NH_2 labelling was performed as a gas-phase reaction. The samples were exposed to PFBA vapour for 20 h. Then the samples were degassed at a pressure of 10^{-5} Pa for 8 h to remove the unreacted PFBA. These labelling reactions were examined with 4,4'diaminodiphenylmethane as model and reached yields in the range of 80%, which are much higher than described in [37].

2.4.3. Derivatization of COOH groups
Carboxylic groups were identified by labelling with vapours of trifluoroethanol in presence of di-*tert*-butyldiimide and pyridine for 8 to 20 h [38]:

$$\text{—COOH} + \text{CF}_3\text{-CH}_2\text{-OH} \xrightarrow[\text{(CH}_3)_3\text{C-N=C=N-C(CH}_3)_3]{\text{C}_5\text{H}_5\text{N}} \text{—COO-CH}_2\text{-CF}_3$$

Afterwards the samples were degassed at a pressure of 10^{-5} Pa to remove non-reacted components. The yield of this reaction was about 90% (see also [38, 39]).

2.5. Self-exciting electron resonance spectroscopy (SEERS)

SEERS and its basic theory are connected with the use of a diode-type radio-frequency reactor. In such a reactor, the self-excited electron vibrations of the plasma depend specifically on electron density and collision rate and can be mathematically described in a theory (developed by ASI for the Hercules sensor system, ASI company, Berlin, Germany). The electron vibrations of the plasma were detected with a sensor in the wall of the reactor and analysed. The SEERS theory is based on the non-linearity of the space charge sheath at the r.f. electrode that provides harmonics with the modulated sheath width and produces high-frequency oscillations in the bulk plasma. The idea of the method is to use the high self-excited (high frequency) damped oscillations of the bulk plasma, which are generated at every peak voltage of the sinusoidal power supply. A computerized hydrodynamic model of the plasma bulk is used, which takes into account the non-linear behaviour of the modulated sheath width. Using this plasma theory it is possible to evaluate the above-mentioned plasma parameters by using the sensor signal. In contrast to Langmuir plasma probes and optical methods, the SEERS method is not influenced by the continuously growing plasma polymer deposit whether on the sensor or at the walls of the reactor. It would be expected that the electron density n_e and electron collision rate μ_e should be closely related to the generation of free radicals which are known to be necessary for further polymerisation reactions in the plasma-off period. The data acquisition time of less than 1 μs for one measurement allows monitoring the relevant plasma parameters also in pulsed plasmas.

3. RESULTS

3.1. Extended chain-building plasma polymers

The general aspects of pulsed plasma polymerisation of vinyl and acrylic monomers have been discussed elsewhere [1, 33, 40, 41]. In this communication the
polymerisation of styrene as a model for a chain extended plasma polymer is considered. Styrene as a vinyl monomer can easily undergo radical gas phase polymerisation in the plasma-off period. This behaviour is shown by the respective XPS
and NEXAFS spectra of pulsed plasma polymerised films (Figs. 3, 4). The spectra
of a commercial polystyrene standard sample and that obtained with pulsed
plasma polymerised styrene are almost identical. As known, the spectra represent
predominantly the primary chemical structure of both polymer materials. Detailed
assignments of peaks and resonance positions were reported in refs. [42, 43]. The
NEXAFS C K-edge spectrum of the pulsed plasma polymerised styrene is slightly
broadened due to branching and crosslinking in the film. Plasma polymers formed
from 1,3-butadiene present a NEXAFS spectrum different from that of the chemically polymerised butadiene. Here, the observed C1s→π* resonance can be assigned to non-saturated carbon species.

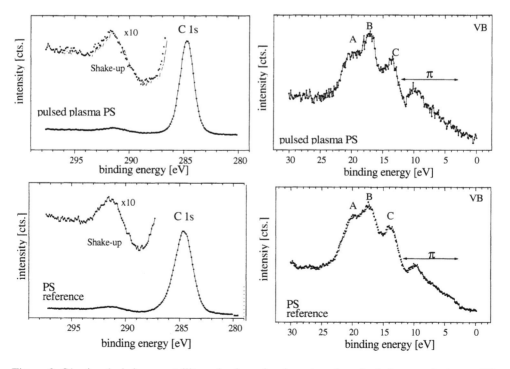

Figure 3. C1s signal, shake-up satellite and valence band spectra of a pulsed plasma polystyrene (30
W, 25 Pa, 10^3 Hz, duty cycle 0.1) and a commercial polystyrene as reference.

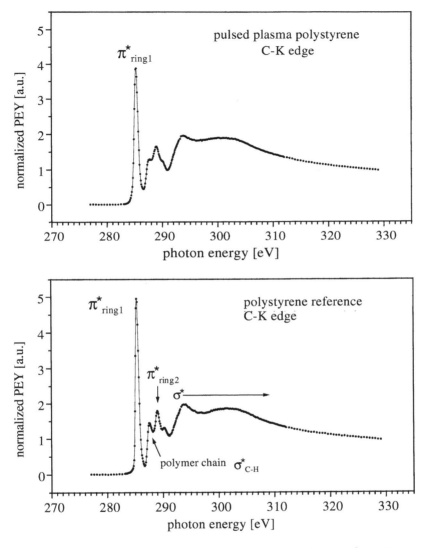

Figure 4. NEXAFS spectra of a pulsed plasma polystyrene (30 W, 25 Pa, 10^3 Hz, duty cycle 0.1) and a commercial polystyrene as reference (PEY–Partial Electron Yield mode).

Pulsed plasma polyethylene (PE) shows differences from both the pulsed plasma polymerised styrene and butadiene as well as from commercial PE. Looking at the valence band spectra, similarities to polypropylene were identified due to the hyperbranching of this polymer. As described earlier, differences between the plasma polymers and the reference materials were also obvious in the IR spectra [41, 42]. More than 30% loss in the absorbance of the vCH_2 and less than 5% in the aromatic ring vibrations and vCH_{arom} vibrations of the pulsed plasma PS

were observed in comparison to commercial PS [1]. Therefore, the most important irregularities in the structure of pulsed plasma polystyrene are localized in the
backbone. Other non-vinyl chain-extending monomers such as 1,3-butadiene and
ethylene show a large number of indications for irregular structures in their IR
spectra also using pulsed or low power plasmas. As described in Section 3.2 they
pick up a large number of oxygen molecules from the air because of the existence
of trapped C radical sites. Other structures can be identified at 3300 ($\nu\equiv$CH) and
2150 cm^{-1} (νC$\equiv$C), at 1900 cm^{-1} (cumulene) as well as at 3060–3160, 1600, 1500,
760, 700 cm^{-1} (aromatic ring formation). However, the pulsed mode produces a
polyethylene whose structure is slightly closer to that of commercial polyethylene. A (very small) ρCH$_2$ vibration at 720 cm^{-1} was detected for the first time, indicating the existence of at least two monomer units (ρCH$_2$ (n$\geq$4)) in the pulsed
plasma polyethylene. So far such a vibration, characteristic of amorphous PE had
not described in the literature for ethylene plasma polymers (cf. [7, 8]). There is a
remarkable difference between the pulsed plasma polymerisation of a vinyl
monomer (styrene) and an olefin (ethylene) based on calculated polymerisation
degrees. Styrene shows a calculated polymerisation degree of $\geq$ 25 (e.g. 2600 Da)
[40, 41] and ethylene of $\approx$ 2.

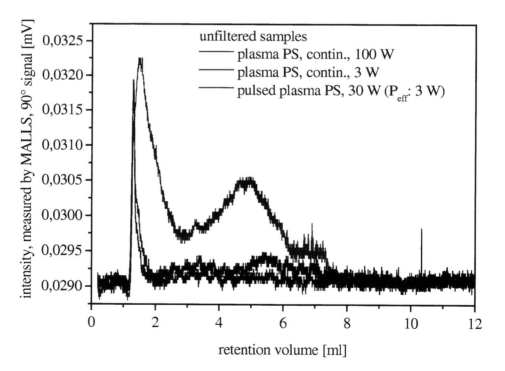

Figure 5. ThFFF elugrams of unfiltered continuous wave and pulsed plasma polymerised polystyrene (MALLS – Multi Angle Laser Light Scattering detector; EASICAL – mixture of polystyrene
standards for calibrating).

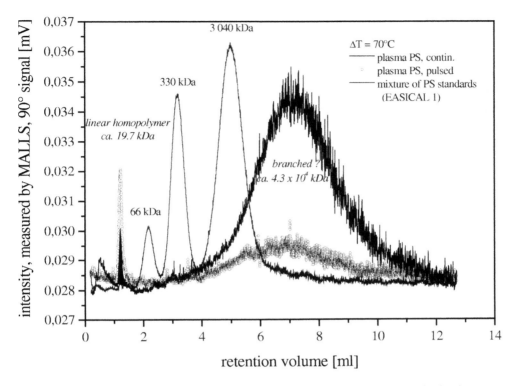

Figure 6. ThFFF elugrams of filtered continuous wave and pulsed plasma polymerised polystyrene (MALLS – Multi Angle Laser Light Scattering detector; EASICAL – mixture of polystyrene standards for calibrating).

These structural differences are represented in thermogravimetric results (TG and DTG). In contrast to the commercial reference polystyrene a very broad range of temperature is observed in which all plasma polymers have degraded. Moreover, the maximum degradation temperature in the DTG curves is shifted to lower temperatures by about 65 K. The range of degradation temperature of pulsed plasma PS is insignificantly smaller than that of the comparable continuous wave plasma PS.

The UV spectra of reference and plasma polymers differ considerably. The slightly yellow plasma polymer solutions absorb much more than commercial polystyrene. Subtracting the curve for pure polystyrene from that of the pulsed plasma polymerised polystyrene a continuous increase of absorbance with decreasing wavelength was observed. Here, in addition to the aromatic rings a large number of conjugated C=C double bonds with different conjugation lengths should exist.

The ThFFF and AFFFF provided more detailed information on molar mass and mass distribution of plasma polymerised styrene (Figs. 5–8). First, the unfiltered ThFFF elugram (also "fractogram") of plasma-polymerised polystyrene at 100 W

continuous wave plasma shows two large peaks (Fig. 5). One peak is situated in
the region of moderate molar mass and has high intensity and moderately broad
molar mass distribution. The second one is located in the very high molar mass
range and represents molar mass $> 10^7$ Da. The other species (P = 3 W continuous
wave and P_{eff} = 3 W pulsed plasma) show intense peaks in the low molar mass
range and diffuse, low intensity peaks in the very high molar mass range. After
calibration with a mixture of polymer standards as reference, the first peak in Fig.
6 can be set to about 19.7 kDa, attributed to linear homopolymers, and the second
one to about 4.3×10^4 kDa (see Fig. 6), due to highly branched or weakly
crosslinked molecules. From AFFFF the very high molar mass fraction cannot be
identified (Fig. 7). However, the low and moderate molar mass fractions are
clearly resolved. In the AFFFF elugram of plasma polymerised polystyrene at 100
W continuous wave plasma the first peak in the low molar mass range could be
split into two components at about 23 and 33 kDa. Furthermore, in the oligomer
fraction an additional low-molar mass fraction of a molar mass ≈ 6 kDa was identified (Fig. 8).

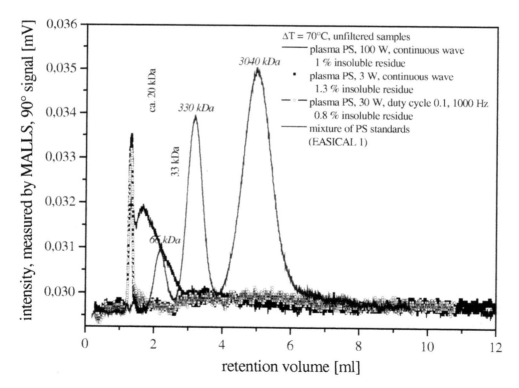

Figure 7. AFFFF elugrams of continuous wave and pulsed plasma polymerised polystyrene
(MALLS – Multi Angle Laser Light Scattering detector; EASICAL – mixture of polystyrene standards for calibrating).

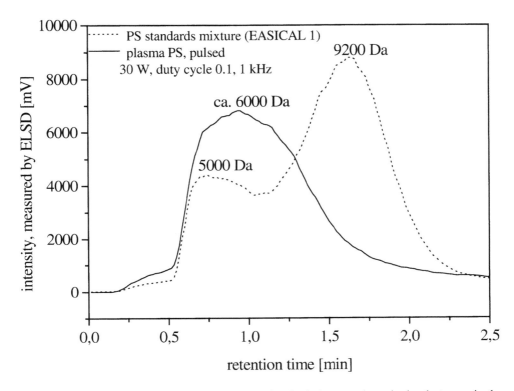

Figure 8. AFFFF elugrams of continuous wave and pulsed plasma polymerised polystyrene in the low-molar mass range (ELSD – Evaporative Light Scattering Detector) and polystyrene standards as reference.

3.2. Ageing behaviour

The plasma polymers from acetylene, ethylene, 1,3-butadiene and styrene pick up a large amount of oxygen from the air during storage in laboratory atmosphere (Fig. 9). This is characteristic of both the continuous wave and pulsed plasma polymerised polymers. The pulsed plasma produced polymers contain about 10% less oxygen than the continuous wave samples. In contrast to the relatively unreactive monomers such as acetylene, ethylene and butadiene both plasma-polymerised products of styrene show a very low tendency for oxygen chemisorption. After 1 month of ageing 15 to 20% oxygen is added in both the continuous wave and the pulsed plasma polymer layers. In situ XPS measurements (cf. Fig. 9) confirm that the oxygen is added successively only by post-plasma auto-oxidation reactions. This mechanism is based on the formation of peroxides/hydroperoxides by the addition of molecular oxygen to the trapped C radicals within the polymer layers [14]. The chemically dominated polymerisation mechanism of styrene (predominantly by radical gas phase polymerisation) is characterized by a small number of remaining radical sites after finishing the plasma polymerisation process. Therefore, a significantly lower post-plasma oxygen intro-

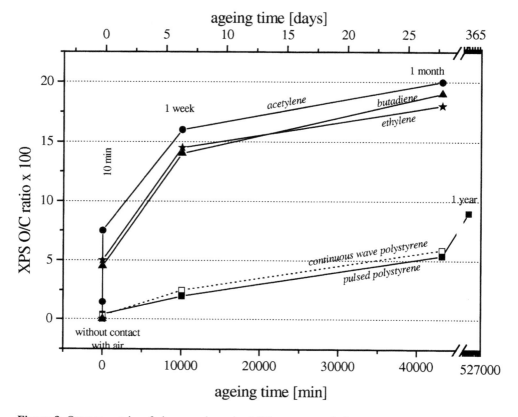

Figure 9. Oxygen uptake of plasma polymerised CH monomers during exposure to ambient air (liquid acetylene was stabilised with acetone, therefore, the resulting plasma polymer contains traces of oxygen under ultra-high vacuum conditions).

duction was observed. Moreover, the possibility of energy dissipation to the phenyl rings may also contribute to the smaller number of radical sites if the low-wavelength UV radiation is considered to be an important factor in C radical production. The non-vinyl monomers show different oxidation stages in the aged layers (primarily acids, ketones/aldehydes and alcohols) as found by fitting the respective C1s signals. In contrast, the pulsed plasma polystyrene shows only the existence of a subpeak at a binding energy of 286.5 eV (alcoholic OH group). After 1 year ageing only a small peak at a binding energy of 287.9 eV appears.

3.3. Plasma polymers bearing functional groups

Allylalcohol, allylamine and acrylic acid were pulsed plasma polymerised to retain the maximum number of functional groups in the resulting plasma polymers. The deposition rates indicated the existence of regular radical gas phase reactions (Table 2). As it is well known from the literature regarding the polymerisation of

Table 2.
Deposition rates of functional groups bearing monomers in pulsed plasma (30 W, duty cycle 0.1, 10^3 Hz, 12 Pa)

Monomer	Deposition rate in the continuous wave plasma (=100)	Deposition rate in the pulsed plasma referenced to that in the continuous plasma
Acrylic acid	100	105
Allylalcohol	100	65
Allylamine	100	65

Table 3.
Retention (yield) of functional groups at the surface of pulsed plasma polymer layers of acrylic acid, allylalcohol and allylamine (X=COOH, OH, NH_2)

Monomer	XPS measured elemental composition (%)	C1s peak fitting (%)	Derivatization results	
			(%)	X per 100 C
Acrylic acid	90 (O)	80	73	≈ 24
Allylalcohol	100 (O)	75	91	≈ 30
Allylamine	>100 (N)	100	55	≈ 18

acrylic and allyl monomers, acrylic acid reacts much faster than the allyl monomers. The respective pulsed plasma polymers were investigated for the retention or yield of functional groups. This retention was primarily measured by XPS including the chemical derivatization of these groups as described in Section 2 (Fig. 10). These results were checked by respective IR spectra. The calculation of the retention degree (yield) was based on the fact that the derivatization included also the bulk of thin layers, i.e. it exceeded the information depth of XPS. Only the PFBA derivatization depth was limited because of steric hindrance but it was more than the XPS information depth (≈ 5 nm). Derivatization and C1s peak fitting was used to calculate the yield of functional groups (Table 3). The plasma polymer from allylalcohol also shows a molar mass distribution in the range of high molar masses as measured by ThFFF (Fig. 11). Nevertheless this polymer was completely soluble in water and tetrahydrofuran.

The allylamine plasma polymer layers showed some ageing effect during storage in ambient air. The higher percentage of N in the freshly prepared polymer layers decreases during the storage to values lower than the expected stoichiometric content of N (Fig. 12). Moreover, a considerable incorporation of oxygen was measured (16 O per 100 C atoms after 40 days storage). The mechanism of the oxygen attack to the neighboured C atom is as follows [44-46]:

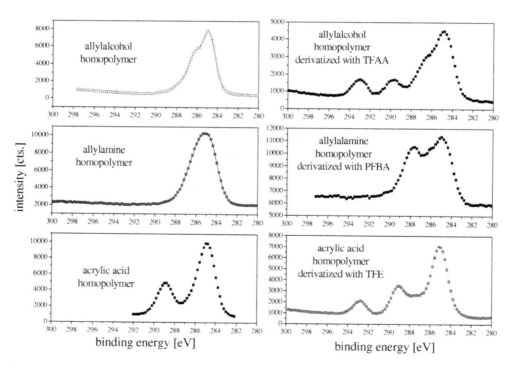

Figure 10. XPS C1s spectra of pulsed plasma polymerised monomers before and after derivatization of functional groups.

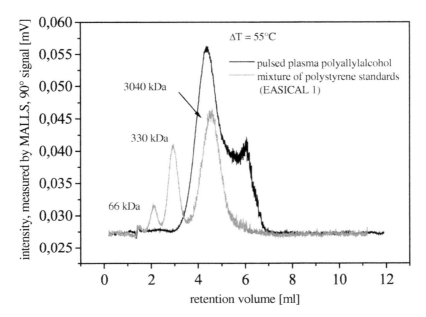

Figure 11. ThFFF elugrams of pulsed plasma polymerised allylalcohol and polystyrene standards as reference.

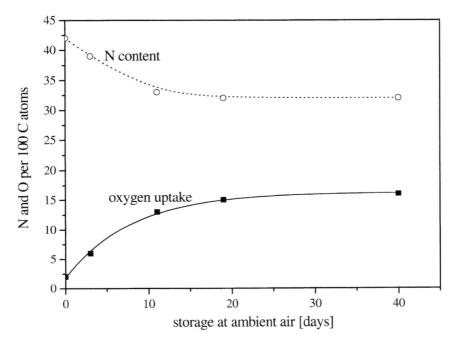

Figure 12. Oxidation of pulsed plasma polymers of allylamine during storage at ambient air as measured by XPS.

3.4. Plasma-activated chemical gas phase copolymerisation of vinyl, acryl, dienes and allyl monomers

The plasma-activated radical gas phase polymerisation as described before should also be applicable to produce copolymers with a variable density of functional groups with high selectivity (Fig. 2). The reaction principle of such copolymerisation was dominated again by the radical polymerisation in the plasma-off period of the pulsed plasma. Therefore, the copolymerisation requires similar chemical reactivities of both the functional groups bearing monomer and the "chain-extending" comonomer that involves at least one reactive double bond (cf. Table 1). Otherwise, a homopolymerisation was observed.

In Fig. 13 the deposition of different copolymerised mixtures of allylalcohol and ethylene are plotted as function of time. In Fig. 14, the deposition rates of these mixtures differ from a linear dependence. Such a non-linear behaviour is characteristic of comonomers with different reactivities (Fig. 14). With butadiene as a comonomer, a considerable lowering of deposition rates for all butadiene-allylalcohol mixtures was observed in comparison to the deposition rate of the allylalcohol homopolymerisation. In contrast, acetylene as a comonomer significantly increases the deposition rate of mixtures.

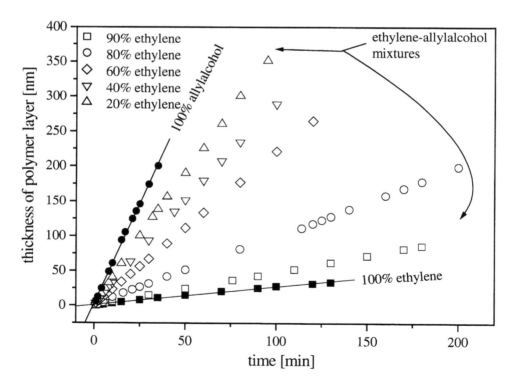

Figure 13. Deposition characteristics of plasma copolymers from allylalcohol and ethylene (P=100 W, P_{eff}=10 W, duty cycle=0.1, f=10^3 Hz, p=26 Pa).

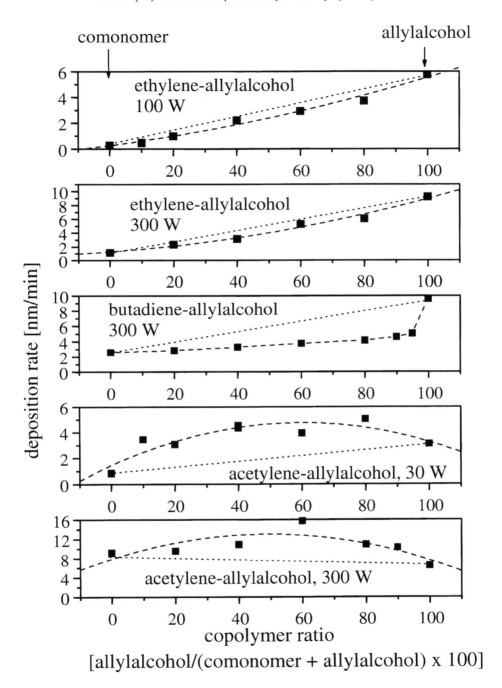

Figure 14. Resulting deposition rates for a variety of "chain-extending" comonomers in the co-polymerisation with allylalcohol (duty cycle=0.1, f=10³ Hz, p=26 Pa).

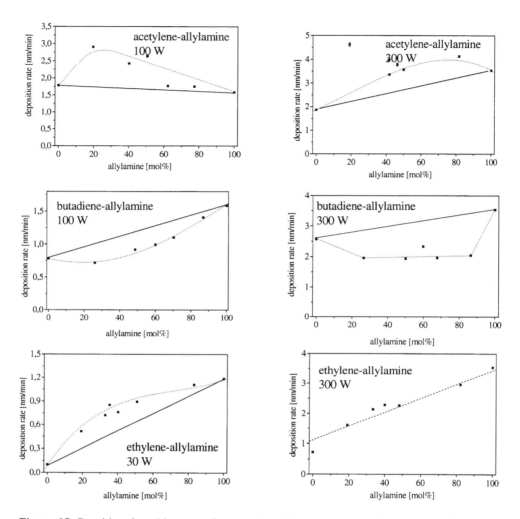

Figure 15. Resulting deposition rates for a variety of "chain-extending" comonomers in the co-polymerisation with allylamine (duty cycle=0.1, f=10^3 Hz, p=26 Pa).

Using allylamine as the functional groups bearing comonomer and butadiene as the chain-extending comonomer the deposition rates of the comonomer mixtures were also lowered (Fig. 15). In contrast, ethylene and acetylene as chain-extending comonomers increase the deposition rates.

The maximum deposition rate of the homopolymers depends on the plasma-on to plasma-off ratio (given as duty cycle). In Fig. 16, this dependence is shown for the ethylene and allylalcohol comonomers. 30 or 100 µs plasma pulses (=plasma-on) were adjusted and then the plasma-off time was successively varied. Allylalcohol shows a maximum deposition rate at very short plasma-off time and ethylene at about 60 µs, which corresponds to a duty cycle of 0.3.

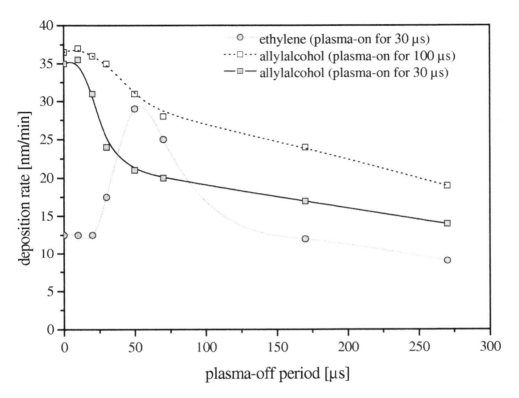

Figure 16. Deposition rates of allylalcohol and ethylene in the pulsed plasma using 30 or 100 μs pulses and varying the plasma-off periods.

3.5. Structure of copolymers

The structures of the copolymers are reflected in the respective C1s-XPS spectra as shown for an ethylene-allylalcohol copolymer (Fig. 17). The spectrum of the copolymer seems to be a linear superposition of the spectra of the homopolymers. To determine the extent of retained functional groups the copolymers were derivatized in the same manner as the homopolymers. In Fig. 18 the elemental composition of an ethylene-allylalcohol copolymer after derivatization with TFAA is plotted. In the range of 0 to 60% ethylene in the gas-vapour mixture of comonomers only homopolymerisation of allylalcohol can be observed. From 60 to <100% copolymerisation occurs. In contrast to the ethylene-allylalcohol system the copolymerisation of butadiene-allylalcohol is dominated by the homopolymerisation of butadiene in the range of 100 to 40% butadiene in the gas/vapour comonomer mixture (Fig. 19). Copolymerisation occurs in the range of <40% butadiene in the gas mixture.

The IR spectroscopy of the ethylene-allylalcohol copolymer confirmed roughly the results of XPS measurements (Fig. 20). With increasing allylalcohol percentage the vOH intensity grows considerably with the concentration of allylalcohol.

However, here, a linear dependence of the vOH intensity on the chemical composition of the gas/vapour mixture is observed. This is reflected in Fig. 20(b). The vOH signal was referenced to the neighbouring vCH vibrations.

As described in Section 3.1 the structure of pulsed plasma ethylene homopolymer differs from that of commercial polyethylene. For example, pulsed plasma polyethylene involves high concentrations of post-plasma oxidised carbon atoms. This is also to be expected when ethylene is used as chain-extending comonomer. In copolymers with ethylene units this behaviour is obvious when measuring the vC=O vibrations and referencing them to the vCH vibrations (Fig. 20). With increasing ethylene percentage the number of >C=O groups increases linearly. It should be noticed that the same tendency was observed when the solubility of copolymers in THF was determined. Pure allylalcohol could be dissolved in both water and tetrahydrofuran but the copolymers with high ethylene percentages could not be completely dissolved.

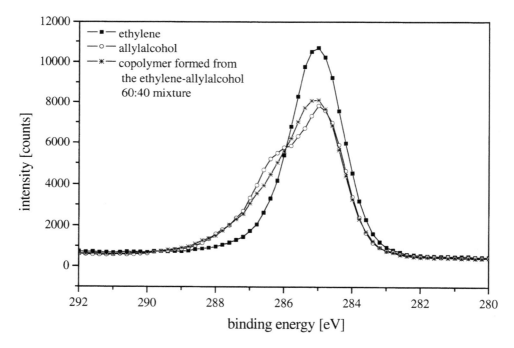

Figure 17. C1s-XPS spectra of allylalcohol, ethylene and an allylalcohol-ethylene copolymer (P=100 W, P_{eff}=10 W, duty cycle=0.1, f=10^3 Hz, p=26 Pa).

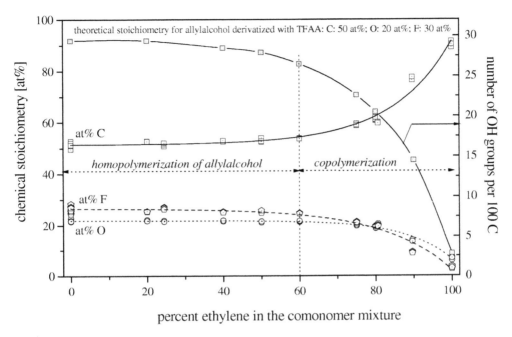

Figure 18. XPS measured elemental composition of a copolymerised ethylene-allylalcohol mixture after derivatization with TFAA (P=100 W, P_{eff}=10 W, duty cycle=0.1, f=10^3 Hz, p=26 Pa).

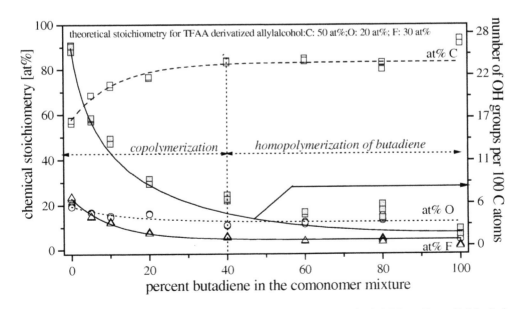

Figure 19. XPS measured elemental composition of a copolymerised 1,3-butadiene-allylalcohol mixture after derivatization with TFAA (P=100 W, P_{eff}=10 W, duty cycle=0.1, f=10^3 Hz, p=26 Pa).

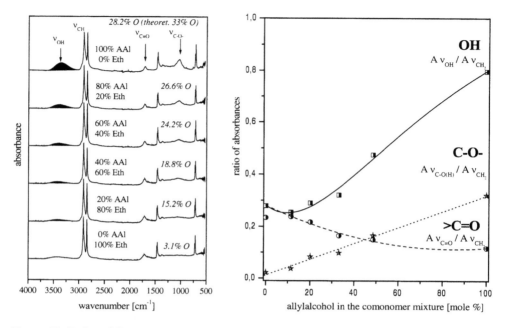

Figure 20. Series of IR spectra (ATR) of all copolymerised ethylene-allylalcohol mixtures (P=100 W, P_{eff}=10 W, duty cycle=0.1, f=10^3 Hz, p=26 Pa) (left) and absorbances (right) of νC=O (1704 cm^{-1}) and νOH (3320 cm^{-1}), νC-O (1032 cm^{-1}) group related vibrations referenced to the $ν_{as}$ CH$_2$ vibration (2925 cm^{-1}).

3.6. Time-resolved SEERS diagnostics during the pulsed plasma polymerisation

Using SEERS, the electron density (n_e), collision rate ($μ_e$) and bias during the r.f. plasma pulses could be measured as function of time for both non-depositing and depositing plasmas. During a single 25 μs pulse in a N_2 plasma both n_e and $μ_e$ strongly increase with both the pressure and the wattage as expected. n_e increases proportionally to the pressure and the number of atoms/molecules in the plasma. The power input has only a weak influence on n_e. The pulse lengths also have an insignificant influence for both the ethylene and the styrene plasmas.

Examining the single 25 μs pulse in styrene plasma in more detail, the decrease in n_e within the first few μs is significant. The electron density is lowered nearly by a factor of 10. The reason is the large electron capture cross section of the styrene molecule (ca. 2.3 nm^2).

3.7. Dependence of Al peel strength on type and density of functional groups at the surface of pulsed plasma polymer coated PP substrates

The aluminium layer (200 nm), which was thermally evaporated onto unmodified polypropylene foils, had very poor adhesion (0 to 10 N/m) (Table 4). Depositing an allylalcohol plasma polymer layer or an allylalcohol-ethylene copolymer layer as an adhesion promoter (150 nm) between Al and PP the peel strength increased significantly (cf. Table 4). The allylalcohol homopolymer layer is soft; therefore, the peel front propagates through the homopolymer layer at low peel strength. The peeled surfaces were checked by XPS (Fig. 21). The C1s signals on both peeled surfaces were identical and corresponded exactly to that of the allylalcohol homopolymer. Increasing the percentage of ethylene in the respective copolymers results in considerably increased peel forces (about 650 N/m). These high peel values are due to the more crosslinked structure of the ethylene-rich copolymers as also indicated by their insolubility. However, the peel front is shifted now partially to the copolymer-PP interface. To avoid the adhesion failure the PP foils were surface-oxidised in an O_2 plasma (10 s). Surprisingly, the measured peel strengths after O_2 plasma pretreatment decrease strongly compared to those of composites with untreated PP (cf. Table 4). The interpretation of this phenomenon is the introduction of a new *Weak Boundary Layer* because of the too long pretreatment (10 s). Another possibility may be the higher input power (100 W in the first series and 300 W in the second series). It is known that higher wattage produces more fragmentation of monomers in the plasma and the resulting plasma polymer has an irregular and highly crosslinked structure. 50 : 50 and 75 : 25 ethylene-allylalcohol copolymer interlayers show the largest adhesion-promoting effect. Ethylene-allylamine copolymers did not show an adhesion-promoting effect as was expected from the possible chemical interactions between the metal and functional groups. The investigations with carboxylic functionalities are in pro-

Table 4.
Types of interlayer and their influence on Al peel strength in the system Al-plasma polymerized interlayer-PP (100 W and 300 W, duty cycle 0.1, 25 Pa)

% ethylene	% allylalcohol	OH density (per 100 C)	Peel strength [N/m]	PP pretreatment	Locus of peel front
100 W					
0	100	29.5	60–100	no	in pp-polyallylalcohol
25	75	29.4	600–700	no	copolymer-PP interface
50	50	28.5	600–700	no	copolymer-PP interface
75	25	26.5	600–700	no	copolymer-PP interface
100	0	3.4	0–10	no	copolymer-PP interface
300 W					
0	100	29.5	50–80	10 s O_2 plasma	in pp-polyallylalcohol
25	75	29.4	40–50	10 s O_2 plasma	copolymer-PP interface
50	50	28.5	90–110	10 s O_2 plasma	copolymer-PP interface
75	25	26.5	100–110	10 s O_2 plasma	copolymer-PP interface
90	10	14.5	50–70	10 s O_2 plasma	copolymer-PP interface
100	0	2.8	0	10 s O_2 plasma	copolymer-PP interface

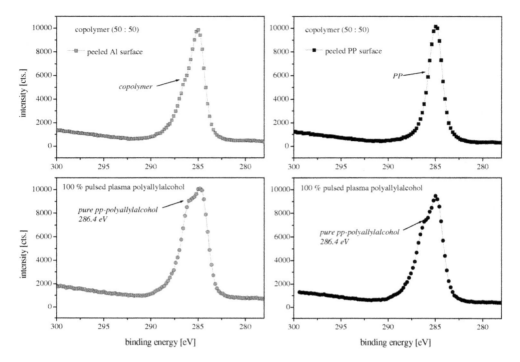

Figure 21. C1s peaks of the peeled Al and PP surfaces with adhesion-promoting copolymer layers from plasma deposited ethylene and allylalcohol (100 W, 1000 Hz, duty cycle 0.1, 25 Pa).

gress. It should be added that the pure pulsed plasma polymerised ethylene did not possess any adhesion-promoting effect (peel strength ≈ 0 N/m, cf. Table 4). Therefore, the small amount of post-plasma introduced oxygen (≈ 3 oxygen atoms per 100 C, cf. Table 4) did not influence the peel strength.

4. DISCUSSION

It was possible to produce OH, NH_2 and COOH functionalised surfaces by deposition, respectively, of pulsed plasma polymerised allylalcohol, allylamine and acrylic acid. Especially the pulsed plasma polymerisation of allylalcohol was very efficient and gave OH densities close to the stoichiometrically expected maximum of OH groups (29.6-measured, 33.3-theoretical). The respective polymer was completely soluble in both tetrahydrofuran and water showing the dominance of the regular radical polymerisation in the plasma-off periods. This was surprising because of the low chemical tendency of allylalcohol to (homo) polymerise in the liquid phase. However, the soft and tacky nature of the plasma polymer layer, characteristic of low-molar mass species, should be due to this chemical behaviour. The allylamine plasma polymer shows a significant instability during long exposure to air, which is characterised by oxidation of the primary amino group (cf. Section 3.3). Therefore, post-plasma chemical processing was performed (derivatization) without long storage of samples in laboratory air. The number of free primary amino groups (unoxidised) was at least 18 NH_2 / 100 C but there were some indications from the literature [38, 39] that the derivatization with PFBA was incomplete ($\approx$ 80% yield was measured). Acrylic acid was easily deposited as plasma polymer layers as predicted by its well-known high chemical polymerisation tendency. However, as reported earlier [40, 41] the percentage of carboxylic groups was reduced during the plasma polymerisation process, also when using the pulsed plasma.

To control the number of functional groups at the surface of the plasma polymer we developed the "plasma-activated chemical" copolymerisation of monomers bearing functional groups and "chain-extending" comonomers. As described before (Section 1) this type of copolymerisation results in a more chemically polymerised copolymer with more defined structure than the simple mixing of chemically unreactive monomers and gases as performed earlier. Four "chain-extending" comonomers were tested. Moreover, the reactivity, characterised by deposition rates in the pulsed plasma, should be in the same range for both the functional group bearing comonomer and the chain-extending comonomer. Therefore, mixtures of allylalcohol-ethylene, allylamine-ethylene or acrylic acid-butadiene (in progress) were well suited for copolymerisation over a broad range of compositions. The most important disadvantage of the chain-extending comonomers such as ethylene and butadiene is the branched and partially cross-linked structure of the resulting homopolymers or blocks/sequences in the copolymers. This was evident by their insolubility and, furthermore, by their picking-up significant percentages of oxygen from the air. Therefore, it is planned to use polystyrene also as a copolymer component.

The chemical composition of pulsed plasma polystyrene was the same as that of atactic polystyrene. The structure of pulsed plasma polystyrene is well defined in the chemical sense but it contains many branched chains with very high molar masses besides homopolymers with low or moderate molar mass. The still existing irregularities in the structure of pulsed plasma PS are shown by the TG/DTG and UV spectroscopy. However, the dominant reaction is the pure chemical (radical) reaction as it was evidenced by the high G-values (number of polymerised monomer molecules per 100 eV power input, G $\approx$ 2500), high average polymerisation degree ($X_w \approx 25$) [40, 41] and the XPS and NEXAFS spectra.

The irregularities within the ethylene (or butadiene) sequences of the copolymers and their tendency to form oxygen containing polar groups by attachment of oxygen from the air may also cause adhesion to metals. However, the pure ethylene pulsed plasma polymer layer did not show any adhesion-promoting effect. Therefore, the strongly increased peel strengths using the 25 : 50, 50 : 50 and 75 : 25 ethylene-allylalcohol copolymer interlayers are due to the higher cohesive strengths of the copolymers as can be shown by comparing them to the low cohesive strength of pure pulsed plasma allylalcohol. The low cohesive strength of the allylalcohol homopolymer was evidenced by the measured low peel strength and the locus of failure, which was within the plasma polymer layer.

5. SUMMARY

The pulsed plasma polymerisation of vinyl, acrylic and allyl monomers is dominated by a chemical (radical) polymerisation in the plasma-off periods. The products are similar to commercial polymers. In addition, pulsed plasma polymers also contain a number of irregularities, especially branched structures. The polymerisation of functional groups bearing monomers is also possible. A high degree of the functional group of the monomer is retained in the plasma polymer (65–95% retention), and OH, NH_2, and COOH groups could be produced. The yields were 30 OH, 18 NH_2, and 24 COOH groups per 100 C atoms. Furthermore, a plasma-initiated copolymerisation can be performed. This process allows to vary the density of functional groups by the addition of a chain-extending comonomer. Olefins, dienes or other hydrocarbon chain-extending comonomers tend to branch and to trap C radical sites (up to 15% of all carbon atoms). Copolymer formation was observed if both comonomers had similar tendencies to polymerise. Self-Exciting Electron Resonance Spectroscopy (SEERS) was used as diagnostic technique for the pulsed plasmas. In Al-PP composites the peel strength of Al proportionally increased with the density of OH functional groups of the adhesion-promoting interlayer produced by the plasma-initiated copolymerisation of allylalcohol and ethylene.

Acknowledgements

We thank the VDI-TZ in Duesseldorf for financing this work under grant 13N7779.

REFERENCES

1. J. Friedrich, I. Retzko, G. Kühn, W. Unger and A. Lippitz, in: *Metallized Plastics 7: Fundamental and Applied Aspects*, K. L. Mittal (Ed.), pp. 117-142, VSP, Utrecht (2001).
2. K. Jesch, J. E. Bloor and P. L. Kronick, J. Polym. Sci., **A1**, 1487 (1966).
3. P. L. Kronick, K. Jesch and J. E. Bloor, J. Polym. Sci., **A4**, 767 (1969).
4. L. F. Thompson and K. G. Mayhan, J. Appl. Polym. Sci., **16**, 2291 (1972).
5. L. F. Thompson and K. G. Mayhan, J. Appl. Polym. Sci., **16**, 2317 (1972).
6. A. R. Denaro, P. A. Owens and A. Crawshaw, Europ. Polym. J., **4**, 93 (1968); **5**, 471 (1969).
7. J. M. Tibbitt, R. Jensen, A. T. Bell and M. Shen, Macromolecules, **10**, 647 (1077).
8. J. M. Tibbitt, M. Shen and A. T. Bell, J. Macromol. Sci., Chem. **A10**, 1623 (1976).
9. G. Rosskamp, PhD thesis, University of Tübingen, Germany (1972).
10. J. Friedrich, PhD thesis, Academy of Sciences, Berlin (1974).
11. J. Friedrich, J. Gähde, H. Frommelt and H. Wittrich, Faserforsch. Textiltechn./Z. Polymerenforsch., **27**, 517 (1976).

12. M. Shen and A. T. Bell, in: *Plasma Polymerization*, M. Shen and A. T. Bell (Eds.), pp. 1-36, ACS Symposium Series 108, Amer. Chem. Soc., Washington, D.C. (1979).
13. H. K. Yasuda, *Plasma Polymerization*, Academic Press, Orlando (1985).
14. A. R. Westwood, Europ. Polym. J., **7**, 363 (1971).
15. F. J. Vastola and J. P. Wightman, J. Appl. Chem., **14**, 69 (1964).
16. J. Meisel and H.-J. Tiller, Z. Chem., **7**, 275 (1972).
17. H. Yasuda and T. Hsu, J. Appl. Polym. Sci., **20**, 1769 (1976); J. Polym. Sci., Polym. Chem. Ed., **15**, 81 (1977).
18. K. Nakajima, A. T. Bell, M. Shen and M. M. Millard, J. Appl. Polym. Sci., **23**, 2627 (1979).
19. J. W. Vinzant, M. Shen and A. T. Bell, ACS Polymer Preprints, **19**, 453 (1978).
20. C. R. Savage and R. B. Timmons, Polym. Mater. Sci. Eng., **64**, 95 (1991).
21. C. R. Savage, R. B. Timmons and J. W. Lin, Chem. Mater., **3**, 575 (1991).
22. H. Schüler and L. Reinebeck, Z. Naturforsch., **6a**, 271 (1951), **7a**, 285 (1952), **9a**, 350 (1954).
23. H. Schüler and M. Stockburger, Z. Naturforsch., **14a**, 981 (1959).
24. H. Schüler, K. Prchal and E. Kloppenburg, Z. Naturforsch., **15a**, 308 (1960).
25. H. König and G. Hellwig, Z. Physik, **129**, 491 (1951).
26. R. Hafer and A. A. Mohamed, Acta Phys. Austriae, **11**, 193 (1957).
27. H. Yasuda and C. E. Lamaze, J. Appl. Polym. Sci., **17**, 1519 (1973).
28. H. Yasuda, H. C. Marsh, M. O. Bumgarner and N. Morosoff, J. Appl. Polym. Sci., **19**, 2845 (1975).
29. H. Yasuda and T. Hirotsu, J. Polym. Sci.: Polym. Chem. Ed., **15**, 2749 (1977).
30. H. Yasuda and T. Hirotsu, J. Appl. Polym. Sci., **22**, 1195 (1978).
31. Th. Gross, A. Lippitz, W. E. S. Unger, J. F. Friedrich and Ch. Wöll, Polymer, **35**, 5590 (1994).
32. I. Koprinarov, A. Lippitz, J. F. Friedrich, W. E. S. Unger and Ch. Wöll, Polymer, **39**, 3001 (1998).
33. I. Retzko, J. F. Friedrich, A. Lippitz and W. E. S. Unger, J. Electr. Spectr. Rel. Phenom. **121**, 111 (2001).
34. St. Weidner, G. Kühn, R. Decker, D. Roessner and J. Friedrich, J. Polym. Sci.: Part A: Polym. Chem. **36**, 1639 (1998).
35. A. Chilkoti and B. D. Ratner, in: *Surface Characterization of Advanced Polymers*, L. Sabbattini, and P. G. Zambonin (Eds.), pp. 221-256, VCH Publishers, Weinheim, Germany (1996).
36. Sh. Geng, PhD thesis, University of Potsdam, Germany (1996).
37. D. E. Everhart and C. N. Reilley, Anal. Chem., **53**, 665 (1981).
38. Y. Nakayama, T. Takayuka, F. Soeda, K. Hatada, S. Nagaoka, J. Suzuki and A. Ishitani, J. Polym. Sci.: Part A, **26**, 559 (1988).
39. M. R. Alexander, P. V. Wright and B. D. Ratner, Surf. Interf. Anal., **24**, 217 (1996).
40. G. Kühn, I. Retzko, A. Lippitz, W. Unger and J. Friedrich, Surface Coatings Technol. **142-144**, 494 (2001).
41. J. F. Friedrich, I. Retzko, G. Kühn, W. E. S. Unger and A. Lippitz, Surface Coatings Technol., **142-144**, 460 (2001).
42. J. F. Friedrich, W. E. S. Unger, A. Lippitz, I. Koprinarov, St. Weidner, G. Kühn, and L. Vogel, in: *Metallized Plastics 5&6: Fundamental and Applied Aspects*, K. L. Mittal (Ed.), pp. 271-293, VSP, Utrecht (1998).
43. W. E. S. Unger, J. F. Friedrich, A. Lippitz, I. Koprinarov, K. Weiss, and Ch. Wöll, in: *Metallized Plastics 5&6: Fundamental and Applied Aspects*, K. L. Mittal (Ed.), pp. 147-168, VSP, Utrecht (1998).
44. R. F. Bartholomew and R. S. Davidson, J. Chem. Soc. (C), 2342 (1971).
45. R. F. Bartholomew and R. S. Davidson, J. Chem. Soc. (C), 2347 (1971).
46. R. F. Bartholomew, R. S. Davidson and M. J. J. Howell, J. Chem. Soc. (C), 2804 (1971).

Polyimides and Other High Temperature Polymers, Vol. 2, pp. 389–406
Ed. K.L. Mittal
© VSP 2003

Metallized polyimide films with high reflectivity and electroconductivity

SAULE KUDAIKULOVA,[1,2] OLEG PRIKHODKO,[3] GALINA BOIKO,[1]
BULAT ZHUBANOV,[1] VANDA YU. VOYTEKUNAS[2] and
MARC J.M. ABADIE[*2]

[1]*Institute of Chemical Sciences, 106 Valihanov, 480100 Alma-Ata, Kazakhstan*
[2]*LEMP/MAO, University Montpellier 2, Place Bataillon, 34095 Montpellier Cedex 05, France*
[3]*Kazakh National University, Physics Department, 85 Tole bi, 480012 Alma-Ata, Kazakhstan*

Abstract—Electroconductive and reflective metallized polyimide films have been prepared by heterogeneous chemical modification of polyimide surface. By carrying out the chemical reactions *in situ* in the modified layers of polyimide surface, a metal phase strongly impregnated into the polyimide surface is obtained. The steps of chemical modification have been studied on the model compound – poly(amic acid) on the basis of isophthaloylchloride and methylenedianthranilic acid which forms insoluble sodium or potassium poly(amic acid) salts (polyamate). Metallization of Kapton® HN & JP (from DuPont) and Upilex® S (from Ube) films has been carried out and the films have been characterized by X-ray diffraction (XRD), X-ray fine diffraction (XRFD), measurements of reflectivity in the visible range and surface resistivity at elevated temperatures. It is shown that reflectivity coefficients of silvered films are 90-92% and surface resistivity is about 0.5 Ω/sq.

1. INTRODUCTION

One of main directions of search for new materials with specific physical properties is metallization of polymer films. Creation of such materials is a difficult task because of low adhesion of metals to polymer surfaces. Such classical methods as vacuum evaporation and electrolytic deposition require special steps for physical or chemical modification of surfaces.

Polyimides (PIs) are widely used for special technological applications because of their high thermal, radiation and chemical resistance, excellent dielectric characteristics and ability to form strong flexible films [1, 2]. To fulfill requirements for different technological applications such as low electrical resistivity and high reflectivity, which are more typical for metals, creation of stable metal-polyimide composites is a necessity.

---

*To whom all correspondence should be addressed. Phone: +33 467 547 825,
Fax: +33 467 144 747, E-mail: abadie@univ-montp2.fr

Thermostable PI films doped with different metals have the potential for use as thin film mirrors/reflectors for use in space, microelectronics, electrochemistry, as well as ferromagnetic materials for magnetic recording, etc. The problem of preparation of strong flexible films with stable electroconductivity can be solved by modification of polyimide (PI) films, particularly by chemical metallization.

The objective of creating reflective, electroconductive metallized PI films is achieved by different methods: homogeneous or heterogeneous. The homogeneous approach [3-6] is to dissolve additives (metal salts and organometallic complexes) into a poly(amic acid) solution (in an amide solvent). The resulting films of pre-polymer upon thermolysis (100-300°C) undergo both cycloimidization and metal-lization. The metallization process includes thermal reduction of metal cations, aggregation of metal clusters with their simultaneous transport, diffusion in the bulk near-surface layer and final appearence of the metal (silver) on the surface of poly-imide film. The ease with which Ag(I) cation is reduced (E° = 0.80 V) raises a synthetic question: Does any Ag(I) compound that is soluble in dimethylacetamide (DMAc) and in the solvent of free poly(amic acid)-polyimide yield essentially the same metallized surface with respect to reflectivity and conductivity [7]? The answer is no; there is clearly a ligand effect on the polymer constrained reduction of Ag(I) and subsequent migration to give metallic surfaces. Preparation of metallized films through Ag(I) acetate and trifluoro acetylacetonate resulted in electroconductive films (0.5 Ω/sq) for a 70 nm silver layer but had poor reflectivity [8]; metalli-zation through Ag(I) acetate 1,1,1,5,5,5-hexafluoro-2,4 pentadione resulted in re-flective (R = 80%) but never conductive films even when heated at 340°C [9]. Replacement of fluorines by hydrogen in one methyl group [10] led to formation of different films: electroconductive (< 0.1 Ω/sq) and reflective (100%) films. The ligand effect plays a dominant role in this type of metallization.

Another approach includes the heterogeneous modification of the films by su-percritical fluid infusion of (1,5-cyclooctadiene-1,1,1,5,5,5-hexafluoro) acetylace-tonate of silver(I) into fully cured polyimide film followed by annealing at 300°C [11]. It was shown that preliminary chemical modification of a polyimide film (on the basis of pyromellitic dianhydride and oxydianiline PMDA-ODA) surface by alkali solution resulted in production of a reflective surface which allowed one to conclude that a coordination site on the polymer backbone was necessary to pro-duce a reflective surface.

It was found [12-15] that introducing organic or inorganic compounds of low molecular weight, into the stretched polymer films, allowed them to fix oriented ordered structures. The elongation of films [(polypropylene, poly(ethylene terephthalate), poly(vinyl chloride)], in the presence of active compounds results in fixed porous structure of films. Loading of the modified polymer films by metal cations followed by their chemical reduction (by sodium borohydride) is one of the efficient ways of preparing metal filled polymer composite films. The size of metal particles (from 4-7 nm up to 132 nm) is controlled by the nature of the polymer matrix as well as by conditions of metal reduction. It is known that interaction of metal salts with reducing agent – sodium borohydride ($NaBH_4$)

aqueous solution – leads mainly to formation of black precipitate, most likely as thin dispersed metal phase with particle size 1-2 nm [16]. Heterogeneous chemical reduction of a metal in the bulk of a polymer matrix can result in electroconductive or reflective films and this is what makes this type of metallization very promising.

A combination of these two approaches is presented in reference [17]. Modified films obtained by homogeneous doping of poly(amide imide)s solution in amide solvent by transition metal salts were reduced in an aqueous solution of sodium borohydride. Thus metallized poly(amide imide) films exhibited low resistivity, around 1-10 Ω/sq and surface morphology was not uniform.

In this work we present formation of a metal phase by carrying out *in situ* chemical reactions in modified polyimide films, as well as results of investigation of their structure and physical properties.

2. EXPERIMENTAL

2.1. Materials

Isophthaloylchloride was prepared according to reference [18] with M.p. 43°C. Neutralization equivalent was determined by saponification followed by potentiometric titration of excess alkali by acid solution and was found to be 101.2 (calculated 101.5).

Methylenedianthranilic acid was prepared from o-aminobenzoic acid and formaldehyde according to reference [18]. Neutralization equivalent determined by reverse titration was found to be 143.5 (calculated – 143.5).

Metal salts $AgNO_3$, AgAc, $CuAc_2$, $CuCl_2$, $Cu(NO_3)_2$, $CoAc_2$, $CoCl_2$, $Co(NO_3)_2$, $NiAc_2$ $NiCl_2$, $Ni(NO_3)_2$ and reducing agent $NaBH_4$ were used as received from Aldrich.

Polyimide films Kapton® HN & JP (from DuPont de Nemours) and Upilex S® (from Ube) were used after washing with aqueous detergent solution and thermally treated at 200°C for 1 hour.

Synthesis of poly(amic acid) based on isophthaloylchloride and methylenedianthranilic acid (PAA-2) was carried out by the reaction of equimolar amounts of methylenedianthranilic acid and isophthaloylchloride in N-methyl-2-pyrrolidone in an inert atmosphere for 2 hours. Films were cast from PAA2 – see Scheme 1 below, solution and dried at room temperature under vacuum. Neutralization equivalent was determined to be 208.1 (calculated 208.1).

2.2. Modification of polyimide films

Hydrolysis of polyimide films was carried out in aqueous, alcohol and alcohol-aqueous solutions of KOH or NaOH. Chelation was carried out by immersion of hydrolyzed films into an aqueous solution of metal salt. Reduction of the chelated films was carried out in an aqueous solution of $NaBH_4$.

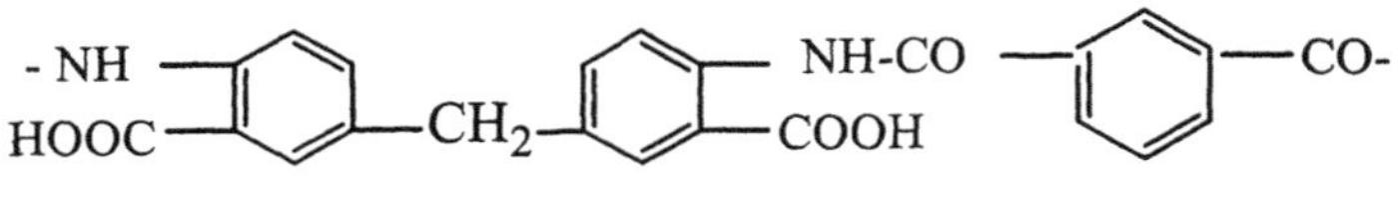

PAA2

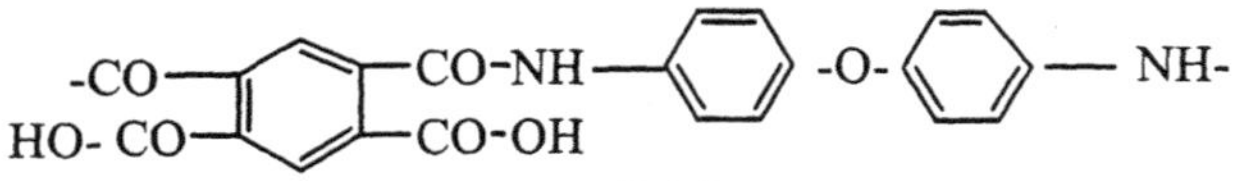

PAA

Scheme 1. Poly(amic acid) structures: PAA2 from methylenedianthranilic acid and isophthaloyl-chloride; PAA from PMDA and ODA.

2.3. Measurements

Metal (M) *content* was determined as follows: M/PAA2 film was burned at 500-600°C and the residue was dissolved in HCl solution. The resulting solution of metal was photocolorimetrically analyzed to determine the amount of metal. Content of potassium during KOH treatment of PAA2 film was determined by the reverse titration method.

IR spectra were recorded on Jasco 810 spectrometer.

Surface electrical resistivity was measured by the two-probe technique with electrodes in contact with the metalized surface of PI film.

Reflectivity spectra were recorded on IR-65V spectrometer. *Scanning electronic microscopy (SEM)* and *X-ray fine diffraction (XRFD)* analyses were carried out on an "SEM-Cambridge 5.360 Instruments" microscope. *X-ray diffractometry* was recorded with a diffractometer "Zeifert analysis".

3. RESULTS AND DISCUSSION

3.1. Heterogeneous chelation of PAA2 films

Surface modification of PI films is widely used to enhance incorporation of metal atoms into the films [19, 20]. One of the main chemical modification routes is hydrolysis of polyimide followed by exchange of K^+ ions by Pd^{2+} cations. This step of metallization process is the most important because it determines the quality and adhesion properties of final metal coating. It is well known that sodium or potassium polyamate dissolves in alkali solutions and fmal hydrolysis represents itself as an equilibrium process of polyimide → polyamate chemical conversion and "washing away" of modified layer. Moreover, PI surface is characterized by well developed system of mesoporous structure formed during the second step of PI formation (transformation of poly(amic acid) film into polyimide with elimination of water and solvent). The diameter of the porous channels achieved is 15-20 nm which increases during hydrolysis in 3M KOH aqueous solution up to 20-

80 nm [21], and the thickness of the modified layer becomes 1-2 μm depending on the temperature of hydrolysis [22, 23]. Evaluation of surface properties of both sides of Kapton type polyimide reference film [24] by contact angle measurements and reflection spectroscopy has shown that the degree of imidization of polyimide films for glass side surface was not as high as that for the air side surface at the same temperature treatment.

In order to exclude the effect of surface morphology and polyimide : poly(amic acid) ratio of different films sides during hydrolysis process, we have taken more hydrolytically stable PAA2 film which can be considered as a model compound of poly(amic acid) (PMDA/ODA). Insolubility of PAA2 polyamate salts allows to achieve practically 100% conversion of PAA/K^+ into $PAA2/M^{n+}$ and to investigate further steps of metallization.

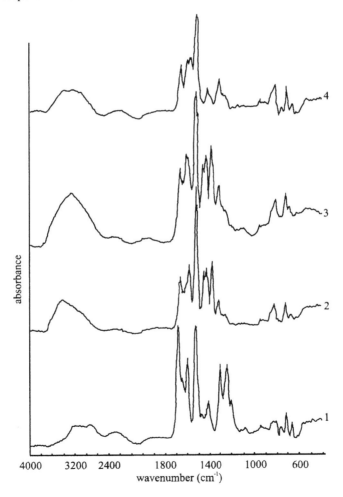

Figure 1. IR spectra of: 1 – PAA2 film; 2 – hydrolysed PAA2 film; 3 – PAA2 film chelated by Cu^{2+}; 4 – PAA2 film chelated by Ag^+.

In Fig. 1 are presented IR-spectra of PAA2 film (curve 1) and hydrolysed PAA2 film in 1% KOH aqueous solution (curve 2). We see that hydrolysis of PAA2 film gives the same characteristic bands in infrared as of polyamate structure of PI based on PMDA and ODA as described in the literature [22]: formation of carboxylate structure, i.e., displacement of 1408 cm^{-1} (carbonyl) band to 1420 cm^{-1} (carboxylate), bands of amide I ($v = 1680$ cm^{-1}, 1670 cm^{-1}) which shifts to 1660 cm^{-1}, 1650 cm^{-1} and band of amide II ($v = 1590$ cm^{-1}) which shifts to 1575 cm^{-1}. The results of the investigation of duration of alkali treatment of PAA2 film on potassium content is presented in Fig. 2. It is seen that hydrolysis of PAA2 film proceeds for 30 minutes and achieves practically 100% conversion.

Exchange of K$^+$ ions with the transition metals takes place following the hydrolysis in aqueous metal salt solution. The spectra of chelated PAA2 films (curves 3 and 4, Fig. 1) are practically the same as for PAA2/K$^+$ except a shift of NH vibration bands ($v = 3330$ cm^{-1} to 3290 cm^{-1}) which points towards the formation of ion-coordination complexes between the NH group of PAA2 potassium salt and the metal cation.

Determination of copper content in the film in the process of ion exchange shows that exchange of K$^+$ by Cu^{2+} cations proceeds nearly with 100% conversion (Fig. 2). This process suggests that there is no need to convert polyamate structure into poly(amic acid) one by additional step of acidic treatment followed by metal loading [20].

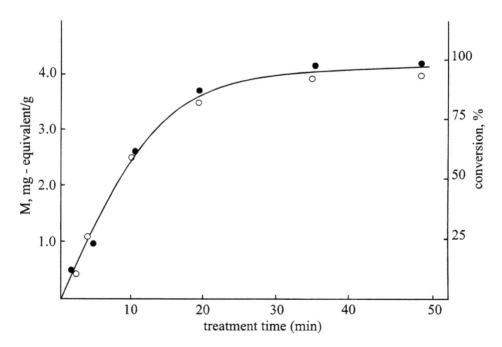

Figure 2. Content M (mg-equivalent/g) of potassium (•) and copper (o) in PAA2 film as a function of treatment duration.

Table 1.
Characteristics of PAA2 films chelated with transition metal cations

Ion M^{n+}	Colour	Metal content, % found	Metal content, % calculated	Nitrogen content, % found	Nitrogen content, % calculated	σ^*, MPa
K	colorless	15.64	15.87	5.34	5.68	90
Cu	green	11.90	13.29	5.26	5.86	115
Co	pink	11.30	12.44	5.43	5.91	110
Ag	colorless	25.20	34.22	5.12	4.44	120

* tensile strength

Because of low degree of dissociation of COOH groups, ion exchange through carboxylate is more efficient and can also be applied for modification of polyimide based on PMDA and ODA. Such a way of modification allows to avoid the acidification step [20].

The data in Table 1 show that potassium forms carboxylate structure with PAA2 in the ratio 2:1 (K^+ : PAA2). Bivalent cations of copper and cobalt form carboxylates with a ratio near 1:1. In spite of monovalence of silver cation, carboxylate structure lies between 1:1 and 1:2 which is probably due to heterogeneity of the process. The tensile strength of chelated PAA2 films increases which shows that metal loading does not result in embrittlement of the films.

3.2. Heterogeneous reduction of chelated films

Thus PAA2 films chelated by cations of transition metals (Ag^+, Cu^{2+}, Ni^{2+}, Co^{2+}) were subjected to reduction by $NaBH_4$, in aqueous or alcohol-aqueous solutions. $NaBH_4$ is a strong reducing agent with reduction potential E = 1.24 V [16]. On immersion of chelated films into alcohol-aqueous solutions of $NaBH_4$, reduction of the incorporated metal cations takes place in 7-10 s and leads to formation of films with metallic lustre. (It should be noted that reactions proceeding during reduction of chelated films are complicated due to heterogeneous nature of the process.) We suggest that reduction of cations in the bulk of PI film proceeds according to the following reactions:

$$4Ni^{2+} + BH_4^- + 8OH^- \rightarrow 4Ni + B(OH)_3 + OH^- + 4H_2O \qquad (1)$$

$$2Co^{2+} + 4BH_4^- + 9H_2O \rightarrow 2Co + B + 3B(OH)_3 + 12.5H_2 \qquad (2)$$

The reactions of soluble nickel [25] and cobalt [16] salts in a basic medium in the presence of different ligands leads to formation of not spongy precipitates, but compact metal coatings with high lustre. Such ligands contain amine and amide groups and are called lustre-forming agents. Their function is to adsorb on the surface of growing metal grains and limit their size. Probably the monomer unit of polyamate network of modified layer plays the role of lustre-forming agent. An

important observation was that all precipitates obtained under conditions of reduction of metals in the aqueous solution were characterized by significant incorporation of metal boride. Later it will be shown that cobalt or nickel coatings obtained by heterogeneous reduction in the bulk of polyimide modified layer do not
contain boron.

The reaction of sodium borohydride with silver results in the formation of pure
metal, independent of the acidity of the reduction medium and presence of ligands
[26, 27].

$$8Ag^+ + BH_4^- + 8OH^- \rightarrow 8Ag + H_2BO_3^- + 5H_2O \qquad (3)$$

Reduction of Cu^{2+}/PAA2 films is complicated and indefinite. The process of
$Cu^{2+} \rightarrow Cu^0$ in solutions goes via different intermediate products and depends on
media conditions as well as reagents ratio. Moreover, reduction of the films is
complicated by the heterogeneous nature of the process, i.e. it is limited by diffusion of the BH_4^- ions into the film. Previously [28] it was stated that interaction of
$NaBH_4$ with $CuCl_2$ resulted in different products depending on the ratio ($\alpha =$
$CuCl_2:NaBH_4$) of the reagents and pH of the solution:

- at $\alpha = 0.1$–0.25; pH 9–9.5 the reaction proceeds as:

$$CuCl_2 + 2NaBH_4 + 6H_2 \rightarrow Cu + 2H_3BO_3 + 2NaCl + 7H_2 \qquad (4)$$

- at $\alpha = 0.25$–0.5; pH 7–9:

$$CuCl_2 + 2NaBH_4 + 6H_2O \rightarrow CuH + 2H_3BO_3 + 2NaCl + 6.5H_2 \qquad (5)$$

- at $\alpha = 1$; pH 5:

$$2CuCl_2 + 2NaBH_4 + 6H_2O \rightarrow CuCl + CuH + 2H_3BO_3 + 2NaCl + HCl + 6H_2 \quad (6)$$

According to reaction (4) metallic copper can be formed in the film under conditions of large excess of BH_4^- ions ($\alpha = 0.1$ to 0.25). This is not acceptable in
conditions of heterogeneous process because the reaction proceeds at large excess
of Cu^{2+} ions in the bulk of chelated modified layer and is limited by diffusion of
BH_4^- ions into the film. The black film formed corresponds to CuH/PAA2 film
according to reaction (5). This is confirmed by chemical conversions of CuH/
PAA2 film which are similar to the conversions of CuH: gas liberation (H_2) is observed in the immersion of the CuH/PAA2 film into NaOH solution; in ammonia
solution the film takes on its initial slightly blue color (Cu^{2+}/PAA2); CuH represents itself as a strong reducing agent. During immersion of CuH/PAA2 black
film into a solution of $AgNO_3$ the intense process of H_2 liberation and formation
of metallized silver surface with mirror reflection takes place.

3.3. Metallized polyimide films

Polyimide films (Kapton and Upilex) were metallized by similar methodology:
hydrolysis, chelation and reduction. All films were covered with metallic lustre

surface, strongly impregnated into polyimide surface with excellent adhesion. (All tests on delamination by using an adhesive tape showed that adhesion was good.)

An SEM micrograph of the surface of metallized (Ag) polyimide Kapton 100HN film is shown in Fig. 3. Surface (dark background) is characterized by a uniform coating consisting of grains with 0.3-0.5 μm in size covered by light spots (silver particles) of size 0.4-0.7 μm, some of them achieve size of 1.2-4 μm. XRFD has shown that black background is silver layer (Fig. 4a) and light spots are silver particles (Fig. 4b). According to XRFD analysis it is seen that the intensity of silver peaks (silver particles, Fig. 4a) is 316 counts while for background silver layer (Fig. 4b) it corresponds to 900 counts. According to these data we estimate the thickness of silver layer as 2.2-2.5 μm. The SEM micrograph of silvered Upilex 25S film (Fig. 5) shows that silver coating is thinner (intensity – 33 counts, Fig. 5a) and uniform (grain size is around 0.3-0.4 μm) and the largest silver particle size is 0.5 μm. (Such metallized surface structure is a result of morphology of original films, which is characterized by more uniform compact surface of Upilex film as compared to Kapton).

We believe that during immersion of chelated film into $NaBH_4$ (pH 10) solution, silver cations due to their high mobility diffuse back to the surface along these channels and are reduced in the near-surface layer. So silver particles on the surface of the film are the result of surface morphology of polyimide films. X-ray diffractogram of silvered Kapton 100HN film (Fig. 6, 1) shows 111 ($\theta = 19.105°$, d = 2.3535) and 200 reflections ($\theta = 22.240°$, d = 2.0352) face centered cubic silver. We believe that silver coating formed by chemical reduction at room temperature and dried at 130°C is characterized by a definite crystal structure. Previously [29] the study of annealing of PMDA-ODA poly(amic acid) doped with silver acetate has shown unclear 111 peaks and the absence of other peaks which were due to two reasons: very thin dispersed form of silver and distortion of crystal lattice. As the cure progresses, silver crystallites grow larger with a corresponding sharpening of the diffraction peaks (Fig. 6) and appearing of new peaks. We can conclude that *in situ* chemical metallization of polyimide surface results in the formation of silver phase with crystal structure. Increase of temperature leads to reverse diffusion of reduced silver to the surface, and growth and perfection of crystal lattice.

According to this methodology cobalt and nickel coated Kapton 100HN composite specimens were also prepared. The structure of these coatings is characterized by larger size of background metal grains (0.8-1.2 μm) and particles 4 μm (Figs. 7-8). X-ray diffractometry has shown that cobalt and nickel metal phases are amorphous (independent of curing). This fact gives reasons to believe that because of significantly lower mobility of Co^{2+} and Ni^{2+} cations compared to Ag^+, they do not move back to the near-surface layer of the film and are reduced in the bulk of modified layer. By electrochemical deposition cobalt and nickel, the layers with lustre can be obtained only in the presence of lustre-forming agents [16]. It is suggested that monomer units of modified polyimide matrix play the role of lustre-forming agent and form metal grains with a size between 800-1200 nm.

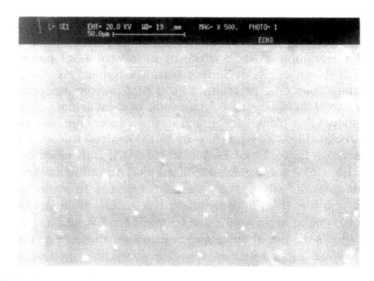

Figure 3. SEM micrograph of Ag/Kapton 100 HN film.

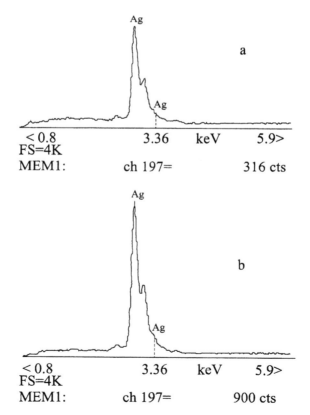

Figure 4. XRFD patterns of Ag/Kapton 100HN film: a – light spot on the surface of the film; b – background of the film (dark surface).

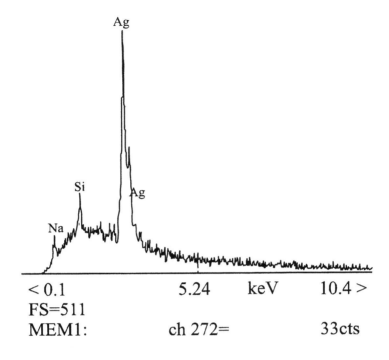

Figure 5. SEM micrograph of Ag/Upilex 25S film (bottom); XRFD analysis (top) of Ag/Upilex 25S film (background).

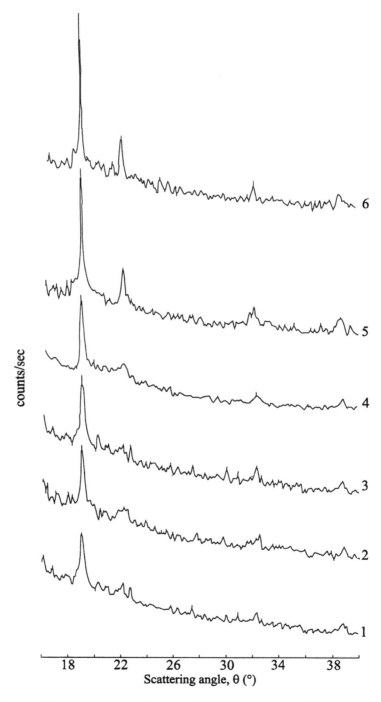

Figure 6. X-ray diffractograms of Ag/Kapton 100HN film cured at different temperatures: 1 – 130°C; 2 – 150°C; 3 – 180°C; 4 – 200°C; 5 – 225°C; 6 – 250°C.

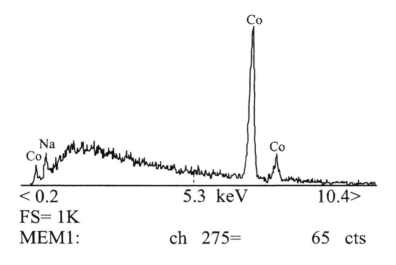

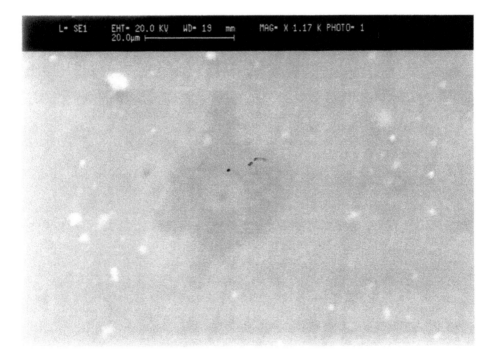

Figure 7. XRFD analysis – (top); SEM micrograph of Co/Kapton 100HN film – (bottom).

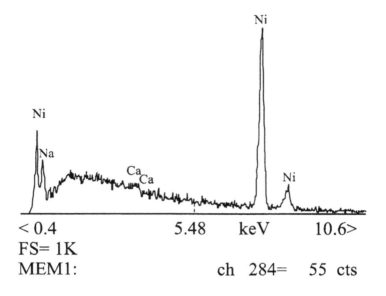

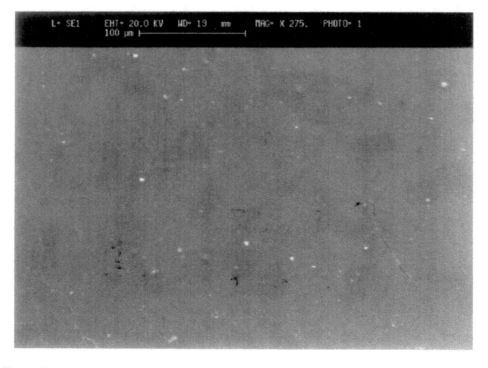

Figure 8. XRFD analysis – (top); SEM micrograph of Ni/Kapton 100HN film – (bottom).

The highly diffusive silver coatings obtained by this technique are not stable at temperatures higher than 250°C. As curing progresses at temperatures higher than 250°C, some whitish haze begins to appear. As this takes place, reflectivity decreases. Light polishing of the surface restores the original optical properties. We assume that this is due to the appearance of silver on the surface of the film and gradual degradation of surface structure followed by conversion of silver into a powder form. The same process was described during curing of silvered polyimide films based on BPDA-ODA [7], however this process takes place at 300°C because of the difference in the microstructure of these films.

To stabilize diffusion of silver at temperatures higher than 250°C and to prevent its appearance on the surface of coating we obtained Ag-Ni and Ag-Co composite coatings on polyimide Kapton HN and Upilex 25S films. Curing up to 300°C showed good stability of silver coating.

3.4. Physical properties of metallized Kapton films

Thus formed silvered polyimide films have good electrical conductivity. In Fig. 9, the temperature dependence of surface resistivity for Ag/Kapton 100HN and Ag-Ni/Kapton 100HN films cured at 200°C is presented. It is found that the relationship is linear. However, different sides of metallized films are characterized by some differences in surface resistivity, which is due to difference in their surface morphology. If we compare the air sides of the films it is seen that silver coating stabilized by nickel is the most conductive, probably because of more deeper incorporation of silver-nickel layer. Glass sides of both films have practically the

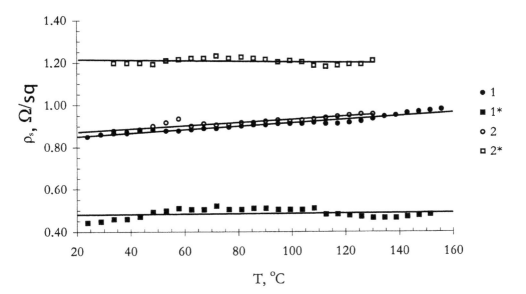

Figure 9. Temperature dependence of surface resistivity (ρ_s) of different sides of metallized Kapton 100HN films: 1 – glass side, 1* air side of Ag-Ni coating; 2 – glass side, 2* – air side of Ag coating.

same resistivity. In the measured temperature range, the surface resistivity practically does not change for all four surfaces. Temperature dependence of surface resistivity $\rho_s(t) = \rho_{0s}(1 + \alpha_s t)$ (where ρ_{0s} and α_s are at 20°C; ρ_s – surface resistivity, α_s – thermal coefficient of resistivity) follows the equations:

1: $\rho_s(t) = 0.85 \ (1 + 9.3 \cdot 10^{-4}t)$ – glass side of Ag-Ni/Kapton 100HN

1*: $\rho_s(t) = 0.48 \ (1 + 1.5 \cdot 10^{-4}t)$ – air side

2: $\rho_s(t) = 0.87 \ (1 + 8.1 \cdot 10^{-4}t)$ – glass side of Ag/Kapton 100HN

2*: $\rho_s(t) = 1.22 \ (1 - 0.8 \cdot 10^{-4}t)$ – air side

It is obvious that temperature coefficients are lower than for pure silver for which $\alpha = 0.0036$ [30, 31]. Possible explanation of such behavior may be the incorporation of silver layer into the polymer matrix. Surface resistivity of Ni/PI and Co/PI films measured at room temperature lies in the range 10-100 Ω/sq.

Fig. 10 shows reflectivity spectra of glass sides of four metallized Kapton 100 HN polyimide films measured in relation to silver reference mirror. Visually the air side for thick Kapton films (500, 300, 200 HN) has slightly lower reflectivity. Reflectivity for both sides of Kapton 100, 75, 50 and 30 HN is practically the same. Reflectivity of the films at $\lambda = 531$ nm (wavelength of the most intense solar irradiation) is 91% for Ag/PI film, 81% for Ag-Ni/PI, 47% for Ni/PI film and 41% for Co/PI film.

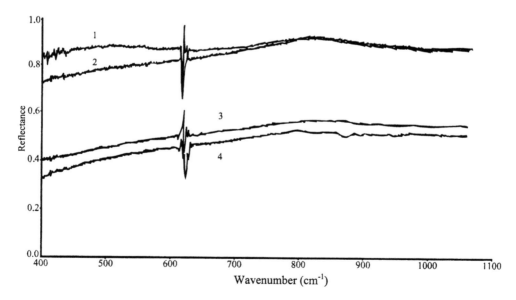

Figure 10. Reflectivity spectra of Kapton 100HN films metallized by: 1 – Ag; 2 – Ag-Ni; 3 – Ni; 4 – Co.

4. CONCLUSION

It can be concluded that heterogeneous chemical modification of polyimide films consisting of consequent steps of hydrolysis in alkali solutions, chelation by metal cations in metal salts solutions, and the following reduction in aqueous solutions of sodium borohydride allows to form a metal phase, strongly impregnated into the near-surface layer of polyimide film. Silver forms crystal structure whereas cobalt and nickel coatings are amorphous. In general, the metallized surface produced by this method represents itself in the background layer of grain structure covered by metal particles of larger size. It is suggested that the grain structure is due to the original morphology of the PI surface characterized by developed mesoporous system.

The preparation of metallized polyimide films by *in situ* chemical modification provides their high electroconductivity and reflectivity.

Acknowledgment

The authors are grateful to the French Embassy in Kazakhstan for support of this research. This work was funded by NATO SfP 97 8013 grant.

REFERENCES

1. C.E. Sroog, J. Polym. Sci., Macromol. Rev., **11**, 161 (1976).
2. M.K. Ghosh and K.L. Mittal (Eds.), *Polyimides: Fundamentals and Applications,* Marcel Dekker, New York (1996).
3. S.A. Ezzell, T.A. Furtsch, E. Khor and L.T. Taylor, J. Polym. Sci. Polym. Chem. Ed., **21**, 865 (1983).
4. R.K. Boggess and L.T. Taylor, J. Polym. Sci. Polym. Chem. Ed., **25**, 685 (1987).
5. R.E. Southward, D.S. Thompson, D.W. Thompson, M.L. Caplan and A.K. St. Clair, in: *Metal-Containing Polymeric Materials,* C.U. Pittman, Jr., C.E. Carraher, B.M. Culbertson, M. Zeldin and J.E. Sheats (Eds.), p. 349, Plenum Press, New York (1996).
6. A.F. Rubira, J.D. Rancourt and L.T. Taylor, in: *Metal-Containing Polymeric Materials,* C.U. Pittman, Jr., C.E. Carraher, B.M. Culbertson, M. Zeldin and J.E. Sheats (Eds.), p. 357, Plenum Press, New York (1996).
7. R.E. Southward, C.M. Boggs, D.W. Thompson and A.K. St. Clair, Chem. Mater. **10**, 1408 (1998).
8. R.E. Southward, D.W. Thompson and A.K. St. Clair, Chem. Mater. **9**, 501 (1997).
9. R.E. Southward, D.S. Thompson, D.W. Thompson and A.K. St. Clair, Chem. Mater. **9**, 1691 (1997).
10. R.E. Southward, D.S. Thompson, D.W. Thompson and A.K. St. Clair, Chem. Mater. **11**, 501 (1999).
11. J. Rosolovsky, R.K. Boggess, A.F. Rubira, L.T. Taylor, D.M. Stoakley and A.K. St. Clair, Polym. Prepr. **38**, 1, 282 (1997).
12. N.I. Nikanorova, Ye.V. Semenova, V.D. Zanegin, G.M. Lukovkin, A.P. Volynsky and N.F. Bakeev, Vysokomolek. Soed. **A34**, 8, 123 (1992).
13. S.V. Stahanova, N.I. Nikanorova, V.D. Zanegin, G.M. Lukovkin, A.P. Volynsky and N.F. Bakeev, Vysokomolek. Soed. **A34**, 133 (1993).
14. N.I. Nikanorova, S.V. Stahanova, A.L. Volynsky and N.F. Bakeev, Vysokomolek. Soed. **A-B39**, 1311 (1997).

15. O.E. Litmanovich, A.G. Bogdanov, A.A. Litmanovich and LM. Papisov, Vysokomolek. Soed. **A-B39**, 1875 (1997).

16. N.N. Maltseva and V.S. Khain, *Sodium Borohydride Properties and Applications*. Nauka, Moscow (1985).

17. C.-J. Huang, C.C. Yen and T.-C. Chang, J. Appl. Polym. Sci. **42**, 2237 (1991).

18. J.K. Stille and T.W. Campbell (Eds.), *Condensation Monomers*, Wiley-Interscience, New York (1972).

19. L.J. Matienzo and W.N. Unertl, in: *Polyimides: Fundamentals and Applications*, M.K. Ghosh and K.L. Mittal (Eds.), p. 629, Marcel Dekker, New York (1996).

20. K.-W. Lee and A. Viehbeck, in: *Polyimides: Fundamentals and Applications*, M.K. Ghosh and K.L. Mittal (Eds.), p. 505, Marcel Dekker, New York (1996).

21. N.B. Melnikova, A.B. Kuzmin, V.B. Shepov, V.R. Kartashov, Z.V. Gerashenko and V.F. Blinov, Plastmassy, **2**, 23 (1988).

22. M.M. Koton, T.A. Redrova, A.N. Krasovsky, K.K. Kalninsh and Yu.N. Sazanov, Reports of USSR AS, **265**, 660 (1982).

23. N.V. Mihailov, V.N. Artemyev, N.V. Kukarkina, A.V. Gribanov, Yu.N. Sazanov and M.M. Koton, Zhurn. Prikladn. Chim., **5**, 1156 (1986).

24. M. Zuo, T. Takeichi, A. Matsumoto and K. Tsutsumi, in *Advances in Polyimides*, H.S. Sachdev, M.M. Khojasteh and C. Feger (Eds.), p. 265, Proceedings of the meeting held in McAffey, New Jersy, October 8-10 (1997).

25. K. Lang, *Die Stromlose Vernicklung - Galvanotechnik*, **56**, 347 (1965).

26. L.F. Hohnstedt, B.O. Miniatos and S.M.C. Waller, Anal. Chem., **37**, 1163 (1965).

27. N.C. Brown and A.C. Boyd, Anal. Chem., **27**, 156 (1955).

28. Z.K. Sterlyadkina and L.S. Alekseeva, Zhurnal Neorg. Chem. **12**, 583 (1967).

29. N.S. Lidorenko, I.Ya. Ravich, L.G. Gindin, A.S. Kagan, A.E. Kovalsky, T.N. Toroptseva and Z.N. Zhigareva, Reports of USSR AS, **182**, 1087 (1968).

30. I.K. Kikoin (Ed.), *Tables of Physical Constants. Reference Book*, p. 306, Atomizdat, Moscow (1976).

31. N.I. Koshkin and M.G. Shirkevich, *Reference Book of Physical Constants*, p. 115, State Publishing House of Physics and Mathematics Literature, Moscow (1962).

Polyimides and Other High Temperature Polymers, Vol. 2, pp. 407–418
Ed. K.L. Mittal
© VSP 2003

RF plasma etching of a polyimide film with oxygen mixed with nitrogen trifluoride

SATORU IWAMORI,[*,1,†] NORIYUKI YANAGAWA,[1]
MITSURU SADAMOTO,[1] RYOUSUKE NARA[2] and SHIGEKI NAKAHARA[3]

[1]*Material Science Laboratory, Mitsui Chemicals, Inc., 580-32 Nagaura, Sodegaura-City, Chiba 299-0265, Japan*
[2]*Functional Materials Laboratory, Mitsui Chemicals, Inc., 580-32 Nagaura, Sodegaura-City, Chiba 299-0265, Japan*
[3]*Mitsui Chemical Analysis and Consulting Service, Inc., 580-32 Nagaura, Sodegaura-City, Chiba 299-0265, Japan*

Abstract—Oxygen mixed with nitrogen trifluoride (NF_3) was used as the gas source for the plasma etching to increase the etching rate of the polyimide (PI) film.
   In order to investigate the effects of NF_3 addition, surfaces of the etched PI films were analyzed with various methods. From the results of x-ray photoelectron spectroscopy (XPS), the chemical bonding state of the etched PI surface with 30% NF_3 / 70% O_2 plasma was similar to that of the surface prepared using 100% O_2 plasma. The results of FT-IR analyses showed that a part of materials deposited on the etched PI film was soluble in chloroform and it contained carbonyl and ether compounds.
   Furthermore, the etching products were analyzed using quadrupole mass spectrometry (QMS) and gas chromatography. The main products were found to be H_2O, HF, CO and CO_2. In addition, CO/CO_2 ratio was found to be related to the etching rate which depended on the NF_3 concentration.

Keywords: Polyimide; etching; nitrogen trifluoride; oxygen; RF plasma.

1. INTRODUCTION

Polyimides (PI) have excellent thermal stability, chemical stability and electrical properties [1, 2]. PI films have been used as an insulating layer in print circuit boards (PCBs) [3]. Oxygen plasma has been used for the surface modification and etching of polymer films [4-10]. Oxygen mixed with carbon tetrafluoride (CF_4) or sulfur hexafluoride (SF_6) has been used as the gas source for the plasma etching to increase the etching rate of the polyimide (PI) film [11-20]. However, there are only a few reports on the PI etching with nitrogen trifluoride (NF_3) [21, 22]. We

---

*To whom all correspondence should be addressed. Phone: +81-76-234-4950,
Fax: +81-76-234-4950, E-mail: iwamori@t.kanazawa-u.ac.jp
†Present address: Department of Human and Mechanical Systems Engineering, Faculty of Engineering, Kanazawa University, 2-40-20, Kodatsuno, Kanazawa 920-8667, Japan.

analyzed the PI surface etched with oxygen mixed with NF_3, the composition of outgas during etching and constructed a chemical reaction model of the PI etching. In this paper, we report on the reason why the etching rate is increased by the addition of fluorine gases such as CF_4, SF_6, NF_3, and furthermore why NF_3 is the most effective gas for increasing the etching rate.

2. EXPERIMENTAL

Kapton-V (0.05 mm thick; Toray-DuPont, Inc.) was used as the PI substrate. Table 1 shows the experimental conditions for plasma etching. Figure 1 shows the etching chamber which is equipped with a pair of parallel electrodes, 13.56 MHz RF generator and quadruple mass spectrometer (QMS).

Table 1.
Etching conditions

Gas	100% O_2	30% NF_3 / 70% O_2	80% NF_3 / 20% O_2
Flow rate (cc/min)	20.0	6.0/14.0	16.0/4.0
Pressure (Pa)	60	60	60
RF power (W)	200	200	200
Etching time (min)	5–20	5–20	5–20

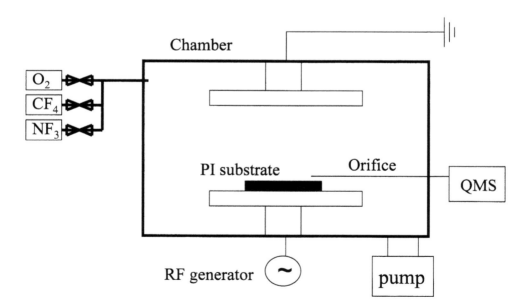

Figure 1. RF plasma system used to etch PI films.

Table 1 shows the etching conditions for the PI film. The etching rates of PI films were measured with a profilometer, Dektak IIA.

The chemical bonding states of the etched PI films were analyzed by x-ray photoelectron spectroscopy (XPS). The fibrils on the etched PI films were analyzed by FT-IR after dissolving them in chloroform and drying. The composition of the outgas during etching was analyzed with QMS and gas chromatography using GC-8A equipment (Shimadzu Co. Ltd.). The columns used for analyses of CO and CO_2 were Unipak 1A and Molecular Sieve 13X (GL Sciences, Inc.), respectively.

The calculations of frontier electron densities and reaction enthalpies were performed using one of the semi-empirical molecular orbital programs, MOPAC^{TM} (Fujitsu, Ltd.).

3. RESULTS AND DISCUSSION

3.1. Surface analyses of etched PI films

Figure 2 shows the etching rate of PI film using oxygen mixed with CF_4 (O_2 / CF_4) and NF_3 (O_2 / NF_3). The horizontal axis represents CF_4 or NF_3 concentration in the inlet gas mixture. Both etch profiles are similar, but the etching rate using O_2 / NF_3 plasma is higher than using O_2 / CF_4 plasma.

Figures 3a, 3b, 3c and 3d show the C(1s) XPS spectra of the PI surface and PI surfaces etched with 100% O_2 plasma, 70% O_2 / 30% NF_3 plasma and 20% O_2 / 80% NF_3 plasma, respectively. After the PI films were etched with 100% O_2 plasma or 70% O_2 / 30% NF_3 plasma, the peaks representing -C-OH (286.9 eV)

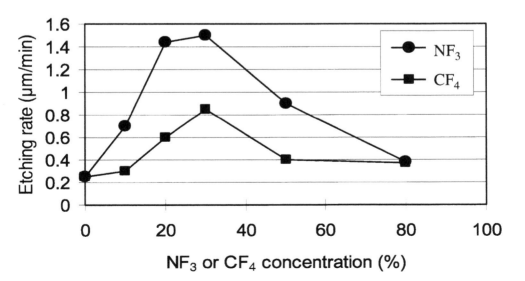

Figure 2. PI etching rate as a function of plasma gas composition.

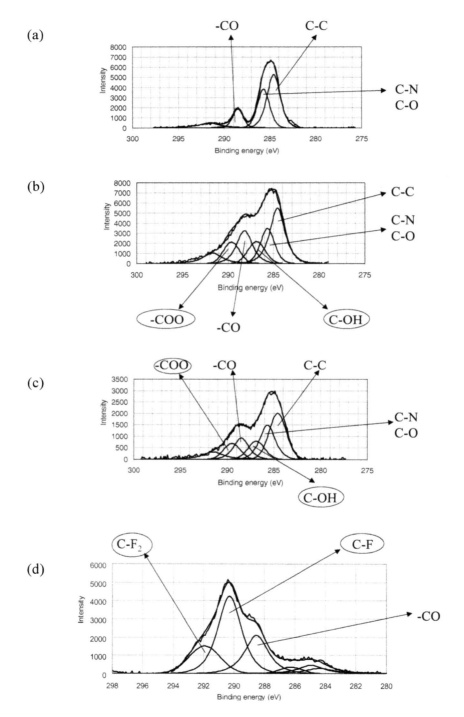

Figure 3. C(1s) XPS spectra of PI (a), etched with 100% O_2 plasma (b), 30% NF_3 / 70% O_2 plasma (c), and 80% NF_3 / 20% O_2 plasma (d).

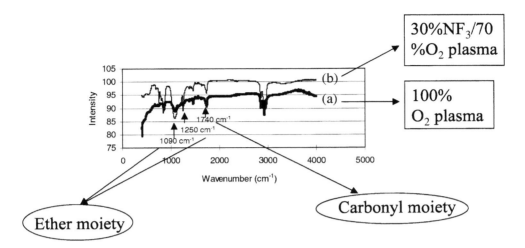

Figure 4. FT-IR spectra of the fibrils at the PI surfaces which were soluble in chloroform after the PI films were etched with 100% O_2 plasma (a), and 70% O_2 / 30% NF_3 (b).

and -(C=O)-O (289.5 eV) moieties appeared at the surfaces. When the PI film was etched with 100% O_2 plasma or 70% O_2 / 30% NF_3 plasma, the amount of oxygen at the surface was increased. When the PI film was etched with 20% O_2 / 80% NF_3 plasma, peaks representing -CF (290.7 eV) and -CF_2 (292.4 eV) moieties appeared at the surface. Even if the PI film was etched with 70% O_2 / 30% NF_3 plasma, fluorine at the surface was small in quantity. However, when it was etched with 20% O_2 / 80% NF_3 plasma, the amount of fluorine at the surface was remarkably increased.

Figure 4 shows the FT-IR spectra of the fibrils at the PI surfaces which were soluble in chloroform after the PI films were etched with 100% O_2 plasma (a), and 70% O_2 / 30% NF_3 (b). The peaks representing carbonyl moiety (1740 cm^{-1}) and ether moiety (1250 cm^{-1} and 1090 cm^{-1}) appeared. The peak intensity of the ether moiety in the fibril at the PI surface etched with 30% NF_3 / 70% O_2 plasma was much larger than that on the surface etched with 100% O_2 plasma.

3.2. Analysis of outgassing from etched PI films

Figures 5a, b and c show the QMS spectra of the outgas during etching with 100% O_2 plasma, 70% O_2 / 30% NF_3 plasma and 20% O_2 / 80% NF_3 plasma, respectively. The main products of 100% O_2 plasma etching were H_2O^+, CO^+ or N_2^+, CO_2^+ or N_2O^+, O_2^+ and O^+ originated from the inlet oxygen. In addition to these peaks, the peak representing HF^+ appeared for etching with 70% O_2 / 30% NF_3 plasma and 20% O_2 / 80% NF_3 plasma.

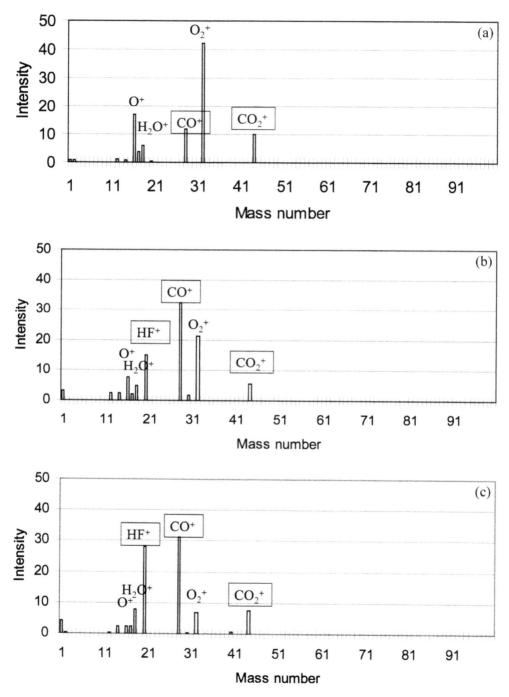

Figure 5. Quadruple mass spectrometry (QMS) spectra of etched PI with 100% O_2 plasma (a), 30% NF_3 / 70% O_2 plasma (b), and 80% NF_3 / 20% O_2 plasma (c).

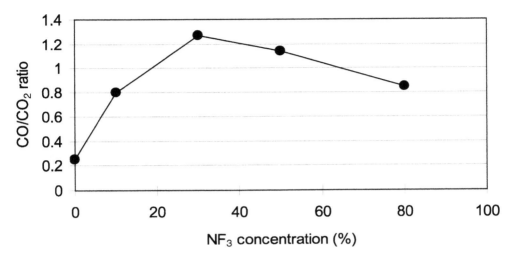

Figure 6. Relationship between the CO / CO_2 ratio and NF_3 concentration in the inlet gas mixture.

CO and CO_2 concentrations during etching were analyzed using a gas chromatograph GC-8A (Shimadzu Co. Ltd.). The rates of formation of CO and CO_2 were 3.6 cc/min and 3.0 cc/min.

If etched PI products are only CO and CO_2, the reaction is:

$$C_{22}O_5N_2H_{10} + z\,O_2 \rightarrow 12CO + 10CO_2 + 5H_2O + y\,NO_x$$

Considering that the PI etching rate was 5.4 mg/min, CO and CO_2 formation rates should be 3.9 cc/min and 3.3 cc/min, respectively.

These results mean that more than 90% of carbon in the PI structure is converted into CO and CO_2 with 70% O_2 / 30% NF_3 plasma etching. These results agree with the QMS analyses in that the main peaks of the carbon compounds in the QMS spectrum shown in Fig. 5b were only CO^+ and CO_2^+.

Figure 6 shows the relationship between the CO / CO_2 ratio and NF_3 concentration in the inlet gas mixture. The CO / CO_2 ratio showed the highest value at the highest etching rate.

3.3. Analysis of CO and CO_2 formation mechanism

In order to investigate the etching mechanism of the PI due to NF_3 / O_2 plasma, the reactivity of the PI with oxygen and NF_3 was evaluated by calculation of reaction enthalpies. Tetracarboxylic anhydride, diamine and benzene were used for model compounds of the PI structure. The main active species are considered to be oxygen and fluorine radicals because the mean free path of oxygen and fluorine ions is considered to be much shorter than the ion sheath at the pressure during the etching. The reaction enthalpies between these compounds (tetracarbox-

ylic anhydride, diamine and benzene) and fluorine and oxygen radicals were calculated.

Figures 7 and 8 show the cleavage schemes of imide ring due to fluorine and
oxygen radicals, respectively. As the enthalpies of the fluorine radical reactions
are negative, a fluorine radical can easily cleave the imide ring. An oxygen radical
can also cleave the imide ring, but this cleavage reaction hardly takes place, because the enthalpy of the formation of the intermediate compound is positive.

The reactions of fluorine radical with a benzene ring are shown in Fig. 9. The
benzene ring is considered to be easily fluorinated by fluorine radicals. The hydrogen in the benzene ring is pulled out by the fluorine radical. The reaction enthalpies of the fluorination of diphenylether were also calculated and are shown in

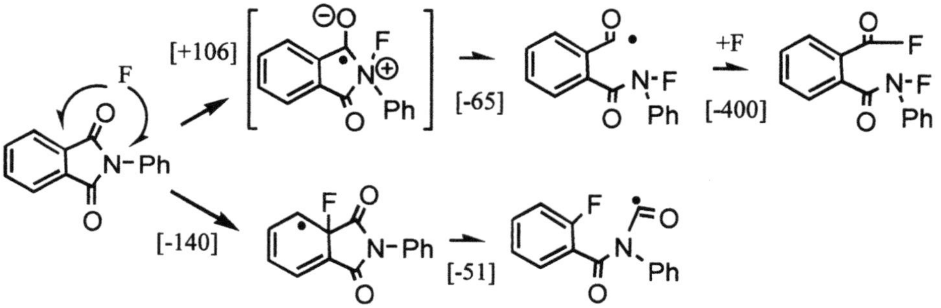

Figure 7. Cleavage scheme of imide ring by fluorine. Reaction enthalpies are shown: [kJ/mol].

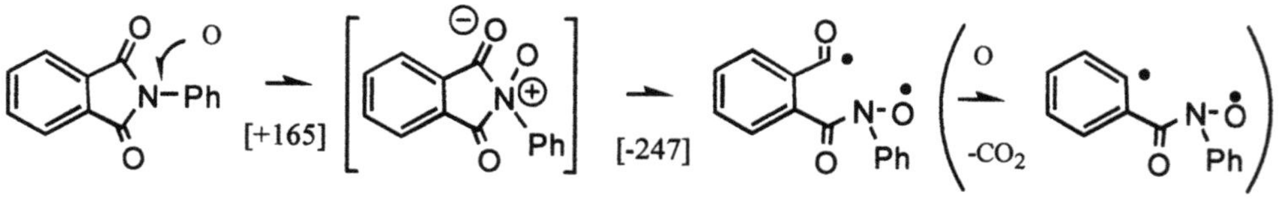

Figure 8. Cleavage scheme of imide ring by oxygen. Reaction enthalpies are shown: [kJ/mol].

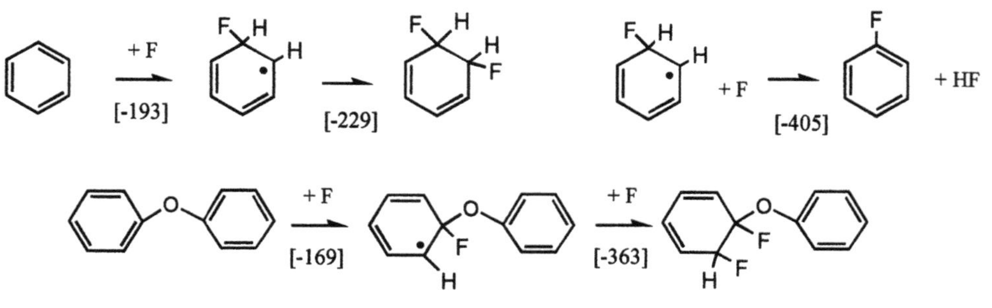

Figure 9. Reactions of fluorine with benzene ring. Reaction enthalpies are shown: [kJ/mol].

Fig. 9. Diphenylether can also be fluorinated by fluorine radicals. These reactions are supported by the results of Fig. 3d and Fig. 5c; large -CF, -CF$_2$ peaks appeared at the PI surface etched with 80% NF$_3$ / 20% O$_2$ plasma, and large hydrogen fluoride peak can also be seen in the QMS spectrum.

The cleavage scheme of fluorinated benzene ring by fluorine radicals is shown in Fig. 10. This cleavage reaction would not take place, because these reaction enthalpies are positive.

Figure 11 shows reactions between a benzene ring and oxygen radicals. The reaction rate of the addition of an oxygen atom to benzene ring is known to be much slower than that to olefin (cyclopentene) [23]. After an oxygen atom is added to a benzene ring, phenol compounds are easily produced. This means that the aromaticity of the benzene ring is not easily lost. In addition, r1 and r2 reactions are also known to be rate-limiting steps [23].

Figure 12 shows reactions between a benzene ring and a diphenylether with oxygen and fluorine radicals. As all of these reaction enthalpies show negative values and the enthalpy of the formation of these allyl radicals is much lower than that of the phenol, −102.3 kJ/mol (Fig. 11), these allyl radicals are more stable than the phenol compounds. This means that a fluorinated benzene ring as well as diphenylether easily break down due to the oxygen radicals.

These results show that the atom contributing to the cleavage of the benzene ring in the PI is oxygen, and the etching rate decreases with decrease of oxygen concentration in the plasma. But the rate-limiting step exists when the PI is etched in 100% O$_2$ plasma. A small quantity of fluorine is necessary for increasing the etching rate, because the benzene ring is easily fluorinated by fluorine radicals

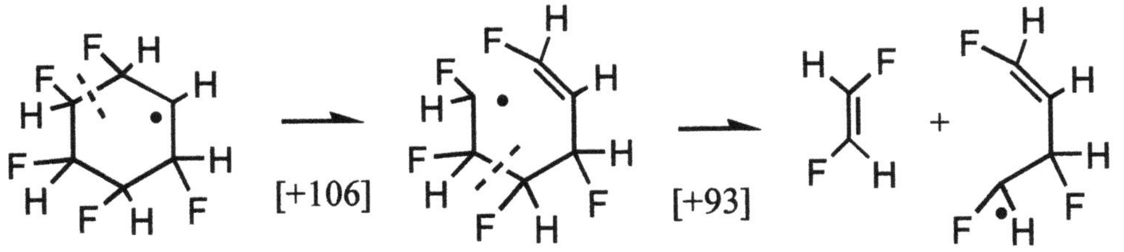

Figure 10. Cleavage scheme of fluorinated benzene ring by fluorine. Reaction enthalpies are shown: [kJ/mol].

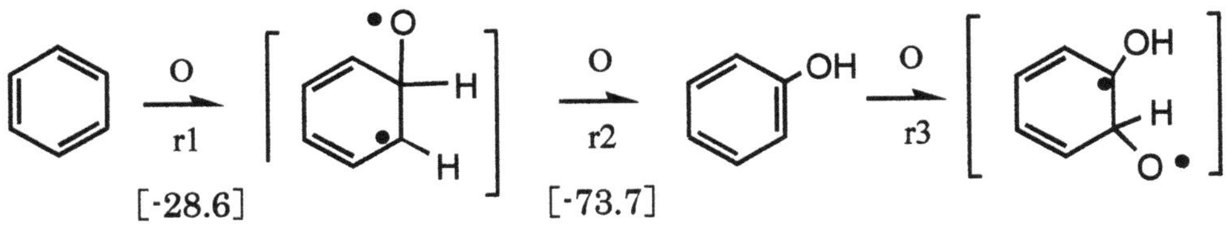

Figure 11. Reactions of oxygen with benzene ring. Reaction enthalpies are shown: [kJ/mol].

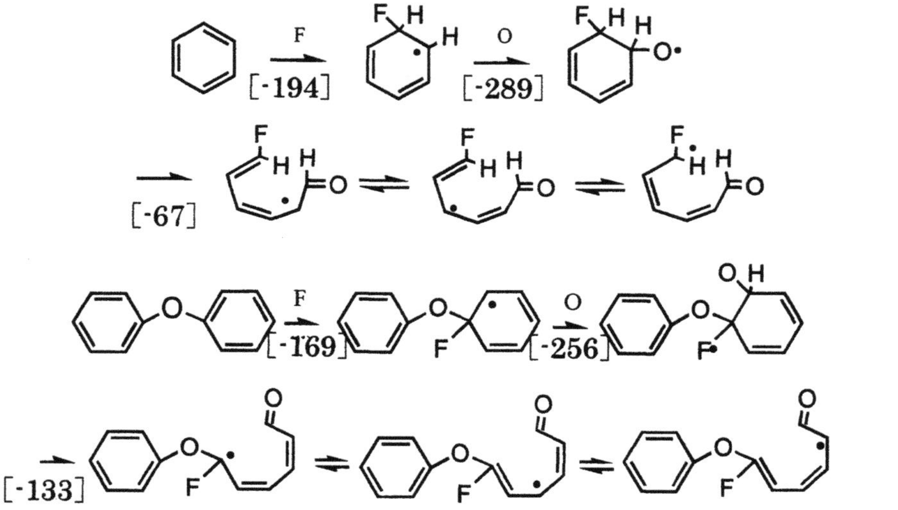

Figure 12. Reactions between a benzene ring and a diphenylether with oxygen and fluorine radicals. Reaction enthalpies are shown: [kJ/mol].

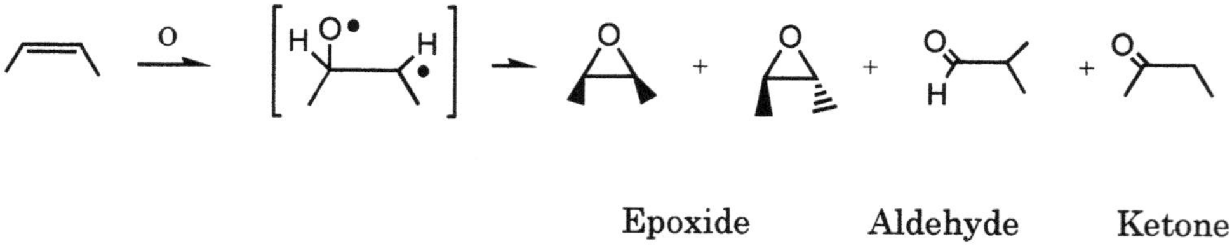

Epoxide   Aldehyde   Ketone

Figure 13. Production of epoxide, ketone and aldehyde compounds by oxygen.

and the fluorinated benzene ring easily breaks down due to the oxygen radicals. This is the reason why the addition of 20–50% NF_3 enhanced the etching rate but 80% NF_3 was not effective in the enhancement.

After the cleavage of benzene ring and imide ring, olefin or aldehyde compounds containing unsaturated carbon are produced. Oxygen radicals are known to react easily with the double bond of an olefin, and produce epoxide, ketone and aldehyde compounds (Fig. 13) [24, 25].

Figure 14 shows reactions of aldehyde and ketone compounds with oxygen radicals. The carbonyl moieties of aldehyde compounds are desorbed and finally converted to CO and CO_2 [26]. The reaction rate constants of olefin and aldehyde compounds with oxygen radicals are known to be much higher than that of benzene ring [27], which means that degradation reactions progress rapidly once the benzene ring is cleaved.

$$MeCHO + O^{\cdot} \rightarrow MeCO^{\cdot} + \,^{\cdot}OH \qquad \text{------ (1)}$$

$$MeCHO + \,^{\cdot}OH \rightarrow MeCO^{\cdot} + \,^{\cdot}OH \qquad \text{------ (2)}$$

$$MeCO^{\cdot} + O \rightarrow MeCOO^{\cdot} \rightarrow Me^{\cdot} + CO_2 \qquad \text{------ (3)}$$

$$MeCO^{\cdot} \rightarrow Me^{\cdot} + CO \qquad \text{------ (4)}$$

Figure 14. Reactions of aldehyde and ketone compounds with oxygen radicals. Me : alkyl group.

$$N_2^{\cdot} + COF_2 \rightarrow N_2 + CO + 2F \qquad \text{------ (1)}$$

$$N_2^{\cdot} + CO_2 \rightarrow N_2 + CO + O \qquad \text{------ (2)}$$

Figure 15. Reaction of nitrogen radical with COF_2 and CO_2.

A nitrogen radical is known to react with COF_2 and CO_2 as shown in Fig. 15 [10]. Nitrogen radicals convert COF_2 and CO_2 to CO. This is one of the reasons why the CO / CO_2 ratio in the etched products shows a high value in 30% NF_3 / 70% O_2 plasma. In addition, nitrogen radicals are predicted to be effective for increasing the concentration of active species for the etching, because they reproduce oxygen and fluorine. This is one of the reasons why NF_3 is more effective in enhancement of etching than CF_4.

These results show that:

(a) rate-limiting steps exist in the case of 100% O_2 plasma etching

(b) the benzene ring is easily fluorinated by fluorine radicals, and the fluorinated benzene ring as well as diphenylether easily break down due to the oxygen radicals

(c) nitrogen converts CO_2 or COF_2 to CO and re-produces atomic oxygen which is the active species for the etching.

4. CONCLUSIONS

(1) The etching rate using O_2 / NF_3 mixture is higher than that with O_2 / CF_4 mixture.

(2) Almost all of the carbon atoms in the PI structure are converted to CO and CO_2 and CO / CO_2 ratio showed highest value at the point of highest etching rate.

(3) The atom contributing to the cleavage of the benzene ring in the PI is oxygen. Rate-limiting steps of the degradation of the benzene ring in the PI structure

exist in the case of 100% O_2 plasma etching. The benzene ring is easily fluorinated by fluorine radicals, and the fluorinated benzene easily breaks down due to the oxygen radicals.

Nitrogen converts CO_2 or COF_2 to CO and re-produces atomic oxygen which is the active species for the etching.

REFERENCES

1. M.K. Ghosh and K.L. Mittal (Eds.), *Polyimides: Fundamentals and Applications*, Marcel Dekker, New York (1996).
2. K.L. Mittal (Ed.), *Polyimides and Other High Temparature Polymers*, Vol.1, VSP, Utrecht (2001).
3. Y. Nakamura, Y. Suzuki and Y. Watanabe, *Thin Solid Films*, **290-291**, 367 (1996).
4. N. Inagaki, S. Tasaka and K. Hibi, *J. Polym. Sci. Part A*, **30**, 1425 (1992).
5. J.E. Klemberg-Sapieha, L. Martinu, E. Sacher and M.R. Wertheimer, *Soc. Plast. Eng. Annu. Tech. Conf.*, **49**, 1093 (1991).
6. E. Occhiello, M. Morra, G. Morini, F. Garbassi and D. Johnson, *J. Appl. Polym. Sci.*, **42**, 2045 (1991).
7. A.D. Katnani, A. Knoll and M.A. Mycek, *J. Adhesion. Sci. Technol.*, **3**, 441 (1989).
8. I.S. Goldstein and F. Kalk, *J. Vac. Sci. Technol.*, **19**, 743 (1981).
9. S. Iwamori, T. Miyashita, S. Fukuda, N. Fukuda and K. Sudoh, *J. Vac. Sci. Technol.*, **B15**, 53 (1997).
10. S. Iwamori, T. Miyashita, S. Fukuda, S. Nozaki, K. Sudoh and N. Fukuda, *J. Adhesion. Sci. Technol.*, **11**, 783 (1997).
11. F.D. Egitto, F. Emmi, R.S. Horwath and V. Vukanovic, *J. Vac. Sci. Technol.*, **B3**, 893 (1985).
12. M. Kogoma and G. Turban, *Plasma Chem. Plasma Proc.*, **6**, 349 (1986).
13. G. Turban and M. Rapeaux, *J. Electrochem. Soc.*, **130**, 2231 (1983).
14. J.A. Folta and R.C. Alkire, *J. Electrochem. Soc.*, **137**, 3173 (1990).
15. A.M. Wrobel, B. Lamontagne and M.R. Wertheimer, *Plasma Chem. Plasma Process.*, **8**, 315 (1988).
16. G. Sauve and M. Moisan, *Appl. Phys. Lett.*, **53**, 470 (1988).
17. V. Vukanovic, G.A. Takacs, E.A. Matuszak, F.D. Egitto, F. Emmi and R.S. Horwath, *J. Vac. Sci. Technol.*, **B 6**, 66 (1988).
18. N.J. Chou, J. Paraszczac, E. Babich, J. Heidenreich, Y.S. Chaug and R.D. Goldblatt, *J. Vac. Sci. Technol.*, **A 5**, 1321 (1987).
19. B. Lamontagne, A.M. Wrobel, G. Jalbert and M.R. Wertheimer, *J. Phys. D*, **20**, 844 (1987).
20. V. Vukanovic, G.A. Takacs, E.A. Matuszak, F.D. Egitto, F. Emmi and R.S. Horwath, *J. Vac. Sci. Technol.*, **A 4**, 698 (1986).
21. A.S. Dawson, M.S.K. Chen and C.C. Dong, *Proc. ISHM*, 512 (1994).
22. J. Leu and K.F. Jensen, *J. Vac. Sci. Technol.*, **A 9**, 2948 (1991).
23. G. Bootcock and J. Cvetanovic, *Can. J. Chem.*, **39**, 2436 (1961).
24. G. Bootcock and J. Cvetanovic, *Can. J. Chem.*, **36**, 623 (1958).
25. R. Klein and M.D. Sheer, *J. Phys. Chem.*, **74**, 613 (1970).
26. R.D. Cadle and J.W. Powers, *J. Phys. Chem.*, **71**, 1702 (1967).
27. C.H. Bamford, *Comprehensive Chemical Kinetics*, p. 111, Elsevier (1976).

Polyimides and Other High Temperature Polymers, Vol. 2, pp. 419–435
Ed. K.L. Mittal
© VSP 2003

Comparison of polyimide film surface properties exposed to real and simulated space environments: Relevance of atomic oxygen effects to wettability in space

MASAHITO TAGAWA,*,1 NOBUO OHMAE,1 YUMIKO NAKATA,2
KEIKO GOTOH3 and MIEKO TAGAWA4

1*Department of Mechanical Engineering, Faculty of Engineering, Kobe University, Rokko-dai 1-1, Nada, Kobe, Hyogo 657-8501, Japan*
2*Seibo Jogakuin Junior College, Fukakusa-tayacho 1, Fushimi, Kyoto 612-0878, Japan*
3*Kyoto University of Education, Fukakusa-fujinomoricho 1, Fushimi, Kyoto 612-8522, Japan*
4*Department of Environmental Engineering, Kanazawa Institute of Technology, Ogi-gaoka 7-1, Nonoichimachi, Ishikawa-gun, Ishikawa 921-8501, Japan*

Abstract—Polyimide surfaces exposed to a real space environment in low Earth orbit (STS-8, STS-46 and SFU missions) are compared to those exposed to a simulated atomic oxygen environment created by a hyperthermal atomic oxygen beam facility. The surface analytical results of the samples exposed to a simulated space environment showed that both the roughness and the oxygen concentration at the atomic oxygen-exposed polyimide surfaces increased with increasing atomic oxygen fluence. The advancing and receding contact angles of water decreased with increasing oxygen concentration at the polyimide surfaces. The Lifshitz-van der Waals component and the acid and base parameters of the surface free energy of polyimide films were calculated from the contact angles of three probe liquids. The base parameter increased with increasing oxygen concentration, whereas the Lifshitz-van der Waals component and the acid parameter did not show any significant change. These analytical results agree with the X-ray photoelectron spectroscopic data showing the formation of surface functional groups due to atomic oxygen exposure. The ground-based experimental results are consistent with the analysis of the three flight samples. It was demonstrated in this study that polyimide surfaces in a low Earth orbit space environment became high-energy surfaces due to the bombardment of atomic oxygen.

Keywords: Polyimide; atomic oxygen; X-ray photoelectron spectroscopy; contact angle; surface free energy; low Earth orbit; space environment.

*To whom all correspondence should be addressed. Phone and Fax: +81-78-803-6126,
E-mail: tagawa@mech.kobe-u.ac.jp

1. INTRODUCTION

The oxidation-induced erosion of polymeric materials in a low Earth orbit (LEO), which is 200-700 km above the sea level, has been one of the major concerns in the field of spacecraft engineering over the last decade. Oxidation of polymers in LEO is due to the collision with ground-state atomic oxygen which is the dominant species in the upper atmosphere of the Earth (>95%) [1]. The flux and collisional energy of atomic oxygen with spacecraft surfaces reach as high as 10^{15} atoms/cm^2/s and 5 eV, respectively, because of the high orbital velocity of spacecraft (8 km/s). Such a high-energy collision with chemically active atomic oxygen leads to the material degradation of the exterior surfaces of spacecraft. Degradation of polymeric materials is so severe that thickness of a few micrometers was lost within only some tens of hours of atomic oxygen exposure in LEO [2]. It has been assessed that after 25 years of operation, a space station will lose 2.5 mm thickness of polymers [3]. Since polyimide has been widely used as a passive thermal control material to cover exterior surfaces of satellites or space structures, polyimide is one of the key materials to be investigated for atomic oxygen reactivity [4]. Some experimental studies have already been carried out both on the ground and as well as during flight [2, 5-13]. However, most ground-based researches have ignored the collisional energy factor because of the difficulty in forming a 5 eV atomic oxygen beam in ground-based facilities. Thus, investigation of the interaction between energetic atomic oxygen and polymer surfaces has not been carried out in detail using ground-based facilities.

On the other hand, while flight experiments have been carried out to study material reactivity with atomic oxygen, they are sometimes disturbed by severe contamination during space flight. Once polyimide surface is contaminated during flight, the atomic oxygen reaction with polyimide is impossible to analyze after retrieval. Moreover, the adsorption of contamination affects the surface properties of polyimide films and influences the function of the passive thermal control system (multilayer thermal insulation, MLI) of spacecraft which uses polyimide at the outermost surface. Due to the requirements for longer operation of spacecraft, avoidance and elimination of contamination of polyimide surfaces becomes more and more important. In general, the surface free energy of a solid is one of the major factors governing the contamination adsorption. Since the polyimide surface is contaminated under atomic oxygen exposure in LEO, the effect of atomic oxygen attack on the MLI polyimide surfaces should also be considered in a ground-based simulation of contamination adsorption. Thus, the evaluation of the surface free energy of polyimide in a simulated LEO space environment is necessary for predicting the contamination behavior at the polyimide surfaces in LEO space environment.

In the present study, we investigated the wettability and surface free energy of atomic oxygen-exposed polyimide surfaces by the sessile drop and the Wilhelmy methods. The laser-detonation atomic oxygen source developed for simulating the LEO space environment was used for this purpose. Chemical and topological

characteristics of the atomic oxygen-exposed polyimide surfaces were characterized using X-ray photoelectron spectroscopy (XPS), Fourier transform infrared
(FT-IR) spectroscopy and atomic force microscopy (AFM). The effect of atomic
oxygen exposure was discussed with respect to contact angles and surface free
energy in conjunction with surface analytical results.

2. EXPERIMENTAL DETAILS

2.1. Materials

Three flight samples were analyzed in this study. A polyimide sample aboard a
space shuttle flight in 1983 (STS-8) was exposed to atomic oxygen environment
for 41.75 hours [11]. Total atomic oxygen fluence was estimated to be 3.5 x 10^{20}
atoms/cm^2. A number of polymeric materials were flown on STS-46 in 1992. In
this flight experiment samples were mounted in a shuttle cargo bay and exposed
to atomic oxygen for 42.25 hours at an altitude of 222 km. Total atomic oxygen
fluence was estimated to be 2.2-2.5 x 10^{20} atoms/cm^2 [12]. A longer exposure was
used on the Space Flyer Unit (SFU) mission in 1995-96. The SFU spacecraft was
launched by a Japanese launch vehicle H-2, and was retrieved by the space shuttle
mission (STS-72). Total atomic oxygen fluence predicted by the MSIS-90 atmos-
pheric model was 5.6-6.2 x 10^{19} atoms/cm^2 [13]. The high orbital altitude of SFU
(482 km) compared with STS-8 (220 km) is responsible for the small atomic oxygen fluence. The polyimide surface located at ram surface of the spacecraft,
where the incident angle of atomic oxygen is 0 degree with respect to surface
normal, was analyzed in this study.

The polyimide samples used in the ground-based experiments in this study
were pyromellitic dianhydride-4, 4'-oxydianiline (PMDA-ODA), commercially
available from DuPont (Kapton-H). The thickness of the polyimide film was 25
µm. Prior to use, the polyimide films were cleaned by ultrasonication in water,
ethanol, and ethyl ether. Water was deionized and doubly distilled using a boro-
silicate glass apparatus.

2.2. Atomic oxygen exposure

In this study, a laser-detonation atomic oxygen source was used to simulate the
atomic oxygen environment in LEO. Figure 1 shows the schematic drawing of the
fast atomic oxygen beam facility including the beam diagnostic system used in
this study [14]. The atomic oxygen source is based on the laser detonation phenomenon and was originally developed by Caledonia *et al.* [15]. This type of
atomic oxygen source uses a pulsed CO_2 laser and a pulsed supersonic valve. The
laser light was focused on the nozzle throat with the concave Au mirror located
50 cm away from the nozzle. The pulsed supersonic valve introduced pure oxygen
gas into the nozzle and the laser light was focused on the oxygen gas in the nozzle. The energy for the dissociation of oxygen molecules into atomic oxygen and

the acceleration of atomic oxygen was provided by the multiphoton absorption process. The atomic oxygen beam, thus generated, was characterized by the time-of-flight (TOF) distribution measured by the quadrupole mass spectrometer installed in the beam line. Translational energies of the species in the beam were calculated using TOF distributions with a flight length of 181 cm. Every mass to charge ratio (m/z) from 1 to 200 was scanned. However, only m/z 16 and 32, which correspond to atomic oxygen and molecular oxygen, respectively, were detected. Figure 2 represents typical translational energy distributions of m/z = 16 and 32. We used the transitional energy distribution relation $P(E) \propto t^2 N(t)$ to calculate the translational energy, where $P(E)$ is the number of atoms with translational energy E, and $N(t)$ the number of atoms arrived at time t (t: the flight time). The mean energy of the hyperthermal atomic oxygen beam was calculated to be 4.7 eV, whereas that of molecular oxygen, which is included in the beam, was 5.0 eV. The atomic oxygen fraction in the beam was approximately 45%, the balance being molecular oxygen (thermal and hyperthermal). The atomic oxygen flux of the beam was measured by a silver-coated quartz crystal microbalance (QCM) with an accommodation coefficient of 1.0. A typical atomic oxygen flux at the sample position, 134 cm away from the nozzle throat, was calculated to be 2.6 x 10^{13} atoms/cm^2/s.

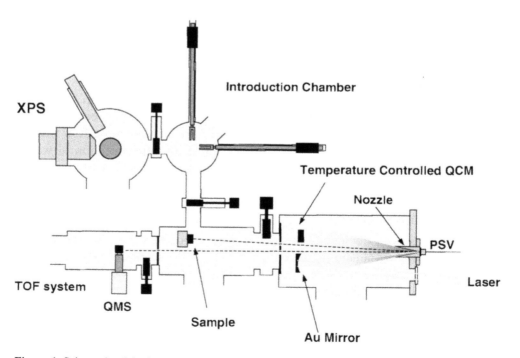

Figure 1. Schematic of the laser detonation atomic oxygen beam facility used in this study.

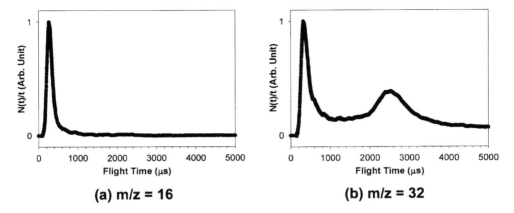

Figure 2. Typical time-of-flight (TOF) spectra of components in the beam. $m/z = 16$ (left panel) and 32 (right panel) correspond to atomic and molecular oxygen, respectively. Mean translational energy of atomic oxygen is 4.7 eV, whereas that of molecular oxygen is 5.8 eV for the peak at 400 µs and 0.1 eV for that at 2500 µs.

2.3. Contact angle measurements

In the present study, the contact angle measurements were carried out using both the sessile drop and the Wilhelmy methods in order to measure the advancing and receding contact angles on each surface of film.

The sessile drop experiments were performed using a video contact angle system. A liquid drop of volume 2-3µl was placed on the PI film surface using a microsyringe [16]. The drop was viewed using a CCD microscope connected to a computer. The drop image was digitized and stored in a computer every second during the measurement. The contact angle and the drop base diameter of each frame were measured by placing markers on the circumference of the drop image on the monitor. The contact angles of water on PI films were plotted as a function of time. The contact angle decreased linearly with time in the initial stage of the measurement. At this stage, the drop base diameter increased and hence water front advanced. Thus, the contact angle extrapolated to time-zero was determined to be the advancing contact angle. After certain period, the drop base diameter decreased due to the evaporation of the liquid, i.e., the water front receded. At this stage, the contact angle decreased and became constant. The contact angle showing the constant value with time was treated as the receding contact angle. The advancing and receding contact angles on pristine polyimide film thus measured (76.9° and 39.5°, respectively), showed a good agreement with those measured by the Wilhelmy method (76.4° and 38.2°). The agreement justifies determination of the advancing and receding contact angles by the sessile drop method.

In the Wilhelmy method, a polyimide film (0.3-0.6 mm in width and 4-5 mm in length) was suspended from the arm of the electrobalance. A beaker containing the probe liquid was raised by an elevator so that the liquid surface reached 2-3 mm higher than the lower edge of the film (immersion) and then moved down to the original position (emersion). A continuous weight recording was made during an immersion-emersion cycle with the three phase boundary velocity of 0.3 mm/min [17]. The contact angle was calculated from the wetting force using the Wilhelmy equation [17]. The effective perimeter of the film was calculated from the wetting force measured at the pentane/air interface. For the atomic oxygen-exposed film, the wetting force on the exposed surface (front side of the sample) was calculated by subtracting the wetting force on the unexposed surface (back side of the sample), using the contact angle on the pristine polyimide, from the total wetting force. From the Wilhelmy method the advancing and receding contact angles on the atomic oxygen-exposed polyimide surface were determined to be 74.4° and 26.3°, whereas using the sessile drop measurement they were 74.0° and 27.6°, respectively. The consistency in the contact angles obtained by the two independent procedures was confirmed.

2.4. Determination of surface free energies of polyimide films

The equation for the work of adhesion, W_A, including the acid-base interaction terms as proposed by van Oss and co-workers [18] can be written as

$$W_A = \gamma_L (1 + \cos \theta)$$

$$= 2 \{(\gamma_S^{LW} \gamma_L^{LW})^{1/2} + (\gamma_S^+ \gamma_L^-)^{1/2} + (\gamma_S^- \gamma_L^+)^{1/2}\}, \tag{1}$$

where θ is the contact angle of the liquid, γ is the surface free energy, the subscripts S and L refer to solid and liquid, respectively, and the superscripts LW, +, and - refer to the Lifshitz-van der Waals component, the Lewis acid parameter, and the Lewis base parameter, respectively [18]. In order to solve the three unknowns for solids, i.e., γ_S^{LW}, γ_S^+ and γ_S^-, the advancing contact angles of three well-characterized liquids (water, diiodomethane, and ethylene glycol) on the polyimide films were measured. The surface free energies of the solids, i.e., γ_S^{LW}, γ_S^+, and γ_S^-, were calculated from equation (1) using the contact angle values, θ, and the published values of γ_L^{LW}, γ_L^+ and γ_L^- [19].

It has been pointed out that equation (1) has the property that negative square roots may appear for particular values of the contact angles [19]. So we took the small negative values of $(\gamma_S^+)^{1/2}$ as effectively zero. On the other hand, the values of $(\gamma_S^-)^{1/2}$ were always positive in this study.

3. RESULTS AND DISCUSSION

3.1. Analysis of flight samples

Polyimide film surfaces exposed to a real space environment on space shuttle missions or other flight programs have been analyzed since the 1980s. It is now

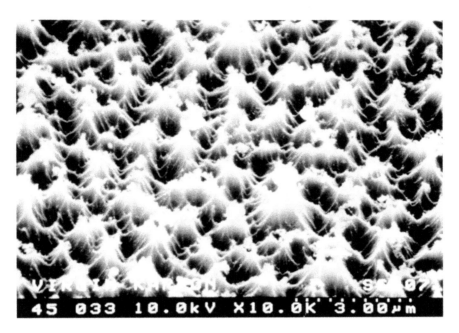

Figure 3. Secondary electron micrograph of the atomic oxygen exposed-polyimide. Atomic oxygen fluence: 3×10^{20} atoms/cm^2.

widely known that the polymer surfaces exposed to atomic oxygen environment in LEO show a characteristic morphological feature called the "shag carpet-like structure" (Figure 3). Due to this rough surface structure, polyimide films lose their transparency, and the optical performance as a thermal control device is affected. The reaction yield of polyimide with atomic oxygen has been studied in many space shuttle flights and the erosion yield of 3.0×10^{-24} cm^3/atom was established. This value is now widely accepted as a standard for measuring atomic oxygen fluence (Kapton equivalent fluence) [2]. The synergistic effects of ultraviolet light and atomic oxygen exposures on the erosion of polyimide have been reported by some researchers in ground-based experiments [20, 21]. However, no significant synergistic effects were observed in flight missions [12]. This point is still unclear at this moment.

The results from XPS analysis on the STS-8 and STS-46 samples are summarized in Table 1. An increase in the oxygen concentration at the atomic oxygen-exposed polyimide was detected, but no significant change in the shape of the C1s core level spectrum was observed for both STS-8 and STS-46 samples compared with the control sample as shown in Figure 4. These analytical results showed that the atomic oxygen reacted with polyimide surfaces and gasified the carbon backbone as CO and/or CO_2. This reaction is responsible for the mass loss of the film. The C1s XPS spectra also suggested that the surfaces themselves were not heavily oxidized even though atomic oxygen attacked the surfaces. However, a decrease

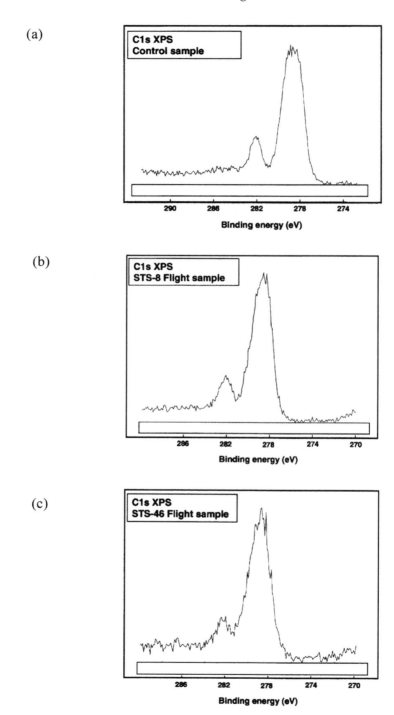

Figure 4. C1s XPS spectra of polyimide surfaces; (a) control sample, (b) STS-8 flight sample, and (c) STS-46 flight sample.

in surface oxygen concentration due to air exposure was observed as described later. Thus, it is natural to consider that the polyimide surfaces in space are much more oxidized due to atomic oxygen exposure.

The contact angles of water on the flight sample surfaces were measured. Although decreases in the contact angles were observed for both STS-8 and STS-46 samples, a significant increase in the advancing contact angle was measured in the case of SFU sample as listed in Table 1. XPS analysis evidences that a heavy silicone contamination, which is supposed to be talc from the FT-IR spectrum (Figure 5), covers the SFU sample surface. The origin of the contamination is not clear, but the silicone-based adhesive which is widely used in spacecraft systems is reported to be one of the possibilities [22].

Table 1.
Atomic concentration and contact angles on the control (not exposed to atomic oxygen) and flight samples

	Atomic concentration (%)			Contact angle (deg.)	
	C	O	N	Advancing	Receding
Control	77.8	15.8	6.4	76.4	38.2
STS-8	73.2	20.1	6.7	71.6	25.4
STS-46	71.2	23.5	5.3	62.6	21.0
SFU*	27.8	41.4	0	109.4	37.0

*balance Si

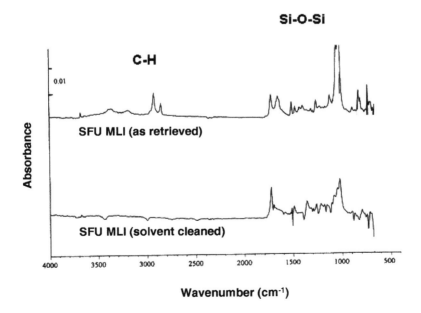

Figure 5. FT-IR differential spectra of SFU MLI samples. Silicone-based contamination was removed by ultrasonication using ethyl ether (bottom, see text).

3.2. Ground-based experimental results

(a) Effect of air exposure on atomic oxygen-exposed polyimide surfaces

An atomic oxygen-exposed polyimide film with an atomic oxygen fluence of 7.6 $\times$ 10^{17} atoms/cm^2 was stored in a Teflon container under a clean air environment of class 10,000, and the XPS spectrum was measured periodically in order to determine the stability of the surface chemistry of the atomic oxygen-exposed polyimide. The results are shown in Table 2. It was observed that N was relatively stable, whereas the atomic concentration of C increased and that of O decreased with air exposure. Namely, the carbon and oxygen concentrations before air exposure were 62.4% and 28.5%, respectively, but, with an air exposure for 220 days, the carbon and oxygen concentrations changed to 69.2% and 23.3%. These values are close to the analytical results on STS-8 and STS-46 flight samples [11, 12].

Figure 6 shows the XPS C1s core level spectra of the atomic oxygen-exposed polyimide films with an atomic oxygen fluence of 7.6×10^{17} atoms/cm^2 (a) without air exposure and (b) with air exposure of 220 days. The XPS spectrum of control sample (not exposed to atomic oxygen) is shown in Figure 4 (a). Note that Figures 6 (a) and (b) were obtained from the same sample. It is obvious from a comparison of Figure 6 (b) and Figure 4 that both flight and laboratory samples exposed to ambient air gave similar C1s XPS spectra. However, before ambient air exposure, the atomic oxygen-exposed surfaces showed the C1s spectrum containing a larger amount of carbonyl (287.7 eV) and ketone (288.9 eV) groups, i.e., a highly oxidized state. These XPS results indicate that the decrease in O shown in Table 2 was due to the decrease of carbonyl and ketone groups at the atomic oxygen-exposed polyimide surfaces. From the experimental results shown in Figure 6, it is suggested that the highly oxidized surfaces produced on exposure to atomic oxygen were deoxidized by ambient air exposure, and the influence of atomic oxygen exposure was lost [23]. Deoxidation process involves removal of volatile products (maybe carbon monoxide and/or carbon dioxide). Diffusion of the high-energy surface functional groups into the bulk is also a possibility.

Table 2.

Effect of air exposure on the surface composition of atomic oxygen-exposed polyimide. Atomic oxygen fluence was 7.6×10^{17} atoms/cm^2

Air exposure (day)	Atomic concentration (%)		
	C	O	N
0	62.4	28.5	9.1
4	64.1	26.6	9.3
10	68.7	25.1	6.2
220	69.2	23.3	7.5

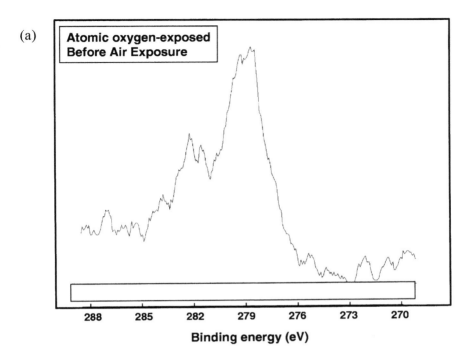

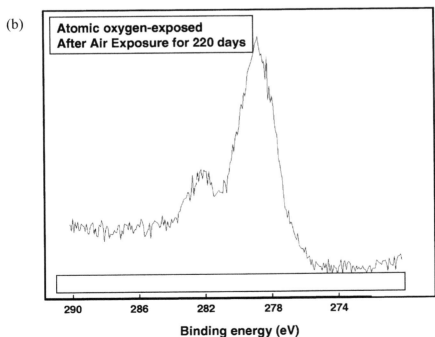

Figure 6. C1s XPS spectra of atomic oxygen-exposed polyimide before (a) and after (b) ambient air exposure for 220 days. Atomic oxygen fluence was 7.6×10^{17} atoms/cm^2.

(b) Effect of atomic oxygen exposure on contact angles and surface free energies
Figure 7 shows the contact angles of water as a function of atomic oxygen (AO)
fluence. With increasing atomic oxygen fluence, both the advancing and receding
contact angles decreased, i.e., the wettability increased. The oxygen concentration
at the polyimide surfaces measured by XPS also increased with atomic oxygen
fluence. The contact angles were replotted against the oxygen concentration, and
the results are shown in Figure 8; both advancing and receding contact angles decreased linearly with increasing oxygen concentration. The behavior of contact
angles as observed in the present study is consistent with the theoretical calculations of the contact angles on heterogeneous surfaces as reported by Johnson and
Dettre [24], i.e., both advancing and receding contact angles decrease with increasing percentage of the high-energy region for heterogeneous surfaces such as
polyimide film surfaces.

Surface roughness is another factor influencing the contact angles. The surface
roughness of the pristine and the atomic oxygen-exposed polyimide films were
examined by AFM. It was observed that the roughness of the polyimide surfaces
increased with increasing atomic oxygen fluence. This is considered to be an early
stage of the formation of the "shag carpet-like structure" which is widely known
as a surface texture of the polyimide exposed to LEO space environment (Figure
3). The roughness of the atomic oxygen-exposed polyimide surfaces was small on

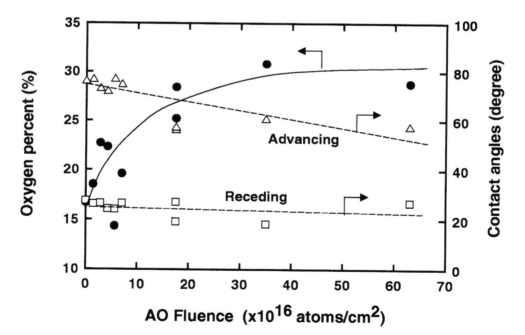

Figure 7. Advancing (△) and receding (□) contact angles of water on atomic oxygen-exposed polyimide films as a function of total atomic oxygen fluence. The oxygen concentration at the surface is
also indicated by ● symbol.

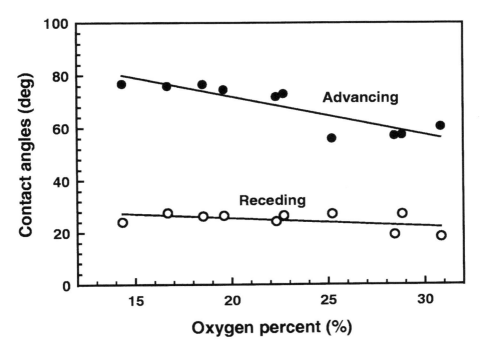

Figure 8. Advancing (●) and receding (○) contact angles of water on polyimide films as a function of surface oxygen concentration.

the order of a few nanometers. Miller *et al*. have demonstrated that the advancing and contact angles on vacuum-deposited poly(tetrafluoroethylene) films observed by the sessile drop method are influenced strongly by nanometer size surface roughness [27]. In the present study, the advancing and receding angles on the atomic oxygen-exposed polyimide measured by the sessile drop method were in fairly agreement with those by the Wilhelmy method, which correspond to the intrinsic angles of the low-energy and high-energy regions on the heterogeneous surface [24, 28]. Such experimental fact seems to show that surface roughness in the range of a few nanometers does not affect the observed contact angles for the oxygen-exposed polyimide films.

The Lifshitz-van der Waals (L-W) component and Lewis acid and base parameters of the surface free energy of polyimide films are plotted on a log scale in Figure 9 as a function of oxygen concentration. The base parameter increased with increasing oxygen concentration whereas the L-W component and acid parameter remained constant. This result may be explained by the introduction of surface functional groups due to atomic oxygen bombardment. The formation of carbonyl and carboxyl groups at the atomic oxygen-exposed polyimide surface was clearly detected by XPS as described earlier. These surface functional groups are expected to play the role of electron donors. Although the acid parameter appears to decrease with increasing oxygen concentration, the tendency may not be

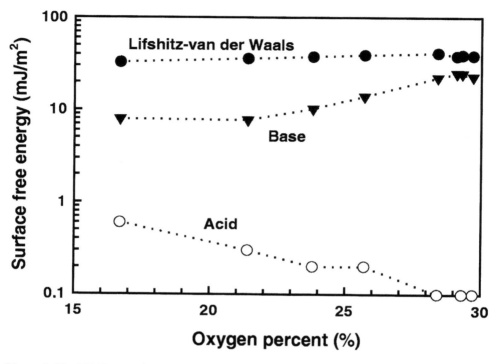

Figure 9. The Lifshitz-van der Waals component (●), the acid parameter (○), and the base parameter (▼) of the surface free energy of polyimide films as a function of surface oxygen concentration.

significant because of their very small values. It is, therefore, concluded that the increase in the wettability of polyimide by the atomic oxygen beam exposure is mainly caused by the increase in the base parameter of the surface free energy due to the formation of carbonyl and/or carboxyl groups.

3.3. Contact angles of water on polyimide surfaces exposed to real and simulated atomic oxygen environments

As shown in the previous sections, XPS results on atomic oxygen-exposed surfaces coincide with the actual flight data. Carbon concentrations at the atomic oxygen beam exposed polyimide surfaces after air exposure (Table 2) are close to those of flight samples (STS-8 and STS-46 in Table 1). The XPS C1s spectra indicated similar surface chemical states between flight and ground-based samples (see Figures 4 and 6). A detailed comparison is reported elsewhere [23]. Additionally, contact angles of water on the polyimide surfaces exposed to atomic oxygen by ground-based facility and by STS flight were compared in this study. The results are shown in Figure 10. Figure 10 is the same plot as Figure 8, but the data points for the flight samples are also added. As shown in Figure 10, both advancing and receding contact angles for the STS-8 and STS-46 samples lie on the

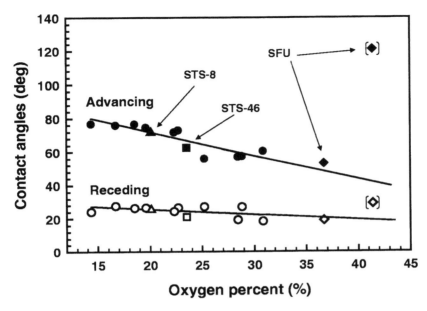

Figure 10. Advancing (●) and receding (○) contact angles of water on polyimide films as a function of surface oxygen concentration (ground-based data). Data from flight samples are added; STS-8 (advancing: ▲, receding: △), STS-46 (advancing: ■, receding: □), and SFU (advancing: ◆, receding: ◇). The data points for the SFU sample before cleaning are shown in brackets.

same line as the ground-based data. In contrast, the data points for the SFU flight sample, which are marked by diamonds in Figure 10, show large contact angles. From the XPS and FT-IR measurements, it was suggested that the surface was covered by a silicone-based contamination, i.e., talc-like material (see Figure 5). The large contact angle and the high oxygen concentration of the SFU sample is due to the silicone-based contamination adsorbed during the space flight. In order to remove the contamination layer from the polyimide surface, the SFU sample was cleaned by ultrasonication in ethyl ether. After the cleaning procedure, advancing and receding contact angles on the SFU sample decreased to 53° and 19°, respectively. The surface oxygen concentration also decreased to 36.7%. The decrease in oxygen concentration is due to the removal of silicone-based contamination consisting of SiOx which may be formed by the atomic oxygen-induced oxidation of adsorbed silicone contamination. These values gave the data plot for the SFU sample on the same line as determined for the ground-based simulated samples. The finding that the original polyimide surface of the SFU sample with the contamination layer showed a high oxygen concentration of 36.7% and small contact angles (53° and 19°) suggested that the polyimide surface was modified by atomic oxygen exposure before being contaminated. Namely, the contamination adsorption on the SFU spacecraft did not occur in the beginning of the mission which is contradictory to the scenario of the vaporization of a silicone-based adhesive to be origin of the contamination. The other possible scenario is that the contamination was due to vaporization of the thermal paint/tile of the orbiter body flap [29].

The experimental results in this study suggest that the increase in wettability of polyimide surfaces due to atomic oxygen bombardment generally occurs in an LEO space environment. Such a hydrophilic surface easily allows for adsorption of contamination. Fluorinated polymers such as PTFE are known to be resistant to atomic oxygen attack as well as being extremely hydrophobic. The surface fluorination of polyimide by plasma treatment [30] or using hyperthermal atom beam technique [31] could be a solution which satisfies the requirements both for protecting from atomic oxygen attack as well as for preventing contamination adsorption.

4. CONCLUSIONS

The polyimide surfaces exposed to an atomic oxygen beam with translational energy of 4.7 eV were characterized by AFM, XPS and contact angle analyses. The AFM and XPS data showed that both the roughness and surface oxygen content at the polyimide surface increased after atomic oxygen exposure. The contact angles of water on the atomic oxygen-exposed polyimide film were found to decrease linearly with increasing oxygen concentration at polyimide surfaces. It was also made clear that the decrease in the contact angle was due to the increase in the base parameter of the surface free energy. These experimental results showed that adsorbed oxygen at the atomic oxygen-exposed polyimide surface formed surface functional groups that behaved as electron donors. It was suggested that the increase in wettability of polyimide surfaces due to atomic oxygen bombardment occurred in LEO space environment. Such a hydrophilic polyimide surface allows for easy adsorption of contamination and may accelerate the performance loss of the thermal control system of spacecraft. Therefore, for the future application of polyimide in space use, the development of atomic oxygen resistant polyimide with a low energy surface for the prevention of contamination adsorption is required.

Acknowledgments

The authors would like to thank Y. Suzuki and K. Miyashita of Nara Women's University, H. Kinoshita of Osaka University for their help with experiments. S.Y. Chung, and T. K. Minton of Jet Propulsion Laboratory, California Institute of Technology, and S. Kibe of National Aerospace Laboratory of Japan are all acknowledged for providing flight samples. A part of this work was supported by the Grant-in-Aid for Scientific Research from the Ministry of Education, Sports, Culture, Science and Technology, Japan; and by the Space Utilization Promotion Fund from the Japan Space Forum.

REFERENCES

1. B. J. Anderson, *NASA Technical Memorandum-4527* (1994).
2. J. C. Gregory, P. N. Peters and J. T. Swann, *Appl. Optics*, **25**, 1290 (1986).
3. J. T. Visentine, *NASA Technical Memorandum -100459*, Vol. 3 (1988).
4. T. K. Minton, *Jet Propulsion Laboratory Publication*, **95-17** (1995).
5. A. F. Whitaker and B. Z. Jang, *J. Appl. Polym. Sci.*, **48**, 1341-1357 (1993).
6. M. A. Golub and T. Wydeven, *Polym. Degradation Stability*, **22**, 325-338 (1998).
7. M. A. Golub, T. Wydeven and R. D. Cormia, *Polym. Commun.*, **29**, 285-288 (1988).
8. J. I. Kleiman, Y. I. Gudimenko, Z. A. Iskanderova, R. C. Tennyson, W. D. Morison, M. S. McIntyre and R. Davidson, *Surface Interface Anal.*, **23**, 335-341 (1995).
9. H. Kinoshita, M. Tagawa, M. Umeno and N. Ohmae, *Trans. Japan Soc. Aeronautics Space Sci.*, **41**, 94-99 (1998).
10. H. Kinoshita, M. Tagawa, M. Umeno and N. Ohmae, *Surface Sci.*, **440**, 49-59 (1999).
11. R. H. Liang, A. Gupta, S. Y. Chung and K. L. Oda, *Jet Propulsion Laboratory Publication*, **87-25** (1987).
12. D. E. Brinza, S. Y. Chung, T. K. Minton and R. H. Liang, *Jet Propulsion Laboratory Publication*, **94-31** (1994).
13. T. Fukatsu, Y. Torii, Y. Koyari, K. Fusegi and M. Ichikawa, *Proceedings of the 7th International Symposium on Materials in Space Environment*, pp. 287-292, Toulouse, France (1997).
14. H. Kinoshita, J. Ikeda, M. Tagawa, M. Umeno and N. Ohmae, *Rev. Scientific Instrum.*, **69**, 2273-2277 (1998).
15. G. E. Caledonia, R. H. Krech and D. B. Green, *AIAA J.*, **25**, 59 (1987).
16. J. Drelich, J. L. Wilbur, J. D. Miller and G. M. Whitesides, *Langmuir*, **12**, 1913-1022 (1996).
17. M. Tagawa, K. Gotoh, A. Yasukawa and M. Ikuta, *Colloid Polym. Sci.*, **268**, 589-594 (1990).
18. C. J. van Oss, R. J. Good and M. K. Chaudhury, *Langmuir*, **4**, 884-891 (1988).
19. R. J. Good, in: *Contact Angle, Wettability and Adhesion*, K. L. Mittal (Ed.), pp. 3-36, VSP, Utrecht, The Netherlands (1993).
20. V. E. Skurat, E. A. Barbashev, I. A. Budashov, Y. I. Dorofeev, A. P. Nikiforov, A. I. Ternovoy, M. van Eesbeek and F. Levadau, *Proceedings of the 7th International Symposium on Materials in Space Environment*, pp. 267-279, Toulouse, France (1997).
21. M. Tagawa, T. Suetomi, H. Kinoshita, M. Umeno and N. Ohmae, *Trans. Japan Soc. Aeronautics Space Sci.*, **42**, 40-45 (1999).
22. R. Yokota, A. Ohnishi, Y. Hashimoto, K. Toki, S. Kuroda, K. Akahori and H. Nagano, *Proceedings of the 7th International Symposium on Materials in Space Environment*, pp. 293-299, Toulouse, France (1997).
23. M. Tagawa, K. Yokota, N. Ohmae and H. Kinoshita, *J. Spacecraft Rockets*, **39**, 447-451 (2002).
24. R. E. Johnson Jr. and R. H. Dettre, in: *Wettability*, J. C. Berg (Ed.), pp. 1-73, Marcel Dekker, New York (1993).
25. R. N. Wenzel, *Ind. Eng. Chem.*, **28**, 988-994 (1936).
26. R. N. Wenzel, *J. Phys. Chem.*, **53**, 1466-1467 (1949).
27. J. D. Miller, S. Veeramasuneni, J. Drelich, M. R. Yalamanchili and G. Yamauchi, *Polym. Eng. Sci.*, **36**, 1849-1855 (1996).
28. A. W. Neumann and R. J. Good, *J. Colloid Interface Sci.*, **38**, 341-358 (1972).
29. C. R. Maag, private communication (1997).
30. M. R. Wertheimer, G. Czeremuszkin, J. Cerny, J. E. Klemberg-Sapieha, L. Martinu and W. Kremers, *Proceedings of the 7th International Symposium on Materials in Space Environment*, pp. 393-402, Toulouse, France (1997).
31. M. Tagawa, unpublished.

Polyimides and Other High Temperature Polymers, Vol. 2, pp. 437–446
Ed. K.L. Mittal

Development and optimization of a laser carbonized polyimide film as a sensor substrate for an all-polymer humidity sensor

JOHN M. INGRAM, JAMES A. NICHOLSON and
AUGUSTUS W. FOUNTAIN III*

Photonics Research Center and Department of Chemistry, United States Military Academy, West Point, NY 10996

Abstract—Kapton® polyimide film was studied as a viable sensor substrate using a capacitive type humidity sensor design. An argon ion laser was used to pyrolyze the Kapton® film to form raised, carbonized filaments. Using a graphics program and a set of computer controlled servo mirrors, an interdigitated circuit was formed on the surface of the polymer. This circuit was coated with a hygrosensitive polymer, HMPTAC (2-hydroxy-3-methacryloxypropyltrimethylammonium chloride), to form an all-polymer humidity sensor. The sensor substrate was optimized for sensitivity by varying pyrolysis wavelength and energy density. The results showed that the polyimide based sensor performed as well as the most commonly used silica/gold and alumina/gold sensor substrates.

Keywords: Polyimide; polymeric sensors; humidity sensor.

1. INTRODUCTION

The problem of developing small, real time point detection systems for chemical agents is being addressed all over the world [1-17]. All proposed approaches use various detection schemes, but virtually all use the same substrate to build their sensing device: a silicon or alumina platform with deposited gold or platinum wiring. This substrate introduces special problems to the sensor design. Silicon substrates with gold wiring are expensive and sometimes need to be post-processed at the manufacturing facility [1-4]. Both silica and alumina substrates have shown adhesion problems [1]. Aluminum is a relatively good thermal conductor, which can lead to sensor film breakdown. Silicon is brittle and alumina is rigid which could lead to sensor damage under torsion or bending stress. While cost issues and fabrication problems are not that critical in the laboratory environment, future

---

*To whom all correspondence should be addressed. Phone: (845) 938-8624, Fax: (845) 938-3062, E-mail: augustus.fountain@usma.edu

research should work in parallel to develop a sensor substrate that would minimize these potential issues in a real-world application.

We believe a carbonized polymer is a candidate for an ideal substrate. Kapton is an example of an insulating polymer that can be laser-carbonized to form conducting filaments. This substrate is flexible, durable, inexpensive and easy to manufacture. While the Kapton polyimide is a good insulator, research has shown that the laser-carbonized filaments are fair conductors [18, 19, 24]. Additional opportunities to increase the effectiveness of some sensor designs lie in the fact that the carbonized filaments are porous and their resistivity can be manipulated easily during processing. Researchers discovered that lines as narrow as 10 μm in width and specific resistance as low as 0.01 $\Omega\cdot$cm could be achieved using cw lasers in the 350–380 nm UV region [22]. Raman spectra of the filaments have shown that the material is principally a "glassy" carbon which is composed of small crystallites [18].

During the last year, our laboratory has developed a capacitive-type humidity sensor prepared on a Kapton substrate using filaments carbonized with UV or visible laser light as electrodes and a coating of 2-hydroxy-3-methacryloxypropyl-trimethylammonium chloride (HMPTAC) as the hygrosensitive dielectric. Research into current humidity sensors has led to the adoption of the general procedure put forth by the Sakai Group [25]. Their procedure used silicon / gold circuit design and coated it with the HMPTAC film. Using a variation of this procedure, our experiment focused on developing a sensor using Kapton as the substrate to see if a viable all-polymer sensor could be developed. Kapton circuits were produced through carbonization of a Kapton film via an argon ion laser operating in the visible and UV regions. The variables involved in making the carbonized circuits were tested to find the lowest resistivity. Substrates were then coated with HMPTAC under a nitrogen atmosphere. The Kapton sensor was then tested using impedance measurements under varying conditions to assess its ability to measure humidity changes. The Kapton-based circuit was able to determine humidity changes in a controlled atmosphere between 5–95% relative humidity. The sensor showed a smooth response to water vapor and displayed no deterioration over four months of use. While research continues with the humidity sensor to optimize the design and sensitivity parameters, we have demonstrated that an all-polymer sensor is feasible.

2. EXPERIMENTAL

2.1. Pyrolysis procedure

A Coherent Innova 200 argon ion laser operating at 514 nm was used to pyrolyze Kapton samples (Figure 1). Pyrolysis took place in a custom-made chamber consisting of a 10×5×8 cm black aluminum box with a removable lid sealed with a rubber gasket. The chamber had a 0.635 cm (1/4") Swagelok inlet and exhaust port to deliver the argon gas flow at around 2 L/min throughout pyrolysis. The front side of the chamber contained a 1 cm thick quartz window. The chamber

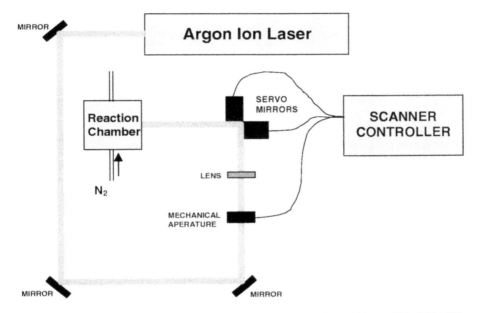

Figure 1. Laser pyrolysis experimental set-up. Argon ion laser is tunable to 333, 364, 488 and 514 nm.

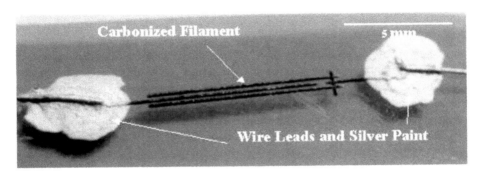

Figure 2. Tongue and fork sensor design. The carbonized filaments are raised above the Kapton surface.

served two purposes: the atmosphere surrounding the Kapton sample could be controlled and the hazardous byproducts of pyrolysis could be exhausted safely.

500HN (127 μm thick) Kapton sheets were used as received from DuPont. The sensor pattern (Figure 2) was laser carbonized onto 3×1 cm piece of Kapton under an argon atmosphere in the pyrolysis chamber. A General Scanning Inc. (Billerica, Massachusetts) DE 2000 scanner was used to control laser pattern on the Kapton surface. Laser beam was focused using a CaF_2 plano-convex lens having a focal length of 20 cm.

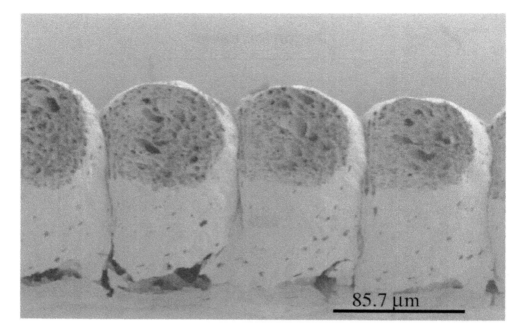

Figure 3. SEM image of a larger carbonized filament. Filament formed with 5 passes of the laser at a scan rate of 111 mm/s and total laser energy of 180 J/cm². The filament exhibited a resistance of 1.29 kΩ·cm.

The laser pyrolysis mechanism of Kapton is described in detail elsewhere [18-24]. Carbonization of a typical sensor was carried out with an energy density of 0.60–1.80 J/cm². Energy density is a function of scan speed, laser power, total number of scans, and the pyrolysis wavelength. Figure 3 is a scanning electron microscope image of a filament formed from multiple scans from a laser with an energy density of 1.80 J/cm² at a pyrolysis wavelength of 514 nm.

2.2. Sensor fabrication

The sensor substrate was taped to a glass slide to provide rigidity throughout fabrication and testing. Wire leads were attached to the ends of the carbonized filaments using silver paint (Figure 2). The substrate was drop coated using a 7.22 mg/mL HMPTAC/methanol solution in a dry nitrogen atmosphere. The sensor was dried at 60°C for 24 hours before testing.

99.9% anhydrous methanol was used as purchased. HMPTAC was purchased from Aldrich. Solid HMPTAC was placed in the methanol at room temperature and sonicated for 2 hours. The saturated solution was separated from any remaining solid material.

2.3. Testing

The test station consisted of a Dell computer running LabVIEW to control and record test conditions. Two computer controlled Aalborg GFC17 mass flow controllers were used to deliver known amounts of dry nitrogen and nitrogen saturated with water vapor to a Plexiglas 15×10×8 cm test box. The glass slide containing the sensor was placed opposite an Omega RH-62C-MV analog hygrometer in the test box. The sensor and the Omega hygrometer were separated by 2 cm. The hygrometer output was sent through a National Instruments SB68 controller to a National Instruments Data Acquisition Card (DAC). The two leads from the fabricated sensor were fed through Kelvin clips into a QuadTech LCR 1970 meter. The LCR meter output was then sent to LabVIEW using a GPIB cable.

3. RESULTS AND DISCUSSION

Impedance measurements of the sensor showed a continuous response over 5–95% relative humidity (Figure 4). The sensor substrate was optimized for dynamic range by varying the filament conductivity, sensor area, and HMPTAC thickness. The sensor was also tested for voltage, temperature and frequency dependence. The optimized sensor was compared to a sensor formed on an alumina/gold substrate fabricated by the Sakai group. Unless otherwise stated, all measurements were carried out at 20°C and 1000 Hz.

3.1. Pyrolysis energy density

By varying the surface pyrolysis energy density, filaments of different size and conductivity can be built (Figure 3). Initial experiments showed that both filament conductivity and size had only a small effect on sensor performance. The ranges of filament conductivity and size were purposely limited for this series of experiments. Further experimentation in this area is still needed. It is hoped that in the future, a greater control over the filaments properties (size, conductivity, porosity) will lead to a more tailored sensor substrate.

3.2. Sensor area

Sensor area is the amount of polymeric surface that interacts with the analyte gas between carbonized filaments. The different combinations of circuit design and filament spacing can be used to manipulate sensor area. The basic sensor design was based on a capacitive type circuit with impedance being the measured circuit parameter. Several circuit designs were used ranging from the tongue and fork design illustrated in Figure 2 to a six fingered comb design. Filament spacing is defined as the distance measured between two filaments. The filament spacing can be viewed as the dielectric distance between two capacitive plates. By varying the design configuration and the filament spacing, the area between filaments was optimized for the widest dynamic sensor response range.

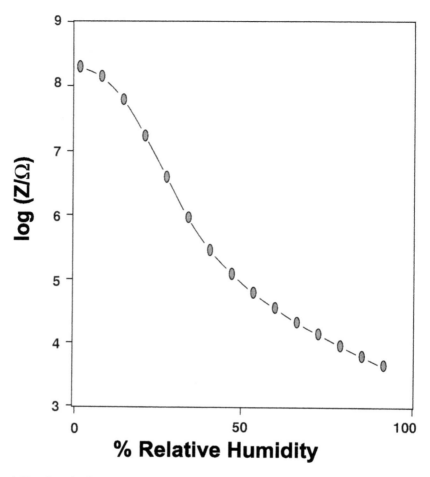

Figure 4. Benchmark all-polymer sensor performance. Sensor was optimized for sensitivity and response time.

A wider dynamic range was shown to correlate to a smaller interaction area. The widest sensor range was found using a simple tongue and fork design (Figure 2) with a 20 μm filament spacing. This behavior should seem intuitive. It follows that the smaller the sensor area, the greater the sensitivity to a change in the environment and the wider the sensor's response. Research using microdots of polymeric sensing materials have shown similar results [5].

3.3. HMPTAC thickness

Both the sensitivity and response time of the sensor were directly related to the thickness of the HMPTAC sensing layer. The thickness was defined as the amount of the HMPTAC polymer drop coated onto the sensor area. For thinly coated sensors a small increase in the dynamic range was observed as the thick-

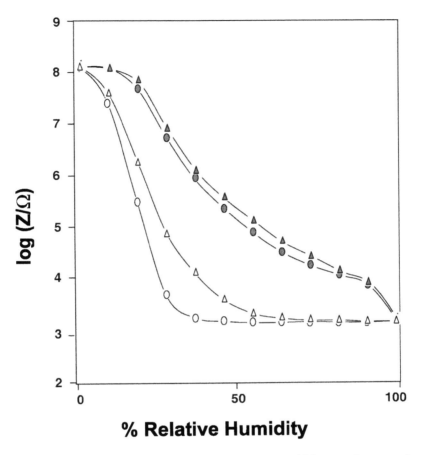

Figure 5. Change in sensor response at various HMPTAC coating thicknesses (in terms of $\mu g/mm^2$) and rates of change in relative humidity. ▲ 34.6 $\mu g/mm^2$ at 1.3% RH/min, ◉ 46.0 $\mu g/mm^2$ at 1.3% RH/min, △ 34.6 $\mu g/mm^2$ at 51% RH/min, ○ 46.0 $\mu g/mm^2$ at 51% RH/min.

ness increased from 5.7 $\mu g/mm^2$ to 17.2 $\mu g/mm^2$. However, at a thickness of 22.9 $\mu g/mm^2$, the sensor exhibited a dramatic increase in dynamic range from 22–95% RH to 5–95% RH. Curiously, this large increase was unique to the small sensor area design. For sensors with wider spaced filaments (60–150 μm) or greater sensor area (four or six fingered comb designs), the range remained constant at 22–95% RH. A thickness of HMPTAC at 22.9 $\mu g/mm^2$ was optimal for the small sensor design. Additional drop coating to increase the HMPTAC layer thickness increased sensitivity only slightly, but greatly increased sensor response time. This effect can be seen in Figure 5. These sensors showed little hysteresis when the relative humidity was changed slowly (1.3% RH per minute) (Figure 6). The sensors demonstrated a response time of 30 seconds for a step increase in relative humidity from either 40% to 90% or from 90% to 40%.

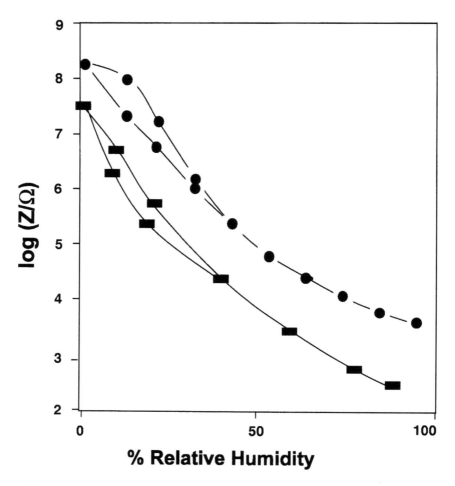

Figure 6. Kapton all-polymer sensor response at 20°C (●) compared with Sakai alumina/gold substrate sensor at 30°C (■). Both sensors demonstrate a similar response and sensitivity, with the differences in actual impedance most likely being a function of temperature.

3.4. Frequency and voltage dependence

Sensor impedance response was measured from 100 Hz to 1 MHz. As shown in Figure 7, the largest linear dynamic range was obtained at a frequency of 100 Hz. While 100 Hz produced a larger linear dynamic range, at low humidity the noise was significant, thus reducing the precision of the measurement. At 1000 Hz the noise contribution is less significant and produces a linear range almost as large. These results correspond to research done on other sensors using alumina or silica substrates [26]. The sensor performance was measured using test voltages from 0.02 to 0.80 V. The sensors were not affected by the changes in the test voltage.

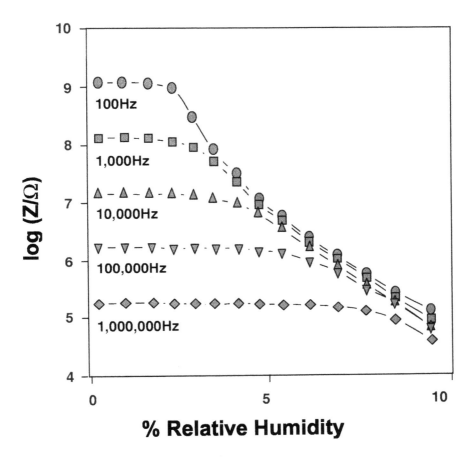

Figure 7. Frequency dependence of the all-polymer sensor response.

3.5. Comparison with alumina/gold substrate

The benchmark sensor was compared directly to results reported by the Sakai research group of Ehime University of Japan [25]. Sakai's sensor was tested at 30°C while ours was tested at 20°C. We tested a sample sensor from 20°C–40°C and observed no difference in dynamic range. The sensor performance compared well with the alumina/gold sensor substrate fabricated by the Sakai group (Figure 6). However, the two sensors were tested at different temperatures and this could have some impact on a direct comparison of the slight hysteresis noticed in the low humidity range.

4. CONCLUSIONS

This research has demonstrated that an all-polymer sensor substrate is a viable substitute for the alumina/gold substrate used in current sensor designs. The dynamic sensing range, response time and hysteresis were comparable to the alumina/gold sensor substrate. Small differences in performance can be attributed to sensor design rather than substrate characteristics. Our future research into chemical sensing will test the versatility of the Kapton substrate in other sensing applications. As a proof of concept, the morphology of the carbonized Kapton filaments and its effect on the sensor's performance is currently being investigated using humidity as the sensed analyte of interest.

Acknowledgements

We would like to thank Dr. Matthew H. Ervin of the Army Research Labs for recording the electron scanning microscope images of our carbonized filaments. We would also like to thank DuPont for generously donating the Kapton.

REFERENCES

1. J.A. Dickson and R.M. Goodman, Proc. IEEE International Symposium on Circuits and Systems, **4**, 341 (2000).
2. A.C. Partridge, M.L. Jansen and W.M. Arnold, Mater. Sci. Eng. C, **12**, 37 (2000).
3. W. Lu and G.G. Wallace, Electroanalysis, **9**, 454 (1997).
4. R.C. Hughes, M.P. Eastman, W.G. Yelton, A.J. Ricco, S.V. Patel and M. W. Jenkins, J. Electrochem. Soc., **146**, 3907 (1999).
5. A.R. Hopkins and N.S. Lewis, Anal. Chem. **73**, 884 (2001).
6. A. Guadarrama, J.A. Fernandez, M. Iniguez, J. Souto and J.A. de Saja, Sensors Actuators B, **77**, 401 (2001).
7. A.L. Jenkins, R. Yin and J.L. Jensen, Analyst, **126**, 798 (2001).
8. G.M. Murray, B.R. Arnold, A.C. Kelly and O.M. Uy, Proc. SPIE, **4206**, 131 (2001).
9. P. Marin-Franch, D.L. Tunnicliffe and D.K. Das-Gupta, Mater. Res. Innovations, **4**, 334 (2001).
10. R.C. Hughes, W.G. Yelton, K.B. Pfeifer and S.V. Patel, J. Electrochem. Soc, **148**, H37 (2001).
11. L. Chen, D.W. McBranch and D.G. Whitten, Proc. Natl. Acad. Sci. (USA), **96**, 12287 (1999).
12. F. Opekar and K. Stulk, Analy. Chimica Acta, **385**, 151 (1999).
13. H.J. Lee, P.D. Beattie and H.H. Girault, J. Electroanalytical Chem., **440**, 73 (1997).
14. G. Gauglitz and A. Bretch, Anal. Chim. Acta., **347**, 219 (1997).
15. P.M. Kramer, B.A. Baumann and P.G. Stoks, Anal. Chim. Acta., **347**, 187 (1997).
16. T.A. Sergeyeva, S.A. Piletsky, A.A. Brovko, E.A. Slinchenko, L.M. Sergeeva and A.V. El'skaya, Anal. Chim. Acta., **392**, 105 (1999).
17. B.J. Doleman, M.C. Longergan, E.J. Severin, T.P. Vaid and N.S. Lewis, Anal. Chem., **70**, 4177 (1998).
18. R. Srinivasan, R.R. Hall, W.D. Wilson, W.D. Loehle and D.C. Allbee, Synth. Metals, **66**, 301 (1994).
19. T. Feurer, R. Sauerbrey, M.C. Smayling and B.J. Story, Appl. Phys. A, **56**, 275 (1993).
20. J. Davenas, Appl. Surf. Sci., **36**, 539 (1989).
21. S.F. Dinetz, E.J. Bird, R.L. Wagner and A.W. Fountain III, J. Anal. Appl. Pyrolysis, **63**, 241 (2002).
22. R. Srinivasan, R.R. Hall and D.C. Allbee, Appl. Phys. Lett., **63**, 3382 (1993).
23. M. Schumann, R. Sauerbrey and M.C. Smayling, Appl. Phys. Lett., **58**, 428 (1991).
24. G.H. Wynn and A.W. Fountain III, J. Electrochem. Soc., **144**, 3769 (1997).
25. Y. Sakai, M. Matuguchi, Y. Sadaoka and K. Hirayama, J. Electrochem. Soc., **140**, 432 (1993).
26. Y. Li, M.J. Lang, N. Camaioni and G. Casalbore-Miceli, Sensors Actuators B, **77**, 625 (2001).

Polyimides and Other High Temperature Polymers, Vol. 2, pp. 447–458
Ed. K.L. Mittal
© VSP 2003

Fluorinated copolyimides for microelectronics applications

J.G. LIU, Z.X. LI, M.H. HE, F.S. WANG and S.Y. YANG*

State Key Laboratory of Engineering Plastics, Center for Molecular Science, Institute of Chemistry, Chinese Academy of Sciences, Beijing 100080, China

Abstract—Fluorinated copolyimides derived from 4,4'-oxydiphthalic anhydride (ODPA) with 4,4'-oxydianiline (ODA) and trifluoromethyl-containing aromatic diamines have been synthesized and characterized. The trifluoromethyl-containing diamines included 2,4-diamino- 3'-trifluoromethyl-azobenzene (TFDA), 2,4-diamino-1-(4'-trifluoromethyl)phenoxy aniline (TFPA), 3,5-diamino-1-(4'-trifluoromethyl)phenoxy benzamide (TFPB), 3,5-diamino-1-(3'-trifluoromethyl) benzamide (TFB), 1,4-bis(4'-aminophenoxy)-2-(3'-trifluoromethylphenyl)benzene (TFPPA), 3,5-diaminotrifluorome-thylbenzene (TFmDA), 4,4'-diamino-4"-(p-trifluoromethylphenoxy) triphenylamine (DATF) and 4-(4'-trifluoromethyl)phenoxy-2,6-bis(4"-aminophenyl)pyridine (TFAP). Copolyimide films, produced by casting the copoly(amic acid) solution followed by thermal imidization, exhibited high thermal stability and high mechanical properties. The copolyimides showed UV-visible absorption cut-off between 330 and 350 nm and pretilt angles as high as 20° for nematic liquid crystals (NLCs), making them highly potential candidates for advanced liquid crystal display (LCD) applications.

Keywords: Fluorinated copolyimides; thermal stability; LCD; pretilt angle.

1. INTRODUCTION

Polyimides have been widely used in advanced microelectronics industry as passivation or stress-relief layers for high density electronic packaging, interlayer dielectrics for wafer-level semiconductor fabrication, or alignment layers for nematic liquid crystals in advanced liquid crystal display devices (LCDs) owing to their outstanding thermal, mechanical and electrical characteristics [1-4]. Because of their portability, low driven voltage, low energy consumption and high resolution quality, LCDs are widely used in notebook computers, mobile telephones, TV sets and so on [5-6], in which polyimides, as the liquid crystal alignment layer on the substrate surface, have played an very important role in the device manufacturing. Several methods are currently used to achieve a uniform alignment of liquid crystals, including oblique evaporation [7], mechanical rub-

---

*To whom all correspondence should be addressed. Phone: +86-10-62564819,
Fax: +86-10-62559373, E-mail: public305@infoc3.icas.ac.cn

bing [8-10] and non-rubbing methods [11-12]. Of these, the mechanical rubbing has been the commonly accepted pathway in the LCD industry for high volume production. The pretilt angle of the liquid crystal is one of the key parameters in an LCD device, which is closely related to the device performance. For instance, a high pretilt angle, which could prevent the creation of reverse tilt disclination in LCDs, is usually required to prevent strip domains in super-twisted nematic LCDs (STN-LCDs) and to properly operate surface-stabilized ferroelectric LCDs (SSFLCDs) [13]. Hence, the generation and control of high pretilt angles in LCD devices have been investigated by many researchers [14-17].

Polyimide coatings rubbed mechanically with cotton have been widely employed as liquid crystal alignment layers in LCD manufacturing due to their uniformity and durability. However, conventional polyimides used in LCDs have some drawbacks, such as poor processability and low pretilt angle. With the rapid development of advanced LCD devices with high density and high performance, the improvement in polyimide features as alignment layers becomes a critical factor. Many attempts have been made to increase the pretilt angle of the nematic liquid crystals (NLCs) on a polyimide surface. It was found that polyimide with long alkyl side chains could generate high pretilt angles by mechanical rubbing [18-21]. However, a polyimide containing alkyl side chains tends to be hydrophobic, which reduces the wettability of liquid crystal [22], and its thermal property, to some extent, is sacrificed. Recently, high pretilt angles of liquid crystals on rubbed polyimide with helical backbone segments and trifluoromethyl moieties were reported [23-26]. And polyimides with trifluoromethyl substituted phenyl or biphenyl groups on the backbone were also reported to show high pretilt angles for liquid crystals [22].

A series of aromatic diamines with trifluoromethyl substituted phenyl groups were synthesized, which were employed to polymerize with an aromatic dianhydride, 4,4'-oxydiphthalic anhydride (ODPA), and another aromatic diamine, 4,4'-oxydianiline (ODA), to afford fluorinated copolyimides. The thermal and mechanical properties of the copolyimides were investigated, and the effects of the chemical structures of polymers on the pretilt angles for liquid crystals are also discussed.

2. EXPERIMENTAL

2.1. Materials

4,4'-Oxydiphthalic anhydride (ODPA, Shanghai Chemspec. Corp.) was recrystallized from acetic anhydride prior to use. 4,4'-oxydianiline (ODA) (Beijing Chemical Reagents Corp.) was recrystallized from ethanol prior to use. Commercially available *N*-methyl-2-pyrrolidinone (NMP), *m*-cresol, *N,N'*-dimethylformamide (DMF) and *N,N'*-dimethylacetamide (DMAc) were purified by vacuum distillation over CaH_2 prior to use.

2.2. Measurements

Differential scanning calorimetry (DSC), thermogravimetric analysis (TGA) and dynamic mechanical analysis (DMA) were performed on a Perkin-Elmer 7 series thermal analysis system at a heating rate of 10°C/min. FT-IR spectra were recorded on a Perkin-Elmer 782 Fourier transform spectrophotometer. Ultraviolet-visible (UV/Vis) spectra were recorded on a Hitachi U-3210 spectrophotometer at room temperature. Copolyimide samples were dried at 100°C for 1 hr before analysis to remove the adsorbed moisture on polyimide films. Inherent viscosity was measured using an Ubbelohde-viscometer with 0.5 g.dl⁻¹ NMP solution at 30°C. The water uptake was determined by immersing the polyimide film $(3.0 \times 1.0 \times 0.05 \text{ cm}^3)$ in water at 25°C for 24 hr, followed by weighing. The electrical properties were measured on a Hewlett-Packard 4284A Presion LCR meter at room temperature.

2.3. Pretilt angle measurements

Copoly (amic acid) solution with 4% solid content in NMP was coated onto indium-tin oxide (ITO) glass by spin-coating at 3000 rpm, which was then thermally imidized at 80°C/1 hr, 180°C/1 hr and 250°C/1 hr. The cured copolyimide film was undirectionally rubbed with a cotton cloth on a rubbing machine with a rubbing pressure R (equation 1) [28], where N is the number of rubbings; M is the length (mm) of rubbing cloth that contacts a certain point of copolyimide film; ω is the speed of rubbing roller (rpm); r is the radius (mm) of the roller and v is the velocity (mm/s) of the substrate stage. The rubbing condition employed were N=1, M=0.3 mm, v=10 mm·s⁻¹ and r=50 mm. The rubbing pressure could be adjusted by changing the ω values from 1000 to 3000 rpm. A sandwich LCD cell was assembled with two pieces of glass substrate, ensuring that the rubbing direction was anti-parallel to each other; the space between the glass substrates was adjusted to be 50 μm by applying a glass fiber-reinforced adhesive at the perimeter of the substrate. The cell was filled with a liquid crystal (E70, Merck, Germany) and sealed with an epoxy adhesive. The pretilt angles and electro-optical properties of the LCD cell were measured with an LCT-5016 spectrometer.

$$R=NM[2\pi r\omega/(v-1)] \tag{1}$$

2.4. Monomer synthesis

2,4-Diamino-3'-trifluoromethylazobenzene (TFDA) [29], 2,4-diamino-1-(4'-tri-fluoromethyl) phenoxyaniline (TFPA) [29], 3,5-diamino-1-(4'-trifluoromethyl) phenoxybenzamide (TFPB) [29], 1,4-bis(4'-aminophenoxy)-2-(3'-trifluoromethyl phenyl)benzene (TFPPA) [30], 4,4'-Diamino-4"-(p-trifluoromethylphenoxy) triphenylamine (DATF) [31] and 4-(4'-trifluoromethyl)phenoxy-2,6-bis(4"-aminophenyl)pyridine (TFAP) [31] were first synthesized in this laboratory. 3,5-

diamino-1-(3'-trifluoromethyl)-benzamide (TFMB) [32] and 3,5-diaminotrifluoro-methylbenzene (TFmDA) [33] were synthesized according to the literature.

2.5. Copolymer synthesis

In a typical experiment, ODPA (3.1022 g, 10 mmol) was added to a stirred solution of ODA (1.0012 g, 5 mmol) and TFAP (2.4876 g, 5 mmol) in NMP (25 ml). The mixture was stirred at room temperature for 24 hr in argon to yield a viscous copoly(amic acid) solution. Copolyimide (PI-8) (see Scheme 1) was obtained by thermally imidizing the copoly(amic acid) solution using the following sequence: 80°C/1 hr, 150°C/1 hr, 250°C/1 hr. The PI-8 film obtained was pale yellow in color.

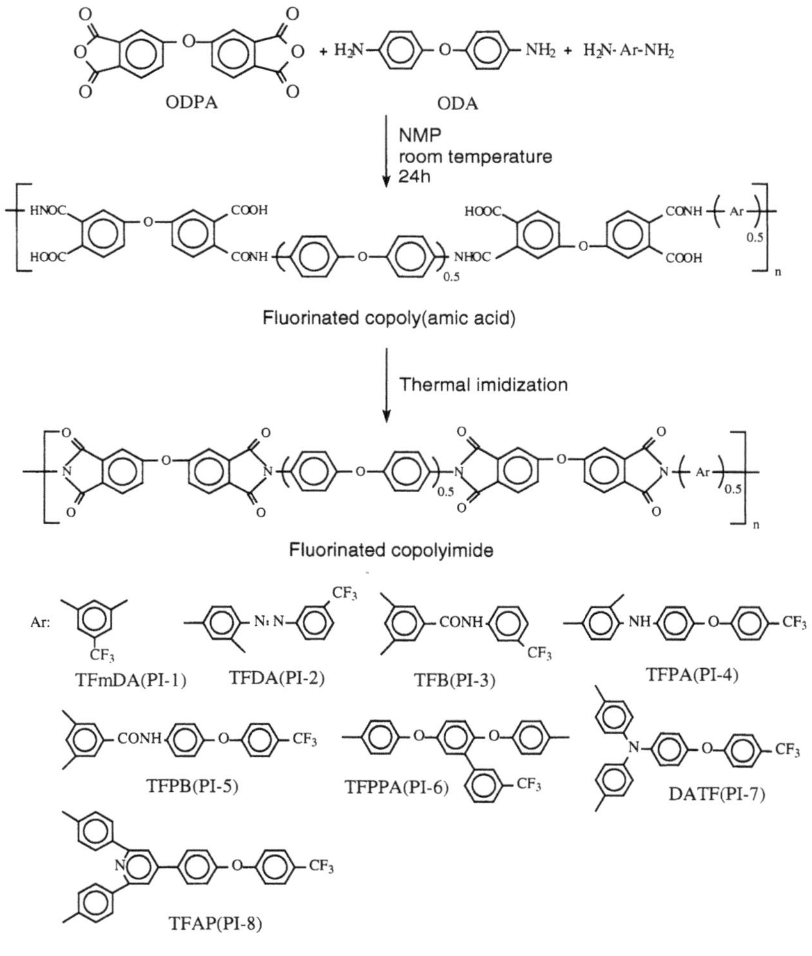

Scheme 1. Synthesis of copolyimides derived from ODPA, ODA and fluorinated aromatic diamines.

PI-1(ODPA/ODA/TFmDA), PI-2(ODPA/ODA/TFDA), PI-3(ODPA/ODA /TFB), PI-4(ODPA/ODA/TFPA), PI-5(ODPA/ODA/TFPB), PI-6(ODPA/ODA /TFPPA) and PI-7(ODPA/ODA/DATF) were prepared at mole ratios of ODPA:ODA: fluorinated diamine = 1:0.5:0.5 in the same manner as described above.

3. RESULTS AND DISCUSSION

3.1. Copolymer synthesis

Copolyimides were prepared by thermal imidization of copoly(amic acid)s, which were prepared by reacting ODPA with ODA and one of the fluorine-containing diamines. The molar ratio of ODPA:ODA: diamine was 1:0.5:0.5, which ensured that the polycondensation reaction proceeded completely to afford high molecular weight copolymers. Scheme 1 shows the synthesis pathway of the copolyimides, in which the trifluoromethyl groups are located at different positions in polymer chains, either in the main chain (PI-1) or in the side phenyl group (PI-2 to PI-8). The inherent viscosity of the copoly(amic acid)s ranged between 0.74 and 1.25 dL/g (Table 1), which was determined by Ubbelohde-viscometer with 0.5 g.dL^{-1} NMP solution at 30°C. It can be seen that PAA-6 showed the highest inherent viscosity (1.25) and PAA-7 showed the lowest value (0.74), implying that the polymer based on TFPPA possesses higher molecular weight than that derived from DATF. Strong and flexible films could be prepared by casting copoly(amic acid) solutions, followed by thermally baking using the following sequence:

Table 1.
Inherent viscosity of PAAs and film properties of copolyimides

PAA	η_{inh} (dL/g)	PI	T_g(°C)		T_d(°C)[a]		$T_{5\%}$[b] (°C)	$T_{10\%}$[b] (°C)	Rw[c] (%)	W_u[d] (%)
			DSC	DMA	T_{d1}[a]	T_{d2}[a]				
PAA-1	1.21	PI-1	271.3	255.3	648.5	–	551.2	608.9	54	0.39
PAA-2	0.86	PI-2	255.9	240.5	456.2	596.1	493.1	533.7	55	0.56
PAA-3	1.02	PI-3	274.1	258.0	441.1	625.9	476.9	523.1	49	0.58
PAA-4	0.89	PI-4	248.5	233.1	463.3	663.6	504.0	578.2	58	0.52
PAA-5	0.96	PI-5	264.6	258.5	443.0	640.6	478.5	538.9	55	0.61
PAA-6	1.25	PI-6	232.4	215.1	598.7	–	534.1	583.6	46	0.42
PAA-7	0.74	PI-7	271.7	256.7	481.0	651.4	542.0	606.1	61	0.44
PAA-8	1.17	PI-8	288.9	273.6	590.0	–	579.9	621.0	60	0.46

a) T_d: Onset decomposition temperature; T_{d1}: The first decomposition temperature; T_{d2}: The second decomposition temperature
b) $T_{5\%}$, $T_{10\%}$: Temperatures for 5% and 10% weight loss, respectively
c) Residual weight retention (Rw %) at 700°C
d) W_u: Water uptake (%)

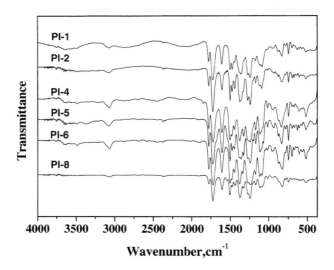

Figure 1. FT-IR spectra of copolyimides derived from different fluorinated aromatic diamines.

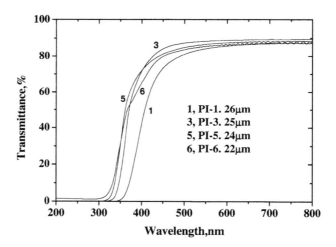

Figure 2. UV-vis spectra of copolyimide films of different thicknesses prepared by ODPA, ODA and different fluorinated aromatic diamines.

80°C/3 h, 120°C/1 h, 200°C/1 h and 250°C/1 h. Figure 1 compares the FT-IR spectra of the copolyimides derived from the different fluorinated aromatic diamines. The absorptions at about 1780 and 1720 cm^{-1}, assigned to the asymmetric and symmetric C=O stretching vibrations of imide groups, respectively, are observed in all of the copolymers, and the absorption at about 1380 cm^{-1} is assigned to the C-N stretching vibration of imide groups. Figure 2 depicts the UV-visible spectra of the copolyimides. It can be seen that the cut-off wavelengths are usually located between 320 and

350 nm, and the transmittance at 450 nm is >80% for most of the fluorine-containing copolyimides. Obviously, the high transparency is attributed to the presence of fluorine-containing groups in the polymer backbone. The copolyimides exhibited low water uptake values in the range of 0.39-0.61%, which might be attributed to the presence of hydrophobic trifluoromethyl substituents in the polymer backbone. PI-1 showed the lowest value, whereas PI-6 had the highest one.

3.2. Thermal properties

Table 1 shows the thermal properties of the copolyimides including onset decomposition temperature (T_d), temperature at 5% weight loss ($T_{5\%}$) and at 10% weight loss ($T_{10\%}$), and the glass transition temperature (T_g). The copolyimides showed different thermal decomposition behaviors (Figure 3). A two-stage decomposition is observed for PI-2 to PI-5 and PI-7, while a one-stage behavior is observed for PI-1, PI-6 and PI-8. In general, the first stage thermal decomposition started at 441.1-481.0°C and the second stage at 596.1-651.4°C for PI-2 to PI-5 and PI-7. The two-stage decomposition behavior might be explained by the presence of weak bonding (-N=N-, -NH- or -CONH-) between the side chain substituents and polymer backbone. The weakly bonded side subsituents would decompose first in the heating process. The glass transition temperatures are located between 232.4-288.9°C as determined by DSC (Figure 4) and 215.1-273.6°C as determined by DMA (Figure 5). For instance, PI-5 has a Tg value of 264.6°C as determined by DSC and 258.5°C as determined by DMA. The decomposition temperatures at 5% and 10% original weight losses are in the ranges of 476.9-579.9°C and 523.1-621.0°C, respectively. And the residual weights of the copolyimides determined by TGA at 750°C in nitrogen are in the range of 46%-61%. Clearly, the copolyimides exhibited high thermal stability.

3.3. Mechanical and electrical properties

Table 2 shows the mechanical and electrical properties of the copolyimides. Thin films (6.35 mm x 127 mm x 22-24 μm) were tested for tensile properties in an Instron universal testing machine (UTM) at 25°C according to GB10410-79. The copolyimide films have tensile strengths of 98.6-137.8 MPa, elongation at break of 8-16%, and tensile modulus of 0.94-1.29 GPa. The surface resistance (R_s) and volume resistance (R_v) are 1.8-7.2 x 10^{14} Ω and 1.2-7.2 x 10^{15} Ω·cm, respectively, indicating that the polyimide films have good electrical insulation properties.

3.4. Pretilt angles, θ_p

Usually, nematic liquid crystal molecules can be aligned on a mechanically rubbed polyimide coating surface along the rubbing direction. TFT-AM-LCDs require that the liquid crystals should have a low value of pretilt angle (1°-5°), however, the liquid crystals in super-twisted nematic LCD (STN-LCD) devices must have higher pretilt angles (4-30°), preferable between 5 and 15° [34]. Table 3

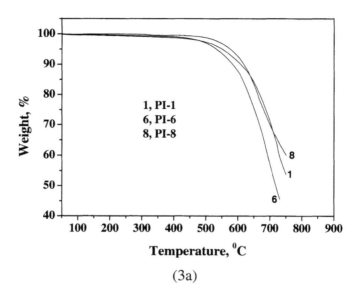

(3a)

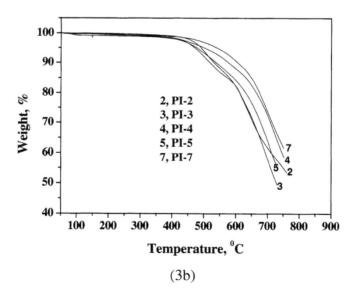

(3b)

Figure 3. TGA curves of copolyimides derived from ODPA, ODA and different fluorinated diamines (In nitrogen, heating rate: 10°C/min); (3a) one-step thermal decomposition behavior; (3b)
two-step thermal decomposition behavior.

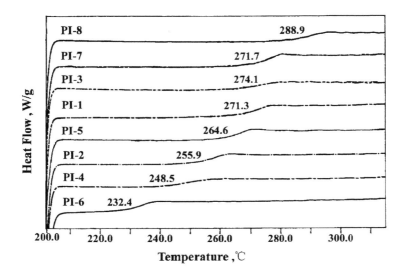

Figure 4. DSC curves of copolyimides derived from ODPA, ODA and different fluorinated aromatic diamines (In air, heating rate: 10°C/min).

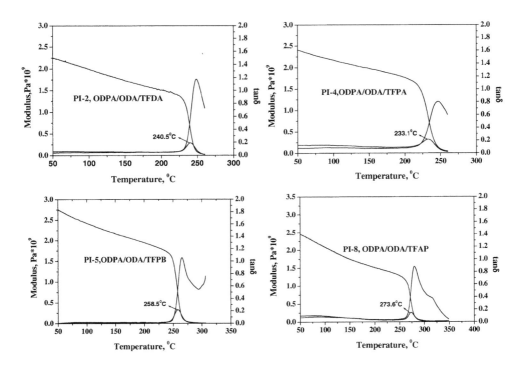

Figure 5. DMA curves of copolyimide films prepared from ODPA, ODA and different fluorinated aromatic diamines (In air, heating rate: 5°C/min).

shows the pretilt angles θ_p of the commercial liquid crystal (E-70) on the rubbed copolyimide coating surfaces. Copolyimides showed different alignment properties with pretilt angles from 2.2-20°. PI-5 exhibited the highest pretilt angle of 20°, followed by PI-4 (10.2°) and PI-8 (9.0°), however, PI-1 and PI-6 had low pretilt angles of only 2-3°. It seems that copolyimides with trifluoromethyl substituted long side substituents in the side phenyl groups, such as PI-4 and PI-5, show high θ_p values and those with short trifluoromethyl substituents in polymer backbone, such as PI-1 and PI-6, have low pretilt angles. This might be interpreted in term of the smaller optical retardation caused by mechanical rubbing for the polymer with short side substituents than those with long trifluoromethyl moieties in polymer backbone. It should be emphasized that thermal annealing the LCD test cells at 150°C for 1 h did not result in obvious change in pretilt angles, indicating that the copolyimides prepared had stable alignment properties. This was mainly attributed to the high thermal-resistance features of the copolyimide backbones.

Table 2.
Mechanical and electrical properties of copolyimides

PI	d^a (μm)	E_B^b (%)	T_S^b (MPa)	Y_M^b (GPa)	R_S^c (Ω) (×10^{14})	R_V^c (Ω·cm) (×10^{15})
PI-1	22	12	117.6	0.98	6.4	7.2
PI-2	24	8	103.4	1.29	2.8	2.6
PI-3	23	11	124.4	1.13	3.2	4.6
PI-4	25	9	98.6	1.09	1.8	1.2
PI-5	22	13	127.6	0.98	7.2	6.8
PI-6	23	14	132.4	0.94	4.7	3.4
PI-7	24	10	104.4	1.04	3.4	2.5
PI-8	22	16	137.8	0.84	4.9	5.4

a) d: Film thickness;
b) E_B: Elongation at Break; T_S: Tensile Strength; Y_M: Young's Modulus;
c) R_S: Surface Resistance; R_V: Volume Resistance

Table 3.
Pretilt angles (°) of E-70 on rubbed copolyimide layers

PI	$\theta_{p\,1}$	$\theta_{p\,2}^a$
PI-1	2.2	2.0
PI-2	4.1	4.0
PI-3	5.2	5.2
PI-4	10.2	10.0
PI-5	20.0	19.8
PI-6	3.0	3.0
PI-7	7.2	7.1
PI-8	9.0	9.0

a) Pretilt angles after thermal treatment at 150°C for 1 h

4. CONCLUSIONS

A series of fluorinated copolyimides derived from 4,4'-oxydiphthalic anhydride (ODPA) with 4,4'-oxydianiline (ODA) and trifluoromethyl-containing aromatic diamines were synthesized and characterized. Strong and flexible copolyimide films could be produced by casting the copoly(amic acid) solution followed by thermal imidization. The copolyimides exhibited high thermal stability and high mechanical properties. The UV-visible absorption cut-off between 330-350 nm and pretilt angles as high as 20° for nematic liquid crystals (NLC) were observed for these copolyimides, and thus are considered good potential candidates for advanced liquid crystal display devices.

Acknowledgment

Funding from the National Natural Science Foundation of China (NSFC) for Distinguished Young Scholars (No. 59925310) is gratefully acknowledged.

REFERENCES

1. M.K. Ghosh and K.L. Mittal (Eds.), *Polyimides: Fundamentals and Applications*, Marcel Dekker, New York (1996).
2. D. Wilson, H.D. Stenzenberger and P.M. Hergenrother (Eds.), *Polyimides*, Chapman & Hall, New York (1990).
3. C. Feger, M.M. Khojasteh and J.E. McGrath (Eds.), *Polyimides: Materials, Chemistry and Characterization*, Elservier, Amsterdam (1989).
4. K.L. Mittal (Ed.), *Polyimides: Synthesis, Characterization and Applications*: Vols 1&2, Plenum, New York (1984).
5. M. Mistutake and M. Yoichi, Proc. Display Device'98, pp. 10-12 (1998).
6. U. Yasuhiro and M. Shoichi, Proc. SPIE, **3143**, 2-8 (1997).
7. J. Janing, Appl. Phys. Lett., **21**, 173 (1972).
8. J.H. Park, J.C. Jung, B.H. Sohn, S.W. Lee and M. Ree, J. Polym Sci., Part A: Polym. Chem. Ed, **39**, 3622-3632 (2001).
9. J.H. Park, B.H. Sohn, J.C. Jung, S.W. Lee and M. Ree, J. Polym Sci., Part A: Polym. Chem. Ed, **39**, 1800-1809 (2001).
10. B.S. Ban, Y.N. Rim and Y.B. Kim, Liquid Crystals, **27**, 125-130 (2000).
11. M. Murata, M. Uchida, Y. Nakajima and K. Saitoh, Jpn. J. Appl. Phys., **32**, L679 (1993).
12. M. Schadt, K. Schmitt, V. Kozinkov and V. Chigrinor, Jpn. J. Appl. Phys., **31**, 2155 (1992).
13. N.A. Clark and S.T. Lagerwall, Appl. Phys. Lett. **36**, 899 (1980).
14. D.S. Seo, K. Muroi and S. Kobayashi, Mol. Cryst. Liq. Cryst. **213**, 223 (1992).
15. L. Li, J. Yin, Y. Sui, H.J. Xu, J.H. Fang, Z.K. Zhu and Z.G. Wang, J. Polym. Sci., Part A: Polym. Chem. Ed, **38**, 1943-1950 (2000).
16. A. Seeboth, Displays, **20**, 131-136 (1999).
17. K.R. Brown, D.A. Bonnell and S.T. Sun, Liquid Crystals, **25**, 597-601 (1998).
18. S.I. Kim, M. Ree, T.J. Shin and J.C. Jung, J. Polym. Sci., Part A: Polym. Chem. Ed, **37**, 2909-2921 (1999).
19. K.W. Lee, S.H. Paek, A. Lien, C. Durning and H. Fukuro, Macromolecules, **29**, 8894 (1996).
20. S.I. Kim, M. Ree, T.J. Shin, J.C. Jung, J. Polym. Sci., Part A: Polym. Chem., **37**, 2909-2921 (1999).

21. D.S. Seo, K. Araya, N. Yoshida, M. Nishikawa, Y. Yabe and S. Kobayashi, Jpn. J. Appl. Phys., **34**, L503 (1995).
22. T. Nihira, Y. Miyamoto, H. Endo and T. Abe, European Patent 0682283 (1995).
23. D.S. Seo and S. Kobayashi, Liquid Crystals, **27**, 883-887 (2000).
24. D.S. Seo and S. Kobayashi, Appl. Phys. Lett. **66**, 1202 (1995).
25. D.S. Seo, M. Nishikawa and S. Kobayashi, Liquid Crystals, **22**, 515-517 (1997).
26. D.S. Seo, S. Kobayashi, M. Nishikawa and Y. Yabe, Liquid Crystals, **19**, 288 (1995).
27. D.S. Seo, Liquid Crystals, **26**, 1615-1619 (1999).
28. D.S. Seo, S. Kobayashi and M. Nishikawa, Appl. Phys Lett. **61**, 2392 (1992).
29. J.G. Liu, M.H. He, Z.X. Li, Z.G. Qian, F.S. Wang and S.Y. Yang, J. Polym. Sci., Part A: Polym. Chem. Ed, **40**, 1572-1582 (2002).
30. H.W. Zhou, J.G. Liu, Z.G. Qian, S.Y. Zhang and S.Y. Yang, J. Polym. Sci., Part A: Polym. Chem. Ed, **39**, 2404-2413 (2001).
31. J.G. Liu, Z.X. Li, J.T. Wu, H.W. Zhou, F.S. Wang and S.Y. Yang, J. Polym. Sci., Part A: Polym. Chem. Ed, **40**, 1583-1593 (2002).
32. K.Y. Choi, M.H. Yi and M.Y. Jin, US patent 6,013,760 (2000).
33. M.K. Gerber, J.R. Pratt, A.K.St. Clair and T.L.St. Clair, Polym. Preprints **31(1)**, 340-341 (1990).
34. B.C. Auman, E. Bohm and B. Fiebranz, US patent 5,759,442 (1998).

Polyimides and Other High Temperature Polymers, Vol. 2, pp. 459–465
Ed. K.L. Mittal
© VSP 2003

Fabrication of thin-film transistors on polyimide films

HELENA GLESKOVA* and SIGURD WAGNER

Department of Electrical Engineering, Princeton University, Princeton, NJ 08544

Abstract—We describe polyimide films as substrates for active thin-film electronics. Amorphous silicon thin-film transistors were fabricated on 51 μm thick polyimide foil (Kapton®E) at a maximum process temperature of 150°C. Kapton E was selected for its chemical stability, high softening or glass transition temperature, relatively low coefficient of humidity expansion, negligible heat shrinkage, low water and oxygen permeability, coefficient of thermal expansion comparable to that of thin-film electronics, and low surface roughness. The fabricated transistors have off-current of ~ 2 x 10^{-12} A, on-off current ratio ~ 10^7, threshold voltage ~ 2 V, electron mobility ~ 0.5 $cm^2V^{-1}s^{-1}$, and subthreshold slope ~ 0.5 V/decade. These performance parameters are similar to those of transistors fabricated at 150°C on glass substrates.

Keywords: Polyimide; flexible electronics; amorphous silicon thin-film transistors.

1. INTRODUCTION

Novel display applications, such as electronic paper, smart labels, and displays for vehicular applications, require flexible thin-film transistor (TFT) backplanes [1-10]. Future sensor skins and electrotextiles will be based on flexible circuits. For these applications, the traditional glass substrate for the thin-film transistors of active-matrix liquid crystal displays must be replaced with foils of organic polymers or metals. Stainless steel foils are suitable as substrates for amorphous [11], nanocrystalline, and polysilicon thin film transistors [12] without much change in the processes used for making them on glass. Since steel is an electrical conductor and its surface roughness is much larger than that of glass, it must be electrically insulated and planarized by using, for example, spin-on-glass.

Another large class of flexible substrates are organic polymers. Their low cost, transparency in the visible part of the light spectrum, and wide variety are attractive attributes. However, in thin-film electronics other characteristics, such as chemical stability, high softening or glass transition temperature, a low coefficient of thermal expansion (comparable to that of the materials used for thin-film elec-

---

*To whom all correspondence should be addressed. Phone: (609) 258-4626,
Fax: (609) 258-3585, E-mail: gleskova@ee.princeton.edu

tronics), negligible shrinkage during circuit fabrication, small coefficient of humidity expansion, low solubility for water, low water and oxygen permeability, and low surface roughness, become important. The standard process temperature for the fabrication of amorphous silicon (a-Si:H) based electronics is ~ 250-350°C. If the process temperature is lowered somewhat, the high-temperature polyimides meet well many of the requirements listed above.

2. SUBSTRATE SELECTION

A key aspect of using organic polymers as substrates for TFT electronics is the initial surface passivation. The passivation layer "seals" the polymer foil and converts the chemistry of the polymer to the chemistry of the passivating material. A standardized substrate chemistry is important, because device fabrication requires complex chemical sequences that cannot be altered to adapt their compatibility with various substrate materials. Semiconductor devices are highly susceptible to contamination, and the passivation layer substantially reduces the possibility of contamination of the TFT layers during their growth, caused by the outgassing, and oxygen and water release from the polymer foil. The passivation layer also serves as an adhesion layer between the TFT layers and the substrate.

During TFT fabrication the polymer foil may be temporarily attached to a rigid substrate, such as a silicon or glass wafer, or used as a free-standing substrate [13-19]. If it is attached, the thermal expansion of the compliant foil substrate during temperature cycling will be constrained by the coefficients of thermal expansion of the stiff glass ($\alpha = 3.76 \times 10^{-6}/°C$ for Corning 1737) or Si wafer ($\alpha = 2.5 \times 10^{-6}/°C$). These are much lower than those of organic polymer substrates and, therefore, are more suitable for the TFT fabrication. While using temporary rigid substrates and bonding agents is acceptable in the laboratory, less labor-intense techniques for carriers will be needed in manufacture. After the fabrication the polymer foil that now carries the TFTs is detached from the rigid substrate. If a free-standing polymer foil is processed, its coefficient of thermal expansion becomes the most important parameter of all. It has to be well matched to the coefficient of thermal expansion of the TFT layers. Also the built-in stress in all TFT layers must be very well controlled. Mismatch in the coefficients of thermal expansion between any TFT layer and the polymer substrate and built-in stress may result in severe substrate curving. Solving these aspects of circuit fabrication on free-standing polymer foils will be important for future roll-to-roll manufacturing.

We selected polyimide Kapton®E as a substrate for a-Si:H TFT fabrication. It is stable to process chemicals, has a glass transition temperature > 350°C, a coefficient of thermal expansion of $12 \times 10^{-6}/°C$ [20], and RMS surface roughness of ~ 30 nm [21]. A 75-μm thick foil shrinks ~ 0.04% after 2 hours at 200°C, has a relatively low humidity expansion coefficient of $9 \times 10^{-6}/\%RH$, a water permeability of 4 g/m^2/day, and oxygen permeability of 4 cm^3/m^2/day [20].

3. EXPERIMENTS

We fabricated amorphous silicon (a-Si:H) thin-film transistors (TFTs) at 150°C on free-standing Kapton E substrates passivated on both sides with SiN$_x$ deposited by plasma enhanced chemical vapor deposition (PECVD) [22]. The standard a-Si:H TFT process temperatures lie between 250°C and 350°C. We have reduced the temperature to 150°C [23] for two reasons: (i) several other polymers can withstand the temperature of 150°C [24] and therefore a 150°C TFT technology can be used on other substrates, and (ii) a-Si:H and SiN$_x$ layers grown by PECVD at 150°C can be grown with a quality comparable to the materials grown at the standard temperature [23].

It is known that the quality of the amorphous silicon (a-Si:H) and silicon nitride (SiN$_x$) deposited by PECVD deteriorates with decreasing deposition temperature. During growth at 150°C, the usual source gases are diluted with hydrogen to assure that the electronic properties of a-Si:H and SiN$_x$ layers are comparable to those of a-Si:H / SiN$_x$ grown at the optimum temperature of 250-350°C [23].

Figure 1 shows a schematic cross section of substrate and thin-film transistor. The TFTs have the inverted, bottom gate staggered geometry with SiN$_x$ backchannel passivation. The polyimide substrate foil was first coated on both sides with a 0.5 μm thick layer of SiN$_x$. All TFTs had the following structure: ~ 100 nm thick Ti/Cr layer as gate electrode, ~ 360 nm of SiN$_x$ as gate dielectric, ~ 100 nm of undoped a-Si:H as channel material, 180 nm of passivating SiN$_x$, ~ 50 nm of highly phosphorus-doped (n$^+$) a-Si:H, and ~ 100 nm thick Al for the source-drain contacts. The TFT channel length was 40 μm and channel width was 400 μm. Fabrication details are given elsewhere [23]. The stress was balanced between the TFT layers such that the foil was flat after fabrication, as shown in Fig. 2(a). Figure 2(b) demonstrates the flexibility of the 75 x 75 mm Kapton substrate bearing 100 a-Si:H TFTs. We have studied the stability of these TFTs under dc electrical bias [25] and carried out extensive experimental and theoretical studies of the bending of TFTs on thin foils [26, 27].

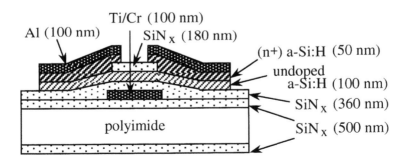

Figure 1. Cross-sectional view of the a-Si:H thin-film transistor showing layer materials and thicknesses.

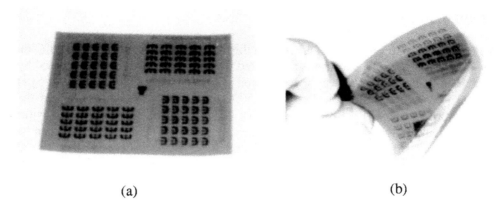

(a)   (b)

Figure 2. Free-standing 51 μm thick Kapton foil with discrete a-Si:H TFTs. After fabrication the foil is flat (a) and in (b) the foil is intentionally bent.

4. RESULTS AND DISCUSSION

The key electrical properties of the TFTs are reflected in the transfer characteristics that depict the drain-to-source current I_{ds} as a function of the gate-to-source voltage V_{gs}. These are plotted in Fig. 3 for drain-to-source voltages $V_{ds} = 0.1$ V and 10 V. At $V_{ds} = 10$ V, the off-current is ~ 2 x 10^{-12} A (~ 5 x 10^{-14} A/μm of gate width) and the on-off current ratio is $\sim 10^{7}$. These characteristics are typical of device quality a-Si:H TFTs grown on glass substrates.

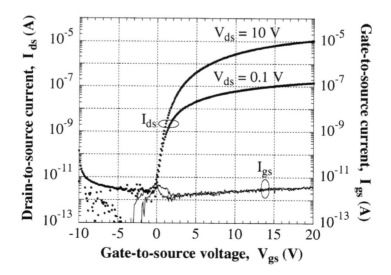

Figure 3. Drain-to-source current I_{ds} and gate-to-source current I_{gs} plotted as functions of the gate-to-source voltage V_{gs}. The drain-to-source voltage V_{ds} was 0.1 V and 10 V. The source was grounded.

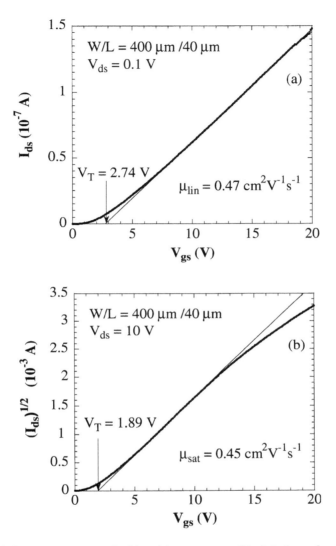

Figure 4. The drain-to-source current I_{ds} (a) and its square root (b) plotted as a function of gate-to-source voltage V_{gs} for the TFT of Fig. 3. The intersection with the x-axis determines the threshold voltage V_T. The electron field-effect mobility μ_n is extracted from the slope of the linear fit.

Another important way of representing the measured transistor performance is shown in Figures 4(a) and (b). The drain-to-source current I_{ds} and its square root are plotted as a function of gate-to-source voltage V_{gs} for the TFT of Fig. 3. These plots are used to determine the threshold voltage V_T and the electron field-effect mobility μ_n in the linear and the saturation regimes, using the following equations originally derived for metal-oxide-semiconductor field-effect transistors (MOSFET) on crystalline silicon [28]:

$$I_{ds(linear)} = \mu_n C_{SiN} \frac{W}{L}\left[\left(V_{gs} - V_T\right)V_{ds} - \frac{V_{ds}^2}{2}\right] \quad \text{(small } V_{ds}) \tag{1}$$

$$I_{ds(saturated)} = \mu_n C_{SiN} \frac{W}{2L}\left(V_{gs} - V_T\right)^2 \quad \text{(large } V_{ds}) \tag{2}$$

where C_{SiN} is the capacitance of the gate insulator, W the channel width, and L the channel length. The dielectric constant ε_r of our 150°C silicon nitride is 7.4 and we calculate $C_{SiN} = \varepsilon_o \varepsilon_r / d_{SiN} = 1.82 \times 10^{-8}$ F cm^{-2}, where ε_o is the permitivity of vacuum and d_{SiN} is the thickness of the gate dielectric. From the plots, we see that both Eqs. (1) and (2) are well obeyed in their respective regions. In the linear current regime (Fig. 4a), we obtain $V_T = 2.84$ V and a mobility of 0.47 cm^{2}V^{-1}s^{-1}. In the saturation current regime (Fig. 4b), we obtain $V_T = 1.89$ V and a saturated mobility of 0.45 cm^{2}V^{-1}s^{-1}. The subthreshold slope, extracted from the exponential region of the curve for $V_{ds} = 10$ V, $S = dV_{gs}/d(\log I_{ds})$, is ~ 0.5 V/decade, and it is typical of a-Si:H TFTs. These values demonstrate that good quality a-Si:H TFTs can be fabricated on thin foils of polyimide.

5. CONCLUSIONS

We fabricated good quality amorphous silicon thin-film transistors on 51 μm thick Kapton E at the maximum process temperature of 150°C. In the TFT fabrication, each layer is optimized for the best electrical performance, and for low built-in mechanical stress. Therefore, the foil is flat after TFT fabrication. The off-current is ~ 2 x 10^{-12} A, the on-off current ratio ~ 10^7, the threshold voltage ~ 2 V, the mobility ~ 0.5 cm^{2}V^{-1}s^{-1}, and the subthreshold slope ~ 0.5 V/decade. Since a-Si:H TFTs are the main building blocks for thin-film electronic circuits, we have demonstrated that amorphous silicon based electronics on organic polymer substrates is feasible.

Acknowledgement

We thank DuPont for the information about Kapton and other polyimides. The authors are grateful to the National Science Foundation, DARPA, and the New Jersey Commission on Science and Technology for financial support.

REFERENCES

1. H. Kuma, T. Iwakuma, F. Moriwaki, M. Fukuda, T. Sekiya, M. Araki, Y. Saito, M. Gomyo and T. Mita, 2001 Digest of Technical Papers of the Society for Information Display **32**, 12 (2001).
2. J.K. Mahon, ibid., p. 22.

3. M.G. Kane, I.G. Hill, J. Campi, M.S. Hammond, B. Greening, C.D. Sheraw, J.A. Nichols, D.J. Gundlach, J.R. Huang, C.C. Kuo, L. Jia, T.N. Jackson, J.L. West and J. Francl, ibid., p. 57.
4. K. Amundson, J. Ewing, P. Kazlas, R. McCarthy, J.D. Albert, R. Zehner, P. Drzaic, J. Rogers, Z. Bao and K. Baldwin, ibid., p. 160.
5. R.G. Stewart, ibid., p. 264.
6. S. Yamamoto, H. Kobayashi, T. Kakinuma, T. Hikichi, N. Hiji, T. Ishii, Y. Harada, M. Koshimzu, K. Maruyama, Y. Niitsu, T. Suzuki, D. Tsuda and H. Arisawa, ibid. p. 362.
7. S. Aomori, T. Maruyama, Y. Nakatani and M. Yoshida, ibid., p. 558.
8. S.K. Park, J.I. Han, W.K. Kim, M.G. Kwak, S.J. Hong and C.J. Lee, ibid., p. 658.
9. Y. Chen, K. Denis, P. Kazlas and P. Drzaic, ibid., p. 157.
10. D. Stryahilev, A. Sazonov and A. Nathan, Proceeding of Asia Display/IDW 2001 Conference, p.1739, Nagoya, Japan (2001).
11. S.D. Theiss and S. Wagner, IEEE Electron Dev. Lett. **17**, 264 (1996).
12. M. Wu, K. Pangal, J.C. Sturm and S. Wagner, Appl. Phys. Lett. **75**, 2244 (1999).
13. G.N. Parsons, C.S. Yang, C.B. Arthur, T.M. Klein and L. Smith, Mater. Res. Soc. Symp. Proc. **508**, 19 (1998).
14. H. Gleskova, S. Wagner and Z. Suo, Mater. Res. Soc. Symp. Proc. **508**, 73 (1998).
15. J.N. Sandoe, Digest of Technical Papers of the Society for Information Display **29**, 293 (1998).
16. E. Lueder, M. Muecke and S. Polach, Proc. of the Int. Display Research Conf., Asia Display '98, 173 (1998).
17. A. Constant, S.G. Burns, H. Shanks, C. Gruber, A. Landin, D. Schmidt, C. Thielen, F. Olympie, T. Schumacher and J. Cobbs, The Electrochemical Society Proceedings **94-35**, 392 (1995).
18. D.B. Thomasson, M. Bonse, J.R. Huang, C.R. Wronski and T.N. Jackson, Technical Digest IEEE Int. Electron Devices Meeting 1998, p. 253 (1998).
19. S.D. Theiss, P.G. Carey, P.M. Smith, P. Wickboldt, T.W. Sigmon, Y.J. Tung and T.-J. King, IEEE Int. Electron Devices Meeting 1998, Technical Digest, p. 257 (1998).
20. J.A. Kreuz, S.N. Milligan and R.F. Sutton, DuPont Films Technical Paper 3/94, Reorder No. H-54504.
21. D.B. Thomasson, M. Bonse, J.R. Huang, C.R. Wronski and T.N. Jackson, International Electron Devices Meeting Technical Digest, pp. 253-256 (1998).
22. H. Gleskova and S. Wagner, IEEE Electron Dev. Lett. **20**, 473 (1999).
23. H. Gleskova, S. Wagner, V. Gašparík and P. Kováč, J. Electrochem. Soc. **148**, G370 (2001).
24. S.M. Gates, Mater. Res. Soc. Symp. Proc. **467**, 843 (1997).
25. H. Gleskova and S. Wagner, IEEE Trans. Electron Dev. **48**, 1667 (2001).
26. H. Gleskova, S. Wagner and Z. Suo, Appl. Phys. Lett. **75**, 3011 (1999).
27. H. Gleskova and S. Wagner, Appl. Phys. Lett **79**, 3347 (2001).
28. S.M. Sze, *Semiconductor Devices – Physics and Technology*, John Wiley & Sons, New York (1985).

Polyimides and Other High Temperature Polymers, Vol. 2, pp. 467–485
Ed. K.L. Mittal
© VSP 2003

Polyimide/polystyrene nanocomposite films as membranes for gas separation and precursors for polyimide nanofoams

ZHI-KANG XU,* ZHEN-MEI LIU and LI XIAO

Institute of Polymer Science, Zhejiang University, Hangzhou 310027, P.R. China

Abstract—A method for the preparation of polyimide (PI) nanofoams with high thermal stability, which can be used as dielectric layers in microelectronics, has been presented in this paper. Polystyrene (PS) and poly[styrene-co-(4-vinylpyridine)] (PSVP) nanospheres were prepared by emulsion polymerization. The diameter of these particles was 30 to 40 nm. PMDA-ODA PI nanocomposite films with various contents of PS and PSVP were obtained by in-situ condensation polymerization in the presence of nanospheres. The distribution of the nanospheres in the films and the morphology of the films were characterized by TEM. Relatively homogeneous distributions were observed for the PI/PSVP nanocomposite films, which can be ascribed to the formation of poly(amic acid) amine salts between pyridinyl and carboxyl moieties. The transport properties of CO_2, O_2, N_2, and CH_4 gases through these nanocomposite films were measured. For the PI/PS-nanosphere composite films, with the increase of PS nanospheres in the films, the gas permeabilities increased and selectivities decreased; these were due to the increase in gas diffusion coefficients and the decrease in diffusion selectivities. For PI/PSVP-nanosphere composite films, increases in both gas permeabilities and selectivities were observed when the PSVP-nanospheres increased from 10 to 20 wt% in the films, while both the diffusion coefficients and diffusion selectivities decreased. It was found that the membrane containing 20 wt% PSVP-nanospheres showed both high gas permeability and high permselectivity toward CO_2. These results can be ascribed to the high solubility of CO_2 and O_2 in the pyridine-containing polymers and the relatively homogenous distribution of PSVP-nanospheres in the PI matrix. Upon thermal treatment, the thermally unstable nanospheres undergo thermolysis, leaving pores with the size and shape dictated by the initial nanosphere morphology. The resulted foams were characterized by TEM, DSC, TG, density measurements and dielectric constant measurements. The results revealed that nanofoams with high thermal stability and low dielectric constant approaching 2.0 could be prepared using PI nanocomposite approach.

Keywords: PI; polystyrene-based nanospheres; composite films; gas separation membranes; nanofoams; dielectric constant.

*To whom all correspondence should be addressed. Phone: (86) (571) 87952131-8218, Fax: (86) (571) 87951773, E-mail: xuzk@ipsm.zju.edu.cn

1. INTRODUCTION

Low dielectric constant materials are widely used as interlayer dielectric films in ultra-large-scale integrated (ULSI) circuit multilevel interconnections [1-2]. The dielectric constant of the interlayer dielectric film controls signal propagation delay time and cross-talk in multilevel interconnections, and the dielectric constant of currently used silicon dioxide (SiO_2) films is too high for the sub-quarter micrometer devices and beyond. Therefore, considerable attention has been focused on the replacement of SiO_2 with new materials having low dielectric constant [1-2]. It is well known that the generation of pores in materials should substantially reduce the dielectric constant because the dielectric constant of air is only unity. However, one of the problems is that it is difficult to maintain the desired thermal and mechanical properties required for the ULSI manufacturing environment. Furthermore, the incorporated air voids must be much smaller than the dimension of microelectronic devices. Recently, Hedrick and co-workers have found that the reduction in the dielectric constant could be simply achieved by incorporation of nano-size air voids into PIs or poly(phenylquinoxalines) matrix [3-6]. The approach they took involved the preparation of block copolymers capable of self-assembly in which the continuous phase was thermally stable polymer and the dispersed phase was thermally unstable. Upon high temperature treatment, the unstable component undergoes thermolysis, leaving behind pores with size and shape dictated by its initial morphology. While this technique has been demonstrated to produce nanofoams with low dielectric constant, the synthesis procedures and processing are relatively complicated. Furthermore, thermal degradation of the labile component reduces the molecular weight as well as certain critical mechanical properties of the resulting nanofoam films.

The use of organic templates to control the structure of inorganic solids has been proven very successful for designing porous materials with pore sizes ranging from angstroms to micrometers [7-9]. In the case of microporous silicates, recent reports illustrate that techniques using latex spheres can be adopted to create silica structures with pore sizes ranging from 5 nm to 1 µm. An improved procedure was used by Park and Xia [10] for the fabrication of three-dimensional microporous films of polyurethane with spherical pores whose dimensions could be precisely controlled in the range from ~ 0.2 to ~ 3 µm. The primary aim of their work was to prepare films with three-dimensionally interconnected networks of pores in the bulk and completely exposed pores on both top and bottom surfaces of the films.

On the other hand, membrane separation of gases has emerged as an important unit operation technique offering specific advantages over more conventional separation procedures (e.g. cryogenic distillation and adsorption) [11-12]. Although polymeric nanocomposite systems containing inorganic nanoparticles are well known as barrier systems [13], however, Kusakabe and co-workers [14] reported that the permeability of CO_2 in a PI/SiO_2 hybrid nanocomposite membrane was 10 times larger than in the corresponding PI membrane.

In the present work, polystyrene-based nanospheres with functional groups on the surface were synthesized and used as templates for PI and then to prepare PI nanocomposite membranes for gas separation. PI nanofoams, preferably having a closed cell structure which is important for minimizing water absorption, were prepared by the thermal decomposition of the PS-based nanospheres. It was expected that the special interaction of the functional groups on the surface with the carboxyl groups of the poly(amic acid) will improve the dispersion homogeneity of the nanospheres in the PI matrix.

2. EXPERIMENTAL

2.1. Materials

Pyromellitic dianhydride (PMDA) and 4,4'-oxydianiline (ODA) were purchased from J&K-ACROS (USA) and sublimed twice prior to use. Styrene was purified by distillation under reduced pressure. 4-Vinylpyridine (MERCK – Schuchardt, Germany) was distilled before use. The cross-linking agents, ethylene glycol dimethacrylate (EGDMA, Aldrich) and divinylbenzene, were used as supplied. The solvent and dispersion medium, N,N'-dimethyl acetamide (DMAc), was dried over CaH_2, distilled under vacuum, purged with nitrogen, and used immediately. Deionized water was purified by double distillation. Other reagents were commercially available and used as received.

2.2. Synthesis of polystyrene (PS) and poly(styrene-co-4-vinylpyridine) (PSVP) nanospheres

Standard microemulsion polymerization technique was used. The reactions were carried out in three-neck flasks at $70 \pm 0.1°C$ equipped with a condenser, mechanical stirrer, and inlets for nitrogen. 80 g water, 2 g sodium dodecyl sulfonate (SDS), 20 g styrene, 0.5 g EGDMA (or divinylbenzene) and 0.172 g potassium persulfate were fed into the reaction vessel to form the monomer emulsion. Prior to polymerization, the reaction mixture was purged with nitrogen, and a small positive pressure of nitrogen was maintained during the synthesis. The mixture was then stirred vigorously and heated at 70°C for 5 hrs. Partially cross-linked polystyrene (PS) nanospheres produced by this procedure yielded particles with average size 35 nm. PSVP nanospheres were prepared by the same process using a mixture of 4-vinylpyridine (5 g) and styrene (15 g) as the monomers. All the resulting nanospheres were freeze dried before further use.

2.3. Dispersion of PS and PSVP nanospheres in poly(amic acid) solution

PMDA-ODA poly(amic acid) (PAA) solutions containing PS and/or PSVP nanospheres were prepared by an in-situ condensation polymerization approach. The predetermined amounts of nanospheres were mixed with 60 ml of DMAc in a 100 ml three-neck flask fitted with a nitrogen inlet and an anhydrous calcium chloride drying tube, and dry nitrogen gas was bubbled through the solution. The mixed system containing nanospheres and solvents was homogenized by intense ultra-sonication and agitation, to form a 5 wt% stable transparent dispersion. After that, the flask was placed in an ice-salt bath and anhydrous ODA was added followed with the addition of PMDA in batches (ODA : PMDA = 2.000 g : 2.2178 g). After all the PMDA was added, the ice-salt bath was taken away, and the solution was stirred at room temperature for 6 h.

2.4. Film fabrication and nanofoam preparation

Nanoporous planar films were prepared by casting the mixture obtained above onto glass plates and drying for 24 h at 80°C under vacuum. The composite films prepared were then thermally cured in a nitrogen atmosphere at 100°C for 60 min, 150°C for 60 min, and 200°C for 4 h. The PI composite films thus obtained were removed from the glass plates and dried for another 24 h at 100°C under vacuum. All composite films were treated according to the same procedure. The thickness of the dry composite films varied from 20 to 30µm. To obtain nanofoam materi-als, these composite films were heated at 380°C until all the thermally unstable nanospheres were thermolysized (about 2 hrs).

2.5. Characterization

The dimensions and size distribution of the nanospheres were measured on a Matec Applied Sciences CHDF-1100 instrument, which separates particles of dif-ferent sizes on the basis of the hydrodynamic effects of particles undergoing Poiseuille flow within a capillary tube. Transmission electron microscopy (TEM) was carried out on a JEOL JEM 1200EX instrument. X-ray photoelectron spec-troscopy (XPS) spectra were recorded on a ESCALABMK II spectrometer and used to extract the surface information of the nanospheres.

2.6. Measurements of gas separation properties

Mean permeability coefficients (P) and effective diffusion coefficients (D_{eff}) were measured at 1 atm and 10 atm for different temperatures between 30°C and 75°C by the time-lag method. The apparatus and experimental procedure employed have been described in previous papers [15-16]. From a single transient gas trans-port experiment, both P and D_{eff} can be obtained. The permeability is defined as:

$$P = J_s \times l / (p_2 - p_1) \qquad (1)$$

where l is the thickness of the film, J_s is the steady-state rate of gas permeation through a unit area of the film, p_2 and p_1 are the upstream and downstream pres-sures of the gas component, respectively. Effective diffusion coefficient can be calculated from the time-lag method using the following equation (2) [15]:

$$D_{eff} = l^2 / 6\theta \qquad (2)$$

where θ is lag time.

The ideal permselectivity of the films toward gas A relative to another gas B can be calculated from the individual pure gas permeabilities:

$$(\alpha_{A/B})^{id} = P_A/P_B \tag{3}$$

where P_A and P_B are the corresponding permeability coefficients of gases A and B, respectively.

The error in the absolute values of the permeability coefficients is estimated to be about $\pm 10\%$, due to uncertainties in the determination of the downstream volume and the effective membrane area and thickness, while the reproducibility was better than $\pm 4\%$.

3. RESULTS AND DISCUSSION

3.1. Synthesis of PS and PSVP nanospheres and fabrication of PMDA-ODA PI/PS or PSVP-based nanospheres composite films

Figure 1 depicts the procedure for the preparation of nanocomposite films and nanofoams studied in this work. The first stage involves the synthesis of the PS or PSVP nanospheres of nanometer size. The ability to tailor the characteristics and structures of emulsion and suspension polymer particles is a direct result of the extensive research efforts that have been carried out over the past 75 years. Numerous techniques have been developed during this time to modify and to control the particle size, degree and type of functionality, as well as the shape and internal structure of emulsion polymer particles.

In this process, the dimensions of the nanosphere templates play an important role in the preparation of the nanofoam materials. The criteria for the selection of template include the dimension of particles, particle monodispersity and the dispersion properties in the PI matrix. An ionic surfactant, SDS, was used in the emulsion polymerization. The cross-linking agent, EGDMA (or divinylbenzene), was added to the system at the maximum concentration of 2 mol% that provided partial cross-linking in the nanospheres. A larger amount of the cross-linking agent led to partial aggregation resulting in the difficulty of thermolysis. Figure 2 shows the TEM micrographs of the nanospheres which were synthesized using the procedure described above. It was found that the average diameter of the spheres was 35 nm, while these nanospheres were nearly monodispersed. The decomposition temperature of the nanospheres is critical for our purpose; it should be high enough to permit standard film preparation, solvent removal and imidization below the T_g of the PI matrix to avoid foam collapse. The TG curves of typical PS and PSVP nanospheres samples in nitrogen are shown in Figure 3. It can be seen that the polystyrene-based nanospheres decomposed rapidly between 300 and 400°C. The XPS spectra of these nanospheres are shown in Figure 4. It can be seen that nitrogen element, which is derived from pyridinyl group, is characterized in the PSVP nanospheres by the peak at approximately 402 eV.

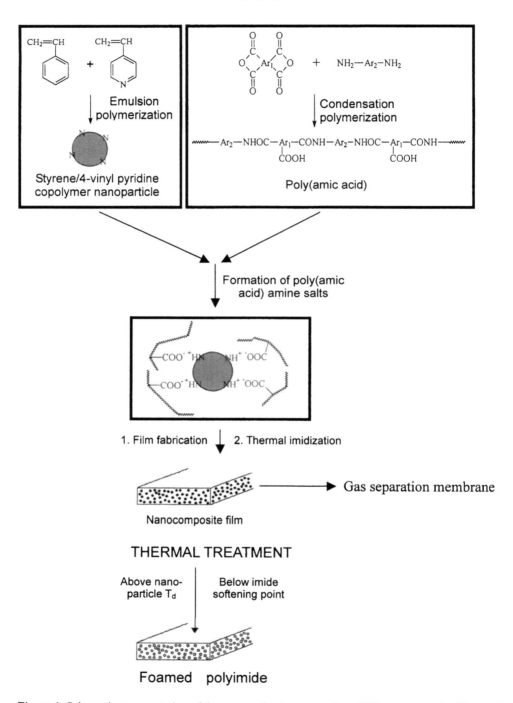

Figure 1. Schematic representation of the process for the preparation of PI nanocomposite films and nanofoams.

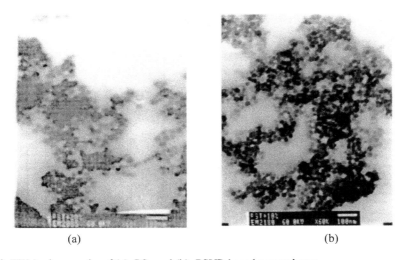

(a)   (b)

Figure 2. TEM micrographs of (a): PS- and (b): PSVP-based nanospheres.

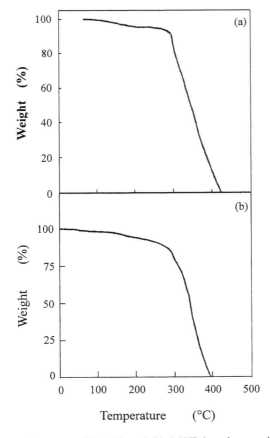

Figure 3. Thermogravimetric curves of (a): PS- and (b): PSVP-based nanospheres.

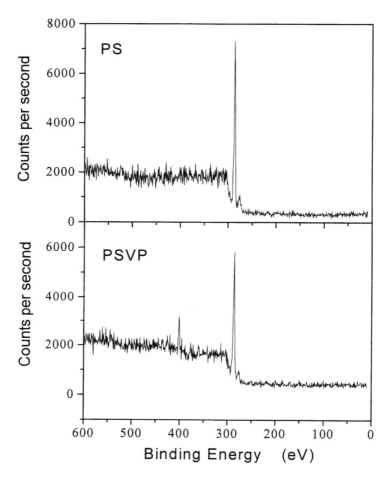

Figure 4. XPS spectra of the polystyrene (PS) and poly(styrene-co-4-vinylpyridine) (PSVP) nanospheres.

As shown in Figure 1, the second stage described in this work involved the dispersion of nanospheres into the PI matrix. The polystyrene-based nanospheres were mixed with DMAc and homogenized by intense ultrasonication and agitation, to form a 5 wt% stable milky suspension with slight sky-blue color. Then the homogeneous dispersions of polystyrene-based nanospheres in poly(amic acid) (PAA) in varying concentrations were obtained by adding PMDA and ODA (in the molar ratio of 1:1) into PS or PSVP/dispersion in designated proportions in ice-salt bath, followed by stirring the mixture continuously for 6 h at 20°C. After that, nanocomposite films of PS or PSVP/PAA were fabricated by the casting method. A typical TG curve for a PAA nanocomposite film containing 20% PSVP in nitrogen is shown in Figure 5(b), and the TG curve of a PAA sample without PSVP is shown in Figure 5(a) for comparison. From these curves, it can

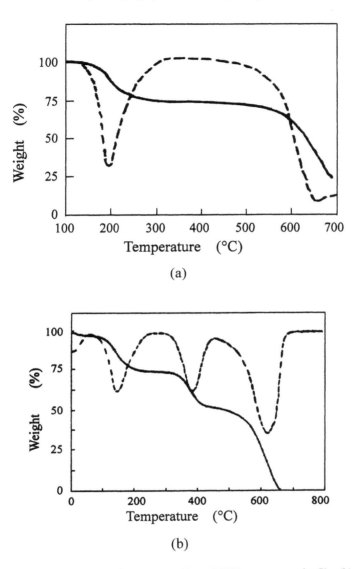

Figure 5. Thermogravimetric curves of PAA (a) and PAA/PSVP nanocomposite film (b). The solid line represents the thermogravimetric loss, and the dashed line is the differential analysis of the solid line.

be seen that the imidization of the pure PAA began at 150°C and ended at 250°C, while the imidization of the PAA in the composite sample began at 110°C and ended at 210°C and the decomposition of the nanospheres began at 320°C and ended at 390°C. In our experiment, thermal imidization at elevated temperature was adapted to prevent the agglomeration of nanospheres in the polymer matrix. All of the nanocomposite films were cured by heating at 100°C for 60 min, 150°C for 60 min, 200°C in nitrogen and holding at this temperature for 4 h.

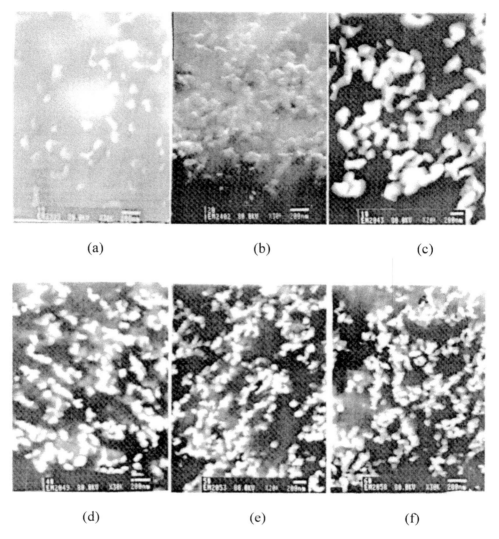

(a)   (b)   (c)

(d)   (e)   (f)

Figure 6. TEM micrographs of PI nanocomposite films. (a): 10 wt% PS, (b): 15 wt% PS, (c): 20 wt% PS, (d): 15 wt% PSVP, (e): 20 wt% PSVP, (f): 25 wt% PSVP.

The TEM micrographs of the PI/PS-based nanospheres composite films containing various nanospheres are shown in Figure 6. It can be seen that nanocomposite films with relatively homogeneous dispersion can be prepared through this process. However, comparing with the original nanospheres, slight agglomeration and distortion of nanospheres in these nanocomposite films can be observed from the TEM pictures. The distortion could be decreased or even eliminated by increasing the degree of cross-linking of the nanospheres. On the other hand, large clusters from the nanospheres were formed in the PI/PS composite film, as can be seen from Figures 6a, 6b and 6c. The relatively homogeneous dispersion of the PSVP nanospheres in the PI matrix might be due to the partial interaction of pyridinyl groups on the surface of the particles with the carboxyl groups of the PAA chains before the PAA was imidized. The surface pyridinyl groups are detected from PSVP nanospheres by XPS (Figure 4). The interaction of tertiary amine with carboxyl group and the formation of amine salt between them has been successfully applied for the preparation of PI Langmuir-Blodgett films [17-18]. Gas transport properties, such as gas diffusion coefficient, of the film might reflect to a certain degree such interaction because the diffusion coefficient depends strongly on the packing density of polymer chains in the film. Specific interactions in the polymer film normally increase the packing density and decrease the gas diffusion coefficient. Therefore, using the time-lag method, the gas transport properties of these PI nanocomposite films were measured.

3.2. Pure gas separation properties of composite films filled with PS-based nanospheres

Gas separation properties of these films for CO_2, O_2, N_2 and CH_4 gases at 35°C and 10 atm are listed in Table 1. For comparison, the gas permeability data obtained under the same conditions in our laboratory for the PMDA/ODA PI film are also included. PIs have good mechanical and thermal properties together with high gas permeabilities, which are greatly affected by the structure of PI molecules; and widely varying gas permeability values are reported in the literatures [19-23]. Hirayama *et al.* [19] have reported the permeabilities of 32 different PIs and examined in detail the relation between gas permeabilities, diffusivities and solubilities and the structures of various PIs. Recently, Alentiev and co-workers [24] presented a novel approach for the prediction of gas transport parameters (permeability and diffusion coefficients) of amorphous PIs. In contrast to the al-

Table 1.
Gas permeabilities (P) and permselectivities (α) of films at 35°C and 10 atm

Film	P_{CO_2}	P_{O_2}	P_{N_2}	P_{CH_4}	α_{CO_2/N_2}	α_{CO_2/CH_4}	α_{O_2/N_2}
PI/10 wt% PS	2.33	0.55	0.085	0.049	27.4	47.5	6.47
PI/15 wt% PS	2.32	0.50	0.090	0.049	25.8	47.3	5.56
PI/20 wt% PS	2.90	1.34	0.91	0.34	3.19	8.53	1.45
PI/25 wt% PS	4.29	1.77	0.91	0.48	4.71	8.94	1.94
PI/10 wt% PSVP	3.58	0.82	0.13	0.10	28.4	35.8	6.31
PI/15 wt% PSVP	3.71	0.81	0.14	0.10	26.5	36.4	5.79
PI/20 wt% PSVP	5.65	0.91	0.15	0.14	38.4	41.2	6.07
PI/25 wt% PSVP	6.55	1.99	1.52	2.19	4.31	3.00	1.31
PI	2.00	0.40	0.063	0.024	31.7	83.3	5.00

P in *Barrers*

ready mentioned approach of Park and Paul [25], this new method is not based on the free volume of the polymers; instead, all the PIs are regarded as alternating copolymers that include dianhydride and diamine sub-units the contributions of which to the gas permeation parameters were deduced. The actually measured gas transport coefficients agreed very well with the calculated values. As regards the PMDA-ODA PI films, the data on gas permeabilities from the literature vary up to an order of magnitude for the same structure due to differences in synthesis method, film preparation, and annealing to remove solvent. Thus one must be selective in choosing data from sources employing similar preparative procedures.

It was found that some PMDA/ODA PI composite films filled with polymeric nanospheres studied in this work had higher permeability and higher selectivity than those of unfilled PMDA/ODA PI film. It can also be seen that the gas permeabilities for both PI/PS and PI/PSVP series of films increase with the increase of nanoparticle fraction from 10 to 25 wt% in the composite films. The gas permeabilities in the PI/PS films are lower than those in the PI/PSVP films. Meanwhile, with the increase of nanoparticle fraction, the selectivities for CO_2/N_2, CO_2/CH_4 and O_2/N_2 decrease from 27.4 to 4.17, 47.5 to 8.94 and 6.47 to 1.94, respectively, for the PI/PS-nanoparticle composite films. These results are consistent with the "trade-off" relationship [26]. However, the PI/PSVP films show increase in both permeabilities and selectivities when the PSVP content increases from 10 to 20 wt%. As can be seen from Table 1, the CO_2 permeability increases from 3.58 to 5.65, while the selectivities for CO_2/N_2 and CO_2/CH_4, respectively, rise from 28.4 to 38.4 and from 35.8 to 41.2 for these films. Concerning the O_2 permeability and the O_2/N_2 selectivity, on the other hand, slight increase in permeability and decrease in selectivity were observed. Further increase in the PSVP content in the composite film decreases the selectivities sharply. This may be ascribed to the defect formation, which was due to the heterogeneous distribution of the nanospheres in the film.

Table 2.

Gas diffusivities (D) and diffusion selectivities of films at 35°C and 10 atm

Film	D_{CO2}	D_{O2}	D_{N2}	D_{CH4}	D_{CO2}/D_{N2}	D_{CO2}/D_{CH4}	D_{O2}/D_{N2}
PI/10 wt% PS	0.30	0.69	0.14	0.030	2.22	10.1	4.93
PI/15 wt% PS	0.42	0.81	0.20	0.057	2.10	7.34	4.05
PI/20 wt% PS	0.50	1.72	2.94	0.14	0.17	3.62	0.58
PI/25 wt% PS	0.76	1.93	6.47	0.73	0.12	1.04	0.29
PI/10 wt% PSVP	0.55	1.13	0.23	0.060	2.37	9.04	4.91
PI/15 wt% PSVP	0.52	1.02	0.23	0.057	2.21	9.10	4.43
PI/20 wt% PSVP	0.24	0.63	0.17	0.033	1.35	7.14	3.70
PI/25 wt% PSVP	0.85	2.91	6.08	–	0.12	–	0.48
PI	0.083	0.39	0.087	0.030	0.21	2.77	4.48

D in $10^{-8} cm^2/s$

It is interesting that, as can be seen in Table 2, with the increase of nanospheres fraction from 10 to 20 (25) wt%, the gas diffusion coefficients of PI/PS film increased while those of the PI/PSVP film decreased. The decrease of diffusion coefficient in the PI/PSVP films can be attributed to the relatively homogeneous dispersion of the nanospheres in the PI matrix, and to the compact packing of PI chains on(in) the nanospheres. These may derive from the partial interaction of pyridinyl groups on the surface of the particles with the carboxyl groups of the poly(amic acid) chains in the film fabrication process.

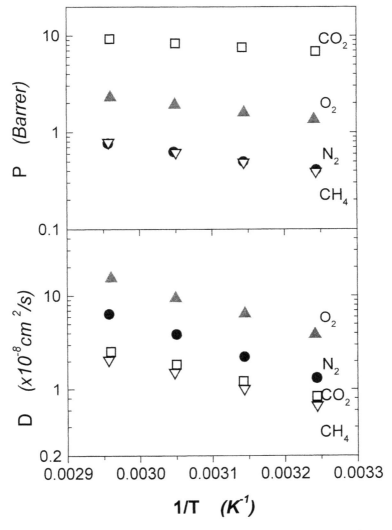

Figure 7. Effects of temperature on the gas permeabilities (top) and diffusivities (bottom) of PI/20 wt% PSVP film at 1 atm.

It is quite surprising that the diffusion coefficient of N_2 for PI/20 wt% PS, PI/25 wt% PS and PI/25 wt% PSVP are much larger than those of O_2 and CO_2. This result might be attributed to the high adsorption of O_2 and CO_2 on the nanospheres surface, which, in turn, would give high solubility of these two gases in the composite films with high fraction of nanospheres. Nevertheless, the exact reason for this phenomenon is not clear at present.

The effects of temperature on permeabilities and diffusivities for CO_2, O_2, N_2, and CH_4 for films containing 20 wt% of PS and PSVP were studied at 1 atm upstream pressure. The results are shown in Figure 7 and Figure 8 respectively. Within a temperature range in which no significant thermal transitions of the polymer occur, the temperature dependence of permeabilities and diffusivity can be described by the Arrhenius expressions:

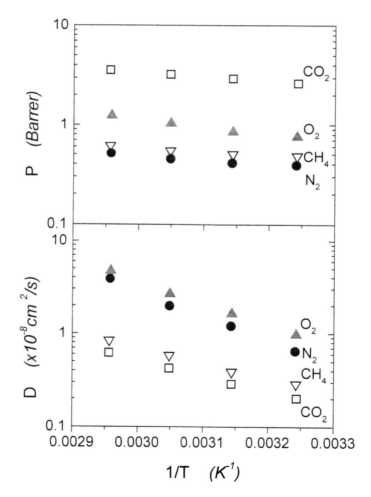

Figure 8. Effects of temperature on the gas permeabilities (top) and diffusivities (bottom) of PI/20 wt% PS film at 1 atm.

Table 3.
Activation energies for gas permeation and diffusion for two composite films

Gas	CO_2	O_2	N_2	CH_4
E_P (kJ/mol)				
PI/20 wt% PS	8.95	15.44	20.48	18.49
PI/20 wt% PSVP	8.59	14.01	7.82	9.12
E_D (kJ/mol)				
PI/20 wt% PS	31.8	40.1	46.6	34.2
PI/20 wt% PSVP	32.5	45.0	54.1	36.4
P_0				
PI/20 wt% PS	224.7	556.1	1143.7	552.2
PI/20 wt% PSVP	75.7	179.1	8.1	15.2
D_0				
PI/20 wt% PS	2.0×10^{-3}	0.244	1.1	4.2×10^{-3}
PI/20 wt% PSVP	6.3×10^{-4}	0.403	8.7	3.7×10^{-3}

E_P: activation energy for gas permeation
E_D: activation energy for gas diffusion
P_0 and D_0: the pre-exponential factors of Arrhenius expressions

$$P = P_0 \exp(-E_P/RT) \qquad (4)$$

$$D = D_0 \exp(-E_d/RT) \qquad (5)$$

where P_0 and D_0 are pre-exponential factors, E_p is the apparent activation energy for permeation, E_d the activation energy for diffusion, R the gas constant, and T is the temperature. It was found that the temperature dependence of P and D for the two composite films could be described by the Arrhenius equations, too. Therefore, activation energies for the gas transport could be obtained and are summarized in Table 3.

3.3. Preparation and characterization of PI nanofoams

Both the PSVP and PS-based nanospheres in the nanocomposite films were thermolysized at 380°C for 2 hrs. The TEM micrographs and the properties of foams prepared through this process are shown in Figures 9, 10 and Tables 4, 5, respectively. The porous structure of the foams is obvious with the white areas representing voids from the PS nanospheres. Comparing with those of the PI nanocomposite, the TEM micrographs showed pores sizes ranging from 30 nm to 100 nm with some interconnections. The larger pore size and the interconnections were mainly due to the agglomeration of the nanospheres in the composite films and partially to the plasticization accompanying the "blowing" effect on the PI at high temperature [27]. It can be seen from Tables 4 and 5 that the volume fraction

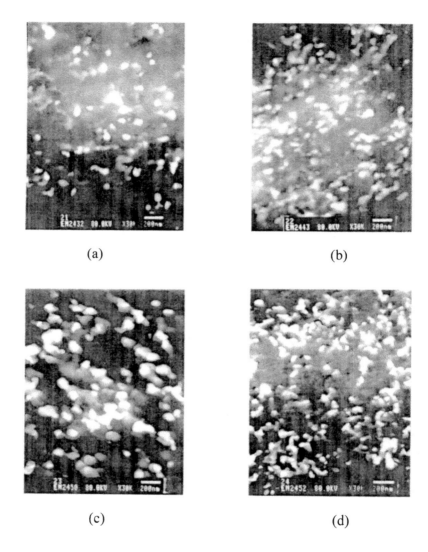

(a)   (b)

(c)   (d)

Figure 9. TEM micrographs of PI foams resulted from (a): 10 wt% PSVP, (b): 15 wt% PSVP, (c): 20 wt% PSVP and (d): 25 wt% PSVP nanocomposite films.

Table 4.
Dielectric constant, ε, of PI foams at 1 MHz

Sample	Thickness (μm)	PSVP nanospheres (wt%)	Void (vol%)	ε
1	31	10	20	3.31
2	25	15	24	2.49
3	22	20	30	2.36
4	23	25	40	2.15

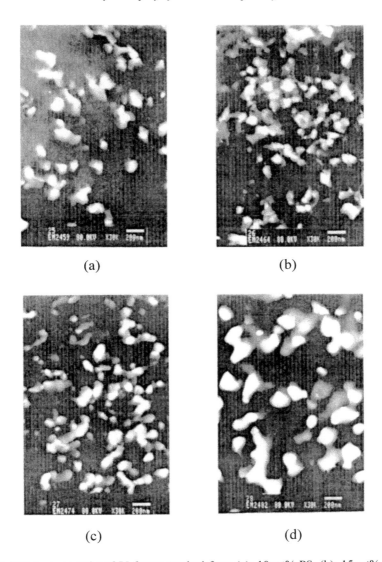

Figure 10. TEM micrographs of PI foams resulted from (a): 10 wt% PS, (b): 15 wt% PS, (c): 20 wt% PS and (d): 25 wt% PS nanocomposite films.

Table 5.
Dielectric constant, ε, of PI foams at 1 MHz

Sample	Thickness (μm)	PS nanospheres (wt%)	Void (vol%)	ε
1	29	10	13	2.60
2	25	15	24	2.32
3	23	20	26	2.21
4	20	25	34	2.02

of voids does not correspond to the volume fraction of polystyrene-based nano-spheres in the initial composite films. In comparison with the initial volume frac-tion of PS-based nanospheres in the composite, the increase in volume fraction of voids incorporated into the matrix can also be understood by considering the "blowing" effect. From the experimental results and corresponding discussion above, we can deduce that the elimination of agglomeration and a relatively low decomposition temperature for the nanospheres will give foams with pores of nanometer size and without interconnection.

4. SUMMARY

PI nanocomposite films with gas permeability and permselectivity regulatable by their composition were prepared in our laboratory by in-situ condensation polym-erization in the presence of PS or PSVP nanospheres as templates. As the contents of PS templates increased, the gas permeabilities increased while permselectivi-ties decreased, compared to the increase of both permeability and permselectivity for PSVP-based PI nanocomposite films. As regard to the dielectric property of PI nanofoams, an increase the content of templates would decrease the dielectric constant. TG analysis showed that the resulted PI nanofoams were thermally sta-ble.

Acknowledgements

This work was supported by the National Natural Science Foundation of China (Grant No. 59703005). Partial financial support for this work from the Foreign Relation Office of the Technical University Berlin is also acknowledged.

REFERENCES

1. T. Homma, *Mater. Sci. Eng.* **R23**, 243 (1998).
2. G. Maier, *Prog. Polym. Sci.*, **26**, 3 (2001).
3. J. L. Hedrick, R. D. Miller, C. J. Hawker, K. R. Carter, W. Volksen, D. Y. Yoon and M. Troll-sas, *Adv. Mater.*, **10**, 1049 (1998).
4. J. L. Hedrick, T. P. Russell, J. Labadie, M. Lucas and S. Swanson, *Polymer*, **36**, 2685 (1995).
5. J. L. Hedrick, T. P. Russell, M. Sanchez, R. DiPietro and S. Swanson, *Macromolecules*, **29**, 3642 (1996).
6. J. L. Hedrick, K. R. Carter, J. Labadie, R. D. Miller, W. Volksen, C. J. Hawker, D. Y. Yoon, T. P. Russell, J. E. McGrath and R. M. Briber, *Adv. Polym. Sci.*, **141**, 1 (1999).
7. L. Manziek, E. Langenmayr, A. Lamola, M. Gallagher, N. Brese and N. Annan, *Chem. Mater.*, **10**, 3101 (1998).
8. B. T. Holland, C. F. Blanford and A. Stein, *Science*, **281**, 538 (1998).
9. F. Caruso, R. A. Caruso and H. Möhwald, *Science*, **282**, 1111 (1998).
10. S. H. Park and Y. Xia, *Chem. Mater.*, **10**, 1745 (1998).
11. W. J. Koros and R. Mahajan, *J. Membrane Sci.*, **181**, 141 (2001).
12. S. A. Stern, *J. Membrane Sci.*, **94**, 1 (1994).
13. P. C. LeBaron, Z. Wang and T. J. Pinnavaia, *Appl. Clay Sci.*, **15**, 11 (1999).

14. K. Kusakabe, K. Ichiki, J. Hayashi, H. Maeda and S. Morooka, *J. Membrane Sci.,* **115**, 65 (1996).
15. Z. K. Xu, M. Böhning, J. Springer, N. Steinhauser and R. Mülhaupt, *Polymer*, **38**, 581 (1997).
16. Z. K. Xu, M. Böhning, J. D. Schultze, G. T. Li, J. Springer, F. P. Glatz and R. Mülhaupt, *Polymer*, **38**, 1573 (1997).
17. M. Iwamoto and M. Kakimoto, in: *Polyimides: Fundamentals and Applications*, M. K. Ghosh and K. L. Mittal (Eds.), p. 815, Marcel Dekker, New York (1996).
18. Z. K. Xu, Z. M. Liu, B. K. Zhu and Y. Y. Xu, *J. Photochem. Photobiol. A: Chem.*, **140**, 81 (2001).
19. Y. Hirayama, T. Yoshinaga, Y. Kusuki, K. Ninomiya, T. Sakakibara and T. Tamari, *J. Membrane Sci.*, **111**, 169 (1996).
20. Y. Hirayama, T. Yoshinaga, Y. Kusuki, K. Ninomiya, T. Sakakibara and T. Tamari, *J. Membrane Sci.*, **111**, 183 (1996).
21. T. H. Kim and K. C. O'Brine, *J. Membrane Sci.*, **37**, 45 (1988).
22. G. R. Husk, P. E. Cassidy and K. L. Gebent, *Macromolecules*, **21**, 1234 (1988).
23. N. Tai-shung Chung, E. Ronald Kafchinski and Paul Foley, *J. Membrane Sci.*, **75**, 181 (1992).
24. A. Yu. Alentiev, K. A. Loza and Yu. P. Yampolskii, *J. Membrane Sci.*, **167**, 91 (2000).
25. J. Y. Park and D. R. Paul, *J. Membrane Sci.* **125**, 23 (1997).
26. L. M. Robeson, *J. Membrane Sci.* **62**, 165 (1991).
27. J. L. Hedrick, C. J. Hawker, R. DiPietro and R. Jerome, *Polymer*, **36**, 4855 (1995).

Polyimides and Other High Temperature Polymers, Vol. 2, pp. 487–497
Ed. K.L. Mittal
© VSP 2003

Amine-quinone polyimides – High temperature polymers that protect iron against corrosion

MIJEONG HAN,[1] HUIMIN BIE,[2] DAVID E. NIKLES[*,1] and
GARRY WARREN[2]

[1]*Department of Chemistry and* [2]*Department of Metallurgical and Materials Engineering and Center for Materials for Information Technology, The University of Alabama, Tuscaloosa, AL 35487-0209*

Abstract—Two dianiline monomers were prepared by the reaction of 1,4-benzoquinone with either 4,4'-methylenedianiline or 4,4'-oxydianiline. These monomers were condensed with either 3,3',4,4'-benzophenonetetracarboxylic dianhydride or 4,4'-(hexafluoroisopropylidene) diphthalic anhydride to make the corresponding poly(amic acid)s. The poly(amic acid)s were converted to the polyimides by thermal imidization at 290°C. The amine-quinone polyimides gave free-standing films with tensile strengths in the range of 90 to 150 MPa and elastic moduli of 0.9 to 1.5 GPa. The thermal decomposition temperature under nitrogen was 440 to 480°C and the glass transition temperature was in the range of 280 to 310°C. Polyimide coatings on iron were exposed to 0.1 M NaCl electrolyte and any degradation was followed by electrochemical impedance spectroscopy. A conventional polyimide, made from 4,4'-methylenedianiline and 3,3',4,4'-benzophenonetetracarboxylic dianhydride, failed after 3 days exposure. The amine-quinone polyimides survived more than one year of exposure. The adhesion between the amine-quinone polyimide coatings and the iron surface was so strong that the coatings could not be delaminated by the electrolyte.

Keywords: Polyimides; iron corrosion; coatings; amine-quinone.

1. INTRODUCTION

Polymers containing the 2,5-diamino-1,4-benzoquinone functional group have been of interest since Erhan and coworkers first reported the synthesis of these polymers by the condensation of a diamine with 1,4-benzoquinone, Fig. 1, in the presence of an oxidizing agent [1]. A number of polymers were prepared using either aliphatic or aromatic diamines [2]. Depending on the choice of diamine, polymers could be prepared that were soluble in organic solvents and films could be cast by solution coating techniques. It was reported that the amine-quinone polymers would displace moisture from the surface of iron, rendering it hydrophobic. This was important, since the rate of iron corrosion greatly increases when surface moisture is present. Erhan reported that the amine-quinone polymer

---

*To whom all correspondence should be addressed. Phone: 205-348-9267, Fax: 205-348-2346,
E-mail: dnikles@mint.ua.edu

coatings inhibited corrosion of iron during exposure to salt spray [3]. Mathias and coworkers [4] provided a simpler synthesis of amine-quinone polymers by the condensation of 2,5-di-n-butoxy-1,4-benzoquinone with 1,6-hexanediamine in amide solvents, thereby eliminating the need for the two, two-electron oxidation steps. Muralidharan and coworkers [5] demonstrated the corrosion inhibition of iron and steel by an amine-quinone oligomer made by electropolymerization. They suggested that the amine-quinone polymers adsorbed onto the metal surface through the imine moiety of the polymer chain and provided FTIR spectroscopic evidence to support this mode of adsorption. Scola and colleagues have reported the use of fluorinated polyaminoquinones as promoters for the adhesion of epoxy to steel [6].

We have synthesized amine-quinone polyurethanes by copolymerizing an amine-quinone diol monomer and an oligomeric diol monomer (e.g., polytetrahydrofuran diol or polycaprolactone diol) with a diisocyanate (e.g., toluene diisocyanate or methylene diphenyl diisocyanate) [7]. Sulfur-quinone polyurethanes were also prepared by using a sulfur-quinone monomer, where the secondary amine groups were replaced with thioether groups [8]. These polyurethanes protected the commercial iron particles used in state-of-the-art metal particle tape against corrosion [9].We have also demonstrated that amine-quinone polyurethanes promote the adhesion of epoxy to steel [10].

Recently we turned our attention to polyimides and questioned whether we could combine the high temperature properties inherent in polyimides with the corrosion protection of metal afforded by the amine-quinone functional group. Two new amine-quinone dianiline monomers (AQMDA and AQODA, Fig. 2) were synthesized and used to prepare a series of amine-quinone polyimides (Fig. 3) [11]. These polymers showed high thermal properties (TGA decomposition temperature above 440°C) and high tensile modulus (E > 1 GPa) one expects for polyimides [12, 13]. One composition (AQPI-2 made from AQMDA and BPDTA) was coated onto iron and shown to protect the iron against attack by aqueous sodium chloride electrolyte [14]. In this paper, we review the thermal and mechanical properties of the amine-quinone polyimides. We also report new results on electrochemical impedance studies of the attack of sodium chloride electrolyte on iron samples protected by amine-quinone polyimides.

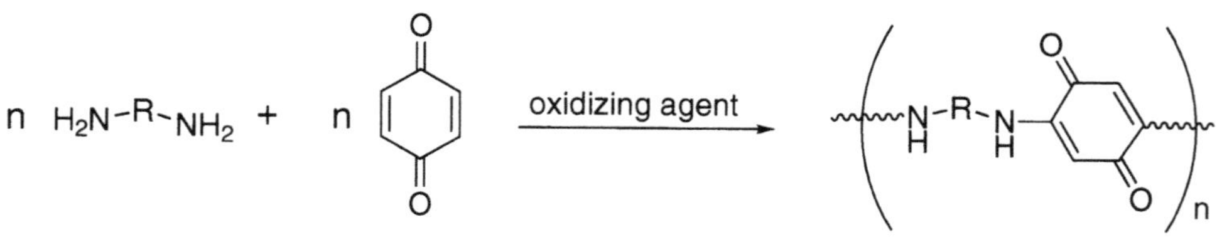

Figure 1. Synthesis of an amine-quinone polyimide by condensation of a diamine with 1,4-benzoquinone.

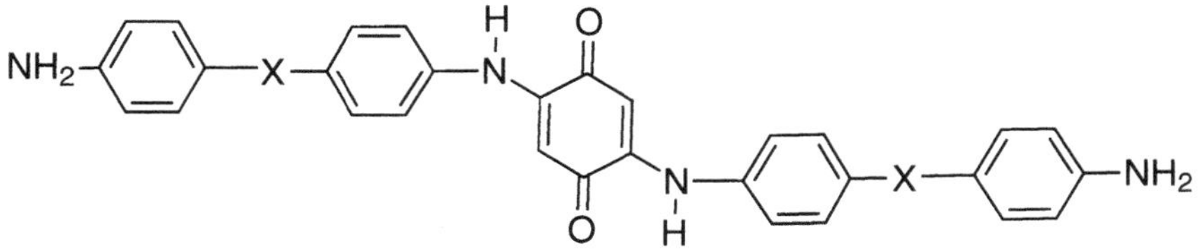

Figure 2. Amine-quinone dianiline monomers, where X = CH$_2$ (AQMDA) or X = O (AQODA).

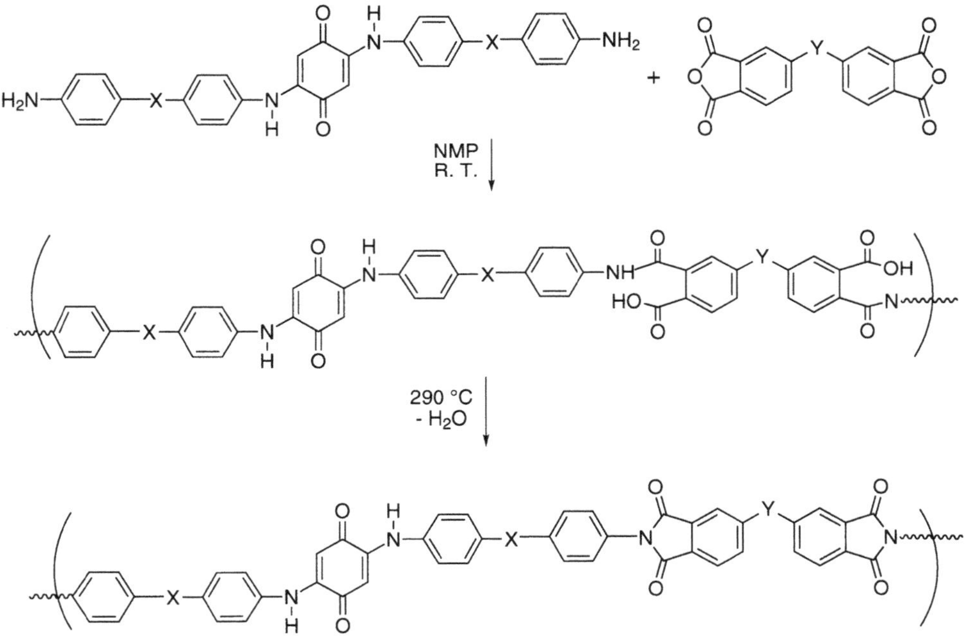

Figure 3. Two-step synthesis of amine-quinone polyimides, where X = CH$_2$ (AQMDA) or X = O (AQODA), and Y is either C=O or C(CF$_3$)$_2$.

2. EXPERIMENTAL

2.1. Materials

The synthesis of the amine-quinone dianiline monomers, AQMDA and AQODA, was described previously [11, 14]. The synthesis of the amine-quinone polyimides was also described earlier [14]. This was a two-stage polymerization where the dianiline monomer was condensed with the dianhydride in dry N-methyl-pyrrolidinone (NMP) to give the poly(amic acid). The poly(amic acid) solution was cast either onto a glass plate by hand draw-down using a wire-wound applicator or onto 50 x 50 mm iron squares by spin coating. The coatings were thermally cured in a vacuum oven by heating to 50°C for 1 h, 150°C for 1 h, 250°C for 1 h, and finally 290°C for 1 h. After cooling to room temperature, the coatings on glass were soaked in water and carefully lifted from the glass to give free-standing films.

2.2. Methods

The intrinsic viscosity, η_{inh} for the poly(amic acid)s was measured at a concentration of 0.5 g/dL in NMP at 30°C using an Ubbelohde viscometer. The thermal decomposition temperature for the polyimides was determined under nitrogen on a TA Instruments Model 2950 thermogravimetric analyzer with the heating rate of 10°C/min. The decomposition temperatures were taken when the weight loss reduced by either 5 or 10% of the original weight. The glass transition temperature was measured on a TA Instruments model 2920 differential scanning calorimeter under nitrogen. The samples were heated to 350°C in nitrogen and cooled to room temperature, and then heated to 400°C. The glass transition temperatures were determined from the second heating. The tensile properties of the polymer films were measured on an Instron Model 4465-STANDARD with 100 N load cell. The samples for the test were prepared by the ASTM method [15] and the crosshead speed was 10 mm/min. Five replicate samples were tested for each polymer and the reported values were the average of the five. The moisture absorption was determined by drying the polyimide films under vacuum at 80°C for 24 hours. The films were weighed and then exposed to 85°C and 85% relative humidity in a Tenney model TH Jr temperature humidity chamber for one week. After exposure the films were weighed again to determine the percent moisture uptake.

3. RESULTS AND DISCUSSION

The viscosity data, Table 1, for the poly(amic acid)s showed that the polymers had very similar molecular weights. The thermal stability of amine-quinone polyimides was determined by measuring the temperature of 5% weight loss ($T_{5\%}$) and percent char yield at 800°C by thermogravimetric analysis under nitrogen, Table 2. All amine-quinone polyimides showed the excellent thermal stability one would expect for polyimides. The polymers exhibited $T_{5\%}$ weight loss temperature ranging from 440 to 479°C. The char yields, residual weight % at 800°C, ranged from 60 to 69%. The thermal stability did not depend on the choice of monomers and all amine-quinone polyimides showed excellent thermal stability. The glass transition temperatures of amine-quinone polyimides were measured by differential scanning calorimetry, Table 2. All polymers showed a glass transition temperature between 279 and 310°C.

Most of the amine-quinone polyimides gave transparent, flexible and tough films suitable for measurements of tensile properties, Table 3. Free-standing films could not be made from AQPI-5, because the poly(amic acid) was always obtained as a gel. In the case of AQPI-4, the film was not very uniform and gave

Table 1.
Composition of the polyimides

Polyimide	Monomers (mole percent)			η_{inh} (dL/g)	Moisture absorption (%)
AQPI-2	AQMDA (50)	BTDA (50)		0.97	0.9
AQPI-3	AQMDA (50)	6FDA (50)		1.18	1.0
AQPI-4	AQODA (50)	BTDA (50)		0.99	0.8
AQPI-5	AQODA (50)	6FDA (50)		$-^a$	$-^a$
AQPI-6	AQMDA (25)	MDA (25)	BTDA (50)	1.02	0.7
AQPI-7	AQMDA (25)	MDA (25)	6FDA (50)	0.96	0.7
AQPI-8	AQODA (25)	ODA (25)	BTDA (50)	1.00	0.8
AQPI-9	AQODA (25)	ODA (25)	6FDA (50)	0.79	1.0
PI	MDA (50)	BTDA (50)		1.01	1.5

[a] The poly(amic acid) from AQODA and 6FDA could not be obtained because it always made a gel.

Table 2.
Thermal properties of polymers

Polymer	$T_{5\%}$ (°C)	Char yield (%)	T_g (°C)
AQPI-2	472	69.0	292
AQPI-3	448	62.5	286
AQPI-4	455	67.6	280
AQPI-6	440	64.2	289
AQPI-7	453	59.8	279
AQPI-8	479	62.8	296
AQPI-9	474	57.9	310
PI	525	61.8	282

Table 3.
Tensile properties of the amine-quinone polyimides

Polyimide	Elastic modulus (GPa)	Tensile strength (MPa)	Elongation at break (%)
AQPI-2	1.1	140	24.4
AQPI-3	1.1	146	24.6
AQPI-4	0.9	92	16.4
AQPI-6	1.0	131	21.6
AQPI-7	1.3	133	13.5
AQPI-8	1.5	161	16.5
AQPI-9	1.2	138	18.3
PI	1.5	169	19.3

poor mechanical properties. For the other amine-quinone polyimides the tensile moduli were between 0.9 and 1.5 GPa and the tensile strengths ranged from 92 to 161 MPa. All polymer films behaved as ductile materials with good modulus, high strength, and moderate elongation at break.

The moisture absorption was determined by exposing the dried polyimide films to 85°C and 85% relative humidity for one week and measuring the percent increase in weight. The amine-quinone polyimides had a moisture absorption in the range of 0.7 to 1.0%, Table 1. The conventional polyimide absorbed 1.5% water. Clearly, the amine-quinone functional groups had very little impact on the affinity of the amine-quinone polyimides for water.

Samples consisting of the polyimide coatings on iron were exposed to NaCl electrolyte and the degradation of the coatings monitored by electrochemical impedance spectroscopy (EIS). In Figures 4 and 5 are shown the Bode plots, i.e. plots of impedance and phase angle as a function of frequency. The initial plot was taken after five minutes exposure. The initial Bode plot for the BTDA/MDA film on iron showed an impedance of 3.9 x 10^5 $\Omega \cdot cm^2$ at low frequency, Fig. 4. The values of the phase angle were less than the $-90°$ as one would expect if the polymer film were acting as a dielectric. After 72 hour exposure to NaCl electro-

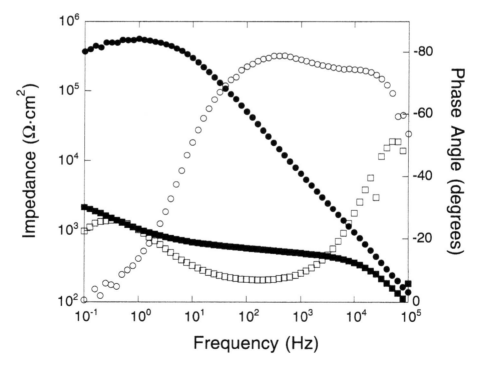

Figure 4. Bode plots of impedance and phase angle as a function of frequency for the conventional polyimide (PI) film on iron. The circles are the initial curves, while the squares are the curves obtained after 72 hours exposure to 0.1 M NaCl eletrolyte. The filled circles and squares are the impedance curves. The open circles and squares are the phase angle curves.

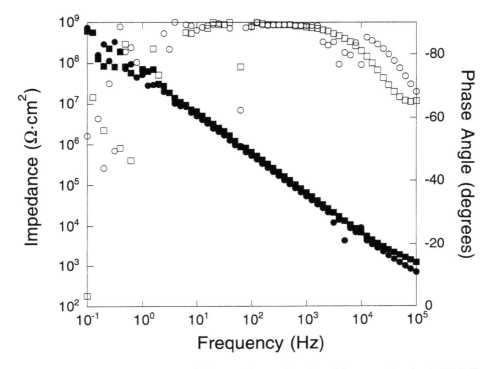

Figure 5. Bode plots of impedance and phase angle as a function of frequency for the AQPI-2 film on iron. The circles are the initial curves, while the squares are the curves obtained after 364 days exposure to 0.1 M NaCl electrolyte. The filled circles and squares are the impedance curves. The open circles and squares are the phase angle curves.

lyte, the impedance dropped to 2×10^3 $\Omega \cdot cm^2$. A new peak appeared at low frequency in the phase angle curve, which was attributed to a double layer capacitance. The electrolyte had diffused through the polymer, thereby lowering the resistance and the capacitance of the coating. Furthermore, the electrolyte had penetrated to the polyimide-iron interface and broke the adhesion bond. When this new feature appeared in the phase angle curve, pitting corrosion soon appeared.

The iron sample coated with AQPI-2 showed a different behavior after exposure to NaCl electrolyte, Fig. 5. The impedance was on the order of 10^8 $\Omega \cdot cm^2$ and the phase angle was near $-90°$ over a wide range of frequencies. The coating was acting as a dielectric. After a year of exposure, there was very little change in the impedance and the phase angle curves, but no peak in the phase angle curve at low frequency appeared.

The EIS spectra were fitted to the equivalent circuit model shown in Figure 6, which consists of a coating resistance (R_c) in parallel with a coating capacitance (C_c), both in series with the solution resistance of the NaCl electrolyte (R_s). After the appearance of the second peak in the phase angle curve, the equivalent circuit

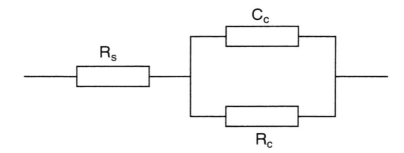

Figure 6. Equivalent circuit for modeling the EIS curves for the polyimides coated iron, where R_s is the NaCl electrolyte solution resistance, R_c is the resistance of the polyimide coating and C_c is the capacitance of the coating.

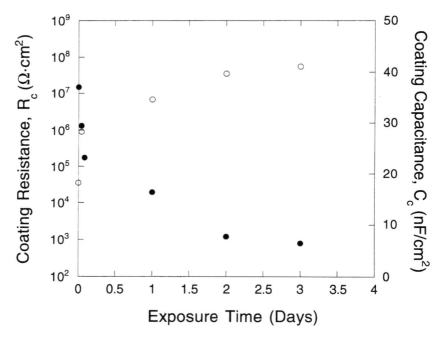

Figure 7. Plot of coating resistance (filled circles) and coating capacitance (open circles) for PI on iron as a function of expoure time to 0.1 M NaCl electrolyte.

in Figure 6 was no longer valid and the experiment was discontinued. In Figure 7 are shown plots of R_c and C_c as a function of exposure time to the electrolyte for the conventional polyimide on iron. The initial coating resistance was low (6 x 10^5 $\Omega \cdot cm^2$) and the initial coating capacitance was high (856 nF/cm²), indicating that the coating was not acting as a pure dielectric. After three days exposure to the electrolyte, the coating resistance dropped, while the coating capacitance in-

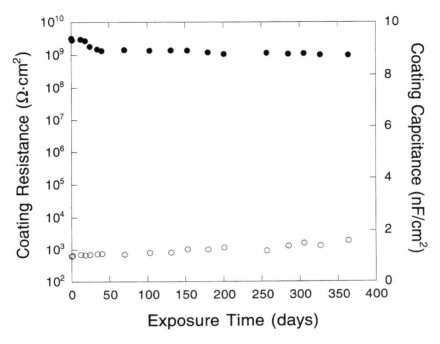

Figure 8. Plot of coating resistance (filled circles) and coating capacitance (open circles) for AQPI-2 on iron as a function of exposure time to 0.1 M NaCl electrolyte.

creased, indicating that the electrolyte was diffusing into the coating. The AQPI-2 coating (Figure 8) had a higher initial resistance (1.2×10^9 $\Omega \cdot cm^2$) and a lower initial capacitance (2.5 nF/cm²), indicating that AQPI-2 was acting as a good dielectric. After a year of exposure to the electrolyte, there was little change in the coating resistance and the coating capacitance for AQPI-2.

A model for the cathodic disbondment of a polymer coating on a metal surface during blister formation includes the disbondment stress and the elastic modulus of the coating [16]. A weakness of this model is that the viscoelastic properties of the polymers are not included. Nevertheless the amine-quinone polyurethanes and the amine-quinone polyimides provide an opportunity to test this model. We have already shown that amine-quinone polyurethanes promote adhesion of epoxy to steel [10]. We speculate that the amine-quinone polymers chemisorb onto the iron surface, providing a strong adhesion bond. The adhesion bond strength should be measured and we expect that it would depend on the nature of the amine-quinone functional group and the number of groups adsorbing onto the iron surface. The elastic moduli for a series of amine-quinone polyurethanes were in the range of 3 to 250 Mpa [10], while the moduli of the polyimides reported here were 0.9 to 1.5 GPa. The range of moduli provided by the amine-quinone polyurethanes and the amine-quinone polyimides offer an opportunity to test the effect of elastic modulus on the rate of blister formation.

4. CONCLUSIONS

An amine-quinone diamine based polyimide showed the high temperature thermal properties typical of conventional polyimides. This amine-quinone polyimide can resist attack by NaCl electrolyte in a manner similar to amine-quinone polyurethanes, thereby protecting iron against corrosion. This raises the possibility of a new class of coatings that protect iron against corrosion in extreme environments.

Acknowledgments

This research was supported, in part, by the corporate sponsors of the Center for Materials for Information Technology and by the NSF Materials Research Science and Engineering Center (award number DMR-9809423). Dr. Mijeong Han was supported by the School of Graduate Studies by a Graduate Council Fellowship.

REFERENCES

1. K. Kaleem, F. Chertok and S. Erhan, *Prog. Org. Coatings* **15**, 63-71 (1987).
2. a) K. Kaleem, F. Chertok and S. Erhan, *J. Polym. Sci., Pt A* **27**, 865-871 (1991);
   b) V. S. Nithianandam, K. Kaleem, F. Chertok and S. Erhan, *J. Appl. Polym. Sci.* **42**, 2893-2897 (1991);
   c) V. S. Nithianandam, F. Chertok and S. Erhan, *J. Appl. Polym. Sci.* **42**, 2899-2901 (1992);
   d) V. S. Nithianandam and S. Erhan, *J. Appl. Polym. Sci.* **42**, 2385-2389 (1991);
   e) V. S. Nithianandam and S. Erhan, *Polymer* **32**, 1146-1149 (1991).
3. V. S. Nithianandam, F. Chertok and S. Erhan, *J. Coatings Technol.* **63** (796), 47-49 (1991).
4. R. F. Colletti, M. J. Stewart, A. E. Taylor, N. J. MacNiell and L. J. Mathias, *J. Polym. Sci., Polym. Chem.* **29**, 1633 (1991).
5. a) K. L. N. Phani, S. Pitchumnani, S. Muralidharan, S. Ravichandran and S. V. K. Iyer, *J. Electroanal. Chem.* **353**, 315-322 (1993);
   b) S. Muralidharan, K. L. N. Phani, S. Pitchumnani, S. Ravichandran and S. V. K. Iyer, *J. Electrochem. Soc.* **142**, 1478-1483 (1995).
6. E. Vaccaro, C. D. Simone and D. A. Scola, *J. Adhesion* **72**, 157-176 (2000).
7. a) D. E. Nikles, J.-L. Liang, J. L. Cain, A. P. Chacko, R. I. Webb and K. J. Belmore, *J. Polym. Sci., Pt A, Polym. Chem.* **33**, 2881-2886 (1995);
   b) D. E. Nikles, A. P. Chacko, J.-L. Liang and R. I. Webb, *J. Polym. Sci., Pt. A, Polym. Chem.* **37**, 2339-2345 (1999);
   c) M. Han and D. E. Nikles, *J. Polym. Sci., Pt A, Polym. Chem.* **38**, 3284-3292 (2000).
8. Y. Hu and D. E. Nikles, *J. Polym. Sci., Pt A, Polym. Chem.* **38**, 3278-3283 (2000).
9. a) J.-L. Liang and D. E. Nikles, *IEEE Trans. Magnetics* **29**, 3649-3651 (1993);
   b) A. P. Chacko and D. E. Nikles, *J. Appl. Phys.* **79**, 4863-4865 (1996);
   c) D. E. Nikles and G. W. Warren, *Polymer News* **23**, 223-231 (1998);
   d) G. W. Warren, R. Sharma, Y. Hu, D. E. Nikles and S. Street, *J. Magn. Magn. Mater.* **193**, 276-278 (1999);
   e) M. Han, A. B. Helms, Y. Hu, D. E. Nikles, J. A. Nikles, R. Sharma, S. C. Street and G. W. Warren, *IEEE Trans. Magnetics* **35**, 2763-2765 (1999).
10. a) K. Vaideeswaran, J. P. Bell and D. E. Nikles, *J. Adhesion Sci. Technol.* **13**, 477-499 (1999);
    b) K. Vaideeswaran, J. P. Bell and D. E. Nikles, *J. Appl. Polym. Sci.* **76**, 1338-1350 (2000).
11. M. Han and D. E. Nikles, *J. Polym. Chem., Pt A, Polym. Chem.* **39**, 4044-4049 (2001).

12. M. J. M. Abadie and B. Sillion (Eds.), *Polyimides and Other High-Temperature Polymers,* Elsevier, Amsterdam (1991).

13. M. K. Ghosh and K. L. Mittal (Eds.), *Polyimides: Fundamentals and Applications*, Marcel Dekker, New York (1996).

14. M. Han, H. Bie, D. E. Nikles and G. W. Warren, *J. Polym. Sci., Pt. A, Polym. Chem.* **38**, 2893-2899 (2000).

15. ASTM Designation D 882-91 "Standard Test Methods for Tensile Properties of Thin Plastic Sheeting".

16. T.-J. Chuang, T. Nguyen and S. Lee, *J. Coatings Technol.* **71***(895)*, 75-85 (1999).

Polyimides and Other High Temperature Polymers, Vol. 2, pp. 499–509
Ed. K.L. Mittal

Semicrystalline DABP-BTDA polyimide modified by fullerenes for wear protection

G.N. GUBANOVA,*,[1] T.K. MELESCHKO,[1] YU.A. FADIN,[2] V.E. YUDIN,[1] YU.P. KOZIREV,[2] A.A. MICHAILOV,[1] N.N. BOGORAD,[1] A.G. KALBIN,[1] G.N. FIODOROVA[1] and V.V. KUDRYAVTSEV[1]

[1]*Institute of Macromolecular Compounds, Russian Academy of Sciences, V.O.Bolshoy pr. 31, 199004 St. Petersburg, Russia*
[2]*Institute of Problems of Mechanical Engineering, Russian Academy of Sciences, V.O.Bolshoy pr. 61, 199178 St. Petersburg, Russia*

Abstract—This work investigates some tribological properties of melt processable semicrystalline polyimide (T_g=250°C, T_m=290°C) based on 3,3'-diaminobenzophenone and 3,3',4,4'-benzophenonetetracarboxylic dianhydride (DABP-BTDA) undoped and doped with fullerenes. The mechanical properties of the obtained polyimide molding were: flexural strength 120 MPa, flexural modulus 2.2 GPa, and failure strain 5.5%. It was shown that the polyimide crystalline structure depended mainly on the synthesis procedure (chemical or thermal imidization) and the kind of solvent. A mixture of fullerenes (C_{60} – 78%, C_{70} – 22%) in the amount of 0.015-1.0 wt.% was added to the polyimide at different stages of its chemical imidization. DSC analysis did not show significant influence of fullerene addition on polyimide melting temperature and heat flow.

Some tribological characteristics were investigated under dry friction condition for different polyimide/fullerene moldings depending on their synthesis procedure. The main parameters studied were the sliding friction coefficient for steel rod on polymer samples and wear at each nominal pressure in tribocontact. It was found that optimal tribological properties were possible to be obtained for polyimide/fullerene samples when fullerene was introduced into the diamine solution during chemical imidization. The sliding friction coefficient of these samples increased from 0.18 to 2.0 at the investigated pressure range (the maximum pressure attained was 8 MPa). These samples also showed less wear rate under experimental conditions as compared to polyimide samples without addition of fullerenes.

Keywords: Semi-crystalline polyimide; moldings; fullerenes; tribology.

1. INTRODUCTION

Polyimides (PI) are among the most promising polymer matrices for composite materials [1] because they combine high thermal stability (up to 500°C) and high

---

*To whom all correspondence should be addressed. Phone: (7) (812) 527 85 71, Fax: (7) (812) 328 68 69, E-mail: medvedll@online.ru

crack resistance with good mechanical properties from cryogenic temperatures to 400°C. Polyimides have received considerable attention from tribologists as a class of thermally stable materials with high mechanical strength. In tribology, PI's are used as coatings, moldings, and matrices for composite materials [2-5].

The class of crystallizable PI exhibiting the properties of a thermoplastic (high crack resistance) and high thermal stability attracts particular attention. Crystallizable PI's display excellent thermal and thermo-oxidative stability, and high mechanical and adhesion properties.

In this work the crystallizable PI's were developed using dianhydride of 3,3',4,4'-benzophenonetetracarboxylic acid (BTDA) and a meta-substituted diamine [6], in particular 3,3'-diaminobenzophenone (3,3' DABP):

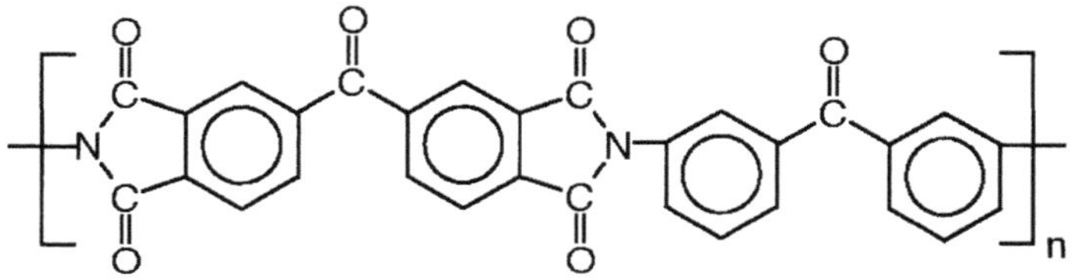

It is important to note that the first investigation of this type of polyimide (a study of calorimetric and rheological behavior) was carried out at the NASA Langley Research Center [7, 8].

We have previously shown that the structural features of this PI were determined, to a considerable extent, by the conditions of its preparation [6]. In particular, BTDA-DABP PI obtained under the conditions of low temperature chemical imidization in solution in an amide solvent is semi-crystalline and its melting temperature is unusually low for this class of polymers (T_{melt} =270-280°C). It was shown that both thermally and chemically imidized PI's could be processed into the matrix of a composite material. Moreover, such processing of chemically imidized PI is more advantageous for this purpose because of low melting temperature which is far lower than the temperature of thermal decomposition of PI. Carbon composites based on this PI exhibit higher interlaminar fracture toughness (750 J/m^2) than similar composites based on thermally imidized semi-crystalline PI and retain high mechanical strength (1 GPa) and thermal stability (250°C) [9]. In the present work, the possibility of processing the BTDA-DABP PI into moldings is investigated and mechanical and tribological properties of moldings are studied. It is known that the addition of fullerenes into the polymers can greatly improve their tribological properties [10]. Therefore, we carried out synthesis and investigated the properties of fullerene-polymer compositions using a semi-crystalline BTDA-DABP PI.

2. EXPERIMENTAL

2.1. Synthesis

BTDA (with T_m=225-226°) recrystallized from dry acetic anhydride and 3,3' DABP were used as monomers [11]. Poly(amic acid) (PAA) was synthesized by introducing a stoichiometric quantity of the dianhydride into the diamine solution of N,N-dimethyl acetamide (DMAc) with stirring at 50°C. The concentration of the PAA solution was 25% (by wt.) and the time of synthesis in DMAc was 4 h.

Powdered PI samples were prepared by chemical imidization in DMAc as described earlier [11].

To obtain fullerene-containing semi-crystalline BTDA-DABP PI, several methods of fullerene introduction into the polymer were employed. A mixture of C_{60}-C_{70} fullerenes in the weight ratio of 78:22% was used. Fullerene was dissolved in a solution of an aromatic diamine (3,3'-diaminobenzophenone – one of the monomers), which was used in the preparation of the poly(amic acid), a precursor of polyimide. We proceeded from the fact that it was possible to distribute fullerene homogeneously and to immobilize it in the polymer bulk by binding its molecules to terminal amino groups of the monomer because fullerene molecules are known to interact actively with amines. In the first experiment (the first method), fullerene was dissolved in DMAc (solution concentration 0.05 mg/ml) and the diamine was introduced into the solution. In the second experiment (the second method), a diamine solution in DMAc was prepared and then fullerene solution in o-dichlorobenzene (solution concentration 20 mg/ml) was introduced into this diamine solution, because fullerene solubility in o-dichlorobenzene is much higher than in DMAc. Therefore, the use of o-dichlorobenzene favored the introduction of a greater amount of fullerene into the polymer than in the case of fullerene solution in DMAc. When o-dichlorobenzene was used, 0.1% of fullerene based on dry polymer weight was introduced. When only DMAc was used as solvent, 0.015% of fullerene was introduced. The mixture of fullerene and 3,3'-DABP in a mixture of solvents, DMAc and o-dichlorobenzene (solvents weight ratio 1:5) was kept for 1 h at room temperature. After this, PAA was synthesized by introducing BTDA, the second monomer, into this solution. Powdered samples of BTDA-DABP PI were subsequently obtained by chemical imidization in solution.

The third way (the third method) of fullerene introduction was to add fullerene solution in o-dichlorobenzene (solution concentration 20 mg/ml) to the solution of the prepared PAA (25 wt.%). The resulting mixture was kept at room temperature for 2 h. Subsequently, just as in the previous cases, PAA chemical imidization in solution was carried out by exposing to the imidizing mixture (consisting of acetic anhydride and triethylamine) for 24 h at room temperature. During imidization, the polymer became insoluble and was precipitated from the reaction mixture in the form of powder.

Finally, in the fourth way (fourth method) of obtaining a fullerene-containing polymer composition, fullerene solution in o-dichlorobenzene (solution concentration 20 mg) was introduced into the polymer solution (3 h after the introduction of the imidizing mixture). In other words, it was done during PAA chemical imidization but before polyimide was precipitated from the reaction mixture.

2.2. Methods

The melt viscosity of PI powders was determined in a rheogoniometer (designed at the Topchiev Institute of Petrochemical Synthesis of Russian Academy of Sciences) with a cone-plane geometry (the cone angle is $1°$, plate diameter 40 mm) at a constant shear rate of 10^{-2} sec^{-1} at a given temperature.

Differential scanning calorimetry (DSC) was performed using the DSM-2M microcalorimeter (designed at the Institute of Biology, Russian Academy of Sciences, Puschino) at a scanning rate of $10°C/min$.

Powders of BTDA-DABP PI obtained by chemical and thermal imidizations were molded as follows: First, discs 20 mm in diameter and 2-3 mm thick were molded at room temperature and a pressure of 200 atm. The disc was placed in a compression mold and heated without pressure to $320°C$ and maintained at that temperature for 1 h. After heating, the compression mold was taken out of the thermostat and the sample in the compression mold was subjected to a pressure of 150 atm. for 2-3 min. The compression mold was cooled to $60\text{-}70°C$ over 40 min at a pressure of 100-150 atm. 5 mm wide, 2 mm thick and 18 mm long samples were cut from molded disc. The samples were subjected to three-point bending tests at a loading rate 1.3 mm/min. Bending strength σ_b, Young modulus E, and failure strain ε_{max} were determined.

The thermogravimetric (TGA) analysis was carried out on a derivatograph (MOM Company, Hungary) in a self-generated atmosphere (in other words, in the atmosphere of volatile products released from the sample during heating); the scanning rate was $10°C/min$.

The tribological characteristics of moldings were determined on a Tribometer, designed at the Institute of Problems of Mechanical Engineering, Russian Academy of Sciences; the testing configuration was "ring-ring". A ring-shaped rotating steel counterbody was in contact with a planar surface of the polymer ring cut from molding. The variable parameters in friction and wear measurements were contact pressure and sliding speed of the counterbody. The contact pressure ranged from 1.7 to 6.7 MPa and the sliding speed of steel counterbody was 0.047 to 0.138 m/sec. The main parameters investigated were the coefficient of sliding friction and wear (mass) for 10 min sliding at each nominal pressure in the tribocontact. The dimentionless wear rate $\overset{\bullet}{W}$ was determined as the volume (V) of material worn during a certain period of time (in our case 10 min) divided by the tribological contact area (S) and the sliding path length (L), i.e.: $\overset{\bullet}{W} = V/SL$ [5]. Polymer particles formed during the experiment were also collected and analyzed using a "Videolab" automated image processing system, designed at Moskow

University [12]. Using this system it is possible to determine the area, perimeter of each wear particle and other parameters.

3. RESULTS AND DISCUSSION

Fig. 1 shows the time dependence of viscosity of the polymer melt for amorphous BTDA-DABP PI samples obtained by chemical and thermal imidizations and for semi-crystalline samples having different melting temperatures (280-290°C for samples obtained by chemical imidization, and 350°C for samples obtained by thermal imidization). It can be seen that the melt viscosity of both amorphous and semi-crystalline samples obtained by chemical imidization is relatively low (Fig. 1, curves 1 and 2) and does not increase for 60 min. The above data indicate that these BTDA-DABP PI samples can be successfully processed into a matrix for composite materials or into a molding under hot pressing conditions.

Polymer samples obtained by thermal imidization (Fig. 1, curves 3 and 4) are characterized by high initial viscosity. During 30 min viscosity increased rapidly. This makes it difficult to process polymers of these structural modifications into a composite matrix or a molding.

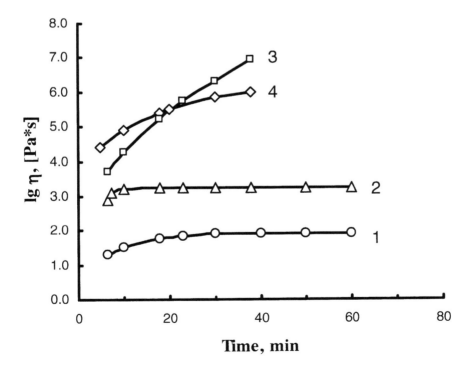

Figure 1. Melt viscosity vs. time at 330°C for amorphous (1,3) and semi-crystalline (2,4) samples of BTDA-DABP PI, obtained by chemical (1,2) and thermal (3,4) imidization.

Table 1 shows the results of bending tests obtained on BTDA-DABP PI moldings. These data show that for a polymer obtained by thermal imidization the bending strength is very low and the sample is brittle. These characteristics of moldings, obtained by thermal imidization, are probably due to the low flowability of melt at T=320°C. As a result, it does not possess the homogeneous structure. Hence, the moldings obtained by chemical imidization, in contrast to its analogs obtained by thermal imidization, can be processed into a molded product at relatively low temperatures, 300-330°C, which are much lower than the PI thermal degradation start temperature 450°C. The BTDA-DABP PI moldings obtained by chemical imidization exhibit relatively high mechanical characteristics, which are comparable to the BTDA-DABP PI amorphous films, obtained by thermal imidization. The mechanical characteristics of this amorphous film at room temperature are as follows: bending strength 100-120 MPa, Young's modulus 3000-3300 MPa, and failure strain 5-6%.

It is for this reason that BTDA-DABP PI obtained by chemical imidization was chosen for fullerene incorporation.

The samples were studied and compared with the samples of semi-crystalline BTDA-DABP PI without fullerene by thermogravimetry (determination of thermal degradation range and thermal stability indices τ_0, τ_5 and τ_{10}, where τ_0 is the temperature of beginning of weight loss, τ_5 and τ_{10} – temperatures of 5% and 10% weight loss respectively) and differential scanning calorimetry (temperature and enthalpies of melting).

The data in Table 2 show that the introduction of fullerene into the reaction solution influenced the process of polymer structure formation. In most cases, the degree of polymer crystallinity decreased, as indicated by decreased melting enthalpy. However, in all cases fullerene-containing polymer has a melting temperature close to that of semi-crystalline BTDA-DABP PI (without fullerene) obtained by chemical imidization. Consequently, it can be processed into a composite matrix or a molding at temperatures of 300-350°C.

A comparison of thermal stability of the samples shows (Table 3) that fullerene introduction into PI increases the thermal stability of BTDA-DABP PI (τ_0 – temperature of the beginning of weight loss): τ_5 and τ_{10} – temperatures of 5% and 10% of weight loss also increase considerably.

Moldings based on BTDA-DABP PI-fullerene compositions were tested for friction behavior. Fig. 2 shows the coefficient of sliding friction vs. pressure in tribocontact for moldings based on polyimide-fullerene compositions and for a molding based on pure BTDA-DABP PI. It should be noted that up to a pressure of 5 MPa the friction coefficient for all moldings (based on polyimide-fullerene compositions, obtained by methods 1-4) remains almost unchanged and is in the range of 0.15-0.2. At contact pressures above 5 MPa, the coefficient of friction stays nearly constant only for the moldings based on polyimide-fullerene compositions obtained by method 2. Hence it is promising for application at contact pressure up to 8 MPa. Moldings, based on polyimide-fullerene compositions, obtained by other methods, are stable only up to 5 MPa.

Table 1.
Mechanical characteristics of BTDA-DABP PI moldings

BTDA-DABP PI	σ_b at 20°C, MPa	E at 20°C, MPa	ε_{max}, %
Semi-crystalline sample (chemical imidization)	118	2150	5.5
Semi-crystalline sample (thermal imidization)	9	820	1.1

σ_b – bending strength, E – Young's modulus, ε_{max} – failure strain.

Table 2.
Melting temperatures and enthalpies of BTDA-DABP PI-fullerene compositions with fullerene introduced in different stages of PI synthesis

BTDA-DABP PI-Fullerene composition	T_g, °C	T_m, °C	ΔH, J/g
Sample 1 – obtained by method 1	245	285	28.0
Sample 2 – obtained by method 2	237	270	17.2
Sample 2'	237	290	9.0
Sample 3 – obtained by method 3	–	273	29.3
Sample 4 – obtained by method 4	237	282	13.7
Sample 5 (without fullerene)	250	290	28.0

Sample 2' was prepared under conditions similar to those for sample 2 but at the end of the synthesis, phthalic anhydride was added.

Table 3.
Thermal stability characteristics of BTDA-DABP PI-fullerene compositions

BTDA DABP PI-Fullerene composition	τ_0 (°C)	τ_5 (°C)	τ_{10} (°C)
Sample 1 – obtained by method 1	465	568	606
Sample 2 – obtained by method 2	480	585	620
Sample 2'	486	550	570
Sample 3 – obtained by method 3	470	555	575
Sample 4 – obtained by method 4	480	550	575
Sample 5 (without fullerene)	450	532	553

Sample 2' was prepared under conditions similar to those for sample 2 but at the end of the synthesis, phthalic anhydride was added.

The molding based on polyimide-fullerene compositions obtained by method 2 is also the best with respect to the second tribological characteristic: wear (Fig. 3). Note that the introduction of fullerene into the monomer, in contrast to the case of its introduction during the stages of PAA synthesis, leads to a considerable decrease in the wear of the molding (Fig. 3, curve 2). For molding without fullerenes, the wear (Fig. 3, curve 5) is much higher than for moldings obtained by method 2 (Fig. 3, curve 2).

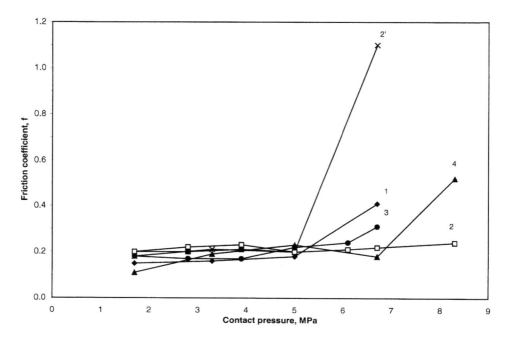

Figure 2. Dependence of average friction coefficient on contact pressure for moldings based on polyimide-fullerene composition. obtained by: 1 – method 1, 2 – method 2, 2' – method 2', 3 – method 3, 4 – method 4.

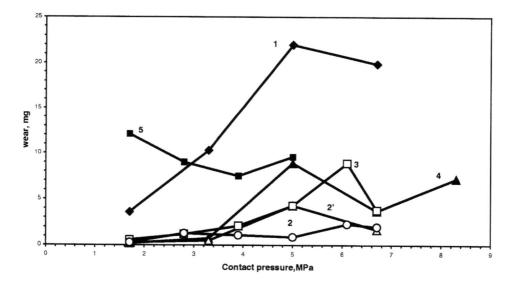

Figure 3. Dependence of mass wear on contact pressure for moldings based on polyimide-fullerene compositions, obtained by: 1 – method 1, 2 – method 2, 2'– method 2', 3 – method 3, 4 – method 4, 5 – molding without addition of fullerene.

The characteristics of wear particles collected in the experiment were also studied. In particular, the mean area of wear particles can indirectly characterize the shear strength of the investigated material. Fig. 4 gives the wear particle sizes (mean area) at different pressures in the tribocontact. These data also confirm the fact that at all nominal pressures the molding based on polyimide-fullerene composition obtained by the method 2, is the best with respect to wear. At all pressures, the mean particle area obtained for this molding is considerably smaller than in all other cases.

Consequently, the sample, based on polyimide-fullerene composition obtained by method 2 is characterized not only by low friction coefficient but also by high shear strength, which is an important parameter of materials.

The dependence of molding dimensionless wear rate on sliding speed was investigated at low sliding speed in the range of $(5-12)10^{-2}$ m/s. The following moldings were chosen: sample 1 with minimum fullerene content (0.015 wt.%), sample 3 with intermediate fullerene content (0.1 wt.%), and sample 3' with maximum fullerene content (1.0 wt.%), (obtained just as in case 3 but with the

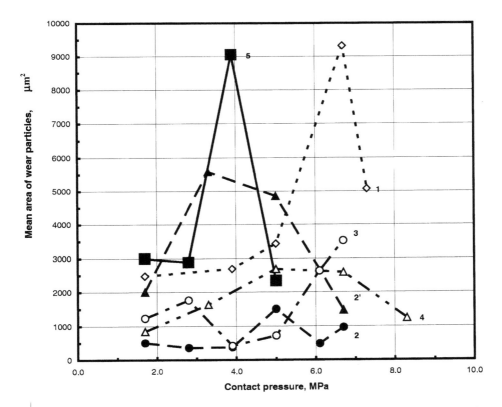

Figure 4. Dependence of mean area of wear particles on contact pressure for moldings based on polyimide-fullerene compositions, obtained by: 1 – method 1, 2 – method 2, 2' – method 2', 3 – method 3, 4 – method 4, 5 – molding without addition of fullerene.

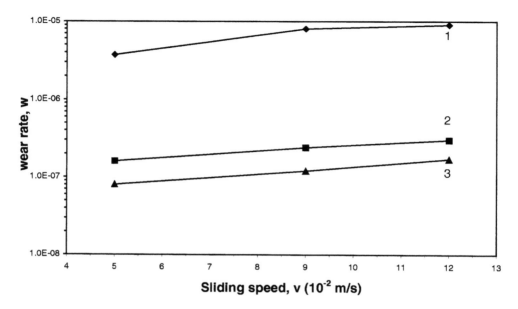

Figure 5. Dependence of dimensionless wear rate at a contact pressure of 3.9 MPa for moldings: 1 – sample 1 (obtained by the method 1), 2 – sample 3 (obtained by the method 3), 3 – sample 3′ (obtained by the method 3) with maximum content of fullerene.

addition of nigner fullerene content in PAA). The data in Fig. 5 show conclusively that wear rate of moldings decreased with increasing fullerene content. When fullerene content was varied from 0.015 to 1.0 wt.%, the wear rate of moldings decreased by two orders of magnitude.

4. CONCLUSIONS

1. BTDA-DABP polyimide obtained by chemical imidization can be processed into a molding at 320-330°C. The resulting molding exhibits bending strength up to 160 MPa, and failure strain of 5-6%. The molding obtained under similar conditions but based on thermally imidized BTDA-DABP PI is very brittle and has low mechanical properties, which is probably due to its low melt viscosity.

2. Several methods of fullerene introduction into BTDA-DABP PI were investigated. It was shown that in all cases fullerene containing polyimide had the crystalline structure with low melting temperature (270-290°C) just as in the case of semi-crystalline PI without fullerene additive. Hence, it could be processed into a molding or a matrix for the composite at 320-330°C.

3. An optimal method for fullerene introduction into BTDA-DABP PI was found, i.e., addition of fullerene in o-dichlorobenzene solution to diamine solution in DMAc. The molding obtained on the basis of this composition (method 2) exhibited tribological characteristics (friction coefficient, wear, and wear rate)

which were superior to those for moldings based on polyimide-fullerene compositions obtained by other methods.

4. The wear properties of moldings improved with increasing mass content of fullerene. For example, when fullerene content was varied from 0.015 to 1.0 wt.%, the wear rate of moldings decreased by two orders of magnitude.

Acknowledgement

This work was carried out with the support of Russian Foundation of Basic Researches (grants 01-03-32415 and 00-15-97297).

REFERENCES

1. M.I. Bessonov, M.M. Koton, V.V. Kudryavtsev and L.A. Laius, *Polyimides-Thermally Stable Polymers*, Consultants Bureau, New York (1987).
2. J.K. Lancaster, Tribology, **6** (6), 219 (1973).
3. S. Bangs, Power Transm. Design, **15** (2), 27 (1973).
4. R.L. Fusaro, ASLE Trans., **25**, 465 (1982).
5. J.C. Anderson, in: *Friction and Wear of Polymer Composites*, K. Friedrich (Ed), pp. 329-362, Elsevier, Amsterdam (1986).
6. S.V. Lukasov, Yu.G. Baklagina, A.G. Kalbin, T.K. Meleschko, T.K Shibaev, E.R. Gasilova, N.N. Bogorad and V.V. Kudryavtsev, in: Polyimides: *Trend in Materials and Applications*, C. Feger (Ed), pp. 251-256, Soc. Plast. Eng., Mid-Hudson Section, Wappingers Falls., NY (1996).
7. H.D. Burks and T.L.St. Clair, SAMPE J., **18** (1), 1-8 (1986).
8. C.E. Sroog, Prog. Polym. Sci. **16**, 561-694 (1991).
9. V.E. Yudin, A.G. Kalbin and T.K. Meleshko, J. Appl. Chem. (Russia), **74**, 1151-1158 (2001).
10. B.M. Ginzburg and D.G. Tochil'nikov, Tech. Phys. **46**, 249-253 (2001).
11. V.V. Kudryavtsev and T.K. Meleshko, J. Appl. Chem. (Russia) **71**, 2035-2040 (1998).
12. Yu.A. Fadin, O.V. Polevaja and I.N. Popov, Tech. Phys. Lett. **19**, 725-726 (1993).

Polyimides and Other High Temperature Polymers, Vol. 2, pp. 511–522
Ed. K.L. Mittal
© VSP 2003

Influence of aramid fabric on fretting wear performance of polyetherimide composite

J. INDUMATHI,[*,1] JAYASHREE BIJWE and ANUP K. GHOSH[2]

[1]*Industrial Tribology, Machine Dynamics and Maintenance Engineering Centre (ITMMEC), Delhi, Hauz Khas, New Delhi 110016, India*
[2]*Centre for Polymer Science and Engineering (CPSE), Indian Institute of Technology, Delhi, Hauz Khas, New Delhi 110016, India*

Abstract—Polyetherimide (PEI) is a high temperature engineering thermoplastic polymer (Tg 218°C and m. p. 380°C) with very good combination of mechanical and electrical properties along with easy injection mouldability. Its fibre reinforced and solid lubricated composites have shown excellent tribological potential in adhesive wear mode. In fretting wear mode, however, only very little is reported in the literature on short fibre reinforced and solid lubricated composites of PEI and not at all on fabric reinforced composites. The authors have reported in their earlier papers on the significant potential of carbon fabric for improving the fretting wear performance of PEI. In this paper efforts have been devoted to explore the tribo-potential of aramid (Kevlar 29) fabric for improving fretting wear performance of PEI. It was observed that aramid fabric (lighter and less expensive than the carbon fabric) improved the fretting wear behavior of PEI significantly. The friction coefficient of PEI (0.4) was independent of operating conditions while that of aramid fabric composite (AFC) very much depended on both load and temperature. Overall, fabric inclusion reduced the friction coefficient of PEI from 0.4 to 0.3 and the specific wear rate to one-third which is very significant improvement. The tribo-performance improvement capability of aramid fabric (AF), however, slightly deteriorated at high operating temperature (150-200°C). The study of worn surfaces by SEM proved to be very helpful in understanding wear mechanisms. It was observed that higher temperature (200°C) was more damaging to the fabric, and hence to the performance of the composite, rather than the higher load (200 N).

Keywords: Polyetherimide; fretting wear; high temperature polymer; aramid fabric reinforced composite.

1. INTRODUCTION

Fretting wear is a type of wear which occurs when there is a small amplitude oscillatory movement between contacting surfaces. This phenomenon generally takes place in the case of bearings; gears; bushes; multilayer leaf springs; palliatives; bolted, riveted and pinned joints; bearing liners; spline couplings; gripped

---

*To whom all correspondence should be addressed. E-mail: jindu_mathi@hotmail.com

components; flanges; seals, etc [1, 2]. As a matter of fact, along with adhesive, abrasive and fatigue mechanisms, the fretting also contributes to a significant extent in the total wear mechanisms, though its extent of contribution depends on the various operating conditions. Polymers and composites are extensively used in the tribo-situations where, most of the time, vibrations are inevitable. The composite or a matrix which performs well in a particular wear mode may not exhibit similar behaviour in other wear modes. In fact, it can also show deterioration in performance. Hence, the tribo-performance has to be evaluated in a particular wear mode. The research on polymer fretting was initiated from the need to enhance the fretting performance of the metals by employing thin coatings or films of polymers on the metal substrates [3, 4]. Kang and Eiss [5] selected a series of siloxane modified polyimide (PI) copolymers with the aim to study the influence of siloxane content, humidity and type of substrate on the fretting wear life of the coatings. Later the efforts were focussed on investigating wear of polymers and their composites in the form of contacting bodies themselves rather than the films [6, 7]. Thus the literature survey on fretting wear of polymers and composites reveals the fact that the influence of fillers in particulate and fibrous forms is more unpredictable as compared to the unidirectional sliding wear mode, because of difference in wear mechanisms.

Polyetherimide (PEI) is a high performance engineering thermoplastic polymer (Tg 218°C and m. p. 380°C) with a very good combination of mechanical and electrical properties along with easy injection mouldability. Its short fibre reinforced and solid lubricated composites have exhibited excellent tribo-potential in sliding wear mode [8] but very little has been reported on fretting wear performance of these composites [9]. Renicke et al. [9] investigated the influence of poly (tetrafluoroethylene) (PTFE) solid lubricant on friction and wear of short glass fibre (30%) reinforced PEI composite. Hardly any literature is available on the fretting wear performance of fabric reinforced composites of PEI. The authors have reported the influence of carbon fabric on fretting wear performance of two-directional composites of PEI [10]. In this paper, the influence of aramid fabric on the fretting wear performance of PEI is presented.

2. EXPERIMENTAL

2.1. Fabrication of composite

The PEI was supplied by GE Plastics USA (properties shown in Table 1) in both granular and moulded (dumbbell shape) forms. High impact aramid fabric (Kevlar 29) was procured from Tory Industries Inc, Germany. The yarn counts in warp and weft were 1300 and 1350, respectively. The aramid fabric composite was prepared using hand lay-up prepreg technique followed by compression moulding. The fabric was cut into 30 cm × 30 cm plies. Twenty four plies of aramid fabric were weighed and PEI resin (40% of the weight of plies) was dissolved in equal amount (vol.%) of solvent (dichloromethane) overnight. The fabric was

Table 1.
Various properties of selected materials

Property	PEI	AFC
% of carbon fabric	0	75 wt.% (74 vol%)
Density (g/cm^3)	1.27	1.33
Hardness (Shore D)	92	90
Tensile strength at break (MPa) (ASTM D 638)	105*	318
Tensile modulus (GPa) (ASTM D 638)	3*	19
Elongation to break (%)	60*	1.8
Toughness (J)	–	14.43
Flexural strength (MPa) (ASTM D 790)	150*	98
Flexural modulus (GPa) (ASTM D 790)	3.3*	11
Inter-laminar shear strength (MPa) (ASTM D 2344)	NA	13
Impact strength (Charpy) (kJ/m^2)	37*	300

*From the supplier's data

stretched and the resin coating was applied using standard hand lay-up process. The pre-pregs were dried for 24 hours in the ambient atmosphere followed by compression moulding at 285°C and 75 MPa pressure with two intermittent breathings.

2.2. Characterisation of the composite

The composite was characterised with various techniques for its composition, physical, thermal and mechanical properties which are shown in Table 1.

2.3. Fretting wear studies

The amplitude of oscillation is the most important parameter which differentiates fretting from reciprocating wear. However, there is no clear demarcation in this respect and an amplitude as high as 2.5 mm has been used to investigate fretting wear [11]. In this work, fretting wear studies were done on SRV Optimol Tester supplied by Optimol Instruments, Germany (Fig. 1), in which a chromium steel ball of diameter 10 mm was fretted against a polymer plate (10 mm × 10 mm × 4 mm). The fabric was always in the direction parallel to that of the fretting direction. The other operating parameters were as follows.

Load: 50, 100, 150, and 200 N; Stroke length (full oscillation width): 1 mm; Frequency: 15, 25, 35, 50 and 75 Hz; Environment: (25°C) except for high temperature studies; Temperature: 25°C, 100°C, 150°C and 200°C; Fretting duration: one hour.

The specific wear rate was calculated using the following equation:

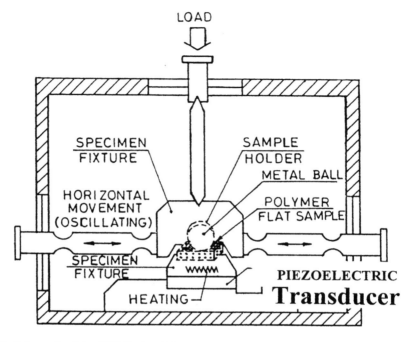

Figure 1. Schematic of the SRV Optimol Tester.

$$K_0 = \Delta V/ \, d \, F_N$$

where, K_0 is the specific wear rate in m^3/N-m, ΔV is the loss in wear volume (m^3), F_N is applied normal load (N) and d is the total sliding distance (m) which is calculated as $2Avt$, where A is the full oscillation width (m), v is the frequency (Hz) and t is the experimental duration (s). Repeatability of the tests was confirmed by carrying out ten tests on each sample and the readings were within 99.5% confidence level.

3. RESULTS AND DISCUSSION

Average friction coefficients (μ) as a function of load and temperature are shown in Fig. 2a and Fig. 2b, respectively. Specific wear rates as a function of load and temperature are plotted in Fig. 3a and Fig. 3b respectively. SEM of the worn composite surfaces under different conditions are shown in Fig. 4. The following are the salient features revealed from these studies.

- Friction co-efficient exhibited by the PEI material was quite high (0.4) and did not show any variation with fretting time, load and temperature. Friction co-efficient of the composite, however, showed increase with increase in load and temperature.

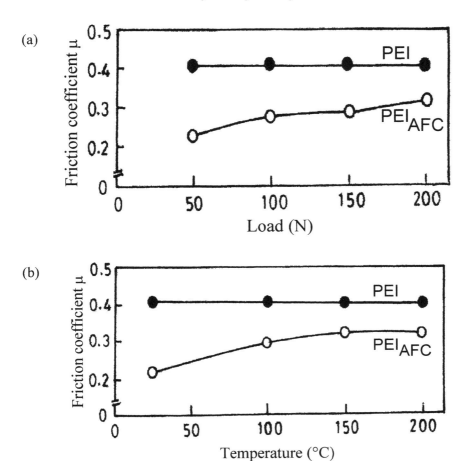

Figure 2. Friction coefficient (μ) for PEI and PEI $_{AFC}$ as a function of (a) load at constant temperature (25°C) and (b) temperature at constant load of 100 N; duration (t) – 1 hr, Amplitude (A) – 1 mm, Frequency (ν) – 50 Hz.

- Aramid fabric inclusion reduced the friction coefficient of PEI from 0.4 to 0.2 at lower loads and at lower temperatures. However, at higher loads and higher temperatures, the friction coefficient of the composite increased from 0.2 to 0.3.
- Specific wear rates of both materials were of the order of 10^{-14} m^3/N-m at all loads (50-200 N). However, fabric reinforcement increased the wear rate of the composite by the factor of 10 at high temperatures [Fig. 3b].
- Specific wear rates of both materials increased with increase in both load and temperature.
- Aramid fabric reinforcement reduced the specific wear rate of PEI to half at lower loads and to one-third at higher loads. However, with increase in temperature the wear performance of AFC was poorer than that of neat polymer.

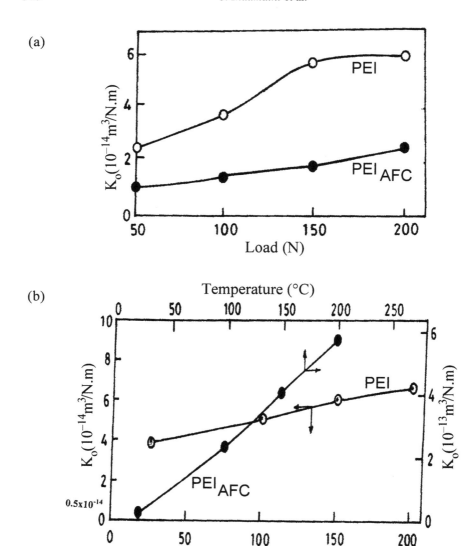

Figure 3. Specific wear rate (K_0) for PEI and PEI $_{AFC}$ as a function of (a) load at constant temperature (25°C) and (b) temperature at constant load of 100 N; duration (t) – 1 hr, Amplitude (A) – 1 mm, Frequency (ν) – 50 Hz.

- Both friction coefficient and wear were sensitive to the selected experimental parameters such as load and temperature.

As seen in Fig. 2a and 2b, the friction coefficient, μ of PEI was found to be independent of load and temperature. The μ was high as compared to the unidirectional sliding mode (μ = 0.3) [8]. In the case of fretting wear, generally high μ values, (> 0.4) are reported for the polymers because of the difference in wear mechanisms as compared to sliding wear. In the authors' laboratory [12] similar values of μ were reported for polyimide (PI) (Vespel SP1), poly(etheretherketone) (PEEK) and polyamide (PA) when evaluated under identical fretting conditions.

In the case of fretting wear, the wear mechanisms are significantly different from those in adhesive wear. Friction and wear mechanisms in the fretting mode are determined by the formation of a third body interphase rather than the two-body contact as in the case of unidirectional sliding. Wear debris in the former mode is of larger size and gets trapped within the contact zone. Hence dominating wear mechanism in fretting wear is always a third body abrasion, though the transfer to the counterfaces also takes place to a slight extent. The wear debris consists of fine powder of metal (due to abrasion), fibres or solid lubricants, if any. In the case of adhesive wear, on the other hand, the extent of abrasion is minimal since the beneficial film transfer on the counterface takes place with the commencement of sliding and thus wear debris, if any, is thrown out of the wear track. The surface temperatures are also lower as compared to the fretting mode since the heat dissipation from the friction zone is more efficient in the unidirectional sliding case.

It is seen from Fig. 2a and 2b that both load and temperature have significantly affected the friction behaviour of the composite. At lower load, μ was around 0.2; with increase in load, μ increased to 0.3. This shows that when load increases sufficiently, aramid fabric melts due to high frictional heating as a result of high contact pressure and this molten material gets transferred to the counterface and could be responsible for increasing the value of μ. At still higher load, however, μ increased from 0.25 to 0.35. This indicated that aramid material is less lubricating at higher loads. Friction coefficient of aramid composite increased with increase in temperature (0.2 to 0.3). Overall the aramid composite showed lubricating effect at lower loads and at lower temperatures.

Aramid fibres consist of highly oriented liquid crystalline polymer. They are soft, flexible and ductile fibres in contrast to glass fibres (GF) and carbon fibres (CF). Under wear loading, they tend to fibrillate and get attached to the counterface. The wear debris of AF does not build up as granules but forms flat thin flakes. Such interface is less mobile than the powdery interface of GF and CF. The AF debris has strong tendency to adhere to one or both sliding surfaces. This patchy and flaky debris thus resists the relative motion between these two bodies leading to high μ. The fretting process thus resembles more to solid body friction rather than third body friction.

The AF debris entrapped within the contact zone is large and flaky. Since AF is also a polymer which melts at moderate temperature ($320°C$), the debris is not abrasive in nature. The debris tends to melt, elongate, fibrillate and gets attached to the counterface. This results in transition of wear mode from third body abrasion to two-body adhesion. The transfer of molten material to the counterface under high load and temperature conditions enhances the specific wear rate very significantly.

Fibre reinforcement generally is very effective in improving the wear resistance of polymers in adhesive wear mode but not necessarily in other wear modes. The available literature on fretting wear studies [11] indicates that the fiber reinforcement is beneficial especially at higher loads since fiber inclusion increases the load carrying capacity, thermal conductivity and resistance to creep and reduces the tendency for material transfer. In the present studies, aramid fiber reinforcement reduced the wear rate to half at lower loads and to one-third at higher loads. However, aramid fabric reinforcement did not help in reducing the wear rate at high temperatures. Schulte et al. [11] reported the fretting wear of unidirectional reinforced composites of epoxy with various fibres such as glass, carbon and aramid. As per their findings, wear resistance of AF+epoxy was better than GF+epoxy; however, it was inferior to CF+epoxy composite which was 3 times higher than that of AF+epoxy composite. Similar trend was observed in authors' studies also which is reported elsewhere [10].

3.1. SEM studies on worn surfaces of AFC

Micrographs of the surface of the AFC pin at various locations worn under different operating conditions are shown in Fig. 4 a-f. Aramid fibre being very sensitive to high temperature, its wear behaviour was significantly influenced by high operating temperature. Micrographs 4 a-c are for the surface worn at high load (200 N, RT) while micrographs 4 d-g are for surface worn at high temperature (200°C) and moderate load (100 N). From a comparison of craters at high load (Fig. 4a) and high temperature (Fig. 4d), a correlation is found between the extent of wear and the appearance of crater. The damage, depth and diameter of the crater at high temperature indicate high wear. Micrographs 4b and 4e show the edges of the craters formed due to high load (4b) and high temperature (4e) on the same magnification. The extent of fibrillation is higher on the surface worn at higher temperature. Both these micrographs 4b (bottom portion) and 4e (extreme left) show the inside of the crater also. The fibre structure in both appears to be completely destroyed because of frictional heating. Moreover, because of higher operating temperature, overlapping molten layers of aramid are apparent in micrograph 4e. The fabric structure outside the crater in both micrographs appears to be intact. Some fibrous debris and flakes transferred after the completion of the experiment also appears on the center portion of the crater.

A molten AF material has been extensively chipped off from the crater zone (4c) which must have been transferred to the ball during fretting. Inside this portion, longitudinal fibres on the verge of losing fibrous structure during frictional heating can be seen. Some fibrils can also be seen on the boundary of this zone. In the case of high temperature (micrograph 4f), however, such features were not observed. A completely molten aramid material in the form of thick overlapping layers elongated during successive traversals can be seen on the surface.

(a)

(b)

(c)

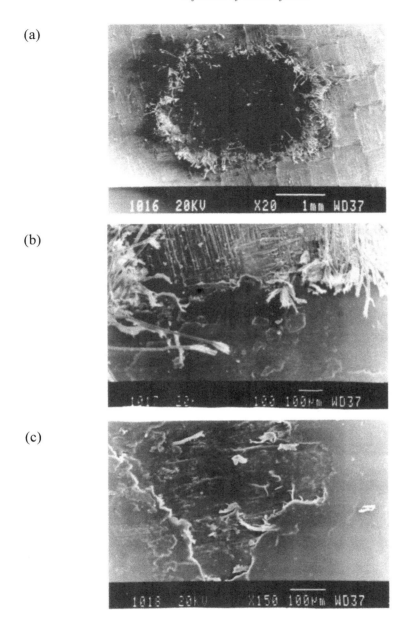

Figure 4. SEM micrographs of worn surfaces of AFC fretted under various conditions showing (a) general view of the crater, fretted under RT and 200 N; (b) Edge of the crater indicating extensive damage to the fibres in the direction anti-parallel to the fretting as well as fibrillation and elongation of aramid fibre; (c) center of crater showing extensive melting of AF and fibrous debris transferred to the surface; (d) pin surface fretted at high temperature 200°C (100 N) showing general view of the crater; (e) edge of the center with chipped off material due to excessive melting of aramid fibre in the crater, as well as fibrillation of aramid fibre at the edge; (f) back transfer of thick molten layer with small fibrous wear debris; (g) extensive melt flow of aramid fibre and back transfer of patches (flaky and fibrilled) of aramid material.

(d)

(e)

(f)

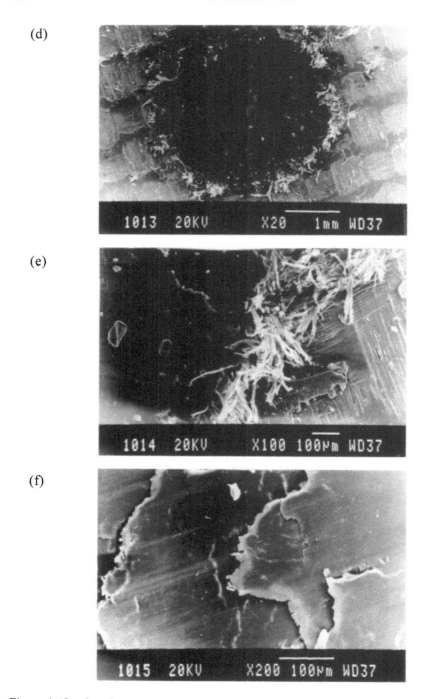

Figure 4. (Continued).

(g)

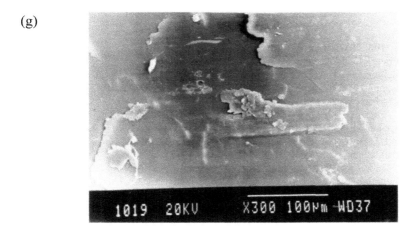

Figure 4. (Continued).

Such layers are due to back transferred material from the counterface. Some small fibrous wear debris also appeared on the surface. Micrograph 4g shows the extensive flow of molten aramid material and its back transfer of wear debris (patchy and fibrous) to the composite pin surface.

4. CONCLUSION

Fretting wear studies on PEI and its aramid fabric reinforced composites under various operating conditions indicated that fabric reinforcement helped in enhancing the friction and wear performance of the composite. The aramid fabric reinforcement reduced the friction coefficient from 0.4 to 0.3 and specific wear rate to one-third which is very significant improvement. However, at higher temperatures, the wear performance of the composite deteriorated because AF melts at 320°C but the debris is not abrasive in nature. The wear debris formed tends to fibrillate, elongate and attach to the counterface. This results in transition of wear mode from third body abrasion to two-body adhesion. Hence, it can be concluded that up to moderate temperatures and high loads AFC proved to be very good material for fretting wear applications.

Acknowledgement

The authors gratefully acknowledge the funding by the Council of Scientific and Industrial Research (CSIR) India for these studies and Mr. M.N. Saraf and Mr. R.K. Gupta from the Defence Materials Stores Research Development Establishment, G.T. Road, Kanpur, India for the help in preparation and characterization of the composites.

REFERENCES

1. R.C. Bill, Wear, **106**, 283-301 (1985).
2. O. Jacobs, K. Friedrich and K. Schulte, Wear, **145**, 167-188 (1991).
3. F.H. Stott, B. Bethune and P.A. Higham, Tribology. Intl., **10**, 211-215 (1977).
4. S. Abrou, D. Play and F.E. Kennedy, ASLE Trans., **30**, 269-281 (1987).
5. C. Kang and N.S. Eiss, Wear, **158**, 29-40 (1992).
6. N. Dahmani, L. Vincent, B. Vannes, Y. Bertheir and M. Godet, Wear, **158**, 15-28 (1992).
7. M. Daoud, A. Chateauminois and L. Vincent, J. Mater. Sci., **34**, 191-194 (1999).
8. U.S. Tewari and J. Bijwe, in: *Polyimides: Fundamentals and Applications*, M.K. Ghosh and K.L. Mittal (Eds.), Chap. 19, Marcel Dekker, NewYork (1996).
9. R. Renicke, F. Haupert and K. Friedrich, Composites Part A, **29A**, 763-771 (1998).
10. J. Bijwe, J. Indumathi, B.K. Satapathy and A.K. Ghosh, Trans. ASME, J. Tribology, in press.
11. K. Schulte, K. Friedrich and O. Jacobs, in: *Advances in Composite Tribology*, Composite Materials Series Vol. 8, K. Friedrich (Ed.), Ch. 18, Elsevier, Amsterdam (1993).
12. J. Bijwe and M. Fahim, in: *Handbook of Advanced Functional Molecules and Polymers*, H.S. Nalwa (Ed.), Ch. 8, pp. 265-321, Gordon and Breach (2001).

Polyimides and Other High Temperature Polymers, Vol. 2, pp. 523–532
Ed. K.L. Mittal
© VSP 2003

Semicrystalline polyimides for advanced composites

V.E. YUDIN,*,[1] V.M. SVETLICHNYI,[1] G.N. GUBANOVA,[1]
A.I. GRIGORIEV,[1] A.L. DIDENKO,[1] T.E. SUKHANOVA,[1]
V.V. KUDRYAVTSEV,[1] S. RATNER[2] and G. MAROM[2]

[1] *Institute of Macromolecular Compounds, Russian Academy of Sciences, St. Petersburg 199004, Russia*
[2] *Casali Institute of Applied Chemistry, The Hebrew University of Jerusalem, Jerusalem 91904, Israel*

Abstract—New semicrystalline polyimides, designed for matrices in composites, were developed. A specific advantage of the proposed polyimides is their ability for crystallization from the melt, thereby retaining their crystallinity through the manufacturing process. The use of semicrystalline polyimides for composites enhances their thermal stability by about 100°C over that of completely amorphous polyimides. The generation of crystallinity after melting, i.e., recrystallization, has been investigated as affected by blending of the polyimides with 5-10% by weight oligoimides of similar chemical structure. Based on thermal analysis and enthalpy measurements, comparative X-ray diffraction analyses and polarized light microscopy of hot-stage controlled crystallization the recrystallization ability is determined for different oligoimides. It is found that in some cases the addition of oligoimides results in complete recrystallization. It is suggested that the main contribution of the oligoimides is through plasticization, allowing segmental chain mobility during crystallization and not via nucleation. A similar effect is obtained by lowering the molecular weight of the polyimide; this, however, causes mechanical property reduction of polyimide matrix in composite materials.

Keywords: Semicrystalline polyimide; oligoimide; recrystallization; carbon fiber; composite.

1. INTRODUCTION

Polyimides (PI) have historically been regarded as an amorphous or mesomorphic class of high-temperature polymers [1, 2], whose incapability to crystallize derives from the high viscosity of their melt. However, because of their typical highly regular structure many of them can be forced to crystallize. New polyimide matrices for advanced composites, which can crystallize after melting, have been developed in recent years, and these are LaRC-CPI [3, 4], LaRC CPI-2 [5], and PI-2 [6]. Their attractive features are excellent thermal and thermooxidative stability, high mechanical and adhesion characteristics and high resistance to solvents, in particular to water and alkalis. Their specific characteristic is the possibility of crystallization from the melt and retention of the crystalline state in the composite. This feature distin-

---

*To whom all correspondence should be addressed. Phone: 7-812-3235065, Fax: 7-812-3286869,
E-mail: yudin@comp.spb.ru

guishes them from the well-known matrix of LaRC-TPI type [7], which after melting of chemically imidized PI powder at T = 270-290°C becomes amorphous and is virtually incapable of passing into the crystalline state as a result of annealing. As pointed out in ref. [8], the only method to crystallize LaRC-TPI from the melt is to boil it additionally in N-methyl-2-pyrrolidone (NMP).

The aim of the present work was to investigate the crystallization and melting process of some new semicrystalline PI's and the effect on this process of such factors as the melting temperature, the time of existence in the melt, annealing and blending with oligoimides. Moreover, the effect of matrix crystalline state on the mechanical properties of carbon fiber reinforced composite (CFRC) has been studied.

2. EXPERIMENTAL

2.1. Materials and preparation

A number of PI semicrystalline binders that could be used as a matrix in CFRC were developed. These semicrystalline binders have the following chemical structures:

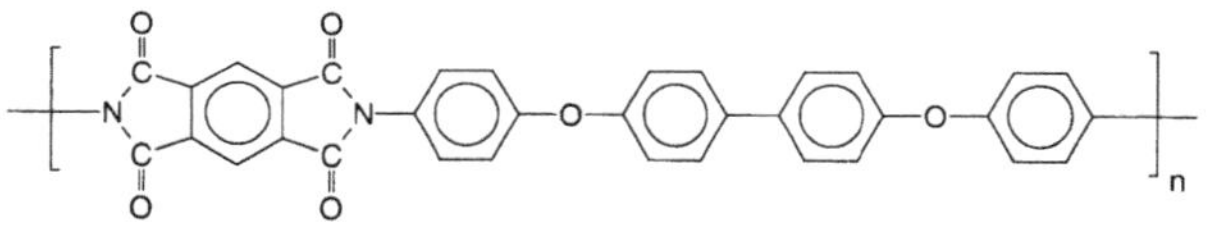

PM-BAPB

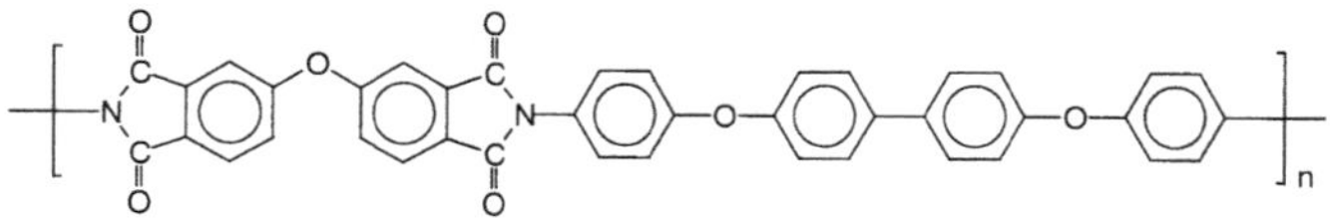

ODPA-BAPB

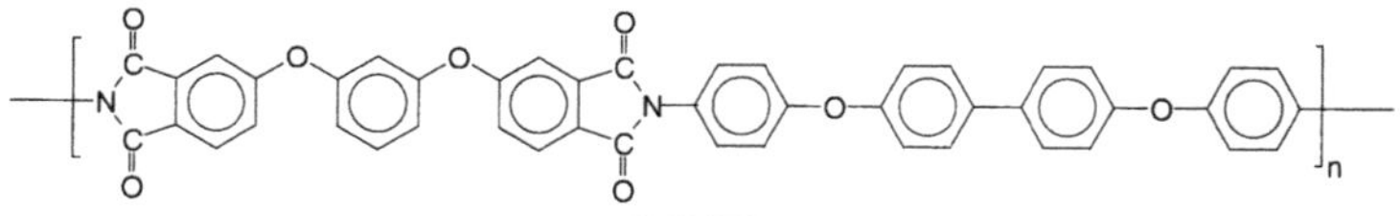

R-BAPB

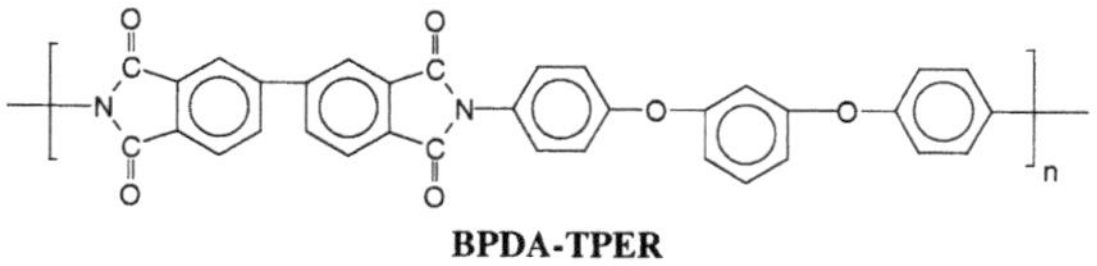

BPDA-TPER

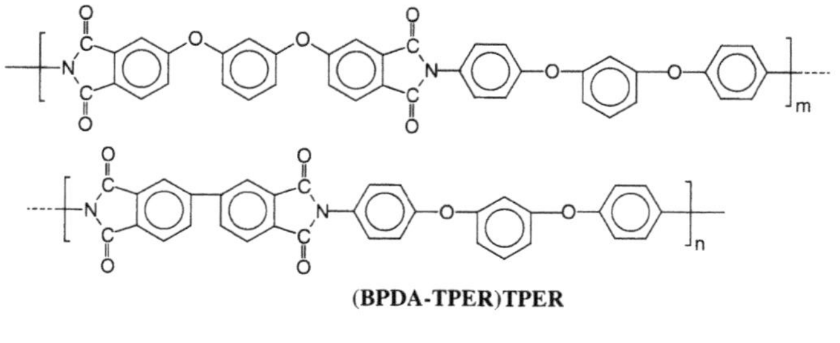

(BPDA-TPER)TPER

PI films, 40-50 μm thick, were prepared from poly(amic acid) (PAA) by casting onto soda lime glass plates and oven curing under air. Polymers of the BPDA-TPER and (BPDA-TPER)TPER types could be obtained in the form of relatively flexible and strong films (BPDA-TPER was an opaque film and (BPDA-TPER)TPER was a transparent film), from NMP or N,N-dimethylacetamide (DMAc) solutions. ODPA-BAPB and R-BAPB did not yield flexible films at all from DMAc. The films of these polymers obtained from DMAc were opaque, brown in color and very brittle after imidization at T = 280°C. According to XRD and DSC data, an active crystallization of these PI's from PAA took place.

The sequence of the preparation of the composite based on semicrystalline PI and low modulus (200 GPa) carbon fiber ELUR (NPO "Chimvolokno", Moscow, Russia) was as follows:

(1) Impregnation of the unidirectional carbon fabric with a 20-30% of the PAA binder;

(2) Prepreg drying for complete solvent removal and thermal treatment up to approximately 100% imidization of the binder;

(3) Prepreg packing into a stack to obtain a composite of the required thickness.

(4) Composite molding above the matrix melting temperature to render the polymer sufficiently fluid for uniform distribution throughout the fibers mass; and

(5) Mold cooling and annealing at temperature below the melting temperature of the polymer and in the vicinity of its crystallization temperature in order to attain the maximum degree of matrix crystallinity in the composite.

2.2. Measurements

DSC was performed using a Mettler DSC-30 Model on 5-10 mg samples contained in aluminum pans with a heating rate of 10-20°C/ min under nitrogen.

Wide-angle XRD measurements were conducted on a diffractometer with a Philips PW-1830 generator. A wavelength of 1.54 Å was generated from a CuKα source, with graphite as the monochromator. Samples were scanned in the angular range of 10-40°.

Interlaminar fracture toughness G_{1C} of the CFRC was measured by the double cantilever beam method on a 1958U-10-1 (Russia) testing machine at a load speed of 10 mm/min.

The storage shear modulus G' and loss modulus G" of the CFRC were determined on a torsion pendulum MK-003 developed at the Institute of Macromolecular Compounds.

3. RESULTS AND DISCUSSION

3.1. Thermomechanical characteristics of PI films

Thermomechanical characteristics of PI films and temperatures of main transitions - glass transition and melting (according to DSC data) of the above polymers are given in Table 1.

The (BPDA-TPER)TPER polymer which is actually a copolymer based on two different dianhydrides of DF and R types was amorphous. It also follows from refs. [9, 10] that PI copolymerization using two different diamines or dianhydrides usually leads to material amorphization. Other polymers listed in Table 1 are semicrystalline and their melting temperature can be determined using DSC. The melting temperature of PM-BAPB is relatively high and is probably close to its thermal degradation temperature. Therefore, it is not possible to determine it precisely using DSC. Nevertheless, according to XRD, this polymer is also semicrystalline. The melting temperature of R-BAPB and ODPA-BAPB can increase by 15-20°C after annealing at a temperature 5-10°C lower than that of the main melting transition. This is probably caused by a certain ordering of the size and shape of crystalline regions in the initial polymer obtained from PAA. Hence, the melting temperature range is given in Table 1 for these polymers. Its lower limit corresponds to the melting temperature of the unannealed polymer, whereas its upper limit corresponds to that of the annealed sample. It should also be noted that according to DSC data, after the first temperature scanning, provided the temperature exceeds the melting temperature of the polymer by 20-30°C, the polymers of the ODPA-BAPB and R-BAPB types are amorphized and are not crystallized during cooling. Thus, they become amorphous, which is confirmed by the results of the second scanning. The BPDA-TPER polymer retains its crystallinity after the first scanning and cooling, but to a much lesser extent, i.e. its melting enthalpy decreases from 22.6 J/g during the first scanning to 2.0 J/g during the second scanning.

Table 1.
Thermomechanical characteristics at T=25°C and main transitions temperatures of semicrystalline PI films obtained from NMP solutions

Polymer	E [GPa]	σ [MPa]	ε [%]	T_g [°C]	T_m [°C]	ΔH [J/g]
PM-BAPB	2.9-3.15	130-135	10-15	300	>450	ND
ODPA-BAPB	3-3.1	90	3-4	245	380-412	28
R-BAPB	2.7-3.3	110	10-15	198-213	317-325	32-45
BPDA-TPER	3.6-3.7	100-110	5	240	387	23
(BPDA-TPER)TPER	3.2	100-110	9-10	202	No	No

E = Young's modulus; σ = tensile strength; ε = elongation at break; T_g = glass transition temperature; T_m = melting temperature; ΔH = melting enthalpy; ND - not detectable.

3.2. Study of the possibility of increasing the degree of crystallinity of the PI matrix in the composite

Preliminary DSC investigations showed that the possibility of crystallinity retention in the matrix was rather small. However, to obtain a homogeneous composite with a uniform matrix distribution throughout the fiber bundle, it is necessary to carry out the binder melting stage in the mold. Moreover, it is desirable to have the mold temperature much higher than the binder melting temperature in order to attain maximum flowability of the polymer melt. Special experiments were carried out to study the possibility of polymer matrix recrystallization after its first melting. The R-BAPB polymer was selected for these experiments because its melting temperature was lower than that of other polymers (Table 1). Therefore, it can be more easily processed into the composite matrix without any danger of thermal degradation which, according to TGA data, starts at 470°C.

The experiment was performed as follows. The PI R-BAPB film obtained from PAA by imidization on glass at T = 300°C and, therefore, being highly crystalline (Table 1), was subjected to melting in a DSC unit for 5 min at a temperature higher than that of polymer melting and subsequently cooled. T_x (see Table 2) is the maximum temperature to which the sample was heated during thermal treatment and at which it started to flow. Subsequently, the sample was cooled to 300°C and was maintained at this temperature for 45 min after which it was cooled to room temperature. After this treatment the sample was heated for the second time and during this second scanning its melting temperature T_{m2} and melting enthalpy ΔH_2 were determined. The results are listed in Table 2.

The DSC results (Table 2) for the R-BAPB polymer show that the maximum temperature to which the polymer can be heated so as to retain crystallinity after cooling must not exceed 320°C, which is quite insufficient for uniform impregnation of fibers. Under the actual molding conditions it is rather difficult to maintain the temperature 5-10°C higher than the melting temperature of the matrix to achieve its flowability state on the one hand and not to exceed too much the upper temperature limit for retaining polymer crystallinity after cooling, on the other. Hence, using a pure polymer of the R-BAPB type as the matrix for composite, it is rather difficult to obtain as a result of melting a composite with a semicrystal-

Table 2.
Melting temperature and melting enthalpy of the R-BAPB film according to DSC data during the second scan

T_x [°C]	T_{m2} [°C]	ΔH_2 [J/g]	T_g [°C]
310	324	24.5	ND
320	314	7.8	ND
330	no	ND	ND
350	no	ND	198

line PI matrix. This is probably due to the fact that after transition into the completely isotropic state polyimide is difficult to crystallize because of its high melt viscosity. This has already been reported [11] for semicrystalline PI's with other chemical structures. Therefore, it was proposed to use the blending of the PI matrix with oligoimides as has already been done by other researchers [12] for increasing the degree of crystallinity of the PI matrix.

3.3. Blending of the PI matrix with oligoimides in order to increase the degree of crystallinity in the composite

A semicrystalline binder of the R-BAPB type was chosen for CFRC because it had been shown to recrystallize after melting. To enhance the recrystallization effect and to increase the degree of crystallinity of the matrix in the composite, R-BAPB polyimide was blended with the following oligoimide (OI):

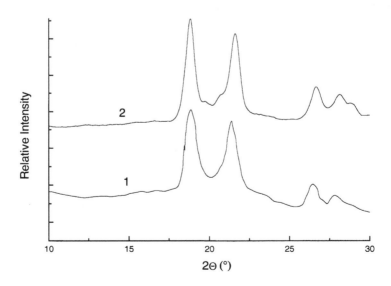

According to XRD this OI is crystalline and its crystalline structure is similar to the crystalline structure of PI R-BAPB (Fig. 1).

The OI was added into the solution of PAA in NMP in weight ratios of PI:OI = 90:10; 80:20; 70:30; 60:40; 50:50. The resulting colloidal suspension was poured onto a glass slide, dried and imidized at 300°C. The films prepared in this way were extremely brittle and broke into small pieces after their imidization on the glass slide.

Figure 1. Wide-angle XRD patterns of the R-BAPB PI film (1) and oligoimide OI powder (2).

The thermal treatment of R-BAPB films with the addition of OI, based on the DSC results was follows. The sample was heated to T = 330°C, i.e. to a temperature at which pure PI melts but does not crystallize upon cooling. The sample was maintained at this temperature (330°C) for 5 min. (time required for composite molding) and then cooled to room temperature. At the second sample scanning to 350°C, its T_m and T_g were also determined.

Table 3 presents the DSC data for PI and OI blends and for pure OI (No. 6). Pure OI crystallizes well and the heat of fusion at the second scanning ($\Delta H = 44.8$ J/g) is higher than at the first scanning ($\Delta H = 32$ J/g). This fact indicates that the degree of crystallinity of OI increases even after the first melting. The T_g for OI is about 50°C lower than that for pure PI and is 154°C.

The PI:OI blends during the first scanning are characterized by two melting peaks, T_{m1} and T_{m2} (Table 3). This is probably caused by the heterophase structure of the film consisting of the PI matrix and small inclusions of OI particles, which did not dissolve in PAA. Only one peak is obtained during the second scanning, i.e. after the first melting and homogeneous OI distribution in the PI matrix on the molecular level.

It is clear that after the first heating not all samples crystallize upon cooling. Moreover, the greater the OI fraction in the PI/OI blends, the more readily the polymer crystallizes and the higher is its degree of crystallinity. It should also be noted that with increasing OI content the T_g of the blends decreases regularly from about 200°C for pure PI to 154°C for pure OI. In this case the melting temperature of the blends changes only slightly from 299 to 308°C (Table 3).

The possibility to obtain a semicrystalline matrix from R-BAPB and OI by chemical imidization was also investigated. According to DSC data, the temperatures of the main endothermic transitions of R-BAPB and OI obtained by thermal and chemical methods coincide. The X-ray diffraction patterns of PI and OI obtained by thermal imidization are characterized by many distinct reflections. PI

Table 3.
DSC data for PI/OI blends and for pure OI

No	PI:OI [%]	1st heating				cooling		2nd heating		
		T_g [°C]	T_{m1}, [°C]	T_{m2} [°C]	$\Sigma\Delta H$ [J/g]	T_c, [°C]	ΔH_c [J/g]	T_g [°C]	T_m [°C]	ΔH [J/g]
1	90:10	191	289	302	28.9	–	–	187	–	–
2	80:20	180	293	307	33.5	–	–	180	308	3.8
3	70:30	–	289	302	41.3	–	–	171	306	6.5
4	60:40	–	287	297	47.4	239	18.0	167	300	24.1
5	50:50	–	288	–	52.4	239	28.6	164	299	30.4
6	0:100	154	272	–	32	247	37.9	–	276	44.8

$\Sigma\Delta H$ = total melting enthalpy for the 1st and 2nd melting peaks; T_c = crystallization temperature; ΔH_c = crystallization enthalpy.

and OI obtained by chemical imidization are characterized by X-ray diffraction patterns with several reflections of low intensity. A comparison of these data suggests that the chemical imidization of PI and OI yields inferior crystalline structures, i.e. more defective structures with smaller crystals.

3.4. Thermomechanical properties of the composites based on the PI matrix alone and on PI with the addition of the oligoimide

CFRC's based on an ELUR carbon fiber were obtained from a pure R-BAPB PI matrix and from those with the addition of the oligoimide OI in the ratio of PI:OI = 60:40 and PI:OI = 90:10. The aim of the experiment was to evaluate the effect of crystallinity on the thermomechanical properties of the polymer. The prepreg was impregnated using PAA solution (with or without OI). Prepregs were dried to remove the solvent and imidized at 280-300°C. Subsequently, they were packed into a pile consisting of 24-25 layers, placed in a mold and molded at 330°C for 5 min. To increase the degree of matrix crystallinity in the CFRC, an additional thermal treatment was also carried out during cooling. It consisted of isothermal maintenance of the CFRC in the mold at $T = 250°C$ for 0.5 hr, i.e. at a temperature close to that of matrix crystallization.

Fig. 2 shows thermomechanical analysis data for CFRC using a torsion pendulum. It can be seen that in the CFRC with a semicrystalline matrix the shear storage modulus G' (Fig. 2, curves 1' and 2') retains a high value at a level of 0.5-1 GPa up to the melting temperature of about 300°C. For the CFRC with an amorphous matrix (pure PI without the addition of OI), a sharp drop in the G' modulus (Fig. 2, curve 3') takes place in the range of glass transition temperature, i.e. at 200°C. Hence the crystallization of PI/OI matrix raises the limit of CFRC performance by almost 100° up to the melting temperature of the matrix.

For additional confirmation of the presence or absence of crystallinity in the PI matrix of the composite, a DSC analysis on the CFRC obtained was performed. In fact, semicrystalline PI matrix with the addition of OI (60:40) in CFRC has a melting temperature of 305°C, i.e. approximately the same as in pure polymer without fibers (Table 3, No. 4). Moreover the melting enthalpy ΔH of CFRC is 8.3 J/g, which, based on the weight of pure matrix in CFRC, is about 25 J/g (volume fraction of the fiber - 0.6, fiber density - 1.75 g/cm^3 and matrix density - 1.4 g/cm^3). This value corresponds to the ΔH value for the matrix in the form of a film (Table 3, No. 4). Consequently, the degree of matrix crystallinity in CFRP approximately corresponds to that of a pure matrix obtained in the form of a film and crystallized under the same conditions. The T_g values for CFRC's with an amorphous matrix also approximately correspond to that of pure PI matrix in the form of films (Table 1) and is 204°C for the amorphous matrix. Slightly higher T_g values for matrices in CFRP can be caused by partial limiting of segmental mobility of the macromolecules in the polymer in the vicinity of carbon fibers or due to possible cross-linking of PI at high temperatures.

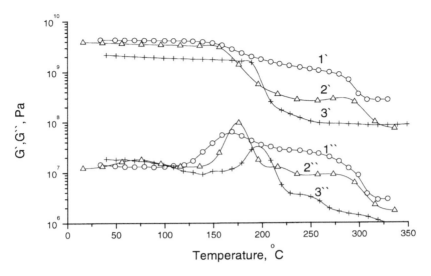

Figure 2. Temperature dependencies of shear storage G' (1',2',3') and loss G" (1",2",3") moduli for CFRC based on blends of PI:OI = 60:40 (1',1") and PI:OI = 90:10 (2',2") and on the pure R-BAPB PI (3',3").

Table 4.
Thermomechanical characteristics of CFRC based on the R-BAPB PI matrix with and without OI

No	Matrix	V_f [%]	σ_f [MPa]	T_g [°C]	T_m [°C]	G'_{25} [GPa]	G_{1C} [J/m^2]
1	Pure PI	48	1050	204	ND	2.1	1600
2	PI:OI = 90:10	57	1230	195	315	3.7	850
3	PI:OI = 60:40	60	1300	169	305	4.3	100

The interlaminar toughness G_{1C} of CFRC's was compared for semicrystalline and amorphous PI matrices. Table 4 gives G_{1C} characteristics obtained for composites with an amorphous matrix (No. 1) and semicrystalline (No. 2 and No. 3) matrices. Furthermore, this table gives the glass transition and melting temperatures of the composites, their shear storage modulus G'_{25} and flexural strength σ_f at room temperature and fiber fraction V_f in CFRC.

An analysis of the data in Table 4 shows that the introduction of OI increases the flowability of the PI matrix, which favors an increase in the fiber volume fraction in CFRC and, hence, its strength, by about 20-25% over that of pure PI. The shear storage modulus of CFRP with a semicrystalline matrix is about twice than that for CFRC with an amorphous matrix. Moreover, this result also follows not only from the higher fiber volume fraction in CFRP with a semicrystalline matrix but also from the higher rigidity of the matrix itself.

The interlaminar toughness of the CFRC with semicrystalline matrix is lower than that for the CFRC with pure amorphous matrix. This probably indicates a high degree of brittleness of the former. The amorphization of the semicrystalline matrix causes increase in G_{1C} up to 200 J/m^2 for matrix PI:OI = 60:40 and to 1300 J/m^2 for matrix PI:OI = 90:10. The lower value of G_{1C} as compared to that of CFRC with a pure amorphous PI matrix is probably due to the presence of a higher quantity of low molecular weight oligoimides in the PI matrix.

4. CONCLUSIONS

New semicrystalline polyimide matrices for composites were developed. During the process of both thermal and chemical imidizations these binders can undergo crystallization. Moreover, after melting they can also crystallize during cooling. The crystallization process during cooling from the melt temperature of the R-BAPB type matrix can be intensified by blending it with oligoimide of the same chemical structure as that of the main PI matrix. It was shown that the higher the oligoimide content in the mixture with the main PI matrix, the greater was the ability of the matrix to crystallize during cooling from the melt temperature. The thermal stability of CFRC with the semicrystalline PI matrix increases by about 100°C over that with the amorphous matrix. Moreover, the shear storage modulus of the CFRC is approximately twice and its mechanical strength increases by 20-25%.

Acknowledgments

The financial support of this work by the Russian Foundation for Basic Research, grant No. 01-03-32415, is gratefully acknowledged.

REFERENCES

1. M.I. Bessonov, M.M. Koton, V.V. Kudryavtsev and L.A. Laius, *Polyimides - Thermally Stable Polymers*, Plenum, New York (1987).
2. C.E. Sroog, Prog. Polym. Sci., **16**, 561 (1991).
3. J.T. Muellerleile, B.G. Risch, D.E. Rodrigues, G.L. Wilkes and D.M. Jones, Polymer, **34**, 789 (1993).
4. P.M. Hergenrother and S.J. Havens, SAMPE J., **24**, 13 (1988).
5. D.K. Brandom and G.L. Wilkes, Polymer, **35**, 5672 (1994).
6. S.Z.D. Cheng, M.L. Mittleman, J.J. Janimak, D. Shen, T.M. Chalmers, H.-S. Lien, C.C. Tso, P.A. Gabori and F.W. Harris, Polym. Int., **29**, 201 (1992).
7. V.L. Bell, B.L. Stump and H. Gager, J. Polym. Sci: Polym. Chem. Ed., **14**, 2275 (1976).
8. J. Wang, A.T. DiBenedetto, J.F. Johnson, S.J. Huang and J.L. Cercena, Polymer, **30**, 718 (1989).
9. V.N. Artemjeva, Yu.G. Baklagina, V.V. Kudriavtsev, V.E. Yudin, N.S. Nesterova, N.V. Kukarkina, Yu.N. Panov and P.I. Chupans, J. Appl. Chem. (Russia), **66**, 1826 (1993).
10. V.N. Artemjeva, L.S. Bolotnikova, T.D. Glumova, M.M. Koton, V.V. Kudriavtsev, N.V. Kukarkina, Yu.N. Panov, N.V. Rumyantseva, P.I. Chupans and V.E. Yudin, J. Appl. Chem. (Russia), **64**, 2405 (1991).
11. A.J. Hsieh, C.R. Desper and N.S. Schneider, Polymer, **33**, 306 (1992).
12. T. Takekoshi, US Patent 4 906 730 (1990).

Milton Keynes UK
Ingram Content Group UK Ltd.
UKHW021926071024
449327UK00022B/1708

9 780367 446789